Industrial Plasma Engineering

Volume 2: Applications to Nonthermal Plasma Processing

Industrial Plasma Engineering

Volume 2: Applications to Nonthermal Plasma Processing

J Reece Roth

Department of Electrical and Computer Engineering
University of Tennessee, Knoxville

CRC Press
Taylor & Francis Group
Boca Raton London New York

CRC Press is an imprint of the
Taylor & Francis Group, an **informa** business
A TAYLOR & FRANCIS BOOK

© IOP Publishing Ltd 2001

British Library Cataloguing-in-Publication Data

A catalogue record for this book is available from the British Library.

ISBN 0 7503 0544 4 hardback
0 7503 0545 2 paperback

Library of Congress Cataloging-in-Publication Data are available

Commissioning Editor: John Navas
Production Editor: Simon Laurenson
Production Control: Sarah Plenty
Cover Design: Victoria Le Billon

Published by Institute of Physics Publishing, wholly owned by The Institute of Physics, London

Institute of Physics Publishing, Dirac House, Temple Back, Bristol BS1 6BE, UK

US Office: Institute of Physics Publishing, The Public Ledger Building, Suite 1035, 150 South Independence Mall West, Philadelphia, PA 19106, USA

Typeset in the UK by Text 2 Text, Torquay, Devon

Contents

Contents
Volume 1: Principles

Preface to Volume 2

Volume 1 of this three-volume work is intended to provide a background in the principles of low temperature, partially ionized Lorentzian plasmas that are used industrially. Volumes 2 and 3 are intended to provide a description of plasma-related processes and devices that are of actual or potential commercial importance. The text assumes that the average student or practicing engineer has not recently taken a course in plasma physics, and possesses a background in physics and calculus that ended at the sophomore level. These three volumes are intended to be used as textbooks at the senior or first-year graduate level by students from all engineering and physical science disciplines, and as a reference source by practicing engineers.

Use of this second volume as a textbook or reference source assumes that the reader is familiar with the material in Volume 1, or has an equivalent background in low-temperature Lorentzian plasma physics. An introduction to plasma physics and the physical processes important in industrial plasmas is contained in the first four chapters of Volume 1. Chapters 5 through 7 of Volume 1 describe the sources of ion and electron beams and ionizing radiation that are used industrially. Chapters 8 through 10 of Volume 1 describe the physics and technology of DC electrical discharges, and chapters 11 through 13 describe the physics and technology of RF plasma sources.

In this second volume, chapter 14 is devoted to some aspects of materials science that are basic to plasma-processing applications. Chapters 15 and 16 are, respectively, devoted to atmospheric and vacuum plasma sources, chapter 17 to the plasma reactors (or plasma 'tools') frequently used industrially, and chapter 18 to specialized methods and devices used in these reactors. Chapter 19 is devoted to the effects of plasma-related parameters on the outcomes of plasma processing. Chapter 20 covers the most frequently used diagnostic methods to measure the independent input variables, the plasma parameters, and the results of plasma processing. Chapters 21 through 25 cover industrial applications categorized as non-thermal plasma processing of materials.

Volume 3 will cover thermal plasma processing and plasma devices.

ix

Chapters in this volume will cover the melting and refining of bulk materials; subsonic and supersonic plasma aerodynamics; the use of plasmas for synthetic and destructive plasma chemistry; electrical sparking, switchgear, and coronas, all important to the electric utilities; plasma lighting devices; the applications of electrohydrodynamics to electrostatic precipitation and paint spraying; and research and development plasmas with potential for industrial applications.

This second volume is not intended to be self-contained. Frequent reference is made to equations, derivations, and data discussed in Volume 1. This volume does not contain derivations from first principles of some advanced material from plasma physics or materials science. Such background can be found in other sources listed among the references at the end of the individual chapters, or in the annotated bibliography that will be included at the end of Volume 3. SI units have been used throughout the text except where conventional usage (e.g., electronvolt, Torr) has become so firmly established that non-SI units are more appropriate.

Available technical dictionaries in the fields of physics, chemistry, and electrotechnology generally have an inadequate coverage of the terminology required to discuss the subject of industrial plasma engineering. In an attempt to deal with this problem and increase the value of this book as a reference source, the technical terminology, jargon, and acronyms used in the field of industrial plasma engineering are not only defined and fully discussed, but also *italicized* and indexed when they first appear in the text. This practice is intended to assist the reader in learning key terminology and concepts, and should provide the practicing engineer with a provisional technical glossary until a proper technical dictionary covering the field of industrial plasma engineering becomes available.

In addition to providing an extensive index, I have attempted to further enhance the book as a reference source by including several appendices at the end of each volume, which provide both a comprehensive listing of the mathematical nomenclature and units used throughout the text, and a collection of frequently used plasma formulae, physical constants, and conversion factors. The index at the end of this second volume includes not only the technical terms used in it, but also those that were used in Volume 1.

In this second volume, which is concerned with specific industrial processing applications of plasma-related technologies, it has been necessary to use trade jargon, some of which includes copyrighted or trademarked trade names. An attempt has been made to indicate the status of such words as copyrighted or trademarked. If any such legally protected terminology has slipped through without the appropriate designation, I apologize to the trademark or copyright holder in advance.

To facilitate the use of this book for classroom instruction, I have prepared an *Instructor's Manual* for both volumes for teachers of the subject, which is available from me for the cost of making a photocopy. This *Manual* includes homework problems and their answers; full-size copies of the figures and tables, from which transparencies can be made; enlarged originals of all the equations in the text for the production of transparencies; and a topical outline of all chapters with pagination keyed to the text.

I would like to express my appreciation to the many individuals and anonymous reviewers whose suggestions and hard work have contributed to the manuscript in its present form. While assuming total responsibility for the contents and correctness of the manuscript myself, I would like to thank my graduate and minicourse students who pointed out errors or opportunities for improvement in early drafts of the manuscripts for both volumes. I especially would like to thank Dr Brian C Gregory and Dr Donald L Smith for their very thorough and helpful review of Volume 1, and I would also like to thank Ms Roberta Campbell who typed the original and most of the later drafts of the manuscript for both volumes, Ms Jenny Daniel, who drafted nearly all the figures in Volume 1, and Mr James Morrison, who drafted the figures in Volume 2.

Finally, I am desirous of establishing contact with the instructors, students, and in-service professionals who use this book in order to improve it, correct it, and answer any questions. Please feel free to contact me with any corrections or comments at (865)-974-4446 Voice, 865-974-5492 FAX (USA); or by e-mail at jrr@utk.edu.

J Reece Roth, PhD
Weston Fulton Professor of Electrical Engineering
University of Tennessee
Knoxville, Tennessee
17 January, 2001

14

Surface Interactions in Plasma Processing

Volume 1 describes the basic principles of low-temperature, Lorentzian, partially ionized plasmas used industrially. This second volume is concerned with industrial processes that use plasmas or plasma-related technologies in applications that have come to be known as *industrial plasma processing*. The first seven chapters of this volume cover fundamentals from materials science, the physics of plasma sources, plasma reactor technology, the kinetic theory and plasma physics of processing plasmas, and diagnostic procedures from the fields of plasma and materials science. The applications to non-thermal plasma processing described in the remaining five chapters have been organized in order of increasing interaction/modification of the surface of the material.

14.1 INDUSTRIAL PLASMA PROCESSING

In this chapter, we consider some basic physical processes from the field of materials science that are of importance in plasma-processing applications.

14.1.1 Industrially Significant Plasma Characteristics

Plasmas are industrially useful because they possess at least one of two important characteristics. The first characteristic is a high power or energy density. Examples include DC electrical arcs or RF inductive plasma torches, in which the plasma power density can range from 100 W/cm^3 to above 10 kW/cm^3. Such plasmas are in or near *thermodynamic* or *thermal equilibrium*, and are used for *thermal plasma processing*. These plasmas are capable of melting or even vaporizing bulk materials, and are used industrially for welding, plasma flame spraying, arc furnaces, and other high-temperature materials-processing applications. Plasma processing with high power density (*thermal*) plasmas will be discussed in volume 3.

The industrial applications to be covered in this volume rely upon the second major characteristic of plasmas: they produce *active species* that are more numerous, different in kind, and/or more energetic than those produced in chemical reactors. These active species make it possible to do things to the surface of materials that can be done in no other way, or which are economically impracticable by other methods. Such industrially useful active species are most frequently produced by *corona* or *glow discharge* plasmas, which have power densities that range from below 10^{-4} to tens of watts per cubic centimeter.

14.1.2   Benefits of Plasma Processing

Minimizing the unfavorable effects of industrial activity on the environment will play a major role in engineering practice for at least the next generation, as was discussed in chapter 1 of volume 1. When judged by broad environmental criteria, plasma-related processes are, in most cases, preferable to conventional processes that accomplish similar results. Thus, adoption of plasma-processing methods is likely to make important contributions to the reduction of carbon dioxide emissions and global warming; to reductions in energy consumption; to increases in the efficiency of energy use; and to the reduction of pollution and environmental contaminants.

Plasma processing has also found increasingly widespread industrial applications because it can produce unique effects of commercial value that can be obtained in no other way. Plasma processing can result in significantly reduced inputs when compared to conventional processes which accomplish the same result; it can reduce occupational hazards at the point of manufacture; and it can reduce unwanted byproducts and/or minimize toxic wastes, compared to conventional methods of accomplishing the same result.

14.1.3   Conventional and Plasma Surface Treatment of Wool

The ability of plasma-based manufacturing methods to improve upon conventional approaches is illustrated by a well-documented example published by Rakowski (1989), who did a paired comparison of conventional and plasma surface treatment of wool. The objective of the treatment processes under study was to achieve printability of patterns on woolen cloth. Rakowski's paper compared the conventional chemical chlorination process used to accomplish this to a newly developed, vacuum plasma treatment process. The conventional chlorination process, like many similar processes in the textile industry, is relatively difficult and expensive to implement; it presents occupational hazards to the workers involved; it produces significant volumes of unwanted or toxic wastes; and the process leaves the wool smelling like chlorine. The conventional process requires 25 MJ/kg of wool, which includes the energy cost of making the input chemicals needed to treat the wool. It therefore requires 840 MW hr (3024 GJ) to treat 120 tonnes of wool, a commercially significant annual production.

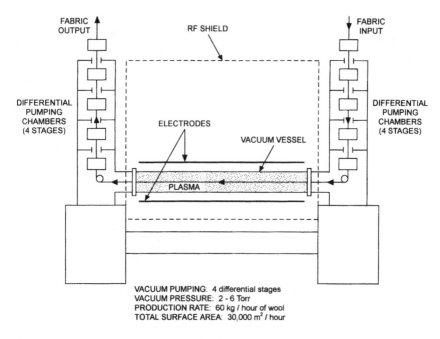

Figure 14.1. Apparatus for continuous low-pressure glow discharge plasma treatment of wool developed by Rakowski (1989).

Rakowski's low-pressure plasma treatment process, which was implemented on a pilot production line, fed wool cloth continuously into a vacuum system where it was exposed to an RF glow discharge plasma operating at pressures from 267 to 800 Pa (2–6 Torr). After plasma treatment, the wool exited to 1 atm, as shown in figure 14.1. The continuous feed of cloth into and out of the vacuum system required four differential stages of vacuum pumping for entrance and exit. The vacuum pumping power was the major energy input and economic expense in this continuous process. The significant energy cost for vacuum pumping results from the total surface area exposed to vacuum in this process. When 60 kg/hr of wool pass through the vacuum system, this represents about 30 000 m^2 of wool fiber surface per hour, all of which is covered with adsorbed monolayers of air molecules. Much of this adsorbed gas causes outgassing and must be pumped away.

Rakowski's low-pressure plasma treatment process required a treatment period in the plasma of 10–15 s, and the woolen cloth moved continuously through the plasma at a rate of 10–15 m/min. The plasma treatment modified several properties of the wool, not just its printability, which was the focus of the study. Compared to the conventional chlorination process, the plasma modification of 120 tonnes per year of wool saves 44 tonnes of sodium hypochlorite, 11 tonnes of sulfuric acid (H_2SO_4), 16 tonnes of bisulfate,

$27\,000$ m^3 of water, and 685 MW hr (2466 GJ) of electrical energy.

Rakowski also found that the power requirement to treat 60 kg/hr of wool using the conventional chlorination treatment (which includes the production of the required chemicals) is 420 kW; the low-pressure plasma treatment process requires only 18 to 36 kW. This is a factor of at least 10 to 20 saving over the conventional process. Most of the remaining energy costs of the low-pressure plasma process are in vacuum pumping requirements. If an atmospheric glow discharge process were used to treat the wool and achieved equivalent results, only the plasma power supply energy would be required. Only 1–4 kW would be required for the plasma power supply, a further factor-of-10 reduction in the power requirements and cost of achieving printability in wool.

14.1.4   Toxic Waste Production in the Microelectronic Industry

Perhaps one of the most striking examples of the role of plasma-related processes in reducing environmental pollution occurred in the microelectronic industry. Here, we compare *dry* or *plasma etching* to its immediate predecessor technology, *wet chemical etching*. The latter technology was widely used until the early 1980s to produce the previous generation of microelectronic circuits with larger component sizes. The use and disposal of chemical stripping and etching fluids, and the wastes from chemical cleaning processes, left the microelectronic industry with a very serious environmental problem (Perry 1993).

Prior to 1982, those working in the microelectronic industry rarely considered the environmental impacts of their processes. This situation changed in January, 1982, when the San Jose, CA *Mercury News* reported that an underground storage tank in Silicon Valley had leaked solvents into a nearby well. This prompted the responsible authorities to investigate 80 other electronic manufacturing facilities, and leaks were found in 85% of the underground tanks that were checked. In the mid-1990s, approximately 150 toxic waste sites were being monitored in Silicon Valley and about 20% of these sites are so seriously contaminated that they are part of the US national Superfund toxic waste cleanup program. Indeed, Silicon Valley has more federal Superfund sites than any other area of its size in the United States, as well as many other toxic waste sites that are being monitored by state and regional agencies.

Figure 14.2, taken from Perry (1993), shows 23 toxic waste sites in Silicon Valley near South San Francisco Bay, a salt-water embayment surrounded by wetlands that nurture fish, birds, and marine mammals. In one of these sites, an underground toxic waste plume is expected to cost over $100\,000\,000$ to clean up. These environmental problems are a legacy of wet chemical etching. Fortunately, further contributions to this problem essentially ceased after 1982, when the microelectronic industry shifted from wet chemical to dry plasma etching, and at the same time took additional steps to greatly reduce the production of chemical wastes from all sources.

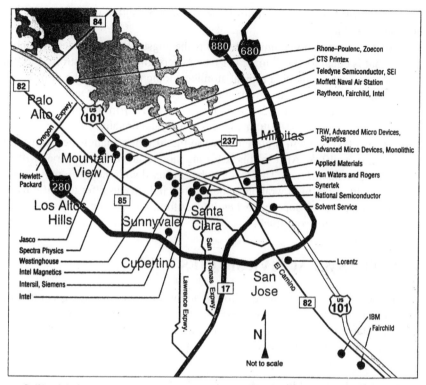

California's Silicon Valley has more Federal Superfund sites than any other area of its size in the nation, plus many other toxic sites that are being monitored by state and regional agencies.

Figure 14.2. A map of Federal Superfund toxic waste sites in Silicon Valley, CA associated with past activities of the microelectronics industry (Perry 1993). ©2001 IEEE.

14.1.5 Relevant Size Scales

The surface properties of materials used commercially are often more important to their function and marketability than their bulk properties. These surface properties are determined by an extremely thin region that can be as small as a few atomic diameters. In considering plasma–surface interactions, it is useful to be aware of the units and size scales involved. The micron, abbreviated μm, is an SI unit equal to 10^{-6} m. It frequently appears in the literature of plasma–surface interactions and the plasma processing of materials. Conventional microelectronic etching technology is accomplished at size scales of about 0.5 μm, and advanced etching technology is carried out at size scales as small as 0.1–0.2 μm. The micron as a unit of length should not be confused with the micron as a unit of vacuum pressure (literally, a difference in the column heights of a mercury U-tube manometer of 10^{-6} m, or 0.13 Pa (10^{-3} Torr)) which is found in the older literature. In this book, the micron will always refer to a unit of length, 10^{-6} m.

Another time-honored unit of length is the ångstrom, abbreviated Å, a non-SI unit that is equal to 10^{-10} m, or 0.1 nanometers (nm). The ångstrom is useful because it is comparable to a typical atomic or ionic radius, and the dimensions between atoms in a solid are conveniently expressed in this unit. One atomic layer of a crystal is approximately 2–3 Å in thickness, and 1 μm therefore represents approximately 3000 to 5000 atomic layers of a crystal or solid.

For this text, we will preferentially use the SI unit of nanometers (nm), for which 1 nm $=$ 10 Å. Also in this text, one atomic layer of a material will be referred to as a *monolayer*. This is not to be confused with the term 'monolayer' which is used in the food wrapping and plastics industry to refer to a film made of a single material, as opposed to a *multilayer* film. Such a single-layer film may be several tens of microns thick.

14.2 PLASMA ACTIVE SPECIES

Plasma-processing effects are due to the action of *active species* generated by the interaction of the working gas(es) with the plasma. These active species are rarely available in purely chemical reactors in the concentrations and active states of excitation found in plasma reactors.

14.2.1 Species Reaching Surface

Two inputs reach the surface of solids exposed to a Lorentzian plasma: the *working gas*, or *feed gas* which might be reflected from, adsorb, absorb, or react chemically with the surface; and the *active species*, which originate directly or indirectly from electron–neutral collisions and subsequent chemical reactions in the plasma. Plasma-generated active species are normally produced in larger numbers and reach higher concentrations than the same species generated in a conventional chemical reactor. The active species usually available from industrial glow discharges and arc plasmas (but not necessarily from dark discharges or corona) include photons, neutral species, and charged particles.

14.2.2 Photons

Photons are available from glow discharge and arc/torch plasmas over a broad spectrum of wavelengths in the electromagnetic spectrum, as summarized in table 14.1. In the infrared portion of the electromagnetic spectrum, the energy of *infrared photons* is too low to interact with the working gas and excite visible radiation from a plasma. These photons possess energies below 1.7 eV, and have, at most, the same general effect as a hot wall or ordinary chemical reactions in inducing plasma-processing effects.

Table 14.1. Photon energies in the electromagnetic spectrum.

Spectral region	Wavelength range (nm)	Energy range (eV)
Infrared	$730 \leq \lambda \leq 10^6$	$0.001\,24 \leq E' \leq 1.70$
Visible	$380 \leq \lambda \leq 780$	$1.59 \leq E' \leq 3.26$
Ultraviolet	$13 \leq \lambda \leq 397$	$3.12 \leq E' \leq 95.3$

Visible photons are more energetic, and have energies ranging from about 1.6 to 3.3 eV. Such photons can break some molecular bonds, and excite atoms with resonances in the visible part of the spectrum. *Ultraviolet photons* are still more energetic, and range from 3.1 to 95 eV. These photons can ionize and excite atoms, scission long hydrocarbon molecules, and break molecular bonds to form smaller molecular fragments.

The energy of photons is given by the *Planck* (Max Planck 1858–1947) *formula*,

$$E = \hbar v \qquad \text{J} \qquad (14.1)$$

or, in electronvolts,

$$E' = \frac{E}{e} = \frac{\hbar v}{e} \qquad \text{eV} \qquad (14.2)$$

where $\hbar$ is Planck's constant, and v the frequency in Hz. The latter is related to the speed of light in free space, c, and the wavelength λ, in meters, of the radiation by

$$v\lambda = c \qquad \text{m/s.} \qquad (14.3)$$

In the ultraviolet, the photon energy is greater than 3 eV. Since photons are uncharged, they are unaffected by electric or magnetic fields that may be present in the sheath above a workpiece, and reach the surface with their original energy.

Ultraviolet and some visible photons may be energetic enough to beak atomic or molecular bonds on a surface, and to produce polymeric free radicals and/or monomers. Sufficiently energetic UV photons from the plasma can modify the molecular state of the surface by the following processes (not necessarily an inclusive list):

(1) *dissociation*, the production of an individual atom or small molecular fragment from a larger molecule, such as a polymer;
(2) *scissioning*, breaking the molecular chain of a polymer; or producing two large molecular fragments from a parent molecule;
(3) *branching*, the production of side chains on a large molecule or polymer; and
(4) *cross-linking*, which produces a two- or three-dimensional molecular matrix by forming bonds between adjacent molecules or polymers.

It should be noted that all these processes can be induced by any sufficiently energetic active species.

Photons can remove electrons from a solid surface by the photoelectric effect, discussed in Volume 1, section 5.2. For this to happen to any significant extent, the photons should have energies in the ultraviolet, above about $E' = 4.5$ eV, the approximate work function of most metals and other materials. The loss of an electron from an electrically insulating solid will leave behind a surface charge.

14.2.3 Neutral Species

Plasmas are capable of producing, through electron–neutral collisions and chemical reactions in the plasma, several kinds of energetic active neutral species capable of interacting strongly with a surface. These neutral species include *reactive atoms*; highly chemically *reactive atomic species* such as H, O, F, Cl, etc; *monomers* which form polymeric chains on surfaces with which they come in contact; relatively *light molecular fragments* such as CH_2, etc; *heavy molecular fragments*, which can form complex compounds on surfaces and/or promote branching and cross-linking of molecules near the surface; *excited atomic or molecular states*, in which excitation of orbital electrons makes a species more chemically reactive than its normal, ground state; and *free radicals*, molecular fragments generated in the plasma with at least one unpaired electron. All such active species may interact strongly with surfaces.

14.2.4 Charged Particles

Industrial partially ionized, Lorentzian plasmas produce, almost entirely by electron–neutral impact ionization, charged particles that may be accelerated to surrounding surfaces by sheath electric fields. Since most industrial glow discharge plasmas float positive with respect to their surroundings, positive ions tend to hit surrounding surfaces, including workpieces, with more energy than electrons or negative ions do. The charged particles available from a plasma include *electrons* produced in the plasma by electron–neutral impact ionization, which typically have kinetic temperatures of 1–10 eV. Such electrons reach surrounding surfaces in numbers equal to the ion flux if that surface is an insulator or is electrically floating. Ions can be positive or negative, but are almost always positive in vacuum glow discharges where the probability of attachment is low. Positive ions are produced by ionization and charge exchange. Negative ions are produced in significant numbers in atmospheric pressure plasmas by attachment of electrons. This process is unlikely under vacuum, because negative ion formation is a three-body process. Molecular ions can also be produced, and may include such charged molecular fragments as OH^-, which may undergo strong chemical reactions in the plasma or on the surface.

14.3 HETEROGENEOUS INTERACTIONS WITH SURFACES

Heterogeneous interactions with surfaces include the chemical reaction or interactions of active species, working gas, energetic individual particles, charged particles, or electromagnetic radiation with surfaces. These interactions may include *heterogeneous chemical reactions* among two or more of the four phases of matter: *solid, liquid, gas*, and *plasma*.

14.3.1 Heterogeneous Interactions in Plasma Processing

Heterogeneous interactions important in industrial contexts usually involve three states of matter, in which neutral gases and plasmas react with a solid surface. Heterogeneous interactions with surfaces are involved in the following industrial processes:

(1) *Plasma cleaning* or *activation* of surfaces, in which exposure to plasma active species results in increases of surface energy, the removal of contaminants, or changes in the chemical structure of surface molecules.

(2) *Plasma thermal diffusion treatment*, as in ion nitriding, boronizing, etc. In this process, a plasma delivers ions to a heated, negatively biased workpiece. The ions are transported by thermal diffusion into the material to form a relatively thick surface layer, which may extend to a millimeter or more in depth.

(3) *Ion-beam implantation*, in which energetic ions in a unidirectional, monoenergetic beam penetrate below the surface of a material and become implanted in a subsurface layer.

(4) *Plasma ion implantation*, in which energetic ions are produced by acceleration across a plasma sheath surrounding the workpiece, and implanted below the surface.

(5) *Sputter deposition of thin films*, in which energetic ions sputter atoms from a target, which are re-deposited as a thin film on a workpiece.

(6) *Plasma Chemical Vapor Deposition (PCVD)*, in which active species from the plasma and the neutral working gas react to form a thin film.

(7) *Plasma etching*, in which ions or other active species from the plasma act as promoters of chemical reactions between a neutral working gas and the substrate.

14.3.2 Characteristic Heterogeneous Interactions

The forms of industrial plasma processing discussed earlier result from two major types of heterogeneous interaction. The first is the interaction of energetic particles with surfaces, in which fluxes of energetic ions, electrons, charge-exchanged neutrals, or photons bombard the surface. The second is the interaction of plasma active species with the surface, including particles in excited states, molecular fragments, free radicals, dissociated atoms, or thermalized charged particles. Some of the more important examples of these heterogeneous interactions encountered in industrial practice include:

(1) *Secondary electron emission*, in which energetic primary species, including ions, electrons, neutrals, or photons knock electrons off the solid surface.
(2) *Sputtering*, in which energetic ions or neutrals knock atoms off the surface of a solid material.
(3) *Erosion*, in which prolonged sputtering results in the removal of a significant depth of surface material.
(4) *Reflection and trapping of ions*, in which ions bounce off the surface (*reflection*) or are retained in the uppermost surface layers of a solid surface (*trapping*).
(5) *Desorption by plasma–wall interactions*, in which monolayers adsorbed on a surface are driven off by bombardment of energetic species or radiation from the plasma.
(6) *Heterogeneous surface chemistry*, in which ions, electrons, excited neutrals, free radicals, or photons from the plasma act as promoters of chemical reactions between a neutral working gas and the solid surface.

In addition to those listed here, many other heterogeneous interactions exist, and can come into play in specific applications of plasma processing.

14.3.3 Energetic Particle-Induced Surface Chemistry

An example of heterogeneous surface chemistry relevant to microelectronic plasma processing is illustrated in figure 14.3, taken from Winters (1980). These data are an example of a heterogeneous *ion-assisted gas–surface chemical reaction*. In this experiment, silicon was exposed to xenon difluoride (XeF_2) gas, with and without a 450 eV argon ion beam bombarding the surface. For the first 200 s of the experiment, the silicon was exposed only to the XeF_2 gas, which etched it at a rate of approximately 0.5 nm/min. When a 450 eV argon ion beam was turned on, the etching rate increased more than a factor of 10, due to ion-assisted surface chemical reactions 'catalyzed' by ion bombardment. After an initial transient period, the etching rate leveled off at approximately 5.5 nm/min. At approximately 640 s, the XeF_2 flow was turned off, leaving the silicon exposed only to sputtering by the argon ion beam. With the XeF_2 entirely removed from the system, the etch rate of the argon ion beam only on the silicon (i.e. the sputtering rate) was about 0.2 nm/min. These data illustrate that, while both the XeF_2 gas alone and the argon ion beam alone were capable of only very small etching rates, the combination of the argon ion beam with the XeF_2 gas produced a very high etching rate.

An example of an electron-assisted gas–surface chemical reaction (Winters 1980) is shown in figure 14.4. These data show the total thickness removed from a silicon dioxide (SiO_2) surface as a function of time. The SiO_2 was exposed to

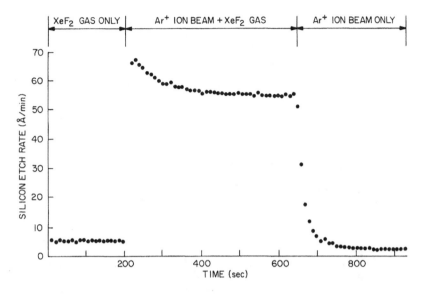

Figure 14.3. Ion-assisted gas–surface chemistry using argon ions and a background gas of XeF_2 on SiO_2. The argon ions had an energy of 450 eV, and a total current of 2.5 μA over an area of 0.1 cm^2 (Winters 1980).

XeF_2 alone for about 800 s, at which time a 1500 eV, 50 mA/cm^2 electron beam was turned on. The electron beam allowed the XeF_2 to etch the SiO_2 surface at a rate of approximately 20 nm/min, a clear-cut example of the catalytic effect of an electron beam on the chemical reaction of XeF_2 with the SiO_2 surface. Such heterogeneous surface catalytic effects have also been observed when an energetic laser beam impinges on a surface in the presence of a suitable etching gas.

14.3.4 Heterogeneous Reactions of Ions with Surfaces

The direct interaction of energetic ions with surfaces is illustrated in figures 14.5 and 14.6, taken from Winters (1980). Figure 14.5 shows the probability of molecular N_2^+ ions dissociating on impact with various metals as a function of the ion energy. These curves indicate that above 100 eV, molecular nitrogen ions completely dissociate on impact, an expected result because the impact energy is well above the dissociation energy of a few eV.

Figure 14.6 shows the nitrogen surface concentration, in atoms/cm^2, as a function of the total dose of incident energetic molecular nitrogen ions. The accumulation of nitrogen was due to very shallow implantation due to the nitrogen energy, and chemical reaction of the nitrogen to form a surface layer of nitride. Formation of this surface nitride layer produced the saturation observed.

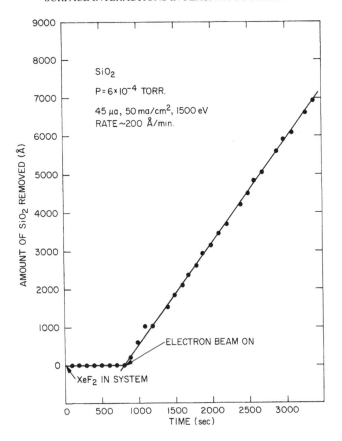

Figure 14.4. Electron-assisted gas–surface chemistry using 1500 eV electrons and XeF$_2$ simultaneously incident on SiO$_2$. Bombardment was conducted at a total pressure of 0.08 Pa (6 × 10^{-4} Torr). The electron beam had a current density of approximately 50 mA/cm^2 (Winters 1980).

14.3.5 Reflection and Trapping on Surfaces

Several processes may occur when an incident ion, atom, or molecule sticks to a surface. *Absorption* is the penetration of an incident particle into the bulk material. When the incident energy becomes sufficiently high, above a few keV, the process of absorption becomes equivalent to the process of *implantation*. *Adsorption* means that the incident particle adheres to the surface only, either by short-range surface forces or by becoming incorporated in the stacked monolayers covering the surface. Particles in the outermost monolayers covering a surface are loosely bound and are generally easy to drive off by heating or bombarding the surface with energetic ions or electrons. Particles in the lowermost monolayers near the surface may be bound with an energy comparable to the work function of

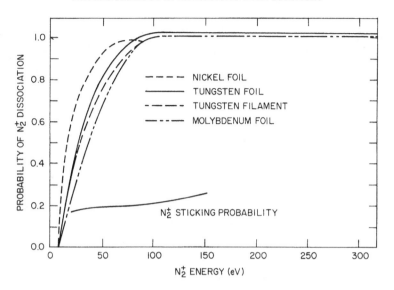

Figure 14.5. Probability that a N_2^+ ion will dissociate upon collision with the four metal surfaces listed, as a function of incident ion energy (Winters 1980).

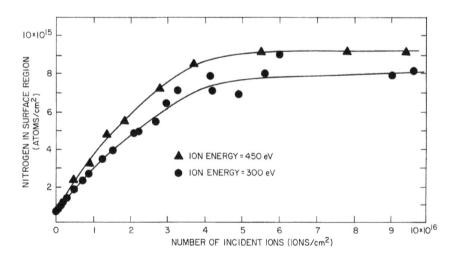

Figure 14.6. The amount of nitrogen contained in the surface region of tungsten bombarded with N_2^+ ions, as a function of the total dose of ions (Winters 1980).

the surface, and are very difficult to dislodge. *Trapping* is the sum of absorption and adsorption.

14.3.5.1   The Reflection Coefficient

The *reflection coefficient*, R, of ions incident on a surface is defined as

$$R \equiv \frac{\text{ions/neutrals reflected from surface}}{\text{ions incident on surface}}. \tag{14.4}$$

The particles reflected from the surface are the same ones that were incident on it, otherwise the process would be sputtering or outgassing, not reflection.

14.3.5.2   The Absorption Coefficient

The *absorption coefficient*, T_b, is defined as

$$T_b = \frac{\text{incident particles absorbed}}{\text{total incident particles}}. \tag{14.5}$$

14.3.5.3   The Adsorption Coefficient

The *adsorption coefficient*, T_d, is given by

$$T_d = \frac{\text{incident particles absorbed}}{\text{total incident particles}}. \tag{14.6}$$

14.3.5.4   The Trapping Coefficient

The *trapping coefficient*, T (also known as the *trapping probability* or the *sticking probability*), is given by

$$T = \frac{\text{incident particles trapped}}{\text{total incident particles}} \tag{14.7}$$

or

$$T = T_b + T_d. \tag{14.8}$$

It follows that the relation between reflection and trapping is

$$T + R = 1. \tag{14.9}$$

14.3.5.5   Characteristic Examples

Selected examples of these coefficients have been taken from the literature and are reproduced in figures 14.7 and 14.8. Figure 14.7, taken from Winters (1980), shows the trapping coefficient, T, for four species of noble gas ions impinging on tungsten. These data illustrate a general trend in which the trapping coefficient increases monotonically with ion energy for most ion–material combinations. Figure 14.8, also from Winters (1980), shows the trapping coefficient, T, of nitrogen gas on tungsten as a function of surface coverage for a range of surface

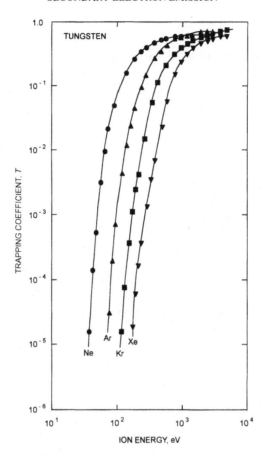

Figure 14.7. The trapping coefficient T as a function of ion energy for neon, argon, krypton, and xenon on tungsten (Winters 1980).

temperatures. These data illustrate the utility of cryogenic baffles, operated at liquid nitrogen temperature (90 K) in vacuum systems, as effective traps of neutral gases. It is also clear that heating the surface will reduce both the loading of trapped particles and the probability that a particle will be trapped.

14.4 SECONDARY ELECTRON EMISSION

Secondary electron emission results from the surface impact of such energetic *primary species* as ions, electrons, neutrals, or photons. In the case of photons, the process is usually referred to as *photoemission*, and the resulting electron as a *photoelectron*. Secondary electron emission is distinct from other physical processes that cause electrons to be emitted from solids, such as field and thermionic emission, discussed in chapter 5 of Volume 1.

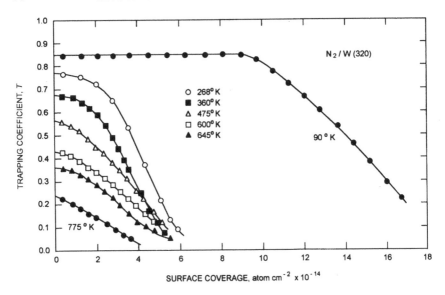

Figure 14.8. The trapping coefficient T as a function of surface coverage for nitrogen incident on the (320) plane of tungsten. The temperature of the tungsten surface is shown as a parameter (Winters 1980).

14.4.1   Secondary Electron Emission Coefficient

The *secondary electron emission coefficient*, γ, is defined as

$$\gamma_j \equiv \frac{\text{No. of electrons leaving surface}}{\text{No. of incident particles of type } j} \tag{14.10}$$

where $j \equiv i$, for incident ions; $j \equiv e$, for incident electrons; $j \equiv n$, for incident neutrals; and $j \equiv p$ for photons. Usually only γ_i is of interest, and we will omit the subscript and assume that γ refers to electron emission resulting from ion bombardment, unless stated otherwise.

14.4.2   Functional Dependence of γ_i

For ion bombardment, the secondary electron emission coefficient is a function of multiple parameters. These parameters include the energy, E_1; the *atomic mass number*, A_1; and the *proton number* (also referred to as the 'atomic number'), Z_1, of the incident ion. The secondary electron emission coefficient also depends on the atomic state of the incident ion, whether excited, ionized, or multiply ionized; the angle of incidence of the particle (unless stated otherwise 90° incidence, normal to the surface, is reported in the literature); the atomic mass number A_2 of the surface material; and the crystallographic nature of the surface (orientation, whether amorphous or crystalline).

14.4.3   Effect on Collected Current

When real currents of positive ions flow to electrodes or surfaces that are monitored by current meters, secondary electron emission can give rise to a false (high) current reading. A current meter cannot distinguish between an ion arriving at an electrode, and γ_i electrons leaving. The relation between the measured current I_m and the true ion current I_i is given by

$$I_m = I_i(1 + \gamma_i) \qquad \text{A.} \qquad (14.11)$$

The coefficient γ_i can be greater than 1.0, so the effect of secondary electron emission can be significant.

14.4.4   Characteristic Data

The secondary electron emission coefficient γ_i as a function of ion energy for various ions incident on aluminum is shown on figures 14.9–14.11. Figure 14.9 shows the secondary electron emission coefficient of carbon, oxygen, helium, and atomic and molecular hydrogen ions incident on aluminum, taken from the Oak Ridge data tables (Thomas 1985). Similar data from Langley *et al* (1984) are shown in figure 14.10, which illustrate the general linear dependence of γ on incident ion velocity. Figure 14.11 includes an estimate of the secondary electron emission coefficient of nitrogen ions incident on aluminum by the present author, in addition to data on the dependence of γ on ion energy from Langley *et al* (1984).

The secondary electron emission coefficient of stainless steel for ions of two different energies, the proton numbers of which are shown on the abscissa, is shown on figure 14.12. The modulation evident in these data is associated with the filling of atomic electron shells in the target material as one moves across the periodic table.

14.5   SPUTTERING

14.5.1   Definitions

Sputtering is the loss of atoms from a surface as the result of energetic bombardment, usually by ions or energetic neutrals. *Erosion* is the loss of a significant thickness of material as the result of the cumulative effect of sputtering. The *sputtering yield* (or *sputtering ratio*) ε is defined as

$$\varepsilon \equiv \frac{\text{No. of sputtered atoms}}{\text{No. of incident particles}}. \qquad (14.12)$$

Electrons (because they are not massive enough) and neutrals (because they are not energetic enough) rarely sputter atoms from surfaces under conditions encountered in industrial practice. Sputtering yields significant enough to be of industrial interest are normally the result of energetic ion bombardment.

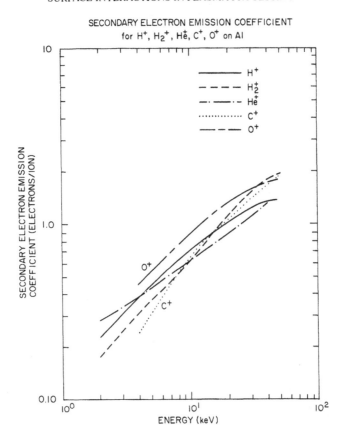

Figure 14.9. The secondary electron emission coefficient γ in electrons per incident ion, for H, H_2, He, C, and O ions on aluminum, and energies from 1 to 100 keV (Thomas 1985).

14.5.2　Functional Dependence of ε

The *sputtering yield* depends upon the energy, E_1, of the incident particle; the mass, M_1, of the incident particle; the angle of incidence of the particle to the surface (normal incidence is customarily reported in the literature); the atomic weight, A_2, of the surface material; and the state and crystalline orientation of the surface material.

14.5.3　Features of Sputtering Yield Curves

A *sputtering yield curve* is a graphical presentation, usually on semilog coordinates, of the sputtering yield ε defined in equation (14.12) as a function of incident particle energy. Such a curve is shown schematically on figure 14.13,

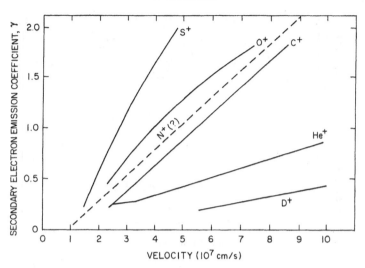

Figure 14.10. The secondary electron emission coefficient γ for deuterium, helium, carbon, oxygen, and sulfur ions on aluminum, with an interpolated estimate for nitrogen ions on aluminum, as a function of velocity in units of 10^7 cm/s (modified from Langley *et al* (1984)).

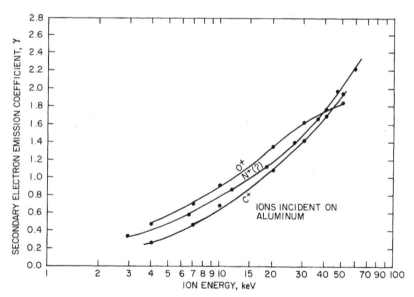

Figure 14.11. The secondary electron emission coefficient γ as a function of energy for carbon, nitrogen, and oxygen ions incident on aluminum (based on Langley *et al* (1984)).

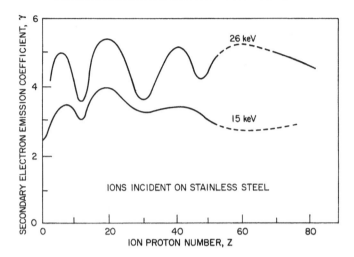

Figure 14.12. The secondary electron emission coefficient γ as a function of the proton number of ions having energies of 15 and 26 keV impacting on stainless steel (Langley *et al* 1984).

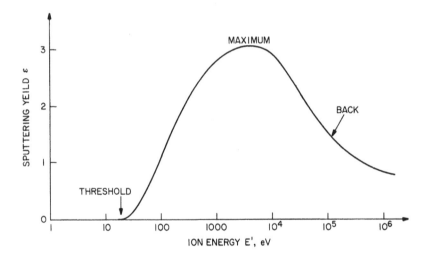

Figure 14.13. Schematic diagram of a sputtering yield curve as a function of ion energy.

and has the following characteristic features: a *threshold*, usually in the range 20–100 eV; a *maximum*, usually between 1 and 10 keV; and the *back*, a region of slow decline from the maximum to energies approaching 1.0 MeV.

Illustrative examples of sputtering data from a very large literature are shown in figures 14.14–14.17. Figure 14.14, taken from Langley *et al* (1984), shows the

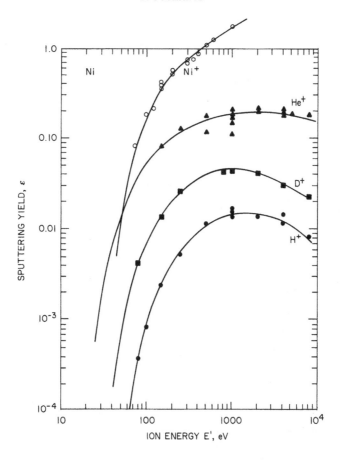

Figure 14.14. The sputtering yield ε as a function of ion energy for protium, deuterium, helium, and nickel ions incident on nickel (Langley *et al* 1984).

sputtering yield at normal incidence of nickel, helium, deuterium, and protium ions on solid nickel as a function of incident ion energy. These data illustrate the increase of the sputtering yield with atomic mass number of the bombarding ion, a dependence expected on the basis that heavier ions do more damage on impact.

Figure 14.15, taken from Wiffen (1984), plots the measured and theoretically calculated sputtering yield of ions of several masses incident on stainless steel as a function of the energy of the incident ion. These data also illustrate that the sputtering yield increases with increasing atomic mass of the incident ion.

Figure 14.16, taken from Winters (1980), shows the sputtering yield as a function of the energy of argon ions incident on silicon and germanium, two materials important in microelectronic circuit fabrication. These data illustrate the dependence of sputtering yield on crystal orientation in single crystal materials.

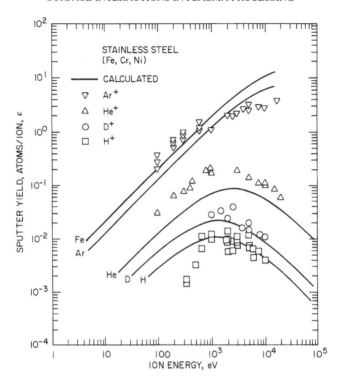

Figure 14.15. The sputtering yield ε for protium, deuterium, helium, and argon ions on stainless steel as a function of energy. Calculated values from an approximate theory are also shown for these ion–surface combinations (Wiffen 1984).

The high sputtering yield of germanium can sometimes render doping of this material problematic, due to rapid erosion of implanted surface layers.

14.5.4   Angular Dependence of Sputtering Yield

Figure 14.17, taken from Molchanov and Tel'kovskii (1961), shows the sputtering yield of 27 keV argon ions on polycrystalline copper, as a function of the angle of incidence. The sputtering yield increases significantly for shallow angles of incidence of the ion on the metal, as the result of a 'snowplowing' effect. The experimental data illustrate an empirical relationship between sputtering yield and incident angle given by

$$\varepsilon(\theta)\cos\theta \approx \varepsilon(\theta = 0) \tag{14.13}$$

that is valid up to angles of $\theta \approx 70°$. This useful approximation is plotted for the yield data in figure 14.17 as the horizontal line.

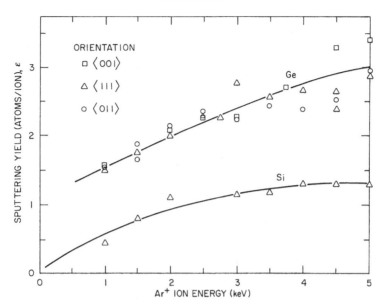

Figure 14.16. The sputtering yield ε as a function of energy for argon ions impacting various lattice planes of silicon and germanium (Winters 1980).

14.5.5 Phenomenological Sputtering Yield

Available theories of sputtering based on the older work of Lindhard and Bohr are described by Winters (1980: 86–100). Unfortunately, these theoretical attempts to describe sputtering are characteristically incorrect by factors of two to five when compared to experimental measurements. As a result, investigators in recent years have tended to use phenomenological correlations of experimental data. Here, we describe such a correlation, developed by Bohdansky *et al* (1980). Their correlation is valid for 'low' energies, below the maximum in the sputtering yield curve of the material of interest.

These authors assume a yield curve with the functional form

$$\varepsilon(E_1) = QY_n(KE_1) \tag{14.14}$$

where Y_n is a dimensionless phenomenological sputtering yield curve obtained from experimental data, and Q and K are constants for each ion–target combination. Values of Q determined experimentally by these authors for a number of ion–target combinations are given in table 14.2.

The constant K appearing in equation (14.14) is given by the approximation (useful for making estimates beyond the database in tables 14.2 and 14.3)

$$\frac{1}{K} = \frac{E_b}{\gamma} \tag{14.15}$$

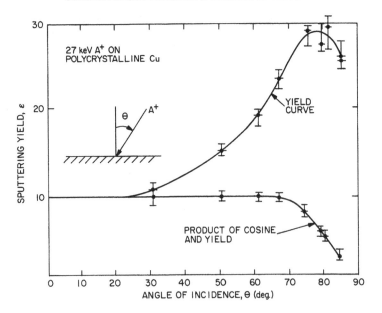

Figure 14.17. The sputtering yield ε as a function of angle of incidence of 27 keV argon ions on polycrystalline copper. The yield curve is shown, as well as the product of the cosine and yield (based on Molchanov and Tel'kovskii (1961)).

where E'_b is the *surface binding energy*, approximately the heat of sublimation of the material, and the *momentum transfer parameter*, γ, which here is equal to

$$\gamma \equiv \frac{4M_1 M_2}{(M_1 + M_2)^2}. \tag{14.16}$$

In this expression, M_1 is the incident particle mass, and M_2 is that of the atoms comprising the solid.

The constant $1/K$ is found by substituting equation (14.16) into equation (14.15), to obtain

$$\frac{1}{K} = \frac{E'_b(M_1 + M_2)^2}{4M_1 M_2} \quad \text{eV}. \tag{14.17}$$

Values of $1/K$ are given in table 14.3, taken from Bohdansky *et al* (1980), for a range of ion–target combinations.

Bohdansky *et al* (1980) have shown that a range of experimental data can be reduced to a common phenomenological scaling law by using the dimensionless energy E_0, given by

$$E_0 = KE_1 = \frac{4M_1 M_2 E_1}{(M_1 + M_2)^2 E_b} = \frac{E_{\max}}{E_b}. \tag{14.18}$$

Table 14.2. Yield factor Q for various target–ion combinations (Bohdansky *et al* 1980).

Target	H	D	^{3}He	^{4}He
		Ion		
Al	0.84	3.2		14
Au	1	2.5	6.1	9
Be	1	2.35		6.4
C	0.68	1.6		5.4
Fe	1.2	2.7		8.5
Mo	0.17	0.52	0.52	2.9
Ni	0.95	2.85		9.5
Si	0.5	1.45		4.9
Ta	0.14	0.38		1.8
Ti	0.25			2.2
V	0.43			4.0
W	0.175	0.37		1.9
Zr				2.8

Target	Ne	Ar	Kr	Xe	Ni
			Ion		
Ni	56	85	114	97	72
W	22	47	80	90	
Mo	33	55	63	67	

Experimental data for 14 elements were correlated with equation (14.14) by Bohdansky *et al* (1980), and the results are shown in figure 14.18. There is generally good agreement between the experimental data and the phenomenological correlation up to dimensionless energies $E_0 = 20$, and a useful similarity up to dimensionless energies $E_0 = 40$.

The full line through the data on figure 14.18 is the Bohdansky *et al* (1980) energy function $Y_n(K, E_1)$, given by

$$Y_n(E_0) = Y_n(KE_1) = 8.5 \times 10^{-3}(E_0)^{1/4}\left[1 - \frac{1}{E_0}\right]^{7/2} \qquad 1 \leq E_0 \leq 40 \text{ keV}.$$
(14.19)

For a dimensionless energy $E_0 = 1.0$, equation (14.18) gives $\gamma E_1 = E_b$. Substituting equation (14.19) into the phenomenological equation (14.14) yields

$$\varepsilon(E_0) = 8.5 \times 10^{-3}Q(E_0)^{1/4}\left(1 - \frac{1}{E_0}\right)^{7/2}.$$
(14.20)

From equation (14.20), $\varepsilon = 0$ when $E_0 = 1.0$. Thus, the threshold energy E_{th},

Table 14.3. Experimentally determined values for $1/K$, in eV, for several target–ion combinations (Bohdansky *et al* 1980).

Target	Ion			
	H	D	^{3}He	^{4}He
Al	53	34		20.5
Au	194	94	60	44
Be	27.5	24		33
C	10	11		16
Fe	64	40		35
Mo	164	86	45	39
Ni	47	32.5		20
Si	24.5	17.5		14
Ta	460	23.5		100
Ti	43.5			22
V	76			27
W	400	175		100
Zr				60

Target	Ion				
	Ne	Ar	Kr	Xe	Ni
Ni	25	31	50	56	34
W	40	45	52	57	
Mo	30	33	43	48	

below which very little sputtering occurs, can be identified with

$$E_0 = 1.0 = K E_{\text{th}} \tag{14.21}$$

or

$$E_{\text{th}} = \frac{1}{K}. \tag{14.22}$$

For combinations of ions and sputtered materials other than those shown in tables 14.2 and 14.3, Bohdansky *et al* (1980) suggest the approximations

$$E_{\text{th}} = \frac{1}{K} = \frac{E_{\text{b}}}{\gamma(1-\gamma)} \qquad M_1/M_2 < 0.3. \tag{14.23}$$

For $M_1 \ll M_2$, equation (14.23) becomes

$$E_{\text{th}} = \frac{1}{K} \approx \frac{E_{\text{b}}}{\gamma} \qquad M_1 \ll M_2. \tag{14.24}$$

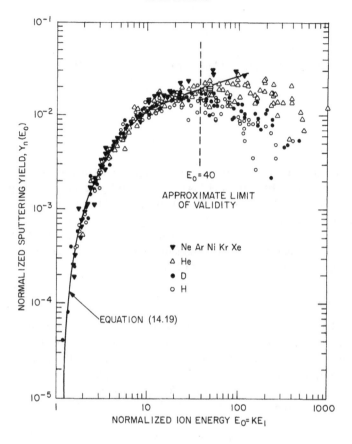

Figure 14.18. The normalized sputtering yield Y as a function of normalized ion energy for hydrogen, deuterium, helium, and several noble gas ions on a range of materials. The full line is equation (14.19) of the text (Bohdansky *et al* (1980)).

When $M_1/M_2 > 0.3$, Bohdansky *et al* (1980) show that the data are consistent with

$$E_{th} = \frac{1}{K} = 8E_B \left(\frac{M_1}{M_2}\right)^{5/2}. \tag{14.25}$$

Equations (14.23)–(14.25) allow one to go beyond the available database to obtain E_{th} or $1/K$, since E_b is known for nearly all materials, as are M_1 and M_2.

The parameter Q for species other than those shown in table 14.2 can be found as functions of M_1 and M_2 in the manner described by Bohdansky *et al* (1980). By plotting Q/M_2 as a function of γ, the authors obtain

$$Q = 0.75M_2\gamma^{5/3} \qquad M_1 < M_2 \tag{14.26}$$

with M_2 in atomic mass units (amu). Substituting equation (14.26) into equation (14.20), one obtains

$$\varepsilon(E_0) = 6.4 \times 10^{-3} M_2 \gamma^{5/3} (E_0)^{1/4} \left[1 - \frac{1}{E_0}\right]^{7/2} \qquad M_1 < M_2, \; E_0 < 40 \tag{14.27}$$

and, using equation (14.21),

$$E_0 = \frac{E_1}{E_{\text{th}}} = KE_1. \tag{14.28}$$

A comparison of the sputtering coefficient predicted by equation (14.27) with some data presented by Wiffen (1984) is shown in figure 14.19. Two points, at deuterium ion energies of 100 and 200 eV, were calculated from equation (14.27) for sputtering on beryllium and carbon. These points are plotted on figure 14.19. The carbon data are about a factor two higher than the experimental data shown on this graph; the beryllium data are a factor two too low, indicating the degree of approximation implicit in equation (14.27).

14.5.6 Erosion

In industrial applications, *erosion* by individual electrons or energetic neutrals is negligible. Erosion normally results from energetic ion beams, or plasma ions accelerated across sheaths with large voltage drops, such as those encountered in plasma ion implantation or magnetron sputtering configurations. The *flux* (also referred to as the *flux density* in the older and/or electrodynamic literature) of sputtered atoms from a surface is

$$\Gamma_s = \varepsilon \Gamma_i \qquad \text{atoms/m}^2\text{-s} \tag{14.29}$$

where Γ_i is the flux of ions reaching the wall. If this flux is a beam of singly charged, monoenergetic ions with a current density J_i A/m^2, the *sputtered atom flux* may be written

$$\Gamma_{sb} = \varepsilon J_i/e \qquad \text{atoms/m}^2\text{-s.} \tag{14.30}$$

If the beam is of area A_b m^2, with a total current I_b A, equation (14.30) becomes

$$\Gamma_{sb} = \frac{\varepsilon I_b}{e A_b} \qquad \text{atoms/m}^2\text{-s.} \tag{14.31}$$

If a surface subject to sputtering is immersed in a plasma, the flux of ions reaching the boundary of a simple Debye sheath is approximately

$$\Gamma_{ip} = \frac{1}{4} n_i \bar{v}_i = n_i \left(\frac{e T_i'}{2\pi m_i}\right)^{1/2} \qquad \text{ions/m}^2\text{-s.} \tag{14.32}$$

The ion flux at the boundary of a Bohm sheath would be greater, since ions enter such a sheath with the Bohm velocity, and not the ion thermal velocity.

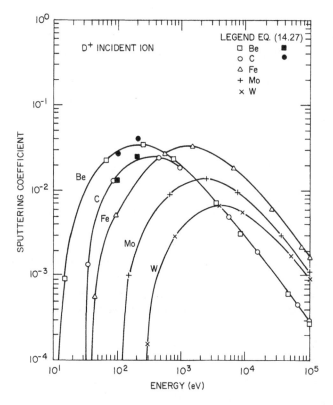

Figure 14.19. The sputtering yield ε as a function of energy for deuterium ions incident on the surface of several materials (Wiffen 1984). The solid symbols are data for beryllium and carbon calculated from equation (14.27).

Substituting equation (14.32) into equation (14.29), a lower bound on the sputtered atom flux due to the presence of the plasma is given by

$$\Gamma_{sp} = \varepsilon n_i \left(\frac{e T_i'}{2\pi m_i} \right)^{1/2} \qquad \text{atoms/m}^2\text{-s.} \qquad (14.33)$$

To calculate the rate at which erosion occurs, or the *erosion velocity*, the atom number density of the sputtered wall material is required, which is given by

$$N_w = \frac{A_0 \rho}{A} \qquad \text{atoms/m}^3 \qquad (14.34)$$

where Avogadro's number $A_0 = 6.02 \times 10^{26}$ atoms/kg-atom, ρ is the density of the sputtered material in kg/m^3, and A is the atomic mass number of the wall material. The *erosion velocity* of a wall under ion-beam bombardment may be found by dividing equation (14.31) by equation (14.34), to yield

$$V_{eb} \equiv \frac{\Gamma_{sb}}{N_w} = \frac{\varepsilon I_b A}{e A_0 \rho A_b} = \frac{\varepsilon J_i A}{e \rho A_0} \qquad \text{m/s.} \qquad (14.35)$$

Similarly, the erosion velocity due to bombardment by ions of mass m_i from a plasma with ion kinetic temperature T_i' eV can be found by dividing equation (14.34) into equation (14.33), to yield

$$V_{eb} = \frac{\Gamma_{sb}}{N_w} = \frac{\varepsilon n_i A}{A_0 \rho} \left(\frac{e T_i'}{2 \pi m_i} \right)^{1/2} \qquad \text{m/s.} \qquad (14.36)$$

In an industrial glow discharge plasma, the ions are likely to be thermalized to room temperature, a kinetic temperature of only 0.025 eV, so equation (14.36) will be negligible compared to equation (14.35), which describes the 'beamlike' ions accelerated across the sheath above a workpiece.

In a geometry in which one of two large parallel plates is sputtered by an energetic ion beam, and the sputtered atoms deposited on the opposite plate, the deposition rate will approximately equal the erosion velocity. The time, in hours, required to erode a thickness L meters of target material is given by

$$T = \frac{L}{3600 V_e} \qquad \text{hr.} \qquad (14.37)$$

For sputtering by monoenergetic ion beams of energy E_i' eV and mass m_i, the ion velocity is found from the kinetic energy,

$$v_i = \sqrt{\frac{2 e E_i'}{m_i}} \qquad \text{m/s.} \qquad (14.38)$$

The current density of an ion beam impinging on a sputtered material is given by

$$J_i = e n_i v_i = e n_i \sqrt{\frac{2 e E_i'}{m_i}} \qquad \text{A/m}^2 \qquad (14.39)$$

where n_i is the beam number density in ions/m^3. Inserting equation (14.39) into equation (14.35), the *erosion velocity* of an ion beam can be expressed as

$$V_{eb} = \frac{\varepsilon J_i A}{e \rho A_0} = \frac{\varepsilon A n_i}{\rho A_0} \sqrt{\frac{2 e E_i'}{m_i}} \qquad \text{m/s.} \qquad (14.40)$$

Although equation (14.40) is expressed in m/s, in industrial applications the erosion/deposition rate is sometimes expressed in μm/hr. In depositing thin films, a deposition rate (which normally cannot exceed the erosion rate of the material that produced it) of the order of 10 μm/hr is of interest in many applications.

14.6 ION IMPLANTATION IN SOLIDS

14.6.1 Ion Implantation Methods

Ion implantation consists of directing ions onto surfaces with enough energy that they penetrate the atomic structure of the material and come to rest many atomic layers below the surface. The ion energies typically used for implantation are in the range 20–300 keV. These energetic ions may originate from space-charge-limited ion sources of the types described in chapter 6, or by extraction from a plasma and acceleration across a sheath to the workpiece, *plasma ion implantation*. The physics and technology of these two implantation methods will be discussed further in chapter 22.

14.6.2 Physical Processes in Ion Implantation

A characteristic implantation process is illustrated in figure 14.20. A beam of energetic ions is directed vertically downward onto a crystalline lattice. A few ions travel long distances through tubes or channels in the orderly structure of the lattice, in *channeled trajectories*. Most ions scatter immediately and perform a random walk into the interior as *dechanneled trajectories*. There is some interaction of the energetic ions with adsorbed surface layers of the surrounding neutral gas (not shown in figure 14.20). This interaction desorbs the outermost monolayers (*outgassing*), but this process reaches a steady state, as discussed in section 21.2.2. In addition to becoming implanted, incident ions can sputter the target surface, driving off ε atoms per incident ion. One normally wishes to avoid sputtering more material than is implanted. This can happen for $\varepsilon > 1.0$, particularly for oblique incidence of energetic ions.

Channeled trajectories occur along clear directions in a crystalline lattice. The channel radii depend on the ion energy through the elastic scattering cross sections, and are comparable to the atomic radii of the lattice atoms. In the ion implantation process, one usually wishes to avoid channeling, in order to implant ions a known, controllable distance below the surface.

14.6.3 Ion Interaction Regimes

As was true in the sputtering process, energetic ions undergoing implantation can interact with a lattice in two ways: (1) at low energies, ions interact with the lattice as a collective whole; or (2) at higher energies, ions can interact by sequential binary collisions with individual atoms of the lattice. It has been the collective experience of individuals in the field of solid state physics that the second, binary collisional model is the most productive theoretically, and the most accurate when compared with experiments.

Binary collisions of ions with atoms in a solid can be of two types:

(1) *Electronic collisions*, in which energy is transferred between the electron

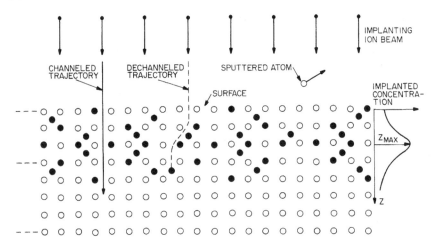

Figure 14.20. Ion implantation of the surface of a crystalline material. Implanted ions are shown as full circles.

clouds of the incident and target atoms. Electrons of either atom can be excited or stripped, causing further ionization. These interactions give rise to electronic stopping.

(2) *Nuclear collisions*, in which the collisions are dominated by the atomic mass of the ions and target atoms, which is concentrated in their nuclei. Such interactions give rise to *nuclear stopping*, which can result in sudden changes in direction, and even *displacements* of target atoms from lattice sites. This terminology is unfortunate since, *in the field of solid state physics*, 'nuclear collisions' do not involve nuclear forces or interactions, the meaning normally understood in fusion research and nuclear physics. In almost all cases, an implanted ion will undergo a large number of collisions before it comes to rest in the lattice.

Interactions of energetic ions with a lattice can be further characterized by two physics regimes: (1) interactions describable by classical mechanics; and (2) interactions requiring a quantum-mechanical description. As will be illustrated later, most industrial applications of sputtering or ion implantation are in the classical physics regime.

The analytical theory of nuclear stopping depends on the *distance of closest approach*, b, given by the point at which an incident ion of energy E_i' eV gives up its maximum kinetic energy to the potential energy of the target atom. In a head-on collision, the initial kinetic energy will be equal to this potential energy,

$$E_1 = eE_1' = \frac{Z_1 Z_2 e^2}{4\pi \varepsilon_0 b} \qquad \text{J.} \qquad (14.41)$$

Equation (14.41) can be solved for the distance of closest approach,

$$b = \frac{Z_1 Z_2 e}{4\pi \varepsilon_0 E_1'} \quad \text{m.} \tag{14.42}$$

The distance of closest approach, b, is to be compared to the *effective atomic radius*, a, of the electron clouds surrounding the interacting nuclei. From scattering theory, discussed in Winters (1980), the effective atomic radius a for an interaction between two atoms with proton numbers Z_1 and Z_2 is given by

$$a = \frac{a_0}{\sqrt{Z_1^{2/3} + Z_2^{2/3}}} \quad \text{m} \tag{14.43}$$

where a_0 is the *Bohr radius*, the classical radius of ground-state electrons in hydrogen, given by

$$a_0 = \frac{h^2}{4\pi^2 e^2 m_p} = 5.2918 \times 10^{-11} \text{ m} \tag{14.44}$$

or

$$a_0 = 0.5918 \text{ Å.} \tag{14.45}$$

When $b < a$, the incident ion undergoes *Rutherford scattering* (a *Coulomb collision*) with an electrostatic potential

$$V \sim \frac{1}{r}. \tag{14.46}$$

When $b > a$, the interaction is screened by the electron clouds, and the repulsive part of the interaction potential is approximated by

$$V \sim \frac{1}{r^n} \tag{14.47}$$

where n is an integer greater than unity. To determine whether the interaction is electronic or nuclear, the *penetration parameter*, ξ, is defined as

$$\xi \equiv \frac{b}{a}. \tag{14.48}$$

If $\xi < 1$, the collision is *nuclear* or *Coulomb-like*. If $\xi \geq 1$, the collision is *electronic*.

To determine whether the collision is *classical* or *quantum mechanical*, the distance of closest approach, b, is compared to the *De Broglie wavelength* of the incident ion,

$$\lambda = \frac{h}{2\pi m_r v_1} \quad \text{m} \tag{14.49}$$

Table 14.4. Slowing down parameters for a 300 keV nitrogen ion implanted in silicon.

Energy E (eV)	χ	$\log \chi$	ξ	$\log \xi$
300 000	141	2.15	0.0273	−1.56
30 000	446	2.65	0.273	−0.564
3 000	1 411	3.15	2.73	0.436
300	4 462	3.65	27.3	1.436
30	14 111	4.15	273	2.436
14	20 655	4.32	585	2.77

where m_r is the reduced mass,

$$m_r = \frac{M_1 M_2}{(M_1 + M_2)} \quad \text{kg} \tag{14.50}$$

and v_1 is the velocity of the incident ion,

$$v_1 = \sqrt{\frac{2e E_1'}{M_1}} \quad \text{m/s.} \tag{14.51}$$

The distance of closest approach, b, and the De Broglie wavelength are used to define the *quantum scale parameter*, χ,

$$\chi \equiv \frac{b}{\lambda}. \tag{14.52}$$

When $\chi \leq 1$, $b \leq \lambda$ and a quantum-mechanical treatment is required. If $a \leq \lambda$ (equivalent to $\xi \leq \chi$) a quantum-mechanical treatment of the electronic interactions is also required. The division of the ξ–χ space into classical and quantum-mechanical regions is indicated in figure 14.21. For *classical interactions*, both $\chi > 1.0$ *and* $\chi > \xi$ must be satisfied simultaneously. For the *nuclear* (or *Coulomb-like*) *stopping* regime, $\xi \leq 1.0$, as indicated in figure 14.21. For the *electronic stopping* regime, $\xi > 1.0$, also indicated in figure 14.21.

The utility of figure 14.21 is that as an ion slows down in a solid, the relation

$$\chi \propto \xi^{1/2} \tag{14.53}$$

represents the slowing down of an ion's velocity. This relation is a straight line of slope $\frac{1}{2}$ which terminates at the upper right-hand corner with the *Frenkel-pair energy*, the energy required to remove a single atom from deep within the lattice. This energy is 14 eV in silicon.

The trajectory of a 300 keV N$^+$ ion slowing down to 14 eV in silicon is plotted in figure 14.21 on the χ–ξ plane, and the individual points are listed in table 14.4. The nitrogen ion begins in the classical nuclear stopping regime at an

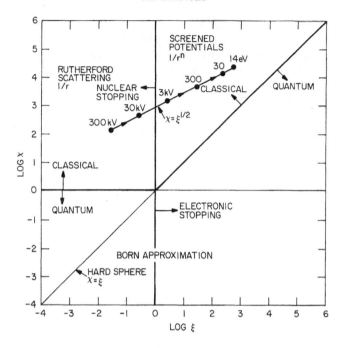

Figure 14.21. The quantum scale parameter ξ plotted as a function of the screening parameter χ.

energy of 300 keV. It moves to the upper right-hand corner through the classical regime along a straight line with a slope of $\frac{1}{2}$, and terminates in the electronic stopping regime. In this case, as with most ion implantations of industrial interest, the interactions remain in a regime described by classical physics during the entire slowing-down process.

REFERENCES

Bohdansky J, Roth J and Bay H L 1980 An analytical formula and important parameters for low-energy ion sputtering *J. Appl. Phys.* **51** 2861–5

Langley R A, Bohdansky J, Eckstein W, Mioduszewski W P, Roth J, Taglauer E, Thomas E W, Verbeek H and Wilson K L 1984 *Data Compendium for Plasma–Surface Interactions* (Vienna: IAEA) ISBN 92-0-139084-X

Molchanov V A and Tel'kovskii V G 1961 Variation of the cathode sputtering coefficient as a function of the angle of incidence of ions on a target *Sov. Phys.–Dokl.* **6** 137–8

Perry T S 1993 Coming clean *IEEE Spectrum* **30** 20–6

Rakowski W 1989 Plasma modification of wool under industrial conditions *Melliand Textilberichte* **70** 780–5

Roth J, Eckstein W, Gauthier E and Laszlo J 1991 Sputtering of low-Z materials *J. Nucl. Mater.* **179–181** 34–6

Thomas E W (ed) 1985 Atomic data for controlled fusion research volume III: particle interactions with surfaces *ORNL Report* 6088/V3

Wiffen F W 1984 Materials requirements and potential solutions for fusion reactors *Proc. IEEE Minicourse on Fusion Experimental/Reactor Systems (17 May, St Louis, MO)* unpublished

Winters H F 1980 *Elementary Processes at Solid Surfaces Immersed in Low Pressure Plasmas (Topics in Current Chemistry 94, Plasma Chemistry III)* ed S Veprek and M Venugopalan (New York: Springer) ISBN 0-387-10166-7, pp 69–125

15

Atmospheric Pressure Plasma Sources

The development of plasma sources for research and industrial applications has a long history, extending back at least to work at the Royal Institution of London in the early years of the 19th century. This work included the electrical arc, developed by Sir Humphry Davy (1778–1829), and the low-pressure glow discharge, developed by Michael Faraday (1791–1867). Two hundred years of plasma-related research and development has left a rich legacy of plasma sources and their variants, a selection of which are discussed in Volume 1 of this book. The many current applications of these sources have been discussed by the US National Research Council (National Academy 1991). Other sources of information on plasma sources and their applications include Boenig (1988), Lieberman and Lichtenberg (1994), Smith (1995), and Lieberman *et al* (1996).

Not all of the plasma sources discussed in the technical literature have proven practicable or economic for industrial applications. For example, sources requiring high magnetic fields (above approximately 50 mT) over a large volume are at a disadvantage for industrial use because of the decreased reliability and increased capital and operating costs associated with electromagnets. Sources that operate under vacuum are at a disadvantage with respect to those that operate at 1 atm because of the increased capital costs and the requirement for batch processing of workpieces associated with vacuum systems.

In the microelectronic industry, plasma sources that operate under vacuum, at low pressures with long ion mean free paths are preferred to higher pressure sources in order to enhance directional etching. Beyond these technical factors, some plasma sources have not been accepted because of unfavorable patent or licensing issues. Only a relatively small number of plasma sources have achieved, or are likely to achieve, widespread industrial acceptance. In this chapter, we emphasize these 'standard' plasma sources, now used in large numbers, as well as selected additional sources that show promise for future development.

The development of atmospheric pressure plasma sources to replace plasma processing in vacuum systems is a current trend in industrial plasma engineering. This trend is likely to continue in the early decades of the 21st century until every possible plasma-processing application involving glow discharges and arcs/torches is conducted at 1 atm, or until it is clear that operation in a vacuum is unavoidable. Thermal plasma processing with arcs or plasma torches, corona treatment, and surface treatment with dielectric barrier discharges has been conducted at 1 atm since their industrial introduction. However, conversion of the industrial plasma-processing applications of vacuum glow discharges to operation at 1 atm has barely begun, and may involve plasma sources discussed in this chapter.

15.1   CHARACTERISTICS OF INDUSTRIAL PLASMA SOURCES

The plasma sources used in industry are described by their manufacturers and users in terms of a small set of characteristics that define their regime of operation. These source characteristics include the plasma type, the nature of its power supply, the operating pressure, and the mode of interaction of the source with a workpiece. Significant differences in these plasma source characteristics can, nonetheless, generate plasmas for which the active species concentrations, electron number densities, electron kinetic temperatures, sheath potential drops, etc, are very similar, while the technical means of generating the plasma are quite different.

In industrial plasma sources, the active-species concentrations and electron-number densities tend to be proportional to the input plasma power density. If the input power density and other plasma parameters are the same in different plasma sources, the usual result is the same plasma-processing effect on the workpiece, achieved by different means.

15.1.1   Plasma Type

The physical fundamentals of plasmas and the physical processes occurring in DC and RF plasma sources of most relevance to industrial applications have been covered in chapters 8 through 13 of Volume 1. The DC electrical discharges of chapters 8 through 10 used in industrial plasma sources can be related to their location on the voltage–current curve of the classical intermediate-pressure electrical discharge, shown in figure 15.1. The *corona source* operates in the *dark discharge regime* on the left-hand side of this diagram, and is discussed in section 15.2. *DC glow discharges* at intermediate currents on this diagram are widely used for sputtering, and their phenomenology is very similar to that of low-frequency RF glow discharges. *DC arcs*, at high currents on the right-hand side of figure 15.1, are high power density discharges capable of damaging sensitive workpieces, but are useful for plasma-processing applications that require the

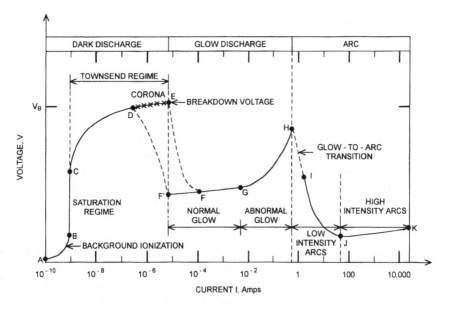

Figure 15.1. The voltage–current characteristic of the classical DC intermediate-pressure electrical discharge tube.

heating or melting of bulk materials. Further discussion of the DC electrical discharge regimes can be found in section 4.9 of Volume 1.

When an RF is applied to generate a plasma, the plasma behaves like a resistive circuit element below the *critical sheath frequency* (see section 19.5.3 of this volume), and like a capacitor with a polarizable dielectric above this frequency. Below the critical sheath frequency at about 1 MHz, the electrode and positive column structures observable along the axis of an *RF glow discharge* resemble an instantaneous snapshot of the structures of a *normal glow discharge* that reverse each half-cycle of the RF. These structures include the *cathode region*, with a linear, *Aston's law* decrease of electric field with distance from the cathode, an *anode dark space*, a *negative glow*, a *Faraday dark space*, and a *positive column* (Ben Gadri 1997, 1999, Ben Gadri *et al* 1994, Massines *et al* 1998).

Above the critical sheath frequency, the sheath structures of the RF glow discharge behave capacitively and become more uniform between the electrodes. The time- and space-resolved sheath characteristics in this regime are not well understood. However, the plasma generation mechanism in RF glow discharges above the critical sheath frequency appears to be the same as that in non-resonant, multimode microwave glow discharges (see figure 13.20 of Volume 1 and the associated discussion). Under a vacuum and in the presence of a magnetic field such that the electron population is *magnetized*, very strong resonant absorption occurs at the electron cyclotron resonance (ECR) frequency. ECR absorption of microwave power is the basis of plasma sources used primarily in the microelectronic industry.

15.1.2   Operating Pressure

The operating pressure of the working gas is a significant factor in choosing a plasma source because it restricts the choice of reactor technology and the type of source used for plasma-processing applications. The most critical choice with respect to pressure is whether the process is to be operated at 1 atm or under vacuum. If a source is operated at 1 atm, as are those discussed in the remainder of this chapter, batch-processing and vacuum equipment are not required; under vacuum both are required, and their cost is relatively insensitive to the vacuum pressure.

Some plasma lighting devices to be discussed in Volume 3 operate at pressures up to several tens of atmospheres. If plasma sources are operated above 1 atm, batch processing of workpieces is required, and above about 3 atm, the enclosures are considered to be pressure vessels with regulatory restrictions that require design according to national boiler codes. For these reasons, plasma processing is rarely conducted above 1 atm.

A few applications of plasma torches and military equipment (i.e. aircraft) require pressures ranging from 1 atm down to 0.1 atm. Other than these, very few plasma sources or industrial plasma-processing applications operate at pressures between 1.3 kPa (10 Torr) and 1 atm.

At 1 atm (101 325 Pa or 760 Torr), which we take to include a barometric range within 5% of the standard atmosphere just quoted, batch processing of workpieces and vacuum pumping equipment are not necessary. This factor leads to significant cost savings in the processing of webs, films, and large numbers of low-value workpieces. Several commercially important plasma sources operate at 1 atm, and are described in this chapter. Chapter 16 covers vacuum plasma sources that operate in the intermediate, low, and high vacuum pressure regimes.

15.1.3   Power Supply Characteristics

Industrial plasma sources operate over a wide range of excitation frequencies, as discussed previously in section 4.10 of Volume 1. Schematic diagrams of DC, power line, and RF plasma source subsystems are shown on figure 15.2. Plasma sources energized by DC power supplies have the advantage that DC power is relatively inexpensive, ranging from $0.10 to $0.50 per watt. DC power supplies may, at most, require a matching network consisting of a simple ballast resistor or inductor to limit the current and stabilize non-thermal arcs or other plasma sources with a negative resistance characteristic. Arc and corona discharges operated at power line frequency (50 or 60 Hz) have a very inexpensive power supply, since the final transformer before the load usually serves as a current limiting ballast.

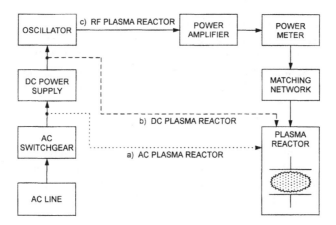

Figure 15.2. Schematic diagram of power supply components needed for plasma sources that require (*a*) AC line voltage only; (*b*) DC electrical power; and (*c*) RF power.

Of plasma sources that require RF power supplies, the dielectric barrier discharge (DBD) and the one atmosphere uniform glow discharge plasma (OAUGDP) operate in the extremely low frequency (ELF) and very low frequency (VLF) range, from approximately 500 Hz to 30 kHz. These frequencies (but not the amplitude of the electric and magnetic fields involved) are below the threshold of regulatory interest in most countries, and plasma sources can be operated over this range without concern with official (the Federal Communications Commission (FCC) in the United States) frequency assignments. Because this is not a widely used frequency band, off-the-shelf equipment is less readily available, and RF power in this range may cost above $1 per watt.

Some RF vacuum glow discharge reactors used for microelectronic deposition or etching, such as triode etching systems, operate the RF power supply for one electrode below the critical sheath frequency, where the plasma appears as a resistive load to the power supply, and the second electrode and power supply above the critical sheath frequency, at 13.56 MHz. The lower range includes portions of the low frequency (LF) and medium frequency (MF) bands, from about 50 kHz to a few MHz. Regulatory restrictions on allowed frequencies and off-site emission intensities must be considered for such sources. Such sources are widely enough used that they can be obtained at a cost of about $1 per watt.

The RF glow discharge plasma reactors used for microelectronic deposition and etching operate above the critical sheath frequency, where the reactor behaves like a capacitive load to its power supply. Such reactors are operated in the high frequency (HF) band (3–30 MHz) at 'sacrifice frequencies' established by regulatory bodies at 13.56 MHz, and at harmonics twice and three times this frequency. These 13.56 MHz RF power supplies have been used in such large numbers for microelectronic and other plasma-processing applications that they are available at a cost of from $0.5 to $1.5 per watt.

Electron cyclotron resonance (ECR) and non-resonant microwave plasma sources operate in the ultra-high frequency (UHF) band (300 MHz to 3 GHz) at the assigned sacrifice frequencies of 915 MHz and 2.45 GHz. The latter frequency is used in domestic microwave ovens. As a result, the technology has been developed to the point that RF power at 2.45 GHz is available at low cost, ranging from $0.20 to $0.50 per watt.

Most plasma sources used for surface treatment, deposition, or etching require only a few hundred watts to a few kilowatts of input power. The capital cost of the plasma source subsystem and its power supply may be such a small fraction of the cost of the complete plasma-processing system or fabrication tool that factors other than cost may dominate the choice of plasma source. For example, some sources are more difficult to operate (involving more independent control variables, higher levels of skill or training, etc) than others, and this can be a determining factor in the choice of plasma source.

In thermal plasma-processing applications requiring arcjets or plasma torches, power levels ranging from a few kilowatts to more than 1 MW may be required. In these applications, efficient and effective use of the plasma power is important. The power supply is a major fraction of the overall cost of the plasma-processing system, and the electrical power is a major contributor to its operating costs.

At frequencies above power line values (50 or 60 Hz), some form of impedance-matching network is desirable. Unlike an antenna or a transmission line, a plasma is a variable load which depends on its physical characteristics (electrical conductivity, number density, electron kinetic temperature, etc). Without impedance matching, a significant fraction of the power supplied will be reflected from an unmatched plasma load, and only a part of the costly total capacity of the power supply is used. At the widely used RF frequencies of 13.56 MHz and 2.45 GHz, impedance-matching circuits are normally provided with the power supply, the most sophisticated of which use feedback circuits to automatically minimize the power reflected from the plasma load.

15.1.4 Plasma Interaction with Workpiece

There are two principal modes by which a plasma can interact with a workpiece during plasma processing. The first is *direct exposure*, in which the workpiece is either immersed in the plasma, or is in direct contact at the plasma boundary. Direct exposure is used when it is desired to maximize the heat flux, as in the heating, welding, or melting of bulk materials, or when it is important to maximize the flux of ions or other active species on the surface of a workpiece. Direct exposure can be accomplished when the workpiece serves as an *active electrode* that carries a real current to a DC or low-frequency power supply. A workpiece can also serve as a *passive boundary* that does not serve as an electrode to generate the plasma, and does not collect a real current.

The second principal mode of exposure, *remote exposure*, is useful in situations for which the workpiece may be damaged by direct exposure to UV radiation, electric fields, ion bombardment, or high thermal fluxes. In a remote exposure configuration, the workpiece is located at a distance from the plasma source, where it is isolated from unwanted forms of plasma interaction. One might convect neutral active species to the remote location with a flow of working gas, thus isolating the workpiece from forms of plasma interaction associated with direct exposure to the plasma.

15.1.5 Problematical Plasma Source Nomenclature

Before proceeding further to discuss industrial plasma sources, it is important to correct two widespread errors of nomenclature that have arisen, particularly in the trade and promotional literature. The first term that needs clarification is *'corona'*. The true corona has been the subject of research and development for at least 150 years and is discussed in chapter 8 of Volume 1, as well as in section 15.2 of this chapter (see Loeb 1965, Cobine 1958). The true corona is generated in the strong electric fields near sharp points or fine wires. The visible portion of the true corona occurs in the region within the *critical radius*, at which the electric field is equal to the breakdown electric field of the surrounding gas. The true corona does not occur between two parallel smooth plates, nor in the presence of an insulating coating over the conductor giving rise to it.

The North American sales and trade literature sometimes confuses the true corona just described with a *dielectric barrier discharge* (DBD), described in section 15.3 of this chapter (see also Eliasson and Kogelschatz 1988, Kanazawa *et al* 1988). The DBD is a *filamentary plasma*, usually generated between parallel-plate or annular cylindrical electrodes, each of which is covered with a dielectric coating (barrier). The active volume within the critical radius of a true corona discharge is much smaller than the volume filled by a DBD discharge. As a result, true corona reactors produce much smaller concentrations and fluxes of active species than the DBD. In the remainder of this text, this distinction will be used, consistent with the history and scientific journal literature of the field.

The second common error in nomenclature in the English-speaking industrial plasma literature is the practice of referring to glow discharge plasmas, particularly atmospheric glow discharge plasmas, as *'silent discharges'*. This appears particularly mysterious, as the discharges in question may operate in a vacuum, where the production of sound outside the vacuum system would not be expected. *Glow discharges* are so called because they emit modest levels of visible light; in the extensive German literature of the 19th and early 20th century, they were referred to as *'glimmentladung'*, literally 'glowing charge'. When English-speaking engineers translated the seminal German plasma papers in the mid-20th century, the technical dictionary of choice was DeVries' *German–English Science Dictionary* (1959). This otherwise excellent dictionary defines 'glimmentladung' (the German term) as 'silent discharge', its supposed English equivalent. This is contrary to the meaning of 'glimmer', which means 'glow' in German. This error has propagated widely in the English industrial plasma literature. To avoid confusion, the term 'silent discharge' will be avoided entirely in these volumes.

The term 'silent discharge' itself apparently arose in the literature of non-equilibrium arcs used as light sources (see section 10.4.1 of Volume 1). There is a mode transition from non-equilibrium arcs with a negative resistance characteristic at low currents, to equilibrium, 'thermal' arcs at higher currents that have a positive resistance characteristic. The non-equilibrium arcs are relatively quiet; above the mode transition, thermal arcs operating in atmospheric air are relatively noisy, and were referred to as 'hissing' arcs (Ayrton 1902) in the early literature.

15.1.6 Plasma Source Applications

Plasma sources have found an enormous range of industrial applications, including many niche applications when physical or chemical-processing methods proved inadequate. New applications are being reported at plasma-related meetings on an almost monthly basis. In terms of sheer numbers, plasma sources are most often found in the well-established areas of surface treatment of materials, microelectronic deposition and etching, ion-beam sources, and the thermal plasma processing of bulk materials.

15.1.6.1 Surface Treatment

Plasma surface treatment involves using direct plasma exposure to change the surface energy, wettability, wickability, and bonding of fabrics, films, and solids, and the sterilization, cleaning, and decontamination of surfaces. To achieve industrially important effects, plasmas used for surface treatment must have a sufficiently high electron number density to provide useful fluxes of active species, but not so high or energetic as to damage the material treated. These constraints rule out dark discharges other than coronas for most applications, because of their low production rate of active species, and arc or torch plasmas, which have power densities and active-species flux intensities high enough to damage exposed material. Glow discharge plasmas, whether operated at 1 atm or under vacuum, possess the appropriate density and active-species flux for nearly all plasma surface treatment applications.

RF glow discharge plasmas can be generated by inductive, capacitive, or microwave excitation, using the physics and technology discussed in chapters 11, 12, and 13 of Volume 1, respectively, while DC corona and glow discharge plasmas may be generated with the respective physics and technology discussed in chapters 8 and 9 of Volume 1. The location of dark and glow discharges on the voltage–current characteristic of the DC low-pressure electrical discharge was discussed in connection with figure 15.1.

15.1.6.2   Deposition and Etching

A relatively few types of plasma source, all operated under vacuum and discussed in chapter 16 of this volume, have been used for microelectronic deposition and etching. In the United States, industrial practice has favored the RF parallel-plate glow discharge operating at a frequency of 13.56 MHz and, more recently, the flat helical coil inductive reactor, also operating at 13.56 MHz. In Japan and other countries, past industrial practice has included extensive use of ECR plasmas generated by microwave power at 2.45 GHz.

Trends within the microelectronics industry to circuit element dimensions below 0.5 μm and larger numbers of circuit elements in a single integrated circuit have created a watershed in the technology of plasma sources. Above this design rule, plasma source technology that operates at intermediate pressures above 13.3 Pa (100 mTorr) is adequate. More advanced etching technologies below a design rule of 0.5 μm, however, require plasma sources that operate in the low-pressure regime below 6.7 Pa (50 mTorr). Several low-pressure plasma sources are under development for microelectronic processing, and are discussed for completeness in chapter 16.

15.1.6.3   Ion-Beam Sources

A selection of ion-beam sources used in aerospace, industrial, and research applications was discussed in chapter 6 of Volume 1. Many of these sources were originally developed for high-energy and solid-state physics research, or as 'ion engines' for electrostatic space propulsion. This technology matured over the period from 1930 to 1970 and was adapted to industrial uses. These uses include ion-beam implantation to improve the hardness and wear (tribological) characteristics of medical prostheses and other high-value items; ion-beam doping of semiconductors and optical components; thin-film deposition by ion-beam sputtering; and ion-beam-enhanced plasma chemical vapor deposition and epitaxy.

These applications of ion beams require long mean free paths, and hence operation in the low-pressure or high-vacuum regimes. At these very low pressures, ionizing collisions are less frequent, and efficient generation of plasmas dense enough to provide the required ion fluxes is difficult. Of the sources discussed in chapter 6 of Volume 1, the *Kaufman* (or *electron bombardment*) *ion source* is preferred when electrical or gas utilization efficiency is important; the *Penning ion source* when it is required to produce ions of a wide range of elements or compounds; and the *Freeman ion source* when physical robustness and reliability are the dominant considerations.

15.1.6.4   Processing Bulk Materials

The oldest industrial applications of plasmas involve the processing of bulk materials using electrical arcs or plasma torches (see Boulos *et al* 1994, Cobine

1958, Hoyaux 1968, Gross *et al* 1969). Plasma torches and arcjets convert electrical power into thermal enthalpy and operate at or near *thermodynamic equilibrium*, in which the ion, electron, and neutral working gas equilibrate to temperatures ranging from 10 000 to 20 000 °C. These factors ensure a high power density and heat flux to a workpiece in direct contact with the plasma.

Bulk materials can be melted at 1 atm of pressure, where plasma torches and arcjets normally operate. These sources make it possible to weld and refine metals, melt scrap metals in arc furnaces, plasma spray thick films of ordinary and refractory materials, and maintain the temperature of ladles and tundishes in metalworking foundries (see Barber 1983, Cary 1979, Paschkis and Persson 1960, Smith and Novak 1991). Plasma torches and arcjets are also used to heat gas flows for energy addition in aerospace wind tunnel applications, and to increase the efficiency of blast furnaces (less coke and oxygen are needed to melt and refine the ore with plasma heat addition).

15.2   ATMOSPHERIC PRESSURE CORONA SOURCES

In section 15.1.5 it was pointed out that a *corona* should not be confused with the *dielectric barrier discharge*, an entirely different type of plasma. There exist at least two broad motivations to study corona in industrial applications. First, a corona can be a regulatory nuisance and safety hazard in the operation of high voltage equipment, and it must be well enough understood to engineer these problems to acceptable levels. Second, a corona can generate relatively low positive and negative ion fluxes that, however, are sufficient for some industrial plasma-processing applications, including plasma surface treatment. Important contributions to the literature of corona discharges have been made by Cobine (1958) and Loeb (1965).

15.2.1   Characteristics of Corona Sources

The physics and phenomenology of corona discharges are discussed in section 8.4 of Volume 1, and the characteristics of corona sources in section 8.5 of that volume. Corona discharges are generated in the regions of high electric field surrounding sharp points, fine wires, edges of sheet metal, etc, as illustrated in figure 15.3. For industrial applications of corona at 1 atm, the fine wire configuration (see section 8.5.2 of Volume 1) is most frequently used.

The coronal electrode is surrounded by an *active region* of *critical radius r_o*, at which the radial electric field falls to the *breakdown electric field* of the working gas. Ionization, excitation, and production of active species take place in the active region, which may be visible at higher currents. A corona characteristically involves voltages from 10 kV to several tens of kV with currents from 1 to 100 mA/m. True corona sources used industrially rarely operate at power levels of more than 1 kW; normal operation may be well below these power levels.

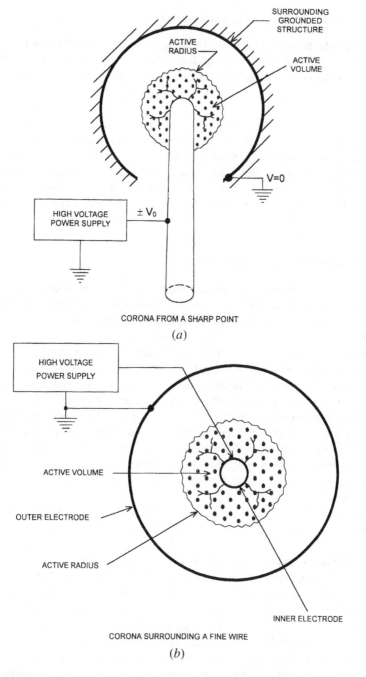

CORONA FROM A SHARP POINT

(*a*)

CORONA SURROUNDING A FINE WIRE

(*b*)

Figure 15.3. Electrode configurations used to produce corona discharges: (*a*) sharp points and (*b*) fine wires.

15.2.2 Phenomenology of Corona Sources

Corona can be generated either by DC excitation or at AC/RF frequencies. The corona around positive wires is uniform, but negative corona may exhibit a *beading instability*, in which the corona forms equally spaced luminous balls along the wire (see Cobine 1958). When a corona is excited at AC or RF frequencies, successive diverging waves of positive and negative thermalized ions propagate away from the source electrode. At atmospheric pressure, the only charged species to escape the active region of corona in any numbers are room-temperature ions, positive or negative, which carry the DC coronal current to the surrounding walls/electrodes. This radially outward flux of ions is known as the *coronal wind* or *ion wind*. The ultraviolet radiation from the active region of corona discharges can change the surface energy of some materials, and affect the ability of some surfaces to hold a static charge.

15.2.3 Regulatory and Safety Issues

Probably no industrial plasma source is surrounded by more potential regulatory and safety issues than coronas. Most of these issues are related to the high voltages at which corona discharges operate, and are particularly vexing because coronas are capable of generating emissions or hazards that fall under the responsibility of two or more distinct governmental regulatory bodies.

High voltages (here, above 220 V) have been used for industrial, small business, and domestic applications for over a century in applications such as neon sign transformers, automotive ignition systems, television sets, etc. It has been found from experience that the social, legal, and practical upper limit in these contexts is about 20 kV, above which various safety, liability, and regulatory issues become too pressing for exposure of the general public. This voltage is approximately the lower limit at which coronas can be generated for industrial applications, thus requiring trained operators and safety monitoring beyond that appropriate for public use.

When corona discharges exist, even at electrode voltages below 20 kV, they can emit broad band RF white noise at and below medium frequencies (MF), causing *radiofrequency interference* (RFI) in the AM band surrounding 1 MHz. This RFI can exceed the limits allowed by national regulatory bodies, including the Federal Communications Commission in the United States. Corona-generated RFI does not extend above about 10 MHz, or to the FM and television bands around 100 MHz. At the same time, corona discharges can generate progressively more energetic x-rays as the operating voltage is increased above this 20 kV threshold, at flux levels that can be hazardous to nearby workers.

Corona discharges at 1 atm also generate acoustic white noise at audio frequencies that can be loud enough to exceed regulatory standards. Such acoustic noise also prompts safety concerns in nearby individuals, since it may be an indicator of the RFI or x-ray emissions discussed previously. In atmospheric air, particularly humid air, plasma chemical reactions in the active region of a corona discharge can produce ozone and nitric acid. These active species are toxic to humans and can corrode or destroy the coronal electrodes and surrounding structures.

Finally, since corona discharges operate at higher voltages than other industrial plasma sources (except ion-beam sources), high voltage and electrical safety are issues. Particularly in humid air, the sharp points or fine wires used to create corona will also form streamers and related phenomena that can cause electrical sparking over large distances. These sparks may be much longer than those appropriate to the breakdown electric field strength in dry air of 30 kV/cm between flat, smooth surfaces. Such sparking can damage solid-state electronic equipment, or electrocute workers unless the high voltage power supply is current limited to 10 mA or less.

15.2.4   Operational Issues of Corona Sources

Corona sources have the advantages of physical robustness, geometric simplicity, ability to operate at 1 atm, and low input power. Their disadvantages include low production rates and fluxes of active species; the inability to produce a range of energetic active species for plasma-processing applications; the production of corrosive active species that attack electrodes, workpieces, and the structure of the corona reactor; and the regulatory and safety issues discussed in the previous section.

15.2.5   Research Issues of Corona Discharges

In spite of the 150-year history of academic research and industrial application of corona discharges, very few areas of industrial plasma engineering offer such a rich selection of poorly understood phenomena or topics for basic research. These phenomena include the physical process responsible for the beading instability of negative corona discharges; the mechanism by which a corona discharge generates broad band white noise in the RF spectrum around 1 MHz; the physical process by which corona discharges generate x-rays with energies above 10 kV, in spite of the high collisionality of the electron population at 1 atm; and the physical process by which a corona discharge at 1 atm produces acoustic noise in the audible range.

Scientific understanding of these physical processes has been retarded by the small size of the active region of corona discharges, where these phenomena are likely to originate. This small size has made it difficult to obtain the time- and space-resolved diagnostic measurements of plasma properties in the active region necessary to shed light on these phenomena.

15.3 ATMOSPHERIC DIELECTRIC BARRIER DISCHARGES (DBDS)

Dielectric barrier discharges (DBDs) that operate at 1 atm have a long history, with patent literature extending back to the 1860s. The DBD is generated in the space between two electrodes, each of which is covered with an insulating, dielectric coating. A DC, AC, or pulsed high voltage is applied to the electrodes, which stimulates electron emission from the instantaneous cathode. These electrons avalanche to form a filament across the gap. When this swarm of electrons reaches the opposite dielectric coating, they spread out over the surface, lowering the local electric field and turning off the avalanche (see Eliasson and Kogelschatz 1988, Kanazawa *et al* 1988).

15.3.1 Applications of Dielectric Barrier Discharges

The first widespread application of DBDs was ozone production for the disinfection of public water supplies. These early patents were issued in Germany in the 1860s, and this remains a major use of this technology. More recently, as the printing industry has had to shift from hydrocarbon-based inks to water-based inks for environmental reasons, DBDs have been used to increase the surface energy (and hence the printability) of papers, films, and polymeric webs. As noted in section 15.1.5, these DBD reactors have been sometimes marketed as 'corona' reactors.

15.3.2 Characteristics of Dielectric Barrier Discharges

DBDs are normally operated in one of the parallel-plate or cylindrical configurations illustrated in figure 15.4. The parallel-plate configuration is used to surface treat fast-moving webs and films, and the annular volume of the cylindrical configuration is used to treat airflows for ozone production. At least one electrode of these geometries is covered with an insulating *dielectric barrier* to prevent real currents from flowing from the discharge volume to the electrodes and power supply.

DBDs used industrially usually operate at 1 atm, with air as the working gas. An exception occurs when air is enriched with oxygen to increase the efficiency of ozone production. Since avalanches in the DBD are very brief, on the order of tens of nanoseconds, the highest production rate of ozone or other active species tends to occur when the applied voltage is pulsed rapidly and repeatedly. The optimum duty cycle depends on the composition of the working gas and the details of the chain of chemical reactions that lead to the desired reaction product. Some commercial ozonizers and DBD reactors for surface treatment are operated at line frequency (50 or 60 Hz), but optimum ozone production appears to occur at RF frequencies on the order of 10 kHz. DBDs characteristically operate at applied

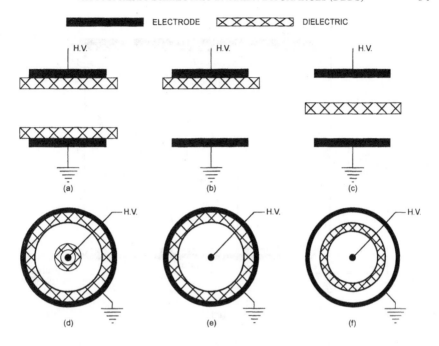

Figure 15.4. Schematic diagram of parallel-plate (a)–(c) and cylindrical (d)–(f) dielectric barrier discharge (DBD) plasma source configurations.

voltages of several tens of kilovolts rms, and at electric fields comparable to the breakdown electric field of the working gas, 30 kV/cm for air.

15.3.3 Dielectric Barrier Discharge Phenomenology

The development of an individual filament in a DBD is illustrated schematically in figure 15.5. An individual filamentary discharge is initiated when a high voltage is applied between the electrodes such that the electric field in the open gap equals or exceeds the breakdown strength of the ambient gas. Electron emission from the surface of the dielectric coating on the instantaneous cathode is stimulated by UV photoemission or ion-induced secondary electron emission.

 These electrons are accelerated in the electric field to energies that equal or exceed the ionization energy of the gas, and create an *avalanche* in which the number of electrons doubles with each generation of ionizing collisions. The high mobility of the electrons compared to the ions allows the electron swarm to move across the gap in durations measured in nanoseconds. The electrons leave behind the slower ions, and various excited and active species that may undergo further chemical reactions.

 When the electron swarm reaches the opposite electrode, the electrons spread out over the insulating surface, counteracting the positive charge on the

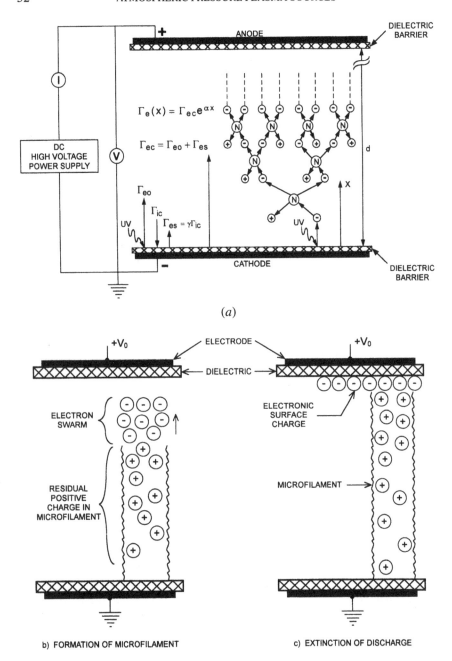

Figure 15.5. Stages in the development of an electron avalanche in a DBD: (*a*) avalanche initiation; (*b*) electron cloud reaches dielectric barrier on opposite electrode; and (*c*) electrons spread out on dielectric surface, leaving behind positive ion cloud.

instantaneous anode. This factor, combined with the cloud of slower ions left behind, reduces the electric field in the vicinity of the filament and terminates any further ionization along the original track in time scales of tens of nanoseconds. If air at 1 atm is the working gas and the duration of the discharge is not prolonged by other means (such as operating the discharge as a one atmosphere uniform glow discharge plasma, see section 15.4), these active species are in a frozen chemical equilibrium, and the DBD preferentially produces ozone as the dominant active species.

The filaments produced by each avalanche characteristically are a few tens to a few hundred nanometers in diameter, and their 'roots' where they contact the dielectric surface can cause pitting or pinholing of a workpiece. Dust or micron-scale asperities on the surface of the insulator can promote the formation of filaments. The overall effect on an exposed workpiece can be quite non-uniform. In some applications, the objective is met by active species produced in the initial avalanche, or by subsequent chained chemical reactions the products of which are well enough mixed with the working gas to provide a uniform effect. The power density of DBDs used industrially is no less than about 50 mW/cm^3 to achieve an adequate effect for surface treatment, and no more than 1 W/cm^3, a limit determined by the heat flux on the dielectric and workpiece surfaces.

15.3.4 Safety and Regulatory Issues of Dielectric Barrier Discharges

The avalanching responsible for the DBD requires an electric field near or above the breakdown electric field for the gas used. This, in turn, requires applied potentials that may be several tens of kilovolts. If such potentials are applied at several kilohertz, streamers and sparking to surrounding grounded surfaces may occur, over distances much greater than would be expected of DC potentials in dry air.

Another safety hazard associated with DBDs operated in air or oxygen-containing gases is ozone production. Ozonizers can generate concentrations of ozone well above the limits for safe 40-hr exposure of workers, and ozone concentrations of regulatory interest are likely to be generated by any DBD operated with oxygen gas present. Ozone is not only a health hazard to humans; it can also damage workpieces and the reactor structure because of its oxidizing power.

Minor regulatory issues that can arise with DBDs include radiofrequency interference (RFI) below a few megahertz, RFI at the operating frequency of the energizing high voltage, and audible noise. The audible noise is rarely loud enough to be of regulatory interest, but has a quality that few human bystanders are willing to tolerate for long durations. Both the RFI and the audible noise are relatively easy to deal with, because the isolation, shielding, and enclosure needed to deal with the high voltage and ozone hazards can reduce these effects to acceptable values.

15.3.5   Operational Issues of Dielectric Barrier Discharges

DBDs would be more widely used for plasma processing if they were capable of producing higher fluxes of active species. They are adequate to raise the surface energy of webs and films to levels above 50 dynes/cm, but have difficulty reaching the higher surface energies required for water-based inks, for example. This limitation on the active-species flux reflects that imposed on the discharge power flux (about 1 W/cm^2) by the presence of a thermally insulating dielectric on the electrodes.

The ozone, nitrogen oxides, and other oxidizing species produced during avalanching can damage everything with which they come into contact, including the workpiece, the dielectric barrier coating on the electrodes, and the structure of the reactor. In some applications, this problem can be avoided by a suitable choice of gases, but in others, the plasma chemical or processing effect depends on these oxidizing species.

The small diameter of the discharge filaments, on the order of tens to a few hundred nanometers, results in a very high local energy deposition where the filaments contact the workpiece or the dielectric barrier coating. This, in combination with the presence of oxidizing species discussed previously, can lead to pinholing of a workpiece or erosive damage of the electrode coating.

Commercial DBDs used for ozone production may be operated at power levels of megawatts, and parallel-plate sources used for surface energy enhancement of webs and films usually operate at the multi-kW level. These relatively high power levels may be accompanied by the safety and regulatory problems discussed earlier, but if a DBD reactor is isolated, shielded, and enclosed to deal with one of these concerns, the others may be dealt with also.

15.3.6   Research Issues of Dielectric Barrier Discharges

As in corona discharges, basic research into the plasma parameters and physical processes of DBDs has been held back by the requirement for time and space resolution of plasma diagnostic measurements at very short durations (nanoseconds), and very small dimensions (nanometers). These experimental difficulties have forced investigators to use a 'black box' approach based on phenomenological relations between input and output and computational modeling, rather than to rely on direct physical measurements of the DBD discharge.

Issues other than time and spatial resolution of the DBD that are either poorly understood or under active investigation include modeling of the plasma chemistry. When DBDs are used as ozonizers or for plasma chemical processing, the energetic-electron-induced chemical reactions during and after the avalanche are modeled in the manner of a chemical reactor. The figure of merit to be optimized is usually the kilograms of product (such as ozone, for example) per kilowatt-hour of energy supplied to the plasma.

Other research issues include understanding the physical processes at the contact point of the filament with the dielectric barrier. This may lead to insight into the self-limiting nature of the avalanche, and to minimize damage due to high particle or energy fluxes on the surface. Finally, experimental observation of DBDs in operation over long periods sometimes reveals a remarkable but poorly understood degree of spatial self-organization of the filaments, in which they occupy patterned, approximately equally spaced locations as though they were the solution to a two-dimensional mathematical eigenvalue problem.

15.4 THE ONE ATMOSPHERE UNIFORM GLOW DISCHARGE PLASMA (OAUGDP)

It has recently become possible to generate large volumes (multiliter) of uniform glow discharge plasma at 1 atm with the one atmosphere uniform glow discharge plasma (OAUGDP). The theory of this reactor was treated in section 12.5.2 of Volume 1, where the physics of the ion-trapping mechanism is analyzed (Roth 1995). Development of the OAUGDP is much more recent than the corona discharge or the DBD discussed previously.

Von Engle *et al* (1933) reported atmospheric pressure glow discharges with bare electrodes in air and hydrogen using both DC and audiofrequency RF driving voltages. This discharge, however, was unstable to the glow-to-arc transition, required specially cooled cathodes, and could only be operated at 1 atm by starting the discharge under vacuum conditions, and slowly raising the pressure to 1 atm. Later work on the OAUGDP included that of Kanazawa *et al* (1988), Yokoyama *et al* (1990), Ben Gadri (1997, 1999), Massines *et al* (1998), Roth (1995, 1997, 1999), and Roth *et al* (1995a, b, 1998). The discharges in the first five of these papers were restricted to working gases with limited industrial applications, such as argon, helium, or helium with an admixture of acetone. However, progress between 1988 and 1999 clearly showed that parallel-plate RF discharges can, under the proper operating conditions, generate a diffuse glow discharge plasma free of filaments in the entire discharge volume, in air, and at 1 atm of pressure.

15.4.1 Characteristics of the OAUGD Plasma Source

A schematic diagram of a parallel-plate OAUGDP reactor system is shown in figure 15.6. A photograph of an OAUGDP source operating in helium is shown in figure 15.7. The reactor volume is bounded by two plane, parallel plates, at least one of which must be covered by an insulating coating. This coating prevents RF arcs and allows the surface charge build-up that carries the discharge from one-half of the RF cycle to the next, as shown by Ben Gadri (1997) and Massines *et al* (1998). The OAUGDP source is highly capacitive as a circuit element, and impedance matching of the power supply to the source is desirable in order to avoid using an expensive, overrated power supply.

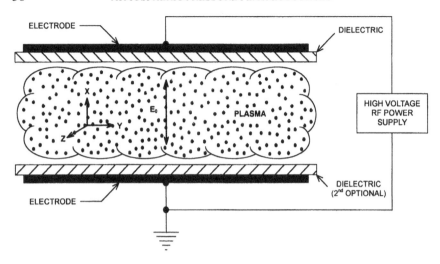

Figure 15.6. Schematic diagram of a parallel-plate OAUGDP reactor, with ion trapping in the RF electric field.

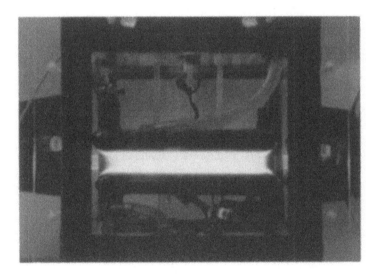

Figure 15.7. Photograph of a parallel-plate OAUGDP operating in helium at 1 atm; vertical height of plasma is 2.5 cm.

For some plasma-processing applications, a relatively thin (a few mm) surface layer of plasma may be more useful than the relatively large volumes of plasma provided by a parallel-plate reactor. The parallel-plate plasma reactor can be geometrically transformed into a planar reactor in the manner illustrated in figure 15.8 (Roth 1997, Roth *et al* 1998). In this configuration, parallel strip

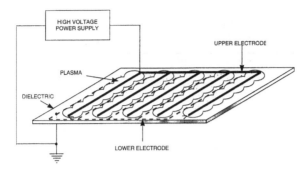

Figure 15.8. The OAUGDP surface-layer plasma source, consisting of a parallel arrangement of electrode strips for the generation of a planar OAUGDP surface layer. Ions are trapped along the electric field lines between adjacent strip electrodes, which are maintained at opposite RF polarities.

electrodes of one RF polarity connected by a bus are located on the upper surface of a dielectric panel, and the electrode of opposite polarity is a conducting sheet below the dielectric. Ions are trapped on the arched electric field lines between the upper and lower electrodes, and a surface plasma layer is generated in the region where ion trapping occurs.

15.4.2   Independent Operating Parameters

The OAUGDP source has been operated at nominal pressures within 5% of one standard atmosphere in the parallel-plate configuration of figure 15.6 over the range of parameters indicated in table 15.1. The power shown in table 15.1 is the *net* power delivered to the plasma, and does not include the *reactive power* (power reflected from the plasma load back to the power supply). The volumetric power densities shown in table 15.1 are far below those of electrical arcs or plasma torches, but much higher than those associated with corona discharges outside their active region.

Such plasma parameters as electron energy and number density in the OAUGDP are poorly known. The energy of ions and active species other than electrons and photons are expected to be close to that of the room temperature gases with which they frequently collide at atmospheric pressure. The electrons remain numerous and energetic enough to excite the neutral background atoms, making this a glow discharge. The existence of excited states that emit visible photons implies that the electron population has a kinetic temperature of at least a few electronvolts during part of the RF cycle. This expectation is consistent with the computational modeling of Ben Gadri (1997, 1999).

Other factors which may affect the production of active species for plasma processing include whether or not the working gas is recycled past the exposed workpiece (some important active species are long lived and their concentration

Table 15.1. Plasma source characteristics for the OAUGDP. Characteristic working gases: He, Ar, N_2O, CO_2, air.

Characteristics	Units	Low	Characteristic	High
		Magnitude		
Frequency, v_0	kHz	0.5	5	40
Voltage, V_{rms}	kV	1.5	7	18
Electrode gap, d	cm	0.20	0.40	2.8
Electric field, E_{rms}	kV/cm	2	8.5 (air)	12
Gas pressure, p	Torr	755	760	775
Power to plasma, p	W	10	200	2000
Power density, $\bar{P}$	mW/cm^2	4	100	600
Plasma volume	liters	0.030	0.20	2.8
Electron density, n_e	no./m^3	?	$\sim 10^{16}$	?
Electron energy, T_e	eV	1 (?)	4	20 (?)
Ion energy, T_i	eV	0.025	0.025	?

builds up upon recirculation); the gas temperature; the temperature of the electrode surfaces; the gas humidity; and the flow rate of the gas past the workpiece. As will be discussed in section 18.6, Lorentzian momentum transfer from the ions to the neutral gas can give rise to important flow acceleration effects.

The RF electric field applied between the electrodes must be strong enough to electrically break down the gas used, about 8.5 kV/cm for air. This electric field is more than a factor of three lower than the sparking electric field for dry air, but is also a factor three higher than the electric field required to initiate an OAUGDP in helium and argon. The RF frequency must be in the optimum range for ion trapping, as discussed in section 12.5.2 of Volume 1. This frequency is characteristically in the kilohertz range. For exploratory research purposes, the two parallel electrode plates illustrated in figure 15.6 may be driven by a power supply capable of operation over the parameter range indicated in table 15.1.

Total power inputs to OAUGDPs can be as little as a few watts, or up to kilowatts for industrial prototype reactors. Plasma power densities in the discharge volume between parallel plates can range from a few mW/cm^3 to 1 W/cm^3, with 0.5 W/cm^3 a nominal value. As long as the minimum electric field for breakdown of the working gas is maintained (about 8.5 kV/cm for air), the gap spacing between the OAUGDP electrodes is determined by geometrical constraints and the maximum voltage that can be conveniently handled in a particular application.

15.4.3   OAUGDP Phenomenology

The range of operating parameters over which an OAUGDP has been operated is indicated approximately in table 15.1. The nominal values indicated are not necessarily simultaneous values. Some observations relating to the operational phenomenology of the OAUGDP are summarized here.

15.4.3.1   Charge Trapping Mechanism

The electric fields employed to create the OAUGDP are normally less than 10 kV/cm even in air, values too low to achieve DC electrical breakdown (sparking) of the operating gas. The OAUGDP is formed, and assumes its normal glow characteristics, when the ions, but not the electrons, are trapped on the electric field lines between electrodes. The electrode gap spacing, electric field, and frequency are adjusted consistent with equation (15.6), such that the mobility drift of the ions in the oscillating RF electric field traps them between the electrodes during an RF cycle. This *ion-trapping mechanism* allows gases such as helium, argon, and air to break down under electric fields as low as 2 or 3 kV/cm for argon and helium, and 8.5 kV/cm for air. During ion trapping, the electrons are free to travel to the electrodes where they are collected, recombine, or build up a surface charge.

  If the RF frequency is so low that both ions and electrons reach the electrodes and recombine, the plasma will either not be initiated or form very few coarse filamentary discharges between the plates. If the applied RF frequency is too high, so that *both* electrons and ions are trapped in the discharge, then the discharge polarizes, and undergoes a *filamentation instability* that forms relatively coarse, large diameter (100 μm) filaments compared to those of a DBD.

15.4.3.2   Trapping Frequencies

In section 12.5.2, it was shown that the rms displacement of an electron or ion during a half-cycle in the parallel-plate geometry of figure 15.6 is given by

$$x_{\mathrm{rms}} = \frac{2}{\pi} \frac{eE_0}{m\omega v_{\mathrm{c}}} \qquad \mathrm{m} \tag{15.1}$$

where E_0 is the maximum electric field during the RF cycle, ω is the RF angular frequency, m the species mass, and v_{c} the collison frequency of the species of interest. If v_0 is the driving frequency in hertz, and V_{rms} is the rms voltage applied to the plates, then the radian RF frequency is given by

$$\omega = 2\pi v_0 \tag{15.2}$$

and the maximum electric field between the plates can be expressed in terms of the peak RF voltage V_0 appearing between them,

$$E_0 = \frac{V_0}{d} = \frac{\pi V_{\mathrm{rms}}}{2d}. \tag{15.3}$$

If the charged particles of interest are confined between the discharge electrode plates during one full cycle, then

$$x_{\text{rms}} \leq \frac{d}{2}.$$ (15.4)

Equation (15.4) states that the rms displacement of a particular charge species must be less than half the clear spacing in order to have a buildup of that species of charge between the plates. In the geometry shown in figure 15.6, the distance d is the clear distance between the electrode plates. Substituting equations (15.2)–(15.4) into equation (15.1) yields the relationship

$$\frac{d}{2} \approx \frac{e V_{\text{rms}}}{2\pi m v_0 v_c d}.$$ (15.5)

By manipulating approximation (15.5), the critical frequency v_0 above which the uniform glow discharge should build up in the plasma volume is given by

$$v_0 \approx \frac{e V_{\text{rms}}}{\pi m v_c d^2} \qquad \text{Hz.}$$ (15.6)

The rms voltage in equation (15.6) is that which bounds the uniform discharge regime. It should be noted that the product $(m v_0)$ in the denominator is smaller for electrons than for ions. The critical frequency for onset of the uniform discharge (ion trapping) therefore is *lower* than the critical frequency for onset of the filamentary discharge (electron trapping). Thus, for operation in the uniform glow discharge regime, the driving frequency should lie between the limits

$$\frac{e V_{\text{rms}}}{\pi m_i v_{ci} d^2} \leq v_0 \leq \frac{e V_{\text{rms}}}{\pi m_e v_{ce} d^2} \qquad \text{Hz.}$$ (15.7)

15.4.3.3 Power Input to Charged Species

Commercial applications of the OAUGDP to the surface treatment of materials require power densities high enough to produce useful fluxes of plasma active species, but not high enough to damage sensitive materials such as fabrics and polymer films. To understand the factors that determine the plasma power density, consider the work done on a single ion or electron between parallel plates by the RF electric field, given by

$$W = \boldsymbol{F} \cdot \mathrm{d}\boldsymbol{x} = e E(t)\, \mathrm{d}x \qquad \text{J.}$$ (15.8)

The power delivered to this ion or electron is

$$p = \frac{\mathrm{d}W}{\mathrm{d}t} = e E(t)\frac{\mathrm{d}x}{\mathrm{d}t} = e E(t)\dot{x} \qquad \text{W/particle.}$$ (15.9)

Taking the derivative of the displacement $x(t)$ given by equation (12.89) in Volume 1, one obtains

$$\dot{x} = C_1 \omega \cos \omega t - C_2 \omega \sin \omega t. \tag{15.10}$$

Substituting equation (12.86) of Volume 1 for $E(t)$ and equation (15.10) into equation (15.9), one obtains

$$p = e E_0 \omega C_1 \sin \omega t \cos \omega t - e E_0 \omega C_2 \sin^2 \omega t \qquad \text{W/particle.} \tag{15.11}$$

The average power per unit volume is found by multiplying equation (15.11) by the electron/ion density, and integrating over a cycle of oscillation, $\omega t = 2\pi$. The first term of equation (15.11) averages to zero, and the remaining term is

$$\bar{P} = n_e \bar{p} = \frac{n_e e E_0 \omega C_2}{2\pi} \int_0^{2\pi} \sin^2 \omega t \, d(\omega t)$$

$$= \frac{n_e e E_0 \omega C_2}{2} \qquad \text{W/m}^3. \tag{15.12}$$

Substituting equation (12.92) of Volume 1 for the integration constant C_2, equation (15.12) becomes

$$\bar{P} = -\frac{n_e e E_0 \omega}{2} \times \frac{v_c e E_0}{\omega m (\omega^2 + v_c^2)} = -\frac{e^2 E_0^2 v_c n_e}{2m(\omega^2 + v_c^2)} \qquad \text{W/m}^3. \tag{15.13}$$

In the OAUGDP reactor, the driving frequency is typically 1–10 kHz, while v_c is GHz for ions, and THz for electrons. Thus, $v_c \gg \omega$, and equation (15.13) may be approximated as

$$\bar{P} \approx -\frac{n_e e^2 E_0^2}{2m v_c} \qquad \text{W/m}^3. \tag{15.14}$$

Using equation (15.3) to write the maximum electric field E_0 in terms of the rms applied voltage in this parallel-plate geometry, one obtains equation (15.14) in the form

$$\bar{P} = -\frac{n_e e^2}{2m v_c} \frac{\pi^2 V_{\text{rms}}^2}{4d^2} = -\frac{n_e \pi^2 e^2 V_{\text{rms}}^2}{8m v_c d^2} \qquad \text{W/m}^3. \tag{15.15}$$

This relation predicts a power density proportional to the square of the applied voltage.

The maximum electric field E_0 can be written in terms of the plasma and reactor parameters by using equations (15.1), (15.2) and (15.4) with $x_{\text{rms}} = d/2$, to obtain

$$E_0 = \frac{v_0 \pi^2 m v_c d}{2e} \qquad \text{V/m.} \tag{15.16}$$

Substituting equation (15.16) into equation (15.14) for the power density yields

$$\bar{P} = \frac{n_e \pi^4 m v_c d^2 v_0^2}{8} \qquad \text{W/m}^3. \tag{15.17}$$

If the plasma number density remains approximately constant, equation (15.17) predicts a power density proportional to the square of the RF frequency, in good agreement with the experimental data illustrated in figure 12.27 of Volume 1.

The relative power input to the ion and electron populations can be calculated from equation (15.15). Under conditions of fixed V_{rms} and electrode spacing d, the relative power delivered to the electron and ion populations is

$$\frac{\bar{P}_i}{\bar{P}_e} = \frac{m \nu_{ce}}{M \nu_{ci}} \tag{15.18}$$

where m and M are the electron and ion mass, and ν_{ce} and ν_{ci} the electron and ion collision frequencies, respectively. Using the collision frequencies at 1 atm for helium gas,

$$\frac{\bar{P}_i}{\bar{P}_e} = \frac{1.8 \times 10^{12}}{4 \times 1837 \times 6.8 \times 10^9} = 0.036. \tag{15.19}$$

Thus for helium, most of the power is delivered to the electron population, where it is most effective in producing ionization and active species.

15.4.3.4 The OAUGDP as a Normal Glow Discharge

The plasma physical characteristics of the OAUGDP in helium were investigated by one-dimensional computer simulations (Massines *et al* 1998) and experimental time-resolved ultra-fast photography (Massines *et al* 1998, Ben Gadri 1997). Short time exposure photographs using a CCD camera (exposure times of 10–100 ns) and computationally modeled characteristics were analogous to those of a normal DC glow discharge in spite of operation at atmospheric pressure. The calculated parameters of the plasma using a one-dimensional fluid model (Massines *et al* 1998, Ben Gadri 1999) included the electric field, as well as the space-charge density.

The results of this model are illustrated in figure 15.9. All the features of the classical normal glow discharge were present between the instantaneous cathode and anode (Ben Gadri 1999). These features included the *cathode fall*, in which the electric field decreases linearly from its maximum value and thus obeys Aston's law; the *negative glow*; the *Faraday dark space*; and finally the *positive column*, where quasi-neutrality holds and the electric field is relatively low and constant. The calculated maximum electron and ion densities are about 3 and 5×10^{11} cm^{-3}, respectively, depending on the plasma operating conditions (Ben Gadri 1997, 1999).

15.4.3.5 Discharge Plasma Chemistry

As it passes through the plasma, the injected gas becomes excited, dissociated, or ionized by electron impact. A wide range of active species are created that can be used for various industrial applications. For air as a working gas, these species

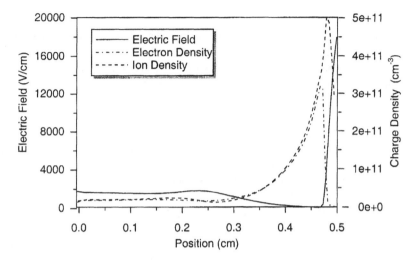

Figure 15.9. Spatial profiles of the electric field and the ion and electron densities at the maximum of the discharge current in a parallel-plate OAUGDP plasma source. Data are from Ben Gadri (1997, 1999) for calculations made in helium with a 10 kHz frequency and a 1.5 kV maximum applied voltage.

include atomic oxygen, ozone, nitrogen oxides, neutral metastable molecules, and free radicals. In the field of non-equilibrium plasma processing, highly reactive species produced in the OAUGDP can be used for dry etching, surface treatment, sterilization, decontamination of surfaces, and other purposes.

15.4.4   Safety and Regulatory Issues of the OAUGDP

Industrial use of the OAUGDP presents safety and regulatory issues that are manageable by appropriate design of the source. Electrical safety is a major issue. Not only is the user faced with the ordinary hazards of high voltages that may be up to 20 kV rms, but also the low-frequency (MF, audio) RF causes electrical sparking over larger separation distances and at lower voltages than would normally be expected. The ion-trapping mechanism that makes the OAUGDP possible contributes to this hazard.

Beyond the enhanced sparking hazard, the 'spark' will, in most cases, be sustained as an RF arc which, like a DC arc, will elongate to distances up to three to five times the distance required to initiate it before being extinguished. In this way, it behaves as an RF version of the DC *expanding arc* discussed in section 10.7.2 of Volume 1. If the arc strikes human skin, it can leave a cauterized burn that can take up to six months to heal. This RF sparking hazard can be alleviated by enclosing components with small radii of curvature in equipotential metal enclosures designed to the corona-shielding standards discussed in section 8.5.4 of Volume 1.

After high voltage safety, other regulatory and safety issues of the OAUGDP include the production of audible noise and toxic and corrosive active species. The audible noise of the OAUGDP originates from the electrohydrodynamic (EHD) coupling of the driving RF voltage first to the plasma, then to the neutral gas by Lorentzian collisions of the ions. This coupling is discussed further in section 18.6 of this volume, and in the plasma aerodynamics chapter of Volume 3. The usual result of this coupling in the OAUGDP is a pure tone at the RF driving frequency (normally in the audible range at a few kilohertz), with the plasma acting as a loudspeaker. This audible noise is characteristically around 50 to 70 dBA, and even if not attenuated is more of a nuisance than a hazard.

OAUGDPs operating in air or oxygen-containing gases produce ozone, and other oxidants such as nitrogen oxides. The ozone concentrations require monitoring in an industrial setting, to ensure concentrations below regulatory limits. The oxidizing active species are often important to surface treatment, sterilization, and other applications of the OAUGDP, and their concentrations are usually maximized. Other hazards include ultraviolet radiation and possible RFI at low frequencies in the ELF and VLF bands. All these potential hazards can be dealt with by suitable isolation, shielding, and enclosure, relatively inexpensive measures that depend on the details of the application.

15.4.5   Operational Issues

As an electrical circuit, the OAUGDP reactor is a variable capacitive load with a resistive component. It has a low power factor when used with conventional power supplies. The resulting impedance mismatch results in a large reflected power from the plasma back to the power supply that does not contribute to plasma formation, may complicate attempts to achieve a uniform discharge, and requires an expensive over-rated power supply. A passive electrical impedance matching network can make the plasma act like a resistor at the resonance frequency, as seen by the transformer secondary. The discharge is stable when impedance matched and produces a normal glow discharge plasma like that modeled by Ben Gadri (1997, 1999).

When operated in air, the OAUGDP is sensitive to humidity in the airflow, which has a significant effect on plasma uniformity. The relative humidity should be below 15% to obtain the desired uniform discharge across the electrode gap. Higher humidity may lead to the formation of filamentary structures, particularly at points, edges or asperities around which high electric fields occur. In addition, forced air flow through the source chamber promotes the formation of a more uniform plasma.

The temperature of the working gas and active species can be controlled to within a few degrees Centigrade of ambient by suitable adjustment of the flow velocity of the working gas and the temperature of the electrode and surrounding surfaces. It has been found that the concentration of active species important to sterilization and surface treatment is sensitive to the temperature of the electrode surfaces with which the OAUGD plasma comes into contact. There appears to be an inverse dependence of active-species concentration on electrode surface temperature.

An OAUGDP operating in air produces ozone and ultraviolet radiation. These may require a protective enclosure through which the workpieces or webs to be processed are fed in a continuous manner. Containment of the ozone produced by OAUGDPs operating with air or significant concentrations of oxygen may be done in a way that also contains other corrosive or toxic species, including ultraviolet radiation. As previously noted, regulatory issues associated with the OAUGDP are manageable, if their requirements are incorporated into the initial reactor design.

15.4.6 Research Issues of the OAUGDP

The computer modeling of Ben Gadri (1997, 1999, Messines *et al* 1998) showed good agreement between the experimentally measured and computationally simulated current waveform and axial structures. However, this work was restricted to a one-dimensional slab plasma consisting of monatomic helium gas. This modeling needs to be extended to two- and three-dimensional computer models with air as the working gas. Such an extension should better accommodate physical processes responsible for the filamentation instability, when electron as well as ion trapping comes into play and the slab plasma becomes polarized.

The origin of the *filamentation instability* that occurs as the RF driving frequency increases and causes electron as well as ion trapping also needs to be investigated experimentally, to provide a database for two- or three-dimensional computer simulations. Other research issues of the OAUGDP include the role of gas flow through the plasma volume in active species formation. For example, in a parallel-plate OAUGDP slab reactor like that illustrated in figure 15.6, does the optimum rate of active species formation occur with the working gas static, flowing normal to the electrodes, or parallel to the electrodes? Other issues that have not been addressed for this relatively recently developed plasma source are the modeling of the plasma chemistry; the effects of surface temperature, gas flow velocity, and gas temperature on active-species formation; and the details of the EHD flow acceleration that occur in the discharge volume.

15.5   ARCJET PLASMA SOURCES

The arcjet plasma sources discussed here are normally operated at 1 atm in industrial settings, and were discussed briefly in section 10.7.3 of Volume 1. These sources operate in the arc regime of the classical $V-I$ curve illustrated in figure 15.1, and are characterized by an intense, dazzling light emission, with very high gas temperatures and power densities. Although this chapter concentrates on DC and AC line frequency arcjet sources, arcjets have been operated at

13.56 MHz in helium (Schutze *et al* 1998), and at the multi-kW level at microwave frequencies (Mitsuda *et al* 1989).

15.5.1   Applications of Arcjets

Arcjet plasma sources are used industrially for plasma spraying of thick films, the heating, melting, or refining of bulk materials, and the epitaxial deposition of bulk materials. In plasma spraying of thick films, powdered material that is to form the desired coating is fed into the plasma jet, where it is melted, and this material is entrained in the flow of the exhaust jet until it condenses on the workpiece. For epitaxial deposition of diamond-like coatings or crystalline materials, deposition at 1 atm with an arcjet plasma source has many economic and operational advantages, including the absence of an expensive vacuum subsystem, and no need for batch processing.

15.5.2   Characteristics of Arcjets

A DC arcjet plasma source used for plasma chemical vapor deposition (PCVD) is illustrated in figure 15.10. *Arcjets* are axisymmetric, and characteristically consist of an uncooled, incandescent axial cathode with a conical tip that faces the converging segment of a water-cooled nozzle that functions as the anode. The carrier and working gases flow axially in the annular space between the cathode and the water-cooled walls. An arc is struck between the anode and cathode. This arc plasma heats the gas as it flows through the converging section of the nozzle. The plasma produced by arc heating is expanded to supersonic velocities in the diverging section of the nozzle, where the resulting *plasma jet* is close to thermodynamic equilibrium.

In this source, the *carrier gas* (usually argon) and *working gases* are fed coaxially through the arcjet, where the working gases may undergo plasma chemistry to produce precursors for deposition. In some cases the same or other working gases may also be injected into the plasma jet or deposition chamber to speed up or otherwise enhance the deposition process. In high power applications, particularly above 50 kW, an axial magnetic field may be applied that causes the root of the arc to rotate on the anode surface due to a $\boldsymbol{J} \times \boldsymbol{B}$ body force (see section 10.3.1 in Volume 1), thus reducing the local average heat flux.

Because the plasma jet is usually close to thermodynamic equilibrium, blackbody radiation losses limit the electron kinetic temperature to 1 or 2 eV, corresponding to gas and electron kinetic temperatures in the range from 12 000 to 24 000 K. Such high temperatures promote a variety of plasma–chemical reactions capable of producing active species and precursors for deposition. These high temperatures also make it possible for the plasma jet to have a sufficiently high energy density to melt, or even vaporize, any element of the periodic table.

The plasma generated by DC arcjet sources is spatially non-uniform. Their electron number densities vary over several orders of magnitude from the plasma

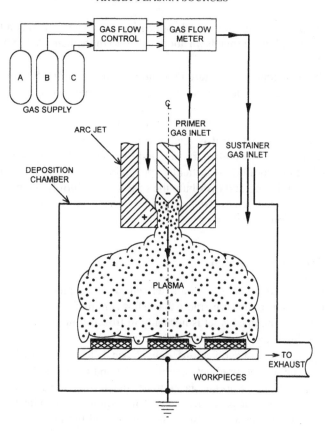

Figure 15.10. Schematic diagram of a DC arcjet used to generate active species from a mixture of injected working gases. The operating pressure can be at or below 1 atm.

jet core at the nozzle throat, to the greatly reduced density immediately above a workpiece on which the jet impinges. The axial electron number density profile in DC arcjet plasma sources may span a range from 10^{21} electrons/m^3 at the core of the plasma jet, down to values of 10^{16} electrons/m^3 or less in the expanded plasma above the workpiece.

15.5.3   Independent Operating Parameters

Arcjet plasma sources can be operated at pressures ranging from a few hundred Pascal (a few Torr) to 1 atm, but most industrial applications are conducted at 1 atm. They are normally operated with direct current (DC), but may be energized at AC, RF, or even microwave frequencies. Power levels for arcjets can range from a few kilowatts to a megawatt or more, depending on the application. The currents drawn range from the lower limit of the thermal arc regime at 20–50 A, to several thousand amperes in megawatt-level applications. Voltages depend on the power level required, and may range from about 100 V to more than 1 kV in high power applications.

The high cathode temperature in an arcjet makes it difficult to operate with oxidizing working gases and, as a result, the use of argon gas is favored. Argon is chemically inert, the least expensive of the 'noble' gases, and is easily ionized, thus leading to electrically stable operation of the arc. Helium gas is also used in the United States, where it is readily available and not too costly. The gas flow rate is adjusted to stabilize the arc, and to produce the gas temperature and power dissipation required for particular applications. When axial magnetic fields are used to spin the arc and distribute the heat load uniformly on the electrodes, they usually are generated by axisymmetric solenoidal electromagnets positioned coaxially at the throat of the nozzle.

15.5.4   Arcjet Phenomenology

The plasma jet from an arcjet is characterized by very high energy densities and power fluxes, but its radius is no more than a few centimeters, limiting its coverage on a workpiece. The plasma jet is normally in frozen-flow chemical equilibrium as it transits the distance from the nozzle to the workpiece.

In plasma-spraying applications, only physical changes (melting) of the feed powder material are required between the arcjet and the workpiece. For PCVD and most epitaxy, the working gases must undergo chemical reactions in the plasma jet to produce the desired species. This requirement can lead to a delicate balancing of the gas flow rate, the power input, and the flow rates of working gases. Workpieces are directly exposed to the plasma jet for plasma spraying and most PCVD applications. However, remote exposure may be used for PCVD when it is desired to reduce the heat flux to the workpiece and the active species reach a frozen flow equilibrium.

15.5.5   Performance Issues

DC arcjet plasma sources are operated at the multi-kW level, with power fluxes easily capable of damaging or melting both the workpiece and the electrodes of the arcjet itself. An arcjet is, by definition, axisymmetric, and the diameter of the plasma jet is only a few centimeters. This concentration of effect generally makes it necessary to have elaborate positioning and fixturing of the workpiece to achieve uniformity of treatment over large areas, or curved workpieces.

Much of the power input to arcjets appears in the form of blackbody ultraviolet and line radiation emission. Both these losses decrease the fraction of the input power available to heat the working gas, produce ionization, melt powders, or create active species for processing. A characteristic power efficiency (power into thermal enthalpy of the plasma/total input power) for a DC arcjet plasma source is about 30%, with the wasted power carried away by radiation, the electrode cooling water, or thermal conduction.

One of the operational problems of arcjet plasma sources is the relatively short lifetime of the electrodes, particularly an uncooled cathode in the presence of oxidizing or chemically reactive working gases. In some applications of arcjets, the uncooled cathode is fed in as it is consumed by erosion and vaporization, much in the manner of a welding rod. The regulatory requirements for the operation of arcjets are very similar to those that apply to arc welding, where protection of workers against high fluxes of ultraviolet radiation and the handling of hot workpieces are the dominant concerns. Some of the research issues in arc technology were reviewed in section 10.8 of Volume 1, and will not be repeated here.

15.6 INDUCTIVELY COUPLED PLASMA TORCHES

Inductively coupled RF plasma torches operate at multi-kW levels in atmospheric pressure applications. The physical processes responsible for plasma heating in these sources have been discussed in chapter 11 of Volume 1. These processes differ from those of the DC and RF plasma sources discussed previously in this chapter, in which the electron population of the plasma is heated by direct acceleration in an applied RF electric field. In inductively coupled sources, the energy is coupled from the power supply to the plasma by a time-varying magnetic field. This time-varying magnetic field induces plasma currents that heat the electron population by ohmic dissipation.

15.6.1 Applications of Plasma Torches

Atmospheric pressure inductively coupled RF plasma torches have been discussed in section 11.4 of Volume 1. This technology is used for continuous-flow plasma chemical reactors, plasma spraying of thick films, surface heat treatment and, more recently, PCVD applications. As continuous-flow chemical reactors, plasma torches are used for destructive plasma chemistry, including the remediation of biohazards and toxic wastes. Heating a plasma by inductive coupling is a technology very similar to the inductive heating of metal ingots, billets, and tundishes widely practiced in foundries and the metalworking industry. This similarity to a familiar technology has led to ready and widespread acceptance of inductive plasma torch technology in these industries.

15.6.2 Characteristics of Plasma Torches

The electron kinetic temperature of inductive plasma torches operated at 1 atm is limited by line and blackbody radiation losses to 1 or 2 eV since, like arcjet plasmas, they are in or close to thermodynamic equilibrium. A single-stage atmospheric pressure *inductively coupled plasma torch* is illustrated in

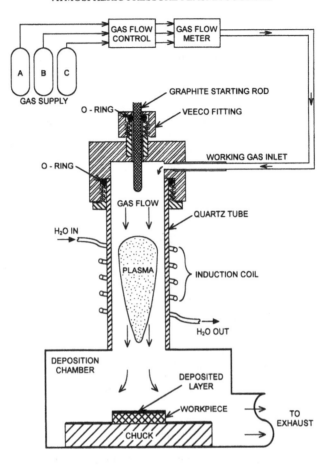

Figure 15.11. Schematic diagram of an axisymmetric inductive plasma torch used as a plasma chemical reactor or for plasma spraying of thick films.

figure 15.11. The inductive current coupling discussed in chapter 11 of Volume 1 creates a multi-kW, electrodeless, intense plasma confined in a quartz tube with a continuous flow of working gases through and around it. Such plasma torches operate close to thermodynamic equilibrium, and at pressures that range from a few kPa to above 1 atm. The plasma is generated by an external RF inductive heating coil operated at frequencies that may range from a few kHz to 13.56 MHz or higher. Because the power levels of inductively coupled atmospheric plasma torches range from several kW to 1 MW or more, the power supply is usually impedance matched to the source.

For applications that do not require plasma–chemical reactions of the feed gases, inert gases such as argon or helium are used. These gases are relatively inexpensive, their low electrical breakdown strength stabilizes the plasma discharge, they do not react chemically with the walls or workpieces, and in plasma-spraying applications, they do not react with the melted powders that form the sprayed coating. The RF current induced in the plasma heats the electron population to kinetic temperatures of 1–2 eV, which, in turn, maintains ionization and produces active species. The gas, electron, and ion temperatures are approximately equal and in the range 10 000–20 000 °C. The maximum electron number density as well as the gas temperature of inductively coupled plasma torches is comparable to that of arcjets, at approximately 10^{21} electrons/m^3.

15.6.3 Plasma Torch Phenomenology

The inductively coupled plasma torch can be used as a high-temperature plasma reactor for synthetic or destructive plasma chemistry, or simply as a source of enthalpy for plasma spraying or heat treatment of workpieces. In the latter applications, inert gases such as helium or argon are used. As indicated in figure 15.11, the working gas in plasma torches flows through the plasma region and becomes entrained in the exiting gas flow. By adjusting the flow rate, the dwell time of the working gases in the plasma can be controlled, and the concentration of the plasma–chemical reaction products varied over a wide range. Workpieces can be exposed to the active species of torch plasmas by *direct contact* with the plasma or by *remote exposure*, in which active species are convected to a workpiece located at a distance from the plasma.

15.6.4 Operational Issues

The axisymmetric coils with which inductively coupled plasmas are generated produce an axisymmetric plasma, which may not be uniform enough across its diameter to provide acceptable processing of a workpiece. In such a case, motional averaging of the deposited layer on the workpiece is required, using one of the methods discussed in section 18.6 of this volume. As noted in chapter 11 of Volume 1, the cross section of the inductive coil need not be circular, and may be formed into rectangular, triangular, and other geometric cross sections to improve depositional or processing uniformity on non-axisymmetric workpieces.

The power efficiency of the atmospheric-pressure inductive plasma source illustrated in figure 15.11 is relatively low, with as little as 10–40% of the input AC electrical line power converted into plasma enthalpy. This relatively low efficiency can be a major issue with plasma torches, since their power level can range from a few kilowatts to several megawatts. Like arcjets, inductive plasma torches are capable of depositing sufficiently high heat fluxes on workpieces to cause damage. This issue is normally dealt with by rapid motion of the workpiece through the plasma and/or vigorous cooling of the workpiece.

The independent parameters used to control the processing of workpieces include the type and mixture ratio of working gases; the flow rate of the working gas; the RF power level; and the coil geometry, including the ratio of the plasma to the coil radius. In addition to these plasma-related parameters, the velocity of the workpiece through the plasma, and hence its duration of exposure, can be adjusted.

The plasma torch may pose significant safety and regulatory issues related to potential radio frequency interference (RFI) from its power supply. In addition, the high power levels and near thermodynamic equilibrium of the plasma jet ensures copious UV emission, which can be hazardous to exposed workers. Except for potential RFI, the safety and regulatory issues of plasma torches are very similar to those of arc welding.

REFERENCES

Ayrton H 1902 *The Electric Arc* (New York: Van Nostrand)

Barber H 1983 *Electroheat* (New York: Granada) ISBN 0-246-11739-7

Ben Gadri R 1997 Numerical simulation of an atmospheric pressure and dielectric barrier controlled glow discharge *PhD Thesis* Order 2644, University Paul Sabatier, Toulouse III, France

——1999 One atmosphere glow discharge structure revealed by computer modeling *IEEE Trans. Plasma Sci.* **27** 36–7

Boenig H V 1988 *Fundamentals of Plasma Chemistry and Technology* (Lancaster, PA: Technomic) ISBN 87762-538-7

Boulos M, Fauchais P and Pfender E 1994 *Thermal Plasma Processing* vol I (New York: Plenum)

Cary H B 1979 *Modern Welding Technology* (Englewood Cliffs, NJ: Prentice-Hall) ISBN 0-13-599290-7

Cobine J D 1958 *Gaseous Conductors* (New York: Dover)

DeVries L 1959 *German–English Science Dictionary* 3rd edn (New York: McGraw-Hill) LCCCN 59-9412

Eliasson B and Kogelschatz U 1988 UV excimer radiation from dielectric-barrier discharges *Appl. Phys.* B **46** 299–303

Gross B, Grycz B and Miklossy K 1969 *Plasma Technology* (New York: Elsevier) LCCCN 68-27535

Hoyaux M F 1968 *Arc Physics* (New York: Springer) LCCCN 68-24015

Kanazawa S, Kogoma M, Moriwaki T and Okazaki S 1988 Stable glow plasma at atmospheric pressure *J. Phys. D: Appl. Phys.* **21** 838–40

Lieberman M A and Lichtenberg A J 1994 *Principles of Plasma Discharges and Materials Processing* (New York: Wiley) ISBN 0-471-00577-0

Lieberman M A, Selwyn G S and Tuszewski M 1996 Plasma generation for materials processing *MRS Bull.* **21** 32–7

Loeb L B 1965 *Electrical Coronas* (Berkeley, CA: University of California Press) LCCCN 64-18642.

Massines F, Rabehi A, Decomps Ph, Ben Gadri R, Segur P and Mayoux C 1998 Mechanisms of a glow discharge at atmospheric pressure controlled by dielectric barrier *J. Appl. Phys.* **83** 2950–7

Mitsuda Y, Yoshida T and Akashi K 1989 Development of a new microwave plasma torch and its application to diamond synthesis *Rev. Sci. Instrum.* **63** 21–30

National Academy 1991 *Plasma Processing of Materials: Scientific Opportunities and Technological Challenges* (Washington, DC: National Academy Press) ISBN 0-309-04597-5

Paschkis V and Persson J 1960 *Industrial Electric Furnaces and Applications* 2nd edn (New York: Interscience) LCCCN 60-11028

Roth J R 1995 *Industrial Plasma Engineering: Volume I, Principles* (Bristol: Institute of Physics Publishing) ISBN 0-7503-0318-2

——1997 Method and apparatus for covering bodies with a uniform glow discharge plasma and applications thereof *US Patent* 5,669,583

——1999 Method and apparatus for cleaning surfaces with a glow discharge plasma at one atmosphere of pressure *US Patent* 5,938,854

Roth J R, Sherman D M and Wilkinson S P 1998 Boundary layer flow control with a one atmosphere uniform glow discharge plasma *AIAA Paper* 98-0328

Roth J R, Tsai P P and Liu C 1995a Steady-state, glow discharge plasma *US Patent* 5,387,842

Roth J R, Tsai P P, Liu C, Laroussi M and Spence P D 1995b One atmosphere, uniform glow discharge plasma *US Patent* 5,414,324

Schutze, Jeong J Y, Babayan S E, Park J, Selwyn G S and Hicks R F 1998 The atmospheric plasma jet: a review and comparison to other plasma sources *IEEE Trans. Plasma Sci.* **26** 1685–94

Smith D L 1995 *Thin-Film Deposition: Principles and Practice* (New York: McGraw-Hill) ISBN 0-07-058502-4

Smith R W and Novak R 1991 Advances and applications in US thermal spray technology I—Technology and Materials *Powder Metall. Int.* **23** 147–55

Von Engle A, Seeliger R and Steenbeck M 1933 On the glow discharge at high pressure Z. *Phys.* **85** 144–60

Yokoyama T, Kogoma M, Moriwaki T and Okazaki S 1990 The mechanism of the stabilization of glow plasma at atmospheric pressure *J. Phys. D: Appl. Phys.* **23** 1125–8

16

Vacuum Plasma Sources

A majority of industrial applications of glow discharge plasmas are conducted under vacuum, at pressures below 1.3 kPa (10 Torr). As has been previously mentioned, the operation of plasma sources under vacuum puts them under the disadvantages of using expensive vacuum equipment (see Kohl 1995), and imposes batch processing as workpieces are cycled between the workplace at 1 atm and vacuum conditions. Operation under vacuum, however, has the important advantage that the isolation, containment, and shielding provided by the vacuum system reduces (but does not eliminate, see Herb (1989)) many of the workplace hazards and regulatory concerns that would otherwise be associated with operation at 1 atm.

This chapter is concerned with the plasma sources most widely used for industrial plasma processing under vacuum or which hold promise for future such applications. Information on vacuum plasma sources and their applications may be found in Batenin *et al* (1992), Boenig (1988), Lieberman and Lichtenberg (1994), Lieberman *et al* (1996), Raizer *et al* (1995), Roth (1995), and Smith (1995).

16.1 INTERMEDIATE-PRESSURE PLASMA SOURCES

In the period from 1960 to 1990, a significant proportion of plasma-processing applications were conducted with glow discharge plasmas at *intermediate pressures*, between 1300 and 13 Pa ($\sim$10–0.1 Torr). In this range, the collisional mean free paths of ions, electrons, and neutral atoms are small compared with the dimensions of the plasma and the reactor. In the next three subsections, the sources most frequently used in this pressure range are discussed.

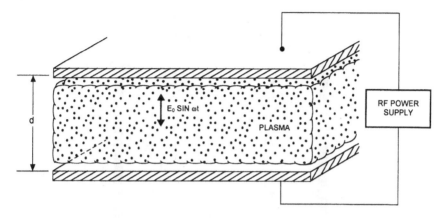

Figure 16.1. A parallel-plate plasma reactor with a slab plasma.

16.1.1 Parallel-Plate Sources

Parallel-plate glow discharge plasma sources are widely used in the United States for microelectronic deposition and etching, and for many plasma-processing applications outside the microelectronics industry. These sources can be energized either by *DC electrical power*, in which the electrodes must collect real currents of impinging ions or electrons, or *RF excitation*, in which the power is coupled to the plasma by displacement currents, without a requirement that real currents flow to the electrodes. The absence of real currents removes a potential source of sputtering and consequent plasma contamination. This is a major factor in the dominance of '*electrodeless*' RF plasma sources for microelectronics applications.

DC and RF glow discharge plasmas have been used for plasma-processing applications in the *intermediate-pressure regime* at pressures above 6.7 Pa (50 mTorr). RF and DC glow discharges can be operated at pressures as high as 1.3 kPa (10 Torr), but their stability above 1 Torr is usually not adequate for production applications. The negative glow plasma of a DC glow discharge tends to have a lower electron number density than the plasma in a characteristic parallel-plate RF glow discharge plasma source operated with the same gas at the same pressure. The DC source is therefore less used for microelectronic plasma processing, because its lower electron number density tends to produce lower fluxes of precursors and active species, leading to longer processing times.

A characteristic parallel-plate plasma source is illustrated in figure 16.1. The workpiece (a wafer in microelectronic applications) to be exposed to active species from the plasma is usually on an electrode. In industrial parallel-plate glow discharge sources, the ratio of electrode separation, d, to plate length, L, is typically $2 \leq L/d \leq 10$. The theory of DC parallel-plate plasma sources, both unmagnetized and magnetized, was discussed in sections 9.2 and 9.5.2, and

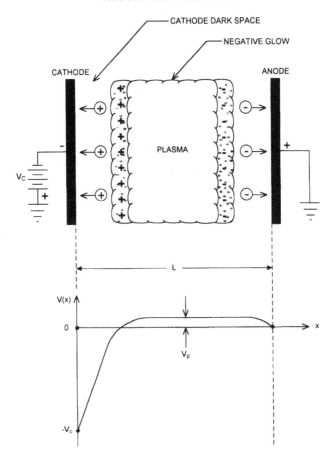

Figure 16.2. The DC obstructed abnormal glow discharge plasma source, with the approximate potential distribution.

unmagnetized and magnetized RF parallel-plate sources in sections 12.1 and 12.2, respectively, of Volume 1.

16.1.1.1 DC Excitation

In the DC *obstructed abnormal glow discharge* illustrated in figure 16.2, real DC currents are collected by the electrodes on which wafers or other workpieces are mounted. Selected references that include parallel-plate DC glow discharge plasma sources are Hirsch and Oskam (1978), Howatson (1976), Lieberman and Lichtenberg (1994), Roth (1995), and Tao *et al* (1996).

For microelectronic plasma etching, the wafers are usually mounted on the cathode in order to take advantage of energetic ion bombardment, which promotes

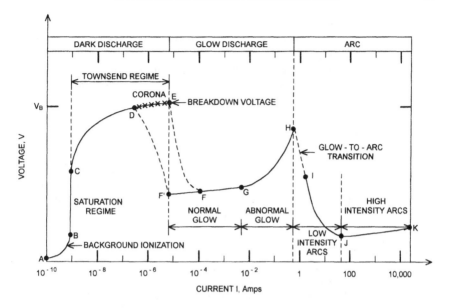

Figure 16.3. A voltage–current characteristic of the classical DC electrical discharge.

directional etching. For deposition applications, workpieces may be mounted on the anode, since production of electrically neutral precursors of the deposited layer does not require the presence of energetic ions or a large sheath voltage drop between the workpiece and the plasma. Many RF-generated plasmas are operated in a hybrid mode, in which the wafer is biased by a DC voltage with respect to the RF-generated plasma potential, in order to control the flux or energy of ions on the wafer. The characteristics of this sheath plasma may behave like that of a DC parallel-plate discharge (see Reinke *et al* 1992).

Figure 16.3 illustrates the classical voltage–current curve for DC electrical discharges. In the *normal glow* regime, between F′ and G in figure 16.3, the *negative glow* plasma may or may not cover the portion of the cathode on which the workpieces are mounted, thus leading to unreliability of exposure and possible non-uniformity of effect. The DC *unobstructed abnormal glow discharge*, discussed in section 9.5.2 of Volume 1, is established between the points G and H on the voltage–current curve of figure 16.3. In the *abnormal glow regime*, whether obstructed or not, the entire cathode, including workpieces mounted on it, is covered by the plasma, ensuring uniformity of effect. The negative glow plasma usually covers the entire anode as well, if the area of the latter is equal to that of the cathode. Unobstructed abnormal glow discharges have a *positive resistance characteristic*, for which a higher applied voltage increases the total current. Also in these discharges, the width d of the cathode region shown on figure 16.2 is equal to the Paschen minimum distance appropriate for the gas and pressure, as discussed in section 9.2 of Volume 1. A positive column and other structures may be present between the negative glow and the anode in these unobstructed discharges.

If an abnormal glow discharge is *obstructed*, the width d of the cathode region is smaller than the Paschen minimum distance, the positive column is absent, and the cathode fall voltage is higher than the Paschen minimum breakdown voltage. This high sheath voltage is useful when energetic ions are required to sputter or etch the cathode surface, or to treat workpieces mounted on it. In some applications where *radiation damage* by ions above a few tens of eV is an issue, one may not wish to operate a DC glow discharge in the obstructed mode, in order to reduce the cathode fall voltage, V_c, to values below the radiation damage threshold. Further discussion of the DC parallel-plate glow discharge may be found in sections 9.2 and 9.5.2 in Volume 1.

16.1.1.2   RF Excitation

The *parallel-plate RF plasma source* operating at 13.56 MHz is widely used in the microelectronic industry in the United States. This source consists of a plane slab of RF glow discharge plasma between two parallel electrode plates, separated by a distance d. These electrodes may not be covered by a dielectric coating. The RF electric field between the plates accelerates electrons that ionize the neutral working gas, as discussed in sections 12.1 and 12.2 of Volume 1. Other references that discuss the parallel-plate RF plasma source include Claude *et al* (1987), Batenin *et al* (1992), Gicquel and Catherine (1991), Godyak *et al* (1991), Lieberman and Lichtenberg (1994), Mutsukura *et al* (1994), Olthoff *et al* (1994), and Raizer *et al* (1995).

When parallel-plate RF plasma sources are used for plasma processing, and particularly for microelectronic etching, one of the three configurations illustrated in figure 16.4 is used. Each is distinguished by the manner in which the RF power supply is connected to the electrodes. In figure 16.4(*a*), when the electrode on which the workpieces are mounted is connected to ground and the opposite electrode to an RF power supply, the *plasma etching* (PE) configuration results. This configuration is preferred for microelectronic deposition, since the large voltage drop between the plasma and the powered electrode may produce active species so energetic that they interfere with the deposition process, or induce radiation damage on the thin film being deposited.

When the electrode on which the workpieces are mounted in figure 16.4(*b*) is connected to an RF power supply and the opposite electrode grounded, the *reactive ion etching* (RIE) configuration results. This configuration is preferred for plasma etching because the voltage drop between the plasma and the powered electrode on which the wafers or workpieces are located is relatively high— typically a few tens of volts. This voltage drop can be adjusted by altering the ratio of the electrode surface areas or DC biasing the powered electrode.

In some microelectronic and other plasma-processing applications, it is

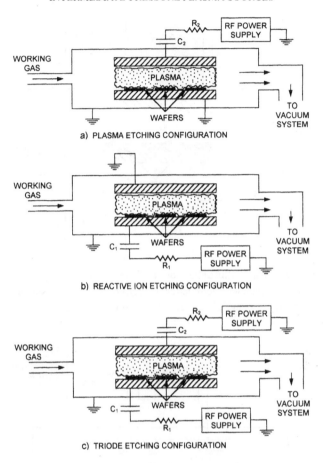

Figure 16.4. Configurations of the RF parallel-plate plasma source: (*a*) the plasma etching configuration; (*b*) the reactive ion etching (RIE) configuration; and (*c*) the triode configuration.

desirable that the workpiece float at or near the *plasma potential*. This is done to reduce 'radiation damage' by ions accelerated in the electric field of the sheath or to prevent contamination of the workpieces by such energetic ions. The floating potential of the workpiece can be matched to the plasma potential by the *triode etching* configuration shown in figure 16.5, in which both electrodes are powered. In a triode etching plasma source, the area of the two electrodes, the blocking capacitances C_1 and C_2, and the resistances R_1 and R_2 can be adjusted until the electrode containing the workpieces floats at the plasma potential.

Finally, some parallel-plate glow discharge RF sources energize each electrode with a different frequency, usually one above and one below the *critical sheath frequency*, discussed in section 19.5.3. This arrangement, illustrated

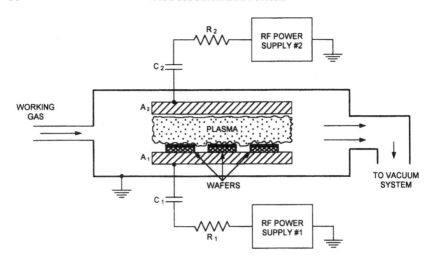

Figure 16.5. Schematic illustration of a triode etching configuration for PCVD applications, in which parallel electrodes are energized by RF power supplies. The circuit elements R_1 and R_2, C_1 and C_2 are adjusted to minimize the voltage drop above the workpiece.

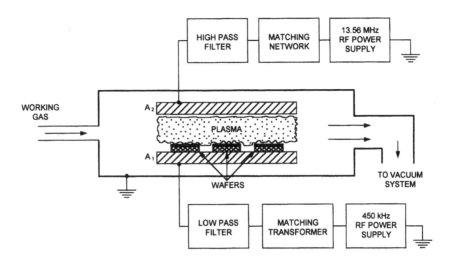

Figure 16.6. Schematic diagram of a parallel-plate RF glow discharge plasma reactor energized by power at two frequencies.

in figure 16.6, generates a plasma in the bulk of which the high-frequency input produces the precursors and active species necessary for plasma-processing applications. The low-frequency input partially shorts out the sheath potential, reduces the ion energy impinging on the workpiece, and minimizes the DC voltage drop across the sheath. This arrangement is intended to minimize any 'radiation damage' that might result from the impact of energetic ions accelerated across the sheath. This short-circuiting of the sheath is accomplished, as shown in figure 16.6, by applying RF power below the critical sheath frequency to the electrode on which the wafer or workpiece is mounted. In such dual-frequency plasma sources, each RF power supply requires its own impedance matching circuit and a filter that blocks the other frequency.

The term 'plasma etching', as applied to the configuration of figure 16.4(a), has both a general and a restricted sense. In the general sense, *plasma etching* (also known as *dry etching* to distinguish it from wet chemical etching) is etching done in a vacuum system in the presence of a plasma. In its restricted sense, plasma etching implies the RF parallel-plate electrode configuration shown in figure 16.4(a).

One of the important advantages of the RIE configuration of figure 16.4(b) is that only the powered electrode containing the wafers is subject to significant fluxes of energetic ions. There is little likelihood that sputtered material from the opposite electrode or surrounding walls will contaminate a workpiece, since ions reaching the grounded electrode will normally have energies below the sputtering threshold. In addition, the ion energy can be controlled during RIE by either adjusting the relative surface area of the electrodes, or by DC biasing the powered electrode with respect to ground. Since the powered electrode in RIE is quite negative, wafers on the powered electrode are not likely to be bombarded either by energetic electrons or negative ions.

16.1.1.3 Glow Discharge Characteristics

The plasma parameters of parallel-plate glow discharges depend on whether the plasma sources are energized by DC or RF power. For most plasma-processing applications, both DC and RF parallel-plate plasma sources are operated from about 133 Pa (1 Torr) down approximately to 6.7 Pa (50 mTorr). At the lower end of this range, active species and precursor fluxes in these sources are reduced to such low levels that deposition times are uneconomically long for most production purposes.

In DC parallel-plate plasma sources, the only plasma usually present is the negative glow of an obstructed abnormal glow discharge. The electron number density of such plasmas tends to be lower than that of RF excited parallel-plate plasma sources, and characteristically spans the range from 10^{15} to 10^{17} electrons/m^3. The electron kinetic temperature in a DC negative glow plasma is usually in the range 2–5 eV. The electron energy distribution function in DC discharges may be non-Maxwellian. When non-Maxwellian, the distribution function may include, in addition to a more numerous but relatively low-temperature Maxwellian component, an additional less numerous population of *primary electrons* with energies from 8 to 20 eV. These primary electrons maintain the ionization of the discharge and are responsible for much of the excitation light that creates the 'glow discharge'.

The electron number density of parallel-plate RF glow discharge sources used for plasma processing in the microelectronic industry is about a factor of 5 to 10 higher than that of a characteristic DC abnormal glow discharge, and may range from 5×10^{15} to 10^{18} electrons/m^3. The electron energy distribution function in RF parallel-plate glow discharges is usually Maxwellian, and the electron kinetic temperature tends to be somewhat higher than those of DC abnormal glow discharges, with characteristic values of 3–8 eV.

The electrode gap distance in parallel-plate plasma sources, both DC and RF, is characteristically from 1 to 3 cm, with the electrode/workpiece diameter being from 10 to 50 times this distance. The plasma volume available to produce active species increases linearly with the gap distance, but the required input power is also proportional to this volume. This gap distance may be determined by the measures required to ensure uniform flow of the working gases, or uniformity of effect on the workpiece.

16.1.1.4 Independent Operating Variables

The independent variables used to control plasma-processing operations include the type of working gas, the neutral gas pressure, the gas flow rate, and the DC or RF power input. Most microelectronic deposition and etching applications use argon as the working gas, with an admixture of working gases appropriate to the specific etching or depositional chemistry desired. The partial pressures and flow rates of these background and working gases are also very useful variables. In microelectronic plasma processing, the gases involved are expensive because they are very pure, and minimizing their use must be considered along with optimizing their effect on uniformity, rate of processing, and other factors. The rate at which a plasma-processing operation proceeds is controlled by the input power to the plasma source. This parameter is often easier to adjust over a wide range in RF plasma sources than in DC abnormal glow discharges.

16.1.1.5 Sheath Potentials

For many plasma-processing applications, particularly in microelectronics, it is important to know the voltage drop across the sheath above the workpiece. This voltage drop indicates whether or not sputtering can be expected, whether ions will be energetic enough to promote etching reactions, whether ions will interfere with deposition reactions, and whether radiation damage may occur.

In DC parallel-plate plasma sources, the workpieces are normally placed on the cathode. The voltage drop across the cathode sheath is known in the classical literature of the field as the *cathode fall voltage*, discussed in section 9.2.1 of Volume 1. Some old data on the cathode fall voltage are available in tables 8.2 and 8.3 of Cobine (1958), but the gases on which these tables are based may have

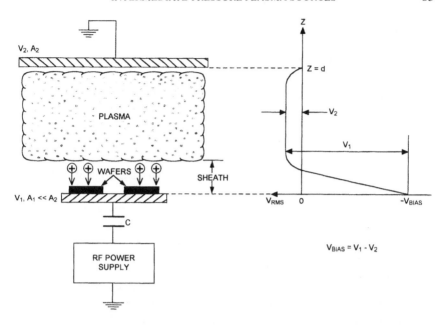

Figure 16.7. Plasma etching with a parallel-plate RF reactive ion etching configuration, with workpieces mounted on an electrode of area A_1, which is much less than the area A_2 of the unpowered, usually grounded electrode.

been contaminated by mercury vapor. The cathode fall voltage depends on the type of gas and the material of the cathode (or its secondary electron emission coefficient). Its magnitude generally lies between the ionization potential of the dominant gas, and about four times this value. AC and low-frequency RF discharges operated below the *critical sheath frequency* tend to behave phenomenologically like DC glow discharges (see Ben Gadri 1997, 1999). The sheath voltage drop therefore is more related to DC discharge behavior than the RF scaling laws discussed later.

In plasma-processing applications, particularly in microelectronic deposition and etching, RF parallel-plate plasma sources with unequal electrode areas are employed, such as that illustrated in figure 16.7. In the configuration shown, the wafer or workpiece is located on the powered RF electrode with a surface area A_1, that is smaller than the grounded electrode of area A_2, which may be the metal wall of the vacuum tank. A characteristic potential profile in such a parallel-plate RF plasma source is shown in figure 16.7. The plasma will float positively with respect to ground by a few volts, V_2, but a greater voltage V_1 will develop between the plasma and the powered electrode on which the wafers are located.

It was observed in section 12.3.2 of Volume 1 that the theory of RF sheaths is not well understood, but a phenomenological relationship relating the sheath voltage drops and the electrode areas has been determined by correlating empirical data,

$$V_1 = V_2 \left(\frac{A_2}{A_1} \right)^q \qquad \text{V} \qquad 1 \leq q \leq 2.5. \qquad (16.1)$$

In this expression, the sheath potential on the left is related only to the plasma potential V_2, the surface area of the RF electrodes on the right, and the power law index q, which is found empirically to range between 1.0 and 2.5. In section 12.3.3 of Volume 1 a current-conserving sheath voltage model was discussed. It was shown that for equal real rms ion currents to each electrode, the exponent q takes the value $q = 2.0$, consistent with the range of q observed in RF glow discharge plasma sources.

For plasma sources used in deposition, it is normally desired to minimize the sheath potential drop V_1 between the plasma and the workpiece. Equation (16.1) suggests that this may most easily be done by increasing the ratio of RF electrode areas above unity, and locating the workpiece on the larger electrode. At the same time, this approach may be inadvisable if the active-species flux on the workpiece is reduced too much by increasing the electrode area. When it is not possible to adjust the electrode area ratio sufficiently, a DC bias power supply may be connected to the powered electrode, to form a hybrid RF–DC plasma source. In such a hybrid configuration, the voltage drop V_1 can be controlled by adjusting the bias voltage on the powered electrode.

16.1.1.6 *Operational issues*

One of the reasons the DC parallel-plate glow discharge is infrequently used for microelectronic plasma processing is that the real currents flowing to the electrodes may introduce contaminants into the plasma or on the surface of the workpiece. These contaminants produce a lower quality product than RF plasma sources that operate on displacement currents. However, DC glow discharge plasmas do find applications to plasma stripping of photoresist (*ashing*), surface treatment of microelectronic wafers between processing steps, and other surface treatment operations outside the field of microelectronics discussed in later chapters of this volume.

Uniformity of effect is necessary, or at least desirable, in most plasma-processing applications, as it is in microelectronic deposition and etching. There are many situations, particularly in etching applications, where it is not possible to achieve this uniformity of effect by the motional averaging discussed in section 18.4, and the workpiece must remain fixed with respect to the plasma source. When the workpiece is fixed with respect to the plasma, uniformity of active-species production and flux is essential, and the RF parallel-plate plasma source is superior to the DC abnormal glow discharge in this respect.

The stronger axial electric fields associated with RF excitation better penetrate the plasma throughout its volume and provide greater radial uniformity of plasma and active species production. In DC discharges, the axial electric field is described by *Aston's law*, which results from the net excess of positive charge through Poisson's equation (see section 9.2.1 of Volume 1). These charge densities have a Bessel function radial profile, however, due to radial diffusion, and this makes it difficult to achieve radial uniformity without using concave electrodes to control the radial distribution of electric field.

DC and RF parallel-plate sources like those shown in figures 16.3 and 16.4 can be stably operated at pressures from 133 Pa (1 Torr) down to 6.7 Pa (50 mTorr). However, they both suffer the disadvantage that the number density of the plasma and the flux of active species generated by the RF plasma decrease with pressure. The time required for a plasma etch may be undesirably long for production applications at the lower end of this range.

The generation of a DC abnormal glow discharge requires only a DC power supply, and no impedance matching. Generating an RF parallel-plate glow discharge requires an RF (usually 13.56 MHz) power supply and an impedance matching network. The resulting cost in dollars per watt may range from two to five times that of a DC supply. Most of the power supplies used for plasma processing to date have been rated at below 1 kW, however, so the relative cost of the power supplies has not dominated other, more important production priorities.

Finally, parallel-plate sources pose few regulatory issues, since the vacuum systems in which they operate provide the isolation, shielding, and confinement required for worker safety. The only major regulatory issue is the use of RF power, but such issues are minimized by operating at the assigned frequency of 13.56 MHz.

16.1.2 Magnetron Plasma Sources

Magnetron plasma sources are characterized by crossed electric and magnetic fields, with the sheath electric field at right angles to a confining magnetic mirror field. The E/B 'magnetron' drift velocity is perpendicular to both the electric and magnetic fields, thus making the plasma itself and its effects on a workpiece more uniform in the $E \times B/B^2$ drift direction. The DC-excited magnetron source has been discussed in section 9.5.6 and the RF-excited magnetron in section 12.4.1 of Volume 1, and additional information can be found in Smith (1995), Vossen and Kern (1978), and Waits (1978).

16.1.2.1 Magnetron Configurations

One of the oldest, and still widely used, magnetron configurations is the *parallel-plate magnetron* illustrated in figure 16.8. This configuration generates a glow discharge between two parallel electrodes, each of which collects a real current of ions or electrons. Magnetron plasma sources usually have permanent magnets (sometimes electromagnets are used) located behind the cathode that generate a magnetic mirror geometry above the cathode surface. This magnetic mirror traps

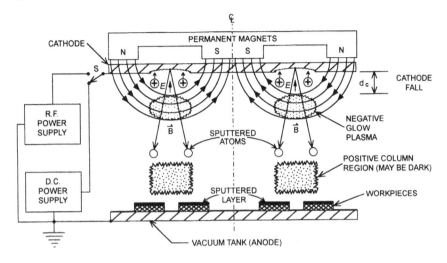

Figure 16.8. Drawing of a parallel-plate magnetron configuration for the deposition of sputtered cathode material on workpieces located on the anode.

the *negative glow* plasma of a DC glow discharge configuration, which is usually operated as an *obstructed abnormal glow discharge.*

The negative glow plasma develops a high potential (the *cathode fall voltage*) between itself and the cathode, which can range from tens to hundreds of volts. Ions accelerated from the negative glow plasma to the cathode arrive with energies high enough to sputter significant fluxes of neutral atoms from the cathode material. A porous fabric, electrically conducting film, or other electrically conducting material may be moved past the parallel-plate magnetron discharge of figure 16.8 at the cathode surface if ion bombardment is desired, or at the anode surface if a flux of sputtered atoms or other active species is required.

The anode location in the parallel-plate magnetron of figure 16.8 is often inconvenient, because any workpieces between it and the plasma must be electrically conducting to allow the passage of a real current. For workpieces which are electrical insulators or in contexts where contamination or radiation damage induced by real currents is not desired, the usual approach is to use the *planar* or *co-planar magnetron* configuration illustrated in figure 16.9. In this configuration, a DC potential is imposed between a planar cathode and a co-planar anode. This geometry gives rise to arched electric field lines, a configuration difficult to analyze analytically. However, this geometry has the important practical advantage that a negative glow plasma can be created above the cathode surface, and real DC currents between the cathode and anode do not have to flow to or through a workpiece on the side of the plasma opposite the cathode.

In the co-planar magnetron configuration of figure 16.8, relatively high cathode fall voltages are developed between the negative glow plasma and the

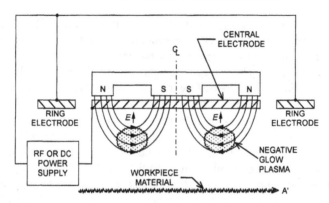

Figure 16.9. Schematic drawing of the co-planar magnetron, in which sputtered material from the cathode is deposited on a workpiece below the magnetron.

cathode. Ions from the negative glow are accelerated to the cathode, and the resulting sputtered atoms are deposited on the workpiece below. This co-planar configuration is particularly advantageous in processing large electrically insulating workpieces such as flat architectural glass, for example. Commercial magnetron sputtering sources are operated in a variety of configurations, of which the parallel-plate and co-planar magnetrons are the most commercially important. Some additional widely used magnetron plasma source configurations will be discussed in chapter 17.

16.1.2.2 Excitation

The *parallel-plate magnetron* illustrated in figure 16.8 and the co-planar magnetron source illustrated in figure 16.9 can be energized either by DC (the usual method) or by low-frequency RF power at frequencies ranging from 1 to 100 kHz. These RF frequencies are below the *critical sheath frequency* (discussed in section 19.5.3) for the plasma and may, in some cases, provide ion trapping along electric field lines, as described in section 12.5.2 of Volume 1. In some configurations, such as the co-planar magnetrons used for coating architectural glass, proprietary combinations of DC and RF excitation may be employed.

16.1.2.3 Characteristics of Magnetron Sources

Because many magnetron plasma sources are operated near the lower limit of the intermediate-pressure regime, the number of ionizing and inelastic collisions that produce active species is reduced. To partially offset this reduction, an attempt is made in magnetron discharges to enhance the number density of charged species by magnetic trapping of electrons in a magnetic mirror geometry. The magnetic field strength of the mirror that traps the negative glow plasma in magnetron discharges is characteristically from 10 to 50 mT. These values are sufficient to magnetize the electron population at the relatively low pressures used in magnetron sputtering applications, but are well below the values required to magnetize the ions. These magnetic fields are usually provided by relatively inexpensive permanent magnets, foregoing the additional degree of operational control (another knob to adjust) that would be afforded by using electromagnets.

The plasma parameters of industrial magnetron sputtering sources have not been well diagnosed with Langmuir probes, microwave interferometers, or other standard plasma diagnostic instruments. The reasons for this include the high degree of spatial non-uniformity and the small size of the negative glow plasmas that need to be measured. It is probably reasonable to assume that the electron number density and electron kinetic temperature of the negative glow plasma in magnetron sputtering discharges is comparable to the parallel-plate glow discharge plasmas discussed previously in section 9.5.2 of Volume 1. The electron number densities of such discharges range from 10^{14} per cubic meter to 5×10^{16} per cubic meter, with electron kinetic temperatures between 2 and 5 eV.

16.1.2.4 *Independent Operating Variables*

Whenever possible, an inert working gas such as argon is used in magnetron plasma sources, particularly in sputtering applications, to avoid oxidation or other chemical reactions between the working gas, the electrodes, or the deposited layer. For sputtering applications in the *low-pressure* regime, argon has the additional advantages of high atomic mass (and hence greater effectiveness as a sputtering agent) and a low binary scattering cross section, with associated long mean free paths.

Any plasma source used for sputtering must trade off the high ion and sputtered atom fluxes possible at the increased ionization rates associated with high operating pressures, against the short mean free paths at high pressures that interfere with sputtered atom transport from the cathode to the workpiece. Magnetron sources used for sputtering characteristically operate at neutral gas pressures down to 6.7 Pa (50 mTorr), where the mean free paths of sputtered atoms are comparable to the distance between the cathode and the workpiece. Electron trapping by the magnetic mirror of the magnetron configuration assists in maintaining the plasma electron number density (and hence the ion and sputtered atom flux) at high levels in spite of low pressures. By contrast, the neutral gas pressure of magnetrons used for PCVD or plasma etching may range up to 133 Pa (1.0 Torr).

Magnetron plasma sources characteristically have power inputs from hundreds of watts to kilowatts. Some of the larger sources used to deposit sputtered coatings on automotive and architectural glass may operate at the kilowatt level. The hybrid DC–RF sources used for sputtering are not usually impedance matched. In magnetron plasma sources used for sputtering, the required characteristics of the sputtered coating determine the cathode material.

In addition to creating a coating with the desired properties (color, infrared reflectivity, etc), the cathode material must have a high sputtering yield at the relatively low energies of argon ions at the cathode fall voltage of the source.

16.1.2.5 Performance Issues

The sputtering yield of neutral atoms in magnetron sputtering plasma sources is a strong function of the energy of the ions (usually argon) that induce sputtering. The higher the ion energy is, the higher will be the flux of sputtered atoms removed from the cathode per unit time. The ion energies reaching the cathode surface are determined by the DC *cathode fall* voltage, which, for obstructed, abnormal glow discharges can be estimated from Child law sheath theory, discussed in section 9.4.5 of Volume 1. In this approach, the cathode sheath thickness is related to the ion current density reaching the cathode by the Child law expression for current density given by equation (9.160) of Volume 1,

$$J_C = \frac{4\varepsilon_0}{9} \left(\frac{2e}{M} \right)^{1/2} \frac{V_c^{3/2}}{S^2} \quad \text{A/m}^3 \tag{16.2}$$

where V_c is the cathode fall voltage; J_C is the Child law current density of ions reaching the cathode; S is the sheath thickness, assumed to be equal to the cathode fall distance d; and M is the mass of the singly charged ions traveling between the negative glow plasma and the cathode. Solving equation (16.2) for the cathode fall voltage V_c, one obtains

$$V_c = \left(\frac{J_C S^2}{4\varepsilon_0} \right)^{2/3} \left(\frac{M}{2e} \right)^{1/3} \quad \text{V.} \tag{16.3}$$

If the Bohm sheath model is used to estimate the current density flowing across the sheath, one obtains from equation (9.161) of Volume 1 an expression for the current density in the cathode fall region,

$$J_C = e n_e v_B = e n_e \left(\frac{e T'_e}{M} \right)^{1/2}. \tag{16.4}$$

In this equation, n_e is the electron number density in the negative glow plasma, and v_B is the Bohm velocity, given by equation (9.146) of Volume 1, which has been substituted into the right-hand side of equation (16.4). Substituting equation (16.4) into equation (16.3) yields, for the cathode fall voltage in a magnetron discharge,

$$V_c = \left(\frac{9e S^2 n_e}{4\varepsilon_0} \right)^{2/3} \left(\frac{T'_e}{2} \right)^{1/3} \quad \text{V.} \tag{16.5}$$

Equation (16.5) implies that the cathode fall voltage, and hence the sputtering yields, can be increased most effectively by increasing the electron number density in the negative glow plasma or by increasing the thickness of the sheath between the cathode and the plasma.

In an obstructed abnormal glow discharge used for magnetron sputtering applications, a characteristic electron kinetic temperature is $T_e' = 3$ eV; a sheath thickness, $S = 1$ cm; and an electron number density $n_e = 10^{15}/m^3$. When substituted into equation (16.5), these values yield a cathode fall voltage $V_c = 290$ V. This voltage is high enough to raise ions to an energy at which significant sputtering will occur, particularly in such materials as aluminum and copper.

An important factor in the utility of magnetron plasma sources for sputtering and other applications is their drift-induced uniformity over large distances. The E/B drift velocity out of the plane of the diagram in figures 16.8 and 16.9 plays an important role in maintaining the uniformity of plasma and effect along the length of magnetron plasma sources that may be 3 or 4 m in length. The usual configuration in planar magnetron sputtering plasma sources is a long, oval racetrack in which the plasma undergoes E/B drift around the racetrack. This drift ensures uniformity of the negative glow plasma in the direction perpendicular to the electric and magnetic fields.

16.1.3 Continuous-Flow Inductively Coupled Plasma Sources

The *continuous-flow inductively coupled plasma source* is illustrated schematically in figure 16.10. It consists of an inductive heating coil wrapped around a quartz or pyrex tube. This source is most frequently, but not exclusively, operated in the intermediate-pressure regime, and is used primarily for applications that require plasma–chemical reactions among the working gases. The configuration and general operation of this source are similar to that of inductive plasma torches operated at atmospheric pressure. However, the plasma is not in thermodynamic equilibrium (although in particular instances it might be in kinetic or plasma–chemical equilibrium), and its power levels in most applications are less than 1 kW, well below that of atmospheric plasma torches. The physical processes responsible for plasma heating in these sources have been discussed in chapter 11 of Volume 1. As was true of atmospheric plasma torches, energy is coupled from the power supply to the plasma by time-varying magnetic fields. These fields induce currents that heat the electron population by ohmic dissipation.

16.1.3.1 *Applications of Inductive Continuous-Flow Sources*

In the intermediate-pressure regime, this technology is used primarily for continuous-flow plasma chemistry and plasma chemical vapor deposition (PCVD) of thin films. In this continuous-flow source of plasma-induced chemical species, the high energy content of the electron population is used in several ways. These include inducing endothermic chemical reactions; breaking high-energy

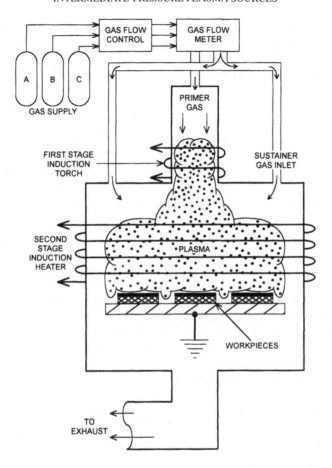

Figure 16.10. A two-stage inductively coupled continuous-flow RF plasma source.

chemical bonds; creating molecular fragments from gases that are precursors for PCVD of thin films; or plasma epitaxy of crystalline materials.

16.1.3.2 Characteristics of Inductive Continuous-Flow Plasma Sources

A common configuration of the *continuous-flow inductively coupled plasma source* is illustrated in figure 16.10. Working gas(es) flow into one end of a quartz or electrically insulating tube, and the plasma is energized by a solenoidal inductive coil wrapped around it. The RF source is usually operated at the standard frequency of 13.56 MHz, and sometimes without impedance matching, since electrical efficiency may not be an issue at the relatively low power levels at which these sources are usually operated. The power levels of inductive continuous-flow plasma sources range from a few tens of watts for plasma–

chemical or small exploratory research investigations, to the kilowatt level in plasma sources intended for synthetic or destructive plasma chemistry in toxic waste remediation.

The plasma in a inductive continuous-flow plasma source may operate at pressures ranging from a few kilopascals ($\sim$10 Torr) down to a few Pascals (a few tens of milliTorr). The working gases are determined by the plasma chemistry needed in a particular application. Because inductively coupled continuous-flow sources are not in thermodynamic equilibrium, the electron kinetic temperatures are in the mid-range of 1–10 eV. The ions and neutral gas are at much lower temperatures, which depend on the degree of disequilibrium of the plasma. These gas temperatures are close to room temperature in the intermediate-pressure regime at which this source is normally operated. The electron number densities are well below those of plasma torches, and are characteristically in the range 10^{16}–10^{19} per cubic meter.

16.1.3.3 Phenomenology of the Inductive Continuous-Flow Source

As indicated in figure 16.10, the working gases in the inductive continuous-flow plasma source flow through the plasma region, where precursors and active species become entrained in the gas flow. By adjusting the flow rate, the dwell time of the working gases in the plasma can be controlled, and the concentration of the plasma–chemical reaction products varied.

Workpieces can be exposed to the active species of inductive continuous-flow plasmas by *direct contact* with the plasma or by *remote exposure*, in which the workpiece is located at a distance from the plasma. This latter configuration is more common in industrial applications, as it reduces or eliminates the possibility of unwanted outcomes that may result from direct plasma exposure.

16.1.3.4 Independent Operating Variables

A variety of independent operating parameters are used to control the processing of workpieces by inductive continuous-flow plasma sources. These include the type and mixture ratio of working gases; the flow rate of the working gas; the operating gas pressure; the RF power level; and the coil geometry, including the ratio of the plasma to the coil radius. In addition to these plasma-related parameters, the motion of the workpiece through the plasma flow, and hence its duration of exposure, can be adjusted.

The choice of working gas is determined by the nature of the plasma–chemical task to be performed. The rate at which the processing task is completed, as well as the composition of the reaction products, may be controlled by adjusting the gas flow rate, the gas pressure, and/or the RF power level. Since the mean free paths of the reaction products are seldom important in deposition applications, the gas pressure can be adjusted freely to obtain a high plasma–chemical reaction rate (at high pressures) or the dominance of a particular reaction product.

Above the *critical sheath frequency*, in the vicinity of 1 MHz for most plasma sources, the plasma parameters and reaction chemistry are not strong functions of RF frequency. As a result, nearly all inductive continuous-flow plasma sources are operated at the standard frequency of 13.56 MHz because such power supplies are readily available, and the regulatory issues surrounding RF interference are minimal at that frequency. Power levels below 1 kW are sufficient for most industrial applications.

16.1.3.5 Performance Issues

The axisymmetric coils with which inductively coupled continuous-flow plasmas are generated produce an axisymmetric plasma, which may not be uniform enough across its diameter to provide acceptable plasma-processing effects on a workpiece. In such a case, motional averaging of the deposited layer on the workpiece is required, using one of the methods discussed in section 18.4 of this volume. As noted previously, the cross section of the inductive coil need not be circular, and may be formed into rectangular, triangular, and other geometric cross sections to improve depositional uniformity.

The inductive continuous-flow plasma source poses few significant safety or regulatory issues. Those surrounding the use of RF power, and RF interference (RFI) from its power supply are normally the most troublesome. Other potential safety and regulatory issues include the production of dangerous levels of ultraviolet radiation, and plasma–chemical reaction products that, if released into the workplace or the environment, may exceed levels considered safe for worker or public exposure.

16.2 LOW-PRESSURE PLASMA SOURCES

After 1990, competitive pressures arose within the microelectronic industry to achieve pattern transfer below a *design rule* of 0.5 μm. This goal motivated the development of several plasma sources that operate in the *low-pressure regime*, between 13 and 0.013 Pa ($\sim$100–0.1 mTorr). In this range, the mean free path of ions and neutrals may be comparable to or greater than the plasma dimensions, and is greater than the sheath thickness between the plasma and the workpiece. The longer the ion mean free path, the less likely it is to undergo a scattering collision in the sheath between the plasma and the workpiece. Such scattering collisions give ions a velocity component parallel to the surface, which in microelectronic etching causes undercutting of the mask and poor pattern transfer.

In addition to operating at low pressure to ensure a vertical etch, competitive pressures have speeded up the etching process in order to fabricate a microelectronic chip in shorter times. This practice more effectively utilizes the expensive clean rooms and cluster tools required for microelectronic circuit fabrication. The requirement for faster etching translates to a requirement for

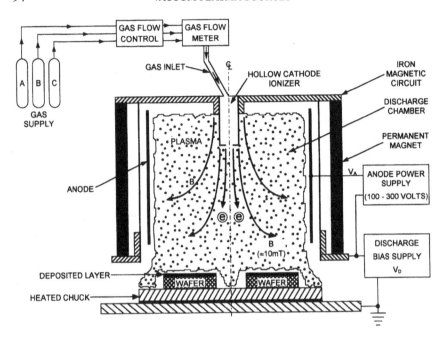

Figure 16.11. Schematic diagram of an electron bombardment plasma source with magnetized plasma electrons. The working gas is admitted to the discharge chamber, preferably through a hollow cathode ionizer.

higher fluxes of active species from the plasma, which result from higher plasma densities.

Four plasma sources are now discussed that are capable of providing simultaneously the ability to operate at background neutral gas pressures below 1.3 Pa (10 mTorr), while generating plasmas with higher average number density than those discussed previously in section 16.1. These include the *electron bombardment (Kaufman) source*, ECR and *non-resonant microwave sources* operating at 2.45 GHz, the *flat-coil inductive RF plasma source*, and the *helicon plasma source*.

16.2.1   The Electron Bombardment Plasma Source

The *Kaufman electron bombardment plasma source* is widely used for plasma chemical vapor deposition (PCVD) in one of two versions illustrated in figures 16.11 and 16.12. This source has been discussed in sections 6.5 and 9.5.4 of Volume 1, and is based on the discharge chamber of the Kaufman electrostatic ion engine developed in the 1960s at the NASA Glenn Research Center for spaceflight applications. This source offers very high electrical and gas utilization efficiencies, as discussed in section 6.5 of Volume 1.

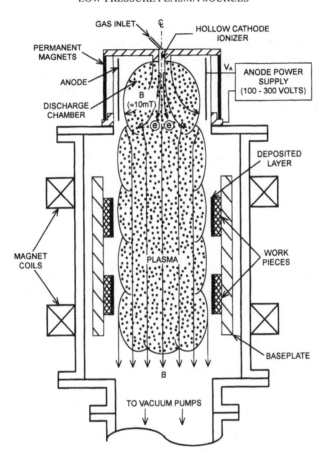

Figure 16.12. A schematic illustration of a remote plasma source configuration, based on the flow of plasma from an electron bombardment source past workpieces that undergo plasma processing by remotely generated active species.

16.2.1.1 Source Characteristics

In the electron bombardment source illustrated in figure 16.11, electrons are confined by a diverging magnetic field created by an iron magnetic circuit that includes permanent magnets. The magnetic induction characteristically ranges from 5 to 20 mT, sufficient to magnetize the electron population. A significant operational restriction on this source is a requirement to operate at neutral gas pressures low enough to magnetize the electron population (below ~1.3 Pa (10 mTorr)) to achieve high electrical and gas utilization efficiencies.

Very high gas utilization efficiencies (above 90%) are achieved in the classical Kaufman electron bombardment source by feeding the working gas into the discharge chamber through a *hollow cathode ionizer* (see section 5.4 of Volume 1, and Kaufman (1961, 1963, 1965)). The dense, small volume plasma generated in the hollow cathode ionizes a large fraction of the gas flowing through it, and produces other active species as well. These active species flow into the discharge chamber and can undergo further plasma chemistry, including the generation of monomers or other precursors to deposition.

Electron bombardment sources used for PCVD applications are most frequently arranged in the *immersion configuration* shown in figure 16.11, in which the discharge chamber plasma is in direct contact with the workpiece. To avoid 'radiation damage' or adverse outcomes of direct plasma exposure of the workpiece the *remote exposure configuration*, shown in figure 16.12, in which the discharge chamber is used as a remote source of active species is also used. These species convect along the magnetic field to a large chamber where they are deposited on surrounding surfaces, including workpieces at the plasma boundary. Either of these configurations may be used for PCVD deposition of microelectronic wafers and other commercial workpieces.

16.2.1.2 Plasma Parameters

Electron bombardment sources characteristically have a low neutral background pressure, below 13 Pa (100 mTorr), and a relatively low electron number density. The resulting active-species fluxes may be inadequate for some industrial PCVD applications. A countervailing factor is that a significant fraction of the (sometimes very expensive) working gas flowing into an electron bombardment source can undergo, in the hollow cathode, relatively efficient conversion into active species. When used for their original aerospace application, these sources have gas utilization efficiencies higher than 90%.

In electron bombardment sources, electron number densities are characteristically 5×10^{16} electrons/m^3, with values that may range between 10^{15} and 10^{17} electrons/m^3. The electron kinetic temperature in these sources ranges from 2 to 5 eV. The electron energy distribution function in this source is usually non-Maxwellian with two components. The first is a less numerous population of energetic primary electrons responsible for ionization and for maintaining the discharge, and a less energetic Maxwellian distribution with the kinetic temperatures given earlier.

16.2.1.3 Independent Operating Variables

The type of gas used in the Kaufman electron bombardment plasma source depends on the application. If used for PCVD, the working gases are those that produce monomers or molecular fragments that are precursors to the deposited material; for etching applications, argon gas would normally be used because it is inexpensive, is easily ionized, and produces a stable plasma.

The neutral gas pressure at which an electron bombardment plasma source is operated must be low enough to magnetize the electron population, and confine the plasma with the magnetic field in the discharge chamber. For any gas, this requires neutral gas pressures below 13 Pa (100 mTorr), and for some gases with high electron scattering cross sections, the required pressures are below a few Pa (~10 mTorr). If the high electrical and gas utilization efficiencies of the Kaufman electron bombardment source are desired, the working gas must be injected into the discharge chamber through a hollow cathode ionizer. The ionizer serves as an electron source as well as a source of ionized, excited, and disassociated active species for the discharge chamber. Without the hollow cathode ionizer, high electrical and gas utilization efficiencies should not be expected.

The total power input into electron bombardment plasma sources in existing commercial applications ranges from a few hundred watts to perhaps a kilowatt. The versions of this source developed for space propulsion systems achieved a very low energy cost for mercury gas in the range 100–150 eV/ion–electron pair. This value is not far above the Stoletow constant for mercury of 50 eV/ion–electron pair (see table 8.2 of Volume 1), the absolute minimum possible energy cost for ionization. This achievement is the basis for the high electrical efficiency of this source. The discharge in the chamber is maintained by an anode power supply, which characteristically operates in the range 100–400 V, depending on the operating gas.

16.2.1.4 Performance Issues

It is possible to achieve a commercially adequate uniformity of effect (less than 5% variation of the deposited layer thickness over the workpiece diameter) with electron bombardment sources. This is done by adjusting the confining magnetic field illustrated in figure 16.12, and by placing the workpieces in a relatively uniform region of plasma remote from the source. The power required to operate an electron bombardment deposition plasma source is characteristically a few hundred watts, including the power required for the hollow cathode ionizer.

The electrical efficiency of generating the plasma, and the utilization efficiency in producing ionized species from the input working gas can both be above 90% (Kaufman 1961, 1963, 1965). The high gas utilization of this source can be particularly valuable if the working gas is very expensive, radioactive, toxic, or has some other property which requires that it be fully used and produce a minimum concentration in the effluent of the plasma reactor.

16.2.2 Microwave Plasma Sources

The 2.45 GHz magnetrons developed for domestic microwave ovens provide an inexpensive source of RF power for industrial plasma sources. Such magnetrons are readily available at power levels up to 1 kW to generate 2.45 GHz microwave glow discharge plasmas in electron cyclotron resonant

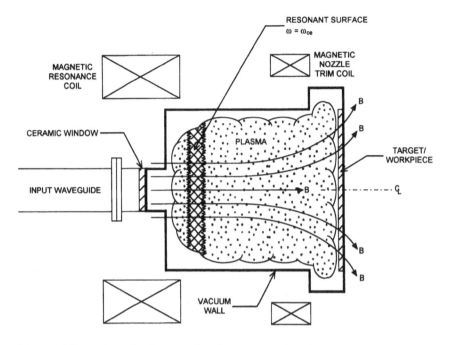

Figure 16.13. Schematic diagram of a plasma source based on an electron cyclotron resonant (ECR) microwave plasma.

(ECR) magnetized plasmas, and in non-resonant, unmagnetized plasmas. The cost of this microwave power can be as little as a few tens of cents per watt, second lowest only to DC electrical power.

A general discussion of microwave plasmas for plasma processing has been published by Claude *et al* (1987), Dusek and Musil (1990), Batenin *et al* (1992), Lieberman and Lichtenberg (1994), and Smith (1995). Selected published work on ECR microwave plasma sources includes Asmussen (1989), Berry and Gorbatkin (1995), Gorbatkin *et al* (1990), Gorbatkin *et al* (1992), Inoue and Nakamura (1995), Reinke *et al* (1992), and Tsai *et al* (1990).

16.2.2.1 Electron Cyclotron Resonant (ECR) Sources

ECR plasma sources can be used for a variety of plasma-processing applications, including the deposition and etching of microelectronic circuits. They are more widely used for these latter applications in East Asia than in the United States. A characteristic ECR microwave plasma source is illustrated in figure 16.13.

Such sources were discussed in section 13.4 of Volume 1, and consist of a 2.45 GHz magnetron; a microwave impedance matching and monitoring circuit; a mode converter from a rectangular to a cylindrical waveguide RF electric field distribution; and a *resonant region* in which the microwave power interacts with a plasma in the presence of a magnetic field of 87.5 mT. This magnetic induction corresponds to the electron gyroresonance frequency at 2.45 GHz. The cylindrical waveguide in the resonant region propagates 2.45 GHz microwave power in an azimuthally symmetric, extraordinary mode (see section 13.2 of Volume 1). The radial electric field vector of the microwave radiation is perpendicular to the axial magnetic field.

When the microwave radiation reaches the point at which the electron cyclotron frequency in the magnetic field equals the 2.45 GHz driving frequency, ECR heating occurs (see section 13.4 of Volume 1). Very strong absorption of the microwave power occurs at the resonant surface indicated in figure 16.13, provided that the electrons are *magnetized*. This power absorption will maintain a plasma, the electron number density of which is at least equal to the *critical electron number density* discussed in section 4.5 of Volume 1. This density for 2.45 GHz microwave power is 7.6×10^{17} electrons/m^3. When ECR plasmas have a density greater than the critical electron number density, the plasma is said to be *overdense*. An *overdense* ECR plasma may have number densities several times 10^{18} electrons/m^3.

The electron number densities characteristic of ECR plasmas are an order of magnitude higher than those characteristic of DC or RF parallel-plate glow discharge plasma sources. As the ECR plasma flows away from the resonant region, however, its density drops substantially. This decrease is often associated with expansion of the plasma in a diverging magnetic nozzle. As the magnetic field weakens, the magnetic field lines diverge from the resonant surface to the wafer, as indicated in figure 16.13.

In order for electrons to undergo RF resonant heating, the electron population must be *magnetized*, that is the two conditions

$$\omega_{ce}\tau_c \gg 1 \tag{16.6}$$

on the *collisionality parameter*, and

$$R_g = \frac{mV_e}{eB} \ll a \tag{16.7}$$

on the *radius of gyration* must be satisfied. Equation (16.6) states that electrons must be able to make several complete gyrations between collisions with the neutral working gas, while equation (16.7) states that the gyroradius of the electrons has to be much smaller than the plasma dimensions.

There is rarely any difficulty satisfying equation (16.7) in industrial plasma-processing applications, but equation (16.6) can best be satisfied for electrons at pressures below 1.3 Pa (10 mTorr). Since this is the pressure regime required for design rules below 0.5 μm and vertical etching in microelectronic applications, a fortuitous compatibility exists between the requirements for ECR plasma generation, and the requirements of advanced microelectronic etching technology.

ECR plasma sources with the resonant surface located within the plasma volume, illustrated in figure 16.13, are *immersed resonant surface plasma*

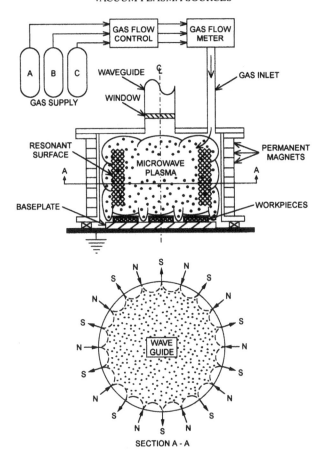

Figure 16.14. Illustration of a plasma source based on ECR resonant surfaces at the boundary of the reactor volume. A section of the reactor is shown in section A–A. The location of the input waveguide on the axis of the reactor is indicated.

sources. The working gas is fed into the relatively dense plasma near the resonant surface, where it is converted into active species for plasma processing. The plasma and active species generated at this surface flow along parallel or diverging magnetic field lines to the workpiece.

ECR plasmas must operate in the low-pressure regime in order for the electron population to be magnetized. It is therefore necessary to have a dielectric 'window' between the microwave waveguide, which is normally at atmospheric pressure, and the vacuum chamber in which the ECR plasma is maintained. This can be done in the manner illustrated in figure 16.13, with a quartz tube leading to a chamber containing the immersed resonant surface where the plasma is generated.

A related ECR plasma source is the *distributed resonant surface plasma source*, which is illustrated in figure 16.14. In this configuration, 2.45 GHz microwave power is supplied by a rectangular waveguide through a vacuum window. The electric field intensity builds up by virtue of wall reflection in the vacuum chamber in which the plasma is generated. Permanent magnets are mounted on the periphery of this source to create a multipolar magnetic field that keeps the plasma away from the walls, where charged species would otherwise recombine. This magnetic confinement at the boundary can as much as double the electron number density of the plasma.

The multipolar permanent magnets serve a second function, however, by providing a magnetic field strong enough to create a distributed electron cyclotron resonant surface around the circumference of the plasma, as illustrated in cross section A–A of figure 16.14. At this surface, a plasma of critical density or an overdense plasma forms thus making active species available for plasma processing of workpieces.

16.2.2.2 Non-resonant Microwave Sources

While ECR microwave plasma sources have found wide use in the microelectronic industry and are becoming more frequently used in other plasma-processing applications, non-resonant microwave plasma sources (those that do not use ECR resonance) have found only a few niche applications. Non-resonant microwave plasma sources have an unfavorable scaling of power coupled with neutral background pressure, an issue to be discussed in section 19.3 of this volume. Non-resonant microwave sources either operate with the plasma electrons unmagnetized, or at a magnetic field strength so low that ECR does not occur anywhere in the plasma volume.

A *non-resonant microwave cavity plasma source* is illustrated in figure 16.15. In this source, 2.45 GHz microwave power is supplied through a vacuum window to a cavity, within which the plasma is generated. The workpieces are mounted for plasma treatment on one of the boundaries. In a *modal microwave plasma source*, the radius R and height D are eigenvalues, related to integer multiples of the wavelength of the microwave radiation, as discussed in section 13.3.2 of Volume 1. In such cavities, the radiation forms modal patterns, with the plasma most dense at the electric field maxima. In a *multimodal microwave plasma source*, the dimensions of the cavity and the radiation propagation characteristics are such that no modal structure forms, and the cavity is filled with a relatively uniform plasma. The 200 liter multimodal Microwave Plasma Facility described in section 13.5 of Volume 1 is an example of such a source.

A second type of non-resonant microwave plasma source is the *continuous-flow tapered waveguide source*, discussed in section 13.5.1 of Volume 1, and illustrated in figure 16.16. In this source, a 2.45 GHz microwave waveguide is tapered to produce a high electric field along the axis of a quartz tube that

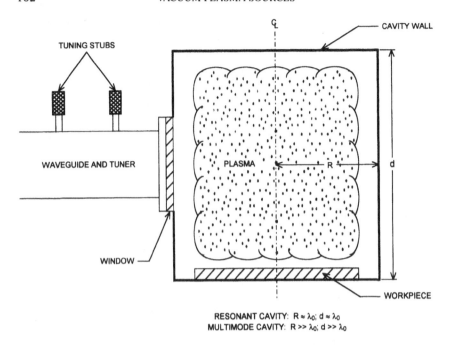

Figure 16.15. A microwave cavity plasma source, in which active species from a microwave-generated plasma impinge on a workpiece located at the wall. Mode structures may exist if the height or radius of the cavity is comparable to the free space wavelength of the microwave power; if the dimensions of the cavity are greater than and/or incommensurate, the free-space-wavelength modal patterns will average out.

passes through the waveguide in a direction parallel to the electric field. The electric field causes electrical breakdown of the gas in the tube and plasma formation, thus stimulating plasma chemistry in the flowing gases. It functions very much like a continuous-flow inductive plasma source, discussed earlier in section 16.1.3. The reaction products from this source can be directed to a workpiece where plasma processing takes place. These sources can operate over a very wide range of pressures, from below 1.3 Pa ($\sim$10 mTorr) to above atmospheric pressure.

16.2.2.3 Plasma Parameters

Microwave plasma sources operate over pressures ranging from a tenth of a mTorr to 1 atm. ECR sources are the most efficient and their electric fields are best coupled to the plasma of the microwave sources, but they must operate at pressures low enough that the electron population of the plasma is magnetized. In non-resonant or unmagnetized microwave plasmas, the coupling of the microwave power to the plasma is not as good as that of ECR resonant plasmas, nor is the

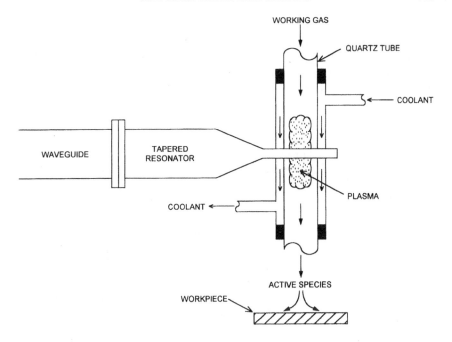

Figure 16.16. A continuous-flow microwave-generated plasma using a tapered waveguide penetrated by a quartz tube. Active species generated in the flowing gas are convected by the flow to the workpiece.

scaling of power density with neutral gas pressure favorable, as will be discussed in section 19.3.

The electron number density of ECR resonant microwave plasmas tends to be higher by at least a factor of 10 than unmagnetized parallel-plate RF glow discharge plasma sources.

The electron kinetic temperature of microwave plasmas, particularly ECR plasmas, is higher than that of other RF and DC glow discharge plasmas, with electron kinetic temperatures ranging from 5 to 15 eV. At the same electron number density, the energetic electrons in microwave sources produce higher fluxes of active species than are found in glow discharges with less energetic electron populations.

16.2.2.4   Performance Issues

The efficiency with which microwave power is transferred to the plasma is relatively high (above 50%) in extraordinary mode ECR plasmas, but is not so high in non-resonant and unmagnetized microwave cavity plasmas. The power actually delivered to the plasma in most industrial ECR plasma sources ranges from a few hundred watts to a kilowatt. Even in deposition and etching plasmas, power levels in excess of 1 kW are sometimes necessary to maintain the high electron number density required to produce an adequate flux of active species.

In ECR microwave plasmas, plasma generation in the resonant region may be non-uniform, and this non-uniformity may propagate as the plasma expands along diverging magnetic field lines from the resonant surface to the workpiece. Industrial experience with ECR microwave plasma sources indicates that it becomes progressively more difficult to produce a uniform plasma across a flat workpiece like a microelectronic wafer, as the diameter is increased from the previous industry standard of 10 cm to the current standard of 30 cm. Other plasma sources, such as the inductive flat-coil or parallel-plate capacitive RF plasma sources, may be better able to provide uniformity of plasma-processing effect as the workpiece diameter increases.

A major disadvantage of ECR microwave plasma sources is their requirement for a magnetic field over a relatively large volume of plasma. The use of electromagnets to generate the required ECR resonant magnetic field is a serious disadvantage in industrial use, since electromagnets require a continuous input of electrical power, usually far more than is required to generate the plasma itself. The requirement for DC electrical power and cooling water for electromagnets also presents reliability and maintenance issues that many industrial engineers would prefer not to deal with if other plasma sources can provide the required active-species fluxes without a magnetic field. An approach to this difficulty is the distributed ECR plasma source illustrated in figure 16.14, in which the magnetic field is generated by permanent magnets.

16.2.3   Planar-Coil Inductive Plasma Sources

The *planar-coil inductive plasma source* shown on figure 16.17 has been used to an increasing extent for plasma processing of microelectronic wafers and other applications. This trend has arisen partly in response to concerns about using electromagnets, and difficulties with the uniformity of effect possible with ECR plasma sources. The physical processes in this source have been discussed in sections 11.4 and 11.5 of Volume 1, and in the journal literature (see Dai and Wu 1995, Li *et al* 1995, Ventzek *et al* 1994, Wainman *et al* 1995).

16.2.3.1   Source Configuration

The *planar-coil inductive plasma source* illustrated in figure 16.17 is energized by 13.56 MHz RF power applied to a planar helical coil. The coupling between the RF power supply and the plasma is achieved by currents induced in a disc at the upper surface of the plasma, below the planar helical coil. As illustrated in figure 16.17, the formation and heating of the plasma occurs in an *interaction layer* of thickness δ, which is approximately equal to the plasma *skin depth*, discussed in section 11.3 of Volume 1. The plasma generated in this interaction layer can reach densities of several times 10^{18} electrons/m^3, in the low-pressure

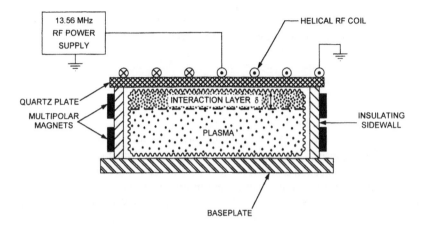

Figure 16.17. A plasma source for etching based on the inductive planar helical coil configuration.

regime at neutral gas pressures below 1.3 Pa (10 mTorr). This plasma convects below the interaction layer to process wafers or other workpieces on the baseplate. The baseplate may have a provision to bias the workpiece, a feature sometimes desirable for microelectronic etching or other plasma-processing operations.

To avoid losses at the cylindrical sidewall of the inductive plasma source shown in figure 16.17, multipolar permanent magnets, discussed in section 3.4.3 of Volume 1, may be mounted on the boundary. The magnetic field magnetizes the electron population only, but is sufficient to significantly reduce plasma bombardment and recombination on surfaces other than the workpiece. These permanent magnets do not pose the reliability or power consumption problems associated with the electromagnets required for ECR plasma sources.

16.2.3.2 Plasma Parameters

The neutral gas pressure at which the planar-coil inductively coupled plasma source can be operated ranges from less than 0.13 Pa (1 mTorr) for etching of microelectronic wafers, up to values in the intermediate-pressure regime for deposition or tasks other than microelectronic plasma processing. The electron kinetic temperature in planar-coil inductive plasma sources can range as high as 5 to 10 eV, and their electron number density can range from values as low as 10^{16} electrons/m^3, to values above 10^{18} electrons/m^3. The power level of the helical coil inductive plasma source illustrated in figure 16.17 can range from a few hundred watts to several kilowatts in microelectronic plasma-processing applications. The electron number density and power input of the planar-coil inductive plasma source is about the same as that of ECR microwave sources.

Its electron number density is about a factor of 10 higher than that generated by parallel-plate RF glow discharge plasma sources.

16.2.3.3 Independent Operating Variables

The active-species flux, duration of exposure, and other effects on plasma processing are controlled by the same easily adjusted parameters as other plasma sources used for processing. For etching applications, argon gas is used as the carrier gas because it is inexpensive, and its electrical characteristics produce a stable, easily controlled discharge. The gas flow rate and working gas pressure are linked in most vacuum systems, and in microelectronic etching applications these are adjusted to produce a satisfactory compromise between low etching rates at low pressure, and undercutting the mask at higher pressure.

The electron number density and active-species production rate are proportional to the power input to the plasma. To achieve the desired plasma-processing effects in acceptable durations, the power level at which microelectronic wafers are treated in planar-coil inductive plasma sources is on the order of kilowatts. Although the efficiency of inductive coupling is only weakly related to the operating frequency above the critical sheath frequency, nearly all planar-coil inductive plasma sources are operated at 13.56 MHz to avoid regulatory problems with RFI.

16.2.3.4 Performance Issues

The flat-coil inductive plasma source is capable of operating over approximately the same range of neutral gas pressures and electron number densities as ECR plasma sources. However, the flat-coil sources do not require electromagnets, and they possess what appears at the time of writing to be a superior capability to achieve uniform plasma densities and processing effects across workpieces up to 30 cm in diameter. By adjusting the radial spacing between individual turns of the planar helical coil shown at the top of the plasma source in figure 16.17, and by adjusting the frequency, the power input, and the operating pressure of the plasma, it should be possible to achieve a virtually flat radial profile of electron number density across the plasma diameter. Such a flat profile should translate into uniformity of processing effect over large diameter wafers.

The axisymmetric coils with which most inductively coupled plasmas are generated usually produce an axisymmetric plasma, which may not be uniform enough across its diameter to provide acceptably uniform processing of a workpiece. In such a case, motional averaging of the workpiece may be required, using one of the methods discussed in section 18.4 of this volume. As noted in chapter 11 of Volume 1, the cross section of the inductive coil need not be circular, and can be formed into rectangular, triangular, and other geometric cross sections to improve depositional uniformity on workpieces with corresponding shapes.

Like other RF-energized plasma sources, the plasma presents a variable load to the power supply. This load changes with operating conditions, so an adjustable impedance matching circuit is necessary. Even with impedance matching, however, the electrical efficiency of the planar-coil inductive plasma source is not likely to be above 50%. The wasted power may appear in ohmic heating of the driving coil, RF radiation from the helical coil acting as an antenna, line radiation from the plasma, and residual impedance mismatch between the planar coil and the plasma.

16.2.4 The Helicon Plasma Source

The *helicon plasma source* has been suggested as a high electron number density, high active-species flux source capable of operation in the low-pressure regime. References on the helicon plasma source include Lieberman *et al* (1996), Smith (1995), and Stevens *et al* (1995).

16.2.4.1 Source Characteristics

The helicon plasma source is illustrated in figure 16.18. It is a cylindrical plasma generated in a magnetic field, and is driven by the RF coil arrangement shown. The helicon plasma source is similar in operation to the *Rotamak* plasma source discussed in section 11.5.2 of Volume 1. In order for the RF power to be resonantly coupled to the plasma, the electrons must be magnetized, and confined throughout the plasma volume by a magnetic induction in the range from 5 to 50 mT. These inductions are generally beyond the reach of permanent magnets and must be generated by external electromagnets if large volumes of plasma are required.

16.2.4.2 Plasma Parameters

The helicon plasma source can operate at pressures below 1.3 Pa (10 mTorr), and number densities as high as 10^{19} electrons/m^3 have been reported. The RF used to energize the source is typically 13.56 MHz, and the coupling to the plasma is both inductive and resonant. Power levels of the helicon plasma source range from a few watts to more than a kilowatt in research applications.

16.2.4.3 Performance Issues

The plasma in a helicon source can operate in the steady state at high electron number density and at low gas pressures. However, its operational requirement for a large volume of magnetic field requiring electromagnets is a serious operational disadvantage. In addition, it is not clear whether helicon plasma sources can provide the radial uniformity of plasma density and active-species fluxes required for application to production-line microelectronic plasma processing that have been demonstrated by other sources.

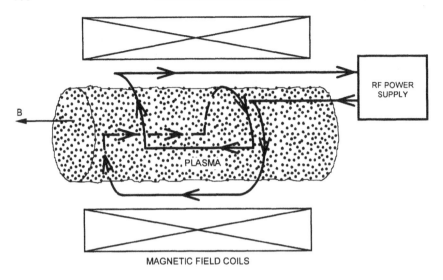

Figure 16.18. A plasma source for plasma etching based on helicon RF excitation in an axisymmetric magnetic field.

16.3   HIGH-VACUUM PLASMA SOURCES

Other than plasma sputtering and a few niche applications requiring long mean free paths, few high-throughput plasma-processing applications are operated under *high vacuum*, at pressures below 0.013 Pa ($\sim 10^{-4}$ Torr). In this regime, the mean free paths of all species are greater than the dimensions of the plasma and workpiece. This is desirable for microelectronic etching applications because it is consistent with vertical etching and good pattern transfer. It is also desirable for evaporative or sputter deposition of workpieces, since vaporized or sputtered atoms will travel directly from the source or target to the workpiece without scattering collisions.

The long mean free paths associated with the high-vacuum regime also have important disadvantages for some plasma-processing applications. Low gas density in the high-vacuum regime reduces the probability of inelastic collisions of plasma electrons that produce active species, and reduces the flux and concentration of active species that result from such collisions. In unmagnetized plasma sources the long mean free paths associated with low pressures make it difficult to conserve electrons, to prevent wall recombination, and to maintain a high enough electron number density to maintain an adequate flux of active species to complete a plasma processing task within an acceptable duration.

Some of the sources available for plasma-processing operations in the high-vacuum regime are discussed in Volume 1, and include *Penning discharges* (section 6.6 of Volume 1), *hollow cathodes* (see section 5.4 of Volume 1, and

Table 16.1. Nominal characteristics of plasma sources used for vacuum plasma processing.

Plasma source	Electron kinetic temperature T_e (eV)	Electron number density n_e (m^{-3})	Background gas pressure P_0 (Torr)	Nominal power input (W)
1. DC glow discharge	2–5	10^{16}	0.1–5	100–300
2. RF glow discharge	3–8	10^{17}	0.05–1.0	200–500
3. ECR plasma	5–15	10^{18}	10^{-4}–0.01	300–1000
4. Inductively coupled	5–15	10^{18}	10^{-3}–0.1	500–2000
5. Helicon	5–15	10^{18}–10^{19}	0.01–0.10	500–2000

Anonymous 1986), *vacuum arcs* (see section 10.3.4 of Volume 1, and Lafferty 1980), and a variety of other specialty sources originally developed for research purposes. These sources, and the high-vacuum regime itself, have associated problems when used for industrial production. The high-vacuum regime requires expensive vacuum pumping equipment, and as the operating pressure drops below 0.013 Pa (10^{-4} Torr), the pump-out time required to reach these pressures lengthens, adding to processing time.

In order to create plasmas of sufficient density to provide adequate fluxes or concentrations of active species in the high-vacuum regime, it is generally necessary to magnetize at least the electron population as is done in ECR microwave, helicon, and Penning sources. This need for a magnetic field normally requires electromagnets, with power requirements that can greatly exceed that of the plasma source itself. The electromagnets also have reliability problems associated with an additional subsystem not needed for the unmagnetized plasma sources operated at higher pressure.

16.4 SUMMARY OF PLASMA SOURCE PARAMETERS

The DC, RF, and flat-coil inductive plasma sources have electron number densities proportional not only to the RF power, but also to background pressure. Because the ECR plasma is generated by a resonant process, its electron number density, while roughly proportional to the power input below the critical density, is not directly proportional to the power input above it. The electron number density in ECR plasma sources is decoupled from the neutral gas pressure, allowing a constant electron number density to be maintained over a wide range of neutral pressures. Such plasma sources allow an additional degree of control over the flux of active species.

Characteristic values of the neutral gas pressure, electron kinetic temperature, and electron number density of plasma sources used for plasma processing are listed in table 16.1. These values are approximations only, as the database of well diagnosed industrial plasmas is small, and the wide range of configurations and operating conditions actually used in industrial applications can lead to large variations of these parameters. The four 'low-pressure' plasma sources are capable of providing plasma parameters almost identical with each other. If the plasma parameters produced by different plasma sources are identical, then the effects associated with etching, deposition, etc also will be the same (Herschkowitz *et al* 1996).

REFERENCES

Anonymous 1986 Board-cleaning technique using hollow cathode plasma discharge *IBM Technical Disclosure Bull.* **29** 1848–50

Asmussen J 1989 Electron cyclotron resonance microwave discharges for etching and thin-film deposition *J. Vac. Sci. Technol.* A **7** 883–93

Batenin V M, Klimovskii I I, Lysov G V and Troitskii V N 1992 *Superhigh Frequency Generators of Plasma* (Boca Raton, FL: Chemical Rubber Company) ISBN 0-8493-9305-1

Ben Gadri R 1997 Numerical simulation of an atmospheric pressure and dielectric barrier controlled glow discharge *PhD Thesis* Order No 2644, University Paul Sabatier, Toulouse III, France

——1999 One atmosphere glow discharge structure revealed by computer modeling *IEEE Trans. Plasma Sci.* **27** 36–7

Berry L A and Gorbatkin S M 1995 Permanent magnet electron cyclotron resonance plasma source with remote window *J. Vac. Sci. Technol.* A **13** 343–8

Boenig H V 1988 *Fundamentals of Plasma Chemistry and Technology* (Lancaster, PA: Technomic) ISBN 87762-538-7

Claude R, Moisan M, Wertheimer M R and Zakrzewski Z 1987 Comparison of microwave and lower frequency discharges for plasma polymerization *Plasma Chem. Plasma Process.* **7** 451–64

Cobine J D 1958 *Gaseous Conductors* (New York: Dover)

Dai F and Wu J C-H 1995 Self-consistent multidimensional electron kinetic analysis for inductively coupled plasma sources *IEEE Trans. Plasma Sci.* **23** 563–72

Dusek V and Musil J 1990 Microwave plasmas in surface treatment technologies *Czech. J. Phys.* **40** 1185–204

Gicquel A and Catherine Y 1991 Plasmas: sources of excited, dissociated and ionized species. Consequences for chemical vapor deposition (CVD) and for surface treatment *J. Physique* IV **1** C2 343–56

Godyak V A, Piejak R B and Alexandrovich B M 1991 Electrical characteristics of parallel-plate RF discharges in argon *IEEE Trans. Plasma Sci.* **19** 660–76

Gorbatkin S M, Berry L A and Roberto J B 1990 Behavior of Ar plasma formed in a mirror field electron cyclotron resonance microwave ion source *J. Vac. Sci. Technol.* A. **8** 2893–9

Gorbatkin S M, Berry L A and Sawyers J 1992 Poly-Si etching using electron cyclotron resonance microwave plasma sources with multipole confinement *J. Vac. Sci. Technol.* A **10** 1295–302

Herb G K 1989 Safety, health, and engineering considerations for plasma processing *Plasma Etching* ed D M Manos and D L Flamm (San Diego, CA: Academic) pp 425–70 ISBN 0-12-469370-9

Hershkowitz N, Ding J, Breun R A, Chen R T S, Meyer J and Quick A K 1996 Does high density–low pressure etching depend on the type of plasma source? *Phys. Plasmas* **3** 2197–202

Hirsh M N and Oskam H J (ed) 1978 *Gaseous Electronics; Vol I, Electrical Discharges* (New York: Academic) ISBN 0-12-349701-9

Howatson A M 1976 *An Introduction to Gas Discharges* 2nd edn (Oxford: Pergamon) ISBN 0-08-020574-7

Inoue M and Nakamura S 1995 Charge separation in an electron cyclotron resonance plasma *J. Vac. Sci. Technol.* A **13** 327–31

Kaufman H R 1961 An ion rocket with an electron-bombardment ion source *NASA Technical Note* TND-585

——1963 The electron bombardment ion rocket *Advanced Propulsion Concepts (Proc. 3rd Symp., October, 1962)* vol 1 (New York: Gordon and Breach)

——1965 Performance correlation for electron bombardment ion sources *NASA Technical Note* TND-3041

Kohl W H 1995 *Handbook of Materials and Techniques for Vacuum Devices* (Woodbury, NY: American Institute of Physics) ISBN 1-56396-387-6

Lafferty J M (ed) 1980 *Vacuum Arcs—Theory and Application* (New York: Wiley) ISBN 0-471-06506-4

Li M, Wu H-M and Chen Y 1995 Two-dimensional simulation of inductive plasma sources with self-consistent power deposition *IEEE Trans.Plasma Sci.* **23** 558–62

Lieberman M A and Lichtenberg A J 1994 *Principles of Plasma Discharges and Materials Processing* (New York: Wiley) ISBN 0-471-00577-0

Lieberman M A, Selwyn G S and Tuszewski M 1996 Plasma generation for materials processing *MRS Bull.* **21** 32–7

Mutsukura N, Fukasawa Y, Machi Y and Kubota T 1994 Diagnostics and control of radio-frequency glow discharge *J. Vac. Sci. Technol.* A **12** 3126–30

Olthoff J K, Van Brunt R J, Radovanov S B, Rees J A and Surowiec R 1994 Kinetic-energy distributions of ions sampled from argon plasma in a parallel-plate, radio frequency reference cell *J. Appl. Phys.* **75** 115–25

Raizer Y P, Schneider M N and Yatsenko N A 1995 *Radio-Frequency Capacitive Discharges* (Boca Raton, FL: Chemical Rubber Company) ISBN 0-8493-8644-6

Reinke P, Schelz S, Jacob W and Moller W 1992 Influence of a direct current bias on the energy of ions from an electron cyclotron resonance plasma *J. Vac. Sci. Technol.* A **10** 434–8

Roth J R 1995 *Industrial Plasma Engineering: Vol I—Principles* (Bristol: Institute of Physics Publishing) ISBN 0-7503-0318-2

Smith D L 1995 *Thin-Film Deposition: Principles and Practice* (New York: McGraw-Hill) ISBN 0-07-058502-4

Stevens J E, Sowa M J and Cecchi J L 1995 Helicon plasma source excited by a flat spiral coil *J. Vac. Sci. Technol.* A **13** 2476–82

Tao W H, Prelas M A and Yasuda H K 1996 Spatial distributions of electron density and electron temperature in direct current glow discharges *J. Vac. Sci. Technol.* A **14** 2113–21

Tsai C C, Berry L A, Gorbatkin S M, Hazelton H H, Roberto J B and Stirling W L 1990

Potential applications of an electron cyclotron resonance multicusp plasma source *J. Vac. Sci. Technol.* A **8** 2900–3

Ventzek P L G, Grapperhaus M and Kushner M J 1994 Investigation of electron source and ion flux uniformity in high plasma density inductively coupled etching tools using two-dimensional modeling *J. Vac. Sci. Technol.* B **12** 3118–37

Vossen J L and Kern W (ed) 1978 *Thin Film Processes* (New York: Academic) ISBN 0-12-728250-5

Wainman P N, Lieberman M A, Lichtenberg A J, Stewart R A and Lee C 1995 Characterization at different aspect ratios (radius/length) of a radio frequency inductively coupled plasma source *J. Vac. Sci. Technol.* A **13** 2464–9

Waits R K 1978 Planer magnetron sputtering *Thin Film Processes* ed J L Vossen and W Kern (New York: Academic) ch II-4, pp 131–73 ISBN 0-12-728250-5

17

Plasma Reactors for Plasma Processing

The two previous chapters were concerned with the *plasma sources* used in industrial applications of plasma processing. These sources are subsystems of a much larger *plasma reactor* or *production tool*, which is the device or apparatus that allows the plasma to process the workpiece in the desired manner. These sources often play a minor role in terms of capital investment and operating costs, but a major role in system reliability, product quality, and engineering manpower requirements.

Once a satisfactory plasma source has been selected for a particular application, it must be incorporated into an appropriate plasma reactor system, or production tool. In this chapter, some of the more widely used plasma reactors are discussed according to their processing application. General references on plasma processing and plasma reactors may be found in Gross *et al* (1969), Flinn (1971), Hollahan and Bell (1974), Boenig (1988), Lieberman and Lichtenberg (1994), Kohl (1995), Madou (1995), Roth (1995), and Lieberman *et al* (1996).

Most industrial glow discharges are operated at pressures below 1.3 kPa (10 Torr). It is therefore necessary to treat webs, films, and individual workpieces exposed to such discharges by *batch processing*, using an apparatus like that shown in figure 17.1. Typical operational steps in plasma processing under vacuum are, with reference to figure 17.1,

(1) Install the feed roll or workpiece.
(2) Close bleed valve, seal off vacuum system.
(3) Open gate valve to vacuum pump.
(4) Pump vacuum system below 1.3 kPa (10 Torr).
(5) Turn on plasma.
(6) Feed material from supply to take-up roll, or expose individual workpiece for the required duration.
(7) Turn off plasma.

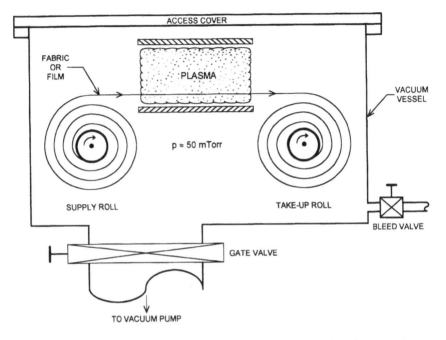

Figure 17.1. Batch processing of fabrics or films in a low-pressure glow discharge plasma.

(8) Close gate valve.
(9) Bleed system up to 1 atm.
(10) Remove take-up roll or workpiece for further processing.

These procedures, and the attendant cost of purchasing and operating vacuum systems, are the principal reasons why low-pressure glow discharges are not more widely used in industry. When such glow discharges are used, it tends to be for the fabrication of high-value items like microelectronic chips or biomedical products, which can be accomplished economically in no other way.

Continuous processing at 1 atm, illustrated in figure 17.2, is much to be preferred in industrial applications. In this case, either a corona discharge, a dielectric barrier discharge (DBD), or a one atmosphere uniform glow discharge plasma (OAUGDP) reactor can provide active species to treat the surface of a workpiece at the plasma boundary. In such continuous processing, the plasma source is usually within a gas-tight safety/environmental enclosure. Such an enclosure is present to control the working gas if other than air, and/or to prevent the outflow of ozone, UV radiation, or other active species that may cause safety concerns.

It is possible to use continuous processing in vacuum glow discharge plasma reactors by passing a fabric web, fiber or film into and out of the vacuum system through one or more stages of vacuum pumping. An example of such a system

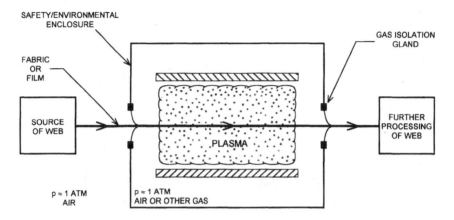

Figure 17.2. Continuous processing of fabric or film through an OAUGDP.

was published by Rakowski (1989), and is illustrated in figure 17.3. In this apparatus, wool toe is fed continuously through a plasma operating at a few hundred Pascals (a few Torr) at speeds, and with results, of commercial interest. This continuous-processing vacuum hybrid has the disadvantages of increased capital and operating costs for the required differential pumping of the vacuum system, as opposed to the smaller capacity vacuum systems that are adequate for batch processing.

17.1 PLASMA REACTORS FOR SURFACE TREATMENT

The industrial plasma treatment of surfaces requires a variety of plasma reactors, the plasma sources for which have been discussed in chapters 15 and 16 of this volume. Additional discussion of plasma reactors for surface treatment may be found in d'Agostino (1990), Coburn (1991), Batenin *et al* (1992), Roth (1999), Montie *et al* (2000), and Roth *et al* (2000). In this section, we survey reactor configurations widely used for plasma surface treatment, and comment on features that qualify them for their applications.

To achieve industrially important effects, plasmas used for surface treatment must have a sufficiently high electron number density to provide useful fluxes of active species, but not so high or energetic as to damage the material being treated. These constraints rule out dark discharges other than coronas for most surface treatment applications, because of their low production rate of active species. These constraints also rule out arc or torch plasmas, which have power densities and active-species flux intensities high enough to damage most exposed materials. Glow discharge plasmas produce the appropriate electron number density and active-species flux for nearly all plasma surface treatment applications.

For the purposes of this section, reactors used for industrial plasma surface

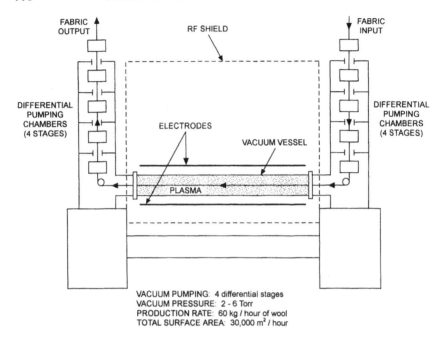

FABRIC OUTPUT

FABRIC INPUT

RF SHIELD

DIFFERENTIAL PUMPING CHAMBERS (4 STAGES)

DIFFERENTIAL PUMPING CHAMBERS (4 STAGES)

ELECTRODES

VACUUM VESSEL

PLASMA

VACUUM PUMPING: 4 differential stages
VACUUM PRESSURE: 2 - 6 Torr
PRODUCTION RATE: 60 kg / hour of wool
TOTAL SURFACE AREA: 30,000 m² / hour

Figure 17.3. Apparatus for continuous low-pressure glow discharge plasma treatment of wool developed by Rakowski (1989).

treatment can be classified according to the type of workpiece they process. These workpieces include thin films and webs processed as rolls for batch processing; and individual solid workpieces such as wafers or three-dimensional objects, the entire surface of which is to be treated.

17.1.1   Plasma Reactors for Films, Webs, and Fibers

Films are two-dimensional continuous sheets of plastics, polymers, or other materials; *webs* are porous, two-dimensional sheets of fabric, either woven or non-woven; and *fibers* are monofilament 'strings' of a polymer, plastic, or natural material. Films, webs, and fibers are normally stored, shipped, and delivered to final production in the form of rolls, and it is these rolls that are normally processed by plasma reactors such as that illustrated schematically by figure 17.1. The usual objective of such plasma processing is to increase the surface energy of the material, which leads to better printing, bonding, wettability, and wickability, as discussed further in chapter 21.

17.1.1.1   Reactors with Atmospheric Pressure Plasma Sources

Three atmospheric pressure plasma sources discussed in chapter 15 lend themselves particularly well to the surface treatment of films, webs, and fibers. These are corona sources, DBDs (sometimes erroneously referred to as corona reactors); and the OAUGDP reactor. Corona sources and DBDs are widely used to increase the surface energy of paper, plastic and polymeric films to moderate levels to improve their printability for use in packaging materials, for example. OAUGDP reactors show promise for increasing the surface energy of materials to levels higher than that possible with either corona or DBDs as the plasma source. These sources are used for films and webs because they can easily be adapted to operate in the parallel-plate configuration illustrated in figure 17.1, which allows treatment of films and webs up to several meters wide.

17.1.1.2   Reactors with Intermediate Pressure Plasma Sources

Before the advent of the OAUGDP source, uniform glow discharges could be generated only under vacuum below 1.3 kPa (10 Torr). Except for large rolls of film or webs, this tended to restrict processing with glow discharge plasmas to high value workpieces for the healthcare or aerospace markets, using the plasma sources described in chapter 16 of this volume.

Parallel-plate plasma sources are preferred to treat films and webs because of their two-dimensional nature. These sources operate in the manner illustrated in figure 17.1, with a film or web passing from roll to roll through a parallel-plate glow discharge. The glow discharge is most frequently energized by RF power at 13.56 MHz in vacuum reactors, but DC obstructed, abnormal glow discharges and parallel-plate DC magnetron plasma sources are used also.

17.1.2   Plasma Reactors for Three-Dimensional Workpieces

The applications of plasma surface treatment to three-dimensional workpieces will be discussed further in chapter 21. These applications include improved bonding of paints, inks, and electroplated layers; improved adhesion of deposited films; and cleaning, sterilization, and decontamination of surfaces. These applications have been demonstrated in the past under vacuum at intermediate pressures, thus requiring batch processing of the workpieces. Although not yet in widespread industrial use, an OAUGDP reactor is capable of treating three-dimensional workpieces that are not too large in one of their dimensions. This reactor can also treat large workpieces with active species convected from the plasma to the workpiece in a remote exposure configuration (Roth *et al* 2000).

17.1.2.1   Reactors with Intermediate Pressure Plasma Sources

A variety of plasma sources similar to those discussed in chapter 16 of this volume have been developed to treat three-dimensional workpieces. Workpieces with a significant extension in the third dimension have required special handling, particularly for those applications for which uniformity of effect is important.

The *DC obstructed abnormal glow discharge* is capable of immersing an entire workpiece of significant size in a large volume of relatively dense and uniform negative glow plasma. *RF-generated parallel-plate plasmas* can be operated with a ratio of parallel-plate separation to electrode dimension great enough to accommodate three-dimensional workpieces, but the uniformity of such plasmas is generally poor. Other intermediate-pressure plasma sources that have been used for the surface treatment of three-dimensional workpieces include the *electron bombardment (Kaufman) plasma source*, and *non-resonant microwave plasma sources*. An attempt is usually made to operate these sources at the high end of the intermediate-pressure range, near 133 Pa (1 Torr). These pressures avoid the longer pump-down times and more expensive vacuum equipment needed to operate plasma sources at lower pressures.

17.1.2.2 *Processing of Three-Dimensional Workpieces*

In any surface treatment process under vacuum, the batch handling of workpieces as they are cycled down to the operating pressure and back up to 1 atm slows the throughput rate. In addition, it is more difficult to manipulate workpieces under vacuum to achieve uniformity of effect by motional averaging than it is at 1 atm. This constraint may lead either to additional complexity and expense, or to a less uniformly treated, inferior product. These factors may be significant enough to make plasma processing under vacuum economically infeasible.

When the workpiece is sufficiently valuable, or there is no other technically feasible alternative to plasma surface treatment, several methods may be used to maintain the throughput of workpieces at a high level. One (*batch processing*) option is to process enough workpieces simultaneously in the vacuum system to provide them at the rate required by a continuous-flow production process. A second option is to have *in-line, quasi-continuous processing*, in which one or several large workpieces are fed into and exit from a vacuum system fitted with one or more airlocks at the entrance and exit. This allows one set of workpieces to be processed under vacuum while the next set in the sequence is pumped down to operating pressure, and the previous set is brought up to atmospheric pressure.

17.1.3 Atmospheric-Pressure Surface Treatment Reactors

The economic and operational advantages of operating at 1 atm have led to the development of a variety of plasma-based surface treatment reactors that function in air and/or at 1 atm. Atmospheric-pressure plasma reactors selected for discussion here include corona, DBDs, and the OAUGDP.

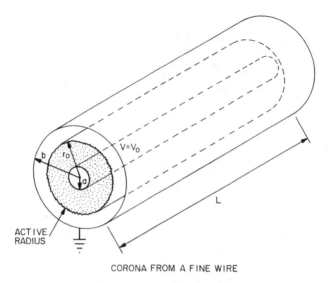

CORONA FROM A FINE WIRE

Figure 17.4. The fine-wire configuration used for corona surface treatment at 1 atm. Each wire in a planar series of fine wires is surrounded by an *active region* of radius r_o.

17.1.3.1 Corona reactors

Corona sources useful for plasma processing have been discussed in section 15.2 of this volume. General references on corona and corona treatment of surfaces include Loeb (1965) and Roth (1995). For corona surface treatment at 1 atm, the fine-wire configuration (see section 8.5.2 of Volume 1) is most widely used. As illustrated in figure 17.4, each wire in a planar series of fine wires is surrounded by an *active region* of radius r_o, at which the radial electric field falls to the *breakdown electric field* of the working gas. Ions or other active species are generated in the active region, which is usually, but not always, visible to the unaided eye.

Coronal currents in surface treatment reactors are small, from a few milliamperes to a few tens of milliamperes per meter of length, and the total rate of generation of active species is correspondingly small. At atmospheric pressure, only ultraviolet photons and room-temperature ions, positive or negative, escape the active region of corona in any numbers. The ions carry the coronal current to the surrounding walls/electrodes and are available for surface treatment. The ultraviolet radiation can change the surface energy of some materials, and also affect the ability of some surfaces to hold a static charge.

Cold ions can be imbedded in molten fibers by the corona reactor shown in figure 17.5 (after Deeds 1992). Here a molten polymer fiber is extruded through a spinneret nozzle, where it encounters a flowing gas containing cold ions generated upstream in the active region surrounding one or more corona

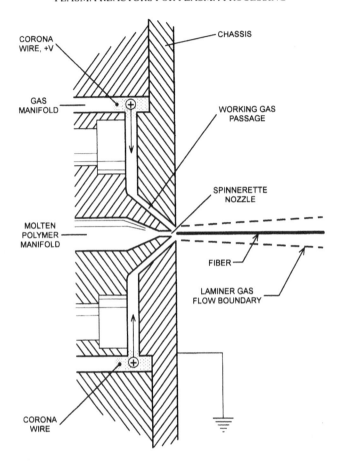

Figure 17.5. Charged polymeric fiber production using a corona discharge (after Deeds 1992).

wires. An arrangement similar to that illustrated in figure 17.5 can be used to produce permanently charged *non-woven fabrics* that are *quasi-electrets* (Tsai and Wadsworth 1995). Figure 17.6 shows a corona reactor for making non-woven, quasi-electret fabrics for filters and face masks. In this reactor, corona wires can be placed in one of three locations near the molten polymer fibers, before or after they impinge on the collector drum and become cross-linked and formed into a fabric. A production corona reactor, with the internal corona wire illustrated in figure 17.5, is illustrated in figure 17.7, after Deeds (1992).

Corona reactors have also been used to treat non-woven webs and polymeric and plastic films to induce higher surface energy and improved wettability, printability, etc. Fabric webs and continuous films can be exposed to corona-generated active species in the configuration illustrated in figure 17.8, in which

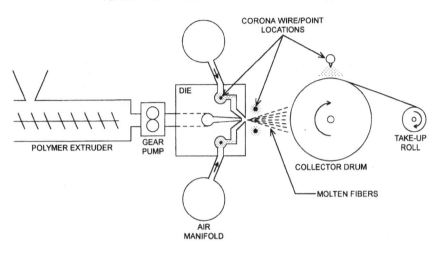

Figure 17.6. Non-woven polymeric meltblown quasi-electret production using an atmospheric corona discharge.

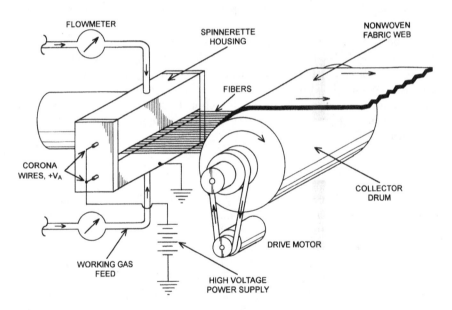

Figure 17.7. Non-woven polymeric meltblown quasi-electret production using an atmospheric corona discharge, isometric view.

the material is pulled, usually over a grounded surface, past an array of fine corona-generating wires. To ensure uniformity of effect across the width of the web, the direction of motion of the web/film is at right angles to the axis of

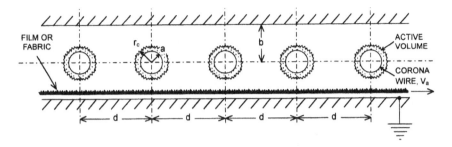

Figure 17.8. Corona wire array for the treatment at 1 atm of thin film or fabric.

the wire(s). Under extreme operating conditions (high voltage, high humidity, etc), a fine corona wire can draw enough current to overheat, expand along its length, and go slack in its mountings. This can short out the power supply. To prevent this, a spring-loaded tensioning assembly like that shown in figure 17.9, may be used (Deeds 1992). Another operational issue is easy replacement of the fragile fine wires, particularly when operated in corrosive or chemically reactive environments.

Corona reactors have the advantages of simplicity, operating at 1 atm, and low input power. Their disadvantages include low formation rates and fluxes of active species, and inability to produce a range of energetic active species that can react chemically with the surface to be treated. In addition, corona reactors pose the regulatory issues discussed in section 15.2.3 of this volume, including radiofrequency interference; x-ray, UV, and ozone production; and high voltage electrical safety.

17.1.3.2   Dielectric Barrier Discharge Reactors

The generation of low power density electrical discharges at 1 atm, other than coronas, is not a recent development. *Filamentary*, or *dielectric barrier discharges (DBD)* between parallel plates or concentric cylindrical electrodes in air at 1 atm have been used in Europe to generate ozone in large quantities for the treatment of public water supplies since the late 19th century. The DBD reactors used for the surface treatment of films and webs referred to in section 15.1.5 are incorrectly called 'corona reactors' in the North American industrial literature. These reactors characteristically use the *DBD plasma source*, the physics of which is discussed in section 15.3 of this volume (see also Ben Gadri 1997).

DBDs used for surface treatment are applied almost exclusively to large surface areas of films, papers, and fabrics, to increase their surface energy. Two of the most common DBD reactor configurations for the surface treatment of films and fibers are illustrated in figure 17.10. Figure 17.10(*a*) is the *planar parallel-plate DBD reactor*, used for fast-moving webs that are not electrostatically charged, and do not adhere to the surface of the electrodes over which they pass.

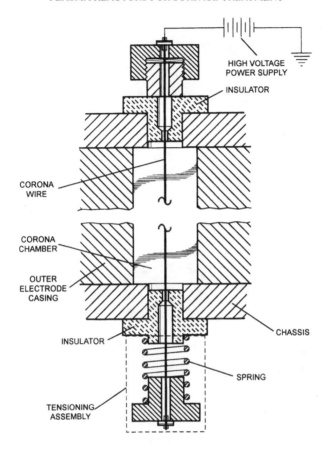

Figure 17.9. Corona wire tensioning assembly (after Deeds 1992).

The *cylindrical annulus DBD reactor* illustrated in figure 17.10(*b*) is used for webs which, because they are electrostatically charged or have high surface energies, adhere to the surface of the DBD electrodes. In this configuration, a rotating drum moves the film or web through an annulus of DBD plasma generated between a cylindrical electrode and the drum, while tension on the film or web is maintained by idler rollers. In these configurations, both electrodes are normally covered by a dielectric barrier, but in some applications, only one may be coated.

DBD reactors are normally operated at atmospheric pressure, with air as the working gas. Sometimes argon may be used because it provides a more stable plasma. The DBD is an efficient generator of ozone when operated in oxygen-containing gas mixtures, including air, and the ozone must be contained or removed by ventilation. A DBD reactor is normally enclosed in a light plastic or sheet metal enclosure, as indicated in figures 10(*a*) and (*b*), in order to provide

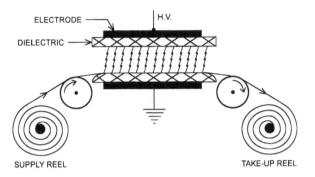

a) PLANAR EXPOSURE CONFIGURATION

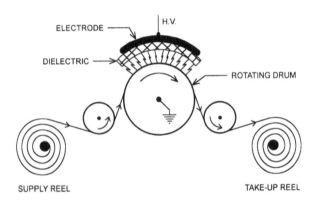

b) ROTATING CYLINDER DBD CONFIGURATION

Figure 17.10. Two DBD reactor configurations for the surface treatment of films and fibers: (*a*) the *planar parallel-plate* DBD reactor, used for fast-moving webs that do not adhere to the surface of the electrodes over which they pass; (*b*) the *cylindrical annulus DBD reactor* used for webs adherent to the surface of the DBD electrodes.

isolation and containment of ozone or other toxic gases from the production line or workplace. A variety of glands and wiping seals have been developed to allow fast-moving webs to enter and exit these enclosures without undue friction or leakage.

At least three operational issues arise in industrial applications of the DBD for its major application, increasing the surface energy of films and webs. One of these is uniformity of effect, the second is damage to the workpiece, and the third is the relatively limited concentration and flux of active species that are available. Macroscopic uniformity of effect over the surface of the workpiece is promoted by motional averaging in the direction of movement of the film or web. Also helpful is careful attention to uniformity of flow of the working gas, uniformity of the electric field, and uniformity of the gap spacing.

Microscopic uniformity of effect is a separate issue raised by the discrete, filamentary nature of the individual DBD discharges discussed earlier in section 15.3.3 of this volume. These filaments characteristically leave small pits on the surface of the workpiece a few tenths to a few microns in diameter. At high power levels, these pits can burn entirely through the workpiece (referred to as *pitting*), leaving holes that can be observed by holding the workpiece up to the light after exposure.

In some applications, a DBD reactor may not be able to provide the high levels of surface energy (above 50 dynes/cm) required to print water-based inks without damaging the workpiece. For other applications, such as increasing the surface energy above 60 dynes/cm, a DBD reactor may not be capable of the areal power densities required, or of producing the necessary active species from air as a working gas. Economic considerations usually limit DBD reactors to air as a working gas. The regulatory issues associated with operating DBD reactors in air or other gases have been discussed previously in section 15.3.4. These issues can be held to a minimal economic and operational impact by designing simple enclosures as an integral part of the reactor design.

17.1.3.3 One Atmosphere Uniform Glow Discharge Plasma Reactors

The properties of a DC glow discharge plasma at 1 atm, using both air and hydrogen, was reported by von Engel *et al* in 1933. Their discharge was initiated at low pressure (thus requiring a vacuum system), required temperature control of the electrodes, and was too unstable with respect to the glow-to-arc transition for routine industrial use.

It has recently become possible to generate large volumes (multiliter) of uniform glow discharge plasma at 1 atm under conditions for which active species are available for surface treatment. The theory of the *OAUGDP* plasma source was treated in section 12.5.2 of Volume 1, where the physics of the ion-trapping mechanism is described. The characteristics of the OAUGDP as a plasma source were described in section 15.4 of this volume. More information on the OAUGDP and its applications, and on related discharges, may be found in Roth *et al* (1995a, b), Ben Gadri (1997, 1999), and Kanazawa *et al* (1988).

The production of a full range of active species from an OAUGDP offers many advantages in an industrial setting. These include: (1) a vacuum system, with its capital and operating costs, is not required; (2) batch processing of workpieces is not required; and (3) continuous surface treatment at the point of manufacture (loom, extruder, paper mill, etc) may be possible.

A schematic diagram of a *parallel-plate* OAUGDP reactor is shown in figure 17.11. The reactor volume is bounded by two insulated, plane parallel plates across which an RF electric field is applied. This electric field characteristically has an amplitude of about 10 kV/cm, and a frequency in the

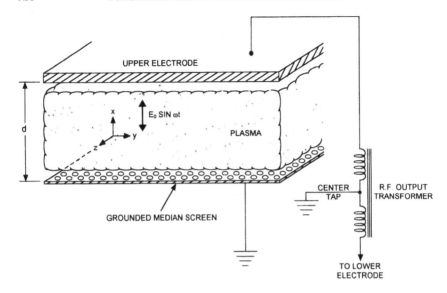

Figure 17.11. Schematic diagram of a parallel-plate OAUGDP reactor, with a center tapped transformer secondary.

audio range. The electric field must be strong enough to break down the gas used, and is about a factor of three lower for helium and argon than for atmospheric air. The RF must be in the correct range for ion trapping, as discussed in section 12.5.2 of Volume 1. If the frequency is too low, the discharge will not initiate, and if it is too high, the plasma polarizes, undergoes a filamentation instability, and forms filamentary discharges between the plates.

The parallel electrode plates may be enclosed in a Plexiglas container, indicated in figure 17.12. The function of the enclosure is to control the composition of the gas used, and to confine potentially harmful active species such as ozone or nitrogen oxides. As illustrated in figure 17.12, the reactor may have a grounded metal median screen located midway between the electrode plates to provide a rigid surface that supports the material to be treated. The median screen may be an uninsulated electrical conductor, and it may be grounded or allowed to float electrically. Holes in the screen allow a gas flow through fabrics or porous workpieces. Keeping the median screen grounded allows treated materials to remain at ground potential. A photograph of a parallel-plate OAUGDP operating in 1 atm of air is shown in figure 17.13.

A useful variant of the parallel-plate plasma reactor, due to Okazaki (1992) is the *cylindrical helix reactor* shown in figure 17.14. Here, two helical strip electrodes are wound on the outside of an insulating tube through which a working gas flows at 1 atm. The helical strips face each other on opposite sides of a diameter, and a plasma is formed in the interior of the tube by the RF electric

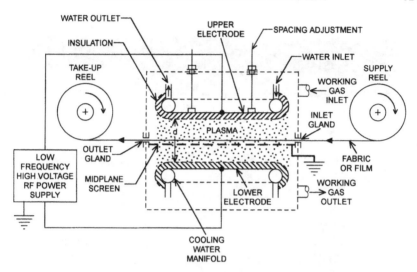

Figure 17.12. Continuous processing of a fabric or film using a parallel-plate OAUGDP reactor.

field. The working gas flows through this plasma, and reacts to produce active species.

For some plasma surface treatment applications, a relatively thin (a few mm) surface layer of plasma may be more useful than the relatively large volumes of plasma provided by a parallel-plate reactor. The parallel-plate plasma reactor can be geometrically transformed into a planar reactor in the manner illustrated in figure 17.15. In this configuration, strip electrodes of opposite RF polarity are located normal to the plane of the diagram on either side of the insulating panel shown in figure 17.15. Ions are trapped on the arched electric field lines between the electrodes, and a surface plasma layer is generated in this region.

The plasma surface layer can take the form of a flat panel, as illustrated in figure 17.16, or a cylinder, as illustrated in figure 17.17. A simple way to produce a cylindrical surface plasma layer is to wind a pair of insulated wires around a cylinder, as shown in figure 17.17. When adjacent wires are energized with high voltage RF at a few kilohertz, ions are trapped on the arched electric field lines between adjacent wires and an OAUGDP surface layer forms. This is illustrated in the photograph of a cylindrical helium plasma in figure 17.18.

For some surface treatment applications, such as anti-static exposure or a temporary increase of surface energy to promote wettability or printability, brief exposure of a moving web or film to a sheet of plasma may be sufficient. Such an OAUGDP may be generated by ion trapping in the rod-and-plane configuration illustrated in figure 17.19, or by the double-rod configuration illustrated in figure 17.20. In these reactor configurations, the rod is usually covered with a glass sleeve to minimize RF conditioning arcs.

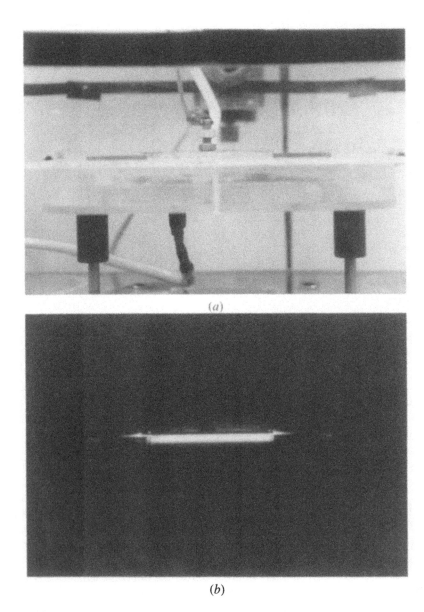

(a)

(b)

Figure 17.13. The MOD-III parallel-plate OAUGDP reactor operating in air. (a) This photograph shows the lower, water-cooled electrode covered by a Pyrex® plate, and an upper metal plate, supported by two microscope slides, operating in air with a uniform glow discharge plasma. (b) This is the same as (a), with the laboratory lights out. Note the uniformity of the plasma.

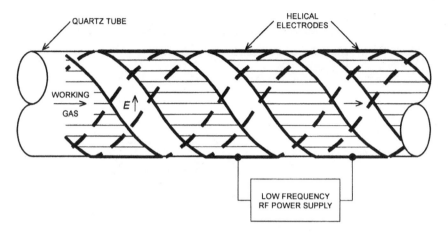

Figure 17.14. Apparatus for generating a 1 atm glow discharge inside an insulating cylinder with a continuous gas flow through the atmospheric plasma (design suggested by Okazaki (1992)).

Figure 17.15. The planar surface layer plasma reactor. Ions are trapped along the electric field lines between adjacent strip electrodes, which are maintained at opposite RF polarities.

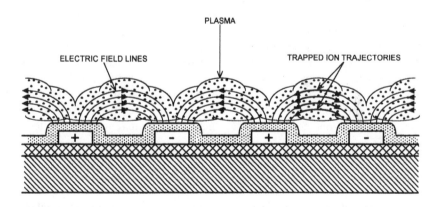

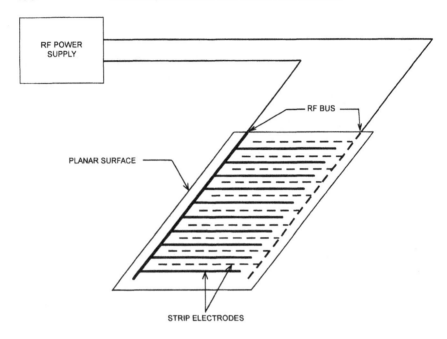

Figure 17.16. A simple strip electrode arrangement for the generation of a planar OAUGDP surface layer.

The active species of an OAUGDP reactor may be used to affect the surface energy and related properties of surfaces. In such applications, the characteristics of the neutral working gas may be as significant as that of the charged species in the plasma. In addition to the type of working gas and its composition (if a mixture), other factors may affect the results of exposure. These factors include whether or not the gas is recycled past the exposed workpiece (some important active species are long lived and their concentration builds up upon recirculation); the gas temperature; the temperature of the surface being treated; the gas humidity; and the flow rate of the gas past the workpiece. An OAUGDP in air produces ozone and ultraviolet radiation, which may require a protective enclosure.

17.1.4 Intermediate-Pressure Glow Discharge Reactors

With rare exceptions, plasma reactors for the surface treatment of materials that operate below atmospheric pressure are glow discharges in the intermediate-pressure regime between 6.7 Pa (50 mTorr) and 1.3 kPa (10 Torr). The intermediate-pressure glow discharge sources used for surface treatment include DC and RF parallel-plate, as well as magnetron plasma sources. These sources have been discussed in section 16.1 of this volume.

Figure 17.17. A planar surface layer plasma reactor on a 5 cm diameter cylinder. A pair of wires is wrapped helically on the surface of the cylinder, providing adjacent electrodes of opposite polarity.

Figure 17.18. The cylinder shown in figure 17.17, operating in helium gas, at a frequency of 3 kHz, and RF voltage of 4 kV, rms.

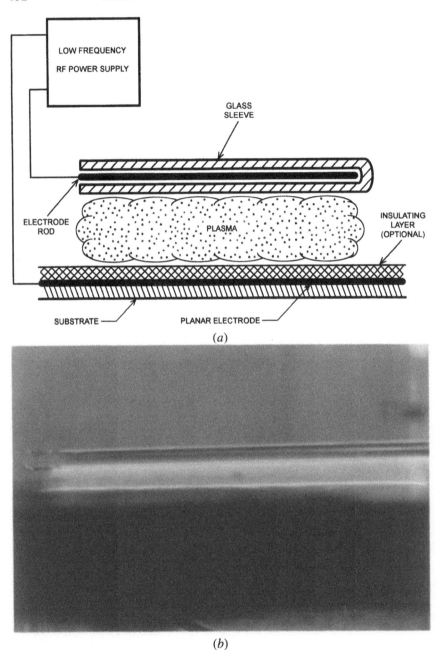

(a)

(b)

Figure 17.19. A single insulated rod and insulated planar electrode configuration for the generation of a sheet of OAUGDP. (a) Schematic diagram of configuration. (b) Photograph of a rod-and-plane OAUGDP configuration operating in helium gas. The rod length is 15 cm.

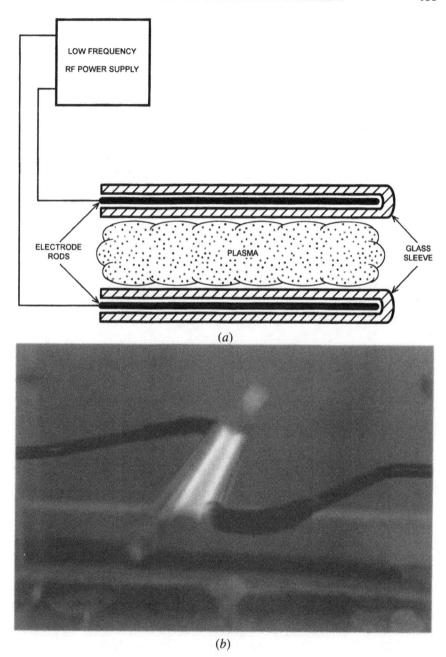

Figure 17.20. A double insulated rod configuration for the generation of an OAUGDP. (*a*) Schematic diagram of configuration. (*b*) Photograph of a double-rod OAUGDP configuration operating in helium gas. The rod length is 15 cm.

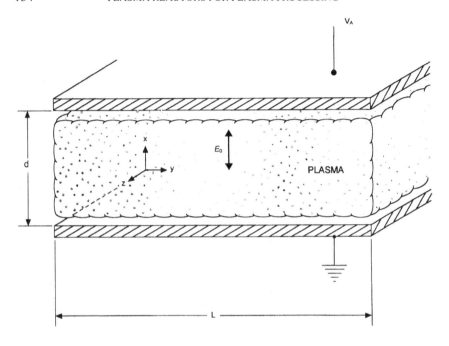

Figure 17.21. A parallel-plate plasma reactor.

17.1.4.1 Parallel-Plate Reactors

Parallel-plate glow discharge reactors intended for surface treatment can be generated by DC or RF capacitive excitation, using the physics and technology discussed in chapters 9 and 12 of Volume 1, and section 16.1.1 of this volume. Additional information on parallel-plate intermediate-pressure plasma reactors may be found in Batenin *et al* (1992) and Raizer *et al* (1995). A characteristic parallel-plate reactor is illustrated in figure 17.21. The workpiece to be exposed to active species from the plasma is usually on or near an electrode. In industrial parallel-plate glow discharge reactors, the ratio L/d of plate width, L, to plate separation, d, may range over values from three to more than 100.

The location of glow discharges on the voltage–current characteristic of the DC low-pressure electrical discharge is discussed in section 4.9 of Volume 1, and is illustrated in figure 17.22. DC glow discharges used in surface treatment characteristically operate between 100 V and 1 kV, and at currents from a few milliamperes to approximately one ampere. The workpieces either are mounted on or comprise the cathode. Most applications require that the plasma be uniform over the surface of the workpiece. To ensure this uniformity, DC glow discharges are normally operated as an *abnormal glow discharge*, as discussed in section 9.5.2 of Volume 1. Such a discharge is established between the points G and H on the voltage–current curve of figure 17.22, where the plasma completely

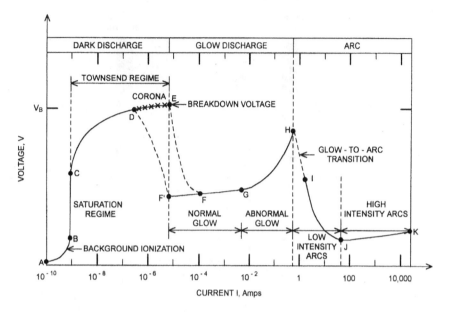

Figure 17.22. The voltage–current characteristic of the classical DC low-pressure electrical discharge tube.

covers the cathode uniformly, and higher voltages are required to further increase the current.

Electron number densities, kinetic temperatures, and active-species production rates and fluxes available from the *positive column* are less than that from the *negative glow*. The positive column therefore is eliminated by operating the discharge as an *obstructed abnormal glow discharge*, leaving only the higher density and more energetic negative glow plasma to produce active species. Obstructed abnormal glow discharges are those for which the width, d, of the cathode region, shown on figure 17.23, is less than the *Paschen minimum distance* appropriate for the gas and pressure, as discussed in section 9.2 of Volume 1. In such obstructed abnormal glow discharges, a positive column and other structures are absent between the negative glow and the anode, and the cathode fall voltage is higher than the Paschen minimum breakdown voltage. This high sheath voltage is useful when energetic ions are required to etch the cathode surface, to sputter atoms off the cathode surface, or to treat workpieces mounted on the cathode.

17.1.4.2 Magnetron Discharges

Magnetron discharges, discussed in section 9.5 of Volume 1, are characterized by crossed electric and magnetic fields in the negative glow plasma, with the sheath (or cathode fall) electric field also at right angles to the magnetic field. When used

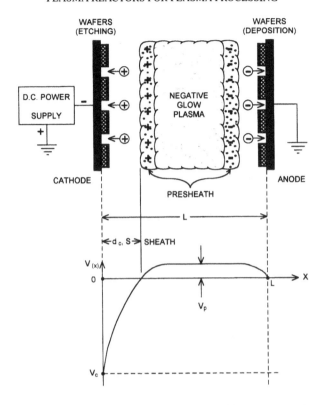

Figure 17.23. The obstructed abnormal glow discharge plasma reactor, with a schematic potential distribution.

for treatment of individual workpieces or electrically conducting surfaces, the *DC parallel-plate magnetron*, illustrated in figure 17.24, is normally preferred. In some applications, supplemental RF excitation is used because it makes possible a higher electron number density in the negative glow, a higher sheath voltage drop, and greater discharge stability. The DC magnetron discharge has been discussed in section 9.5.6 and the RF magnetron in section 12.4.1 of Volume 1.

The magnetic field employed in magnetron discharges is usually generated by inexpensive permanent magnets, strong enough to magnetize and trap the negative glow electrons, but not the ions, in a local magnetic mirror above the cathode surface. This magnetic mirror trapping increases the electron number density and the active species production rate. The E/B 'magnetron' drift is into or out of the plane of the diagram in figure 17.24, thus making the plasma itself and its effects on a workpiece more uniform in the $E \times B/B^2$ drift direction. A porous fabric, electrically conducting film, or other electrically conducting material may be moved past the magnetron discharge at the electrode surface position A if ion bombardment is desired, or at the anode surface position A′ if a

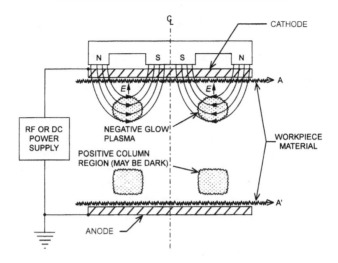

Figure 17.24. A low-pressure parallel-plate magnetron reactor. Fabrics or films may be exposed either at the position A or A'.

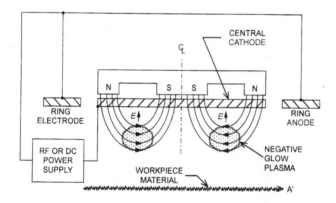

Figure 17.25. A schematic diagram of the low-pressure co-planar magnetron reactor. A fabric or film (usually electrically insulating) passes by the plasma at the position A' for sputtering or other plasma surface treatment.

flux of sputtered atoms or other active species is required.

If the workpiece material is an insulator, or if charge build-up on the workpiece is undesirable, the *co-planar magnetron* geometry shown in figure 17.25 can be used with either RF or DC excitation. Placement of the workpiece material at the position A' allows it to receive a flux of active species, without having to interrupt real or displacement currents, or be immersed in a sheath.

17.1.4.3   Microwave Energized Reactors

Microwave plasma reactors may be used to generate active species needed for the plasma treatment of surfaces. They have the advantage of being 'electrodeless', and thus produce a plasma with fewer sputtered impurities than result from ion bombardment of the cathode of DC glow discharges or the powered electrode of RF glow discharges. A survey of microwave plasmas in surface treatment technologies has been published by Dusek and Musil (1990). Other more specialized references include Asmussen (1989) and Batenin *et al* (1992). Microwave glow discharge reactors used for surface treatment characteristically operate at the standard frequency of 2.45 GHz, and at power levels of a few hundred watts. The free-space wavelength at this frequency is 12.24 cm, a circumstance that may lead to the formation of undesirable modal patterns in the power density of such reactors.

The *waveguide-coupled reactor* has been discussed in section 13.5.1 of Volume 1, and is illustrated in figure 17.26. In this configuration, a microwave-generated plasma is created inside a quartz or dielectric tube inserted through a tapered waveguide. A feed gas flows through the plasma, and the resulting active species are convected to the workpiece by the gas flow. This arrangement is usually used at operating pressures below 13.3 kPa (100 Torr). At atmospheric pressure, such high power densities ($0.1–10$ kW/cm^3) are required to maintain a microwave plasma that the power flux of active species is likely to damage the workpiece.

A second microwave reactor is the *microwave cavity reactor*, illustrated in figure 17.27, and discussed in section 13.5 of Volume 1 and in section 16.2.2. This reactor resembles a domestic microwave oven, except that it characteristically operates below 1.33 kPa (10 Torr), rather than at 1 atm. The microwave radiation arrives through a waveguide maintained at 1 atm, and passes into the evacuated reactor cavity through a ceramic window. A good impedance match—which yields a minimum of reflected power—is obtained by adjustment of tuning stubs in the input waveguide.

In the *resonant cavity microwave reactor*, also illustrated by figure 17.27, the free-space wavelength may be comparable to the radius or the axial dimension of the cavity, or a small integer multiple of these. Such a resonant cavity is undesirable for many applications, for reasons similar to those that apply to domestic microwave ovens. The modal concentrations of microwave power density produce a non-uniform plasma, which yields a non-uniform flux of active species and effect on a workpiece.

More satisfactory for applications requiring uniform surface treatment is the *non-resonant multimode cavity reactor*, an example of which was discussed in section 13.5.3 of Volume 1. In these reactors, the wavelength is smaller than any dimension of the cavity, thus resulting in a multimode, uniform plasma and a uniform flux of active species on workpieces.

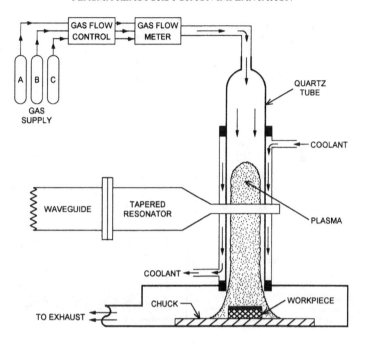

Figure 17.26. A microwave waveguide coupled reactor, in which the active species of a microwave-generated plasma are convected to a workpiece by the neutral gas flow.

17.1.5 Operational Issues

The intermediate-pressure glow discharge reactors discussed here operate below 1.33 kPa (10 Torr), and require vacuum systems and batch processing of treated materials. The requirement of operation in a vacuum system reduces the attractiveness of these plasmas for applications requiring continuous processing. If vacuum exposure and batch processing of the material are acceptable, however, these reactors offer a relatively simple way to generate the active species required with off-the-shelf hardware.

17.2 PLASMA REACTORS FOR ION IMPLANTATION

DC abnormal glow discharges have been used for *low-energy plasma thermal diffusion treatment* (nitriding, carbonizing, etc) for nearly a century. The use of glow discharges for *plasma ion implantation* as a competing technology to *ion-beam implantation*, however, is as recent as the past two decades (Conrad 1988). Each of these ion-implantation technologies requires its own characteristic plasma source, the subject matter of this section.

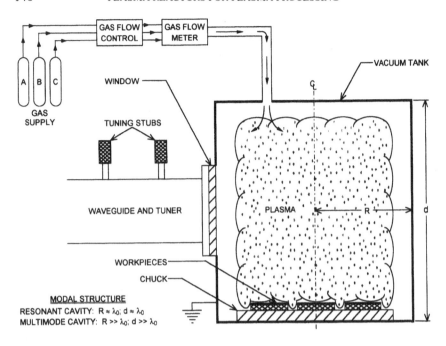

Figure 17.27. A microwave cavity reactor, in which active species from a microwave-generated plasma impinge on a workpiece located at the wall.

17.2.1 Reactors for Low-Energy Plasma Thermal Diffusion Treatment

Low-energy plasma thermal diffusion treatment is most frequently accomplished by inserting a metallic workpiece into an intermediate-pressure DC abnormal glow discharge plasma, and using the workpiece as the cathode. The workpiece (cathode) is heated, either by ion bombardment or by placing the entire workpiece in an external furnace. Ions that reach the heated surface thermally diffuse into the interior of the workpiece over periods of time from tens of minutes to several hours, and reach depths up to one or two millimeters. The addition of such thermally diffused atoms can greatly improve the hardness and wear characteristics of ferrous workpieces. References relating to this process include Grube and Gay (1969), Hollahan and Bell (1974), Penfold and Thornton (1975), and Roth (1997).

The region of the classical DC electrical discharge $V-I$ characteristic in which low-energy plasma thermal diffusion treatment is accomplished is shown in figure 17.22. In the *abnormal glow* region from G to H on the voltage–current curve, the plasma completely covers the cathode, exposing the entire cathode surface to a flux of ions.

Low-energy plasma thermal diffusion treatment utilizes apparatus like that illustrated in figure 17.28, taken from Hollahan and Bell (1974). In this example,

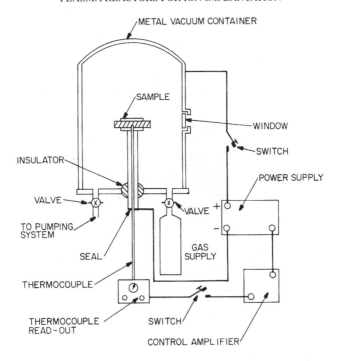

Figure 17.28. A typical system for the low-energy plasma thermal diffusion treatment of metals (Hollahan and Bell 1974).

the workpiece is mounted in the center of an evacuated metal bell jar and is connected as a cathode to a DC power supply. The workpiece is maintained at a potential from 500 V to a few kV. Ions arrive at the surface of the cathode, with a flux determined by the random ion flux. Sometimes the current density of bombarding ions is sufficient to raise the workpiece to temperatures adequate for thermal diffusion of the ionic species into the surface. The apparatus illustrated in figure 17.28 is an example that does not require an external furnace. The temperature of the workpiece is monitored with a thermocouple, and controlled by switching the discharge voltage on and off. In many applications, however, the workpiece must be heated by operating the entire discharge in a furnace, as illustrated in figure 17.29 (Grube and Gay 1978).

Low-energy plasma thermal diffusion treatment has the important advantage that, if it is operated in the abnormal glow discharge regime, the surface of complex workpieces with compound curvature in three dimensions can be covered with plasma. The entire surface can be treated simultaneously to improve its hardness or related properties. This process is used, for example, to harden the complex molds used to fabricate automotive bumper and fender panels.

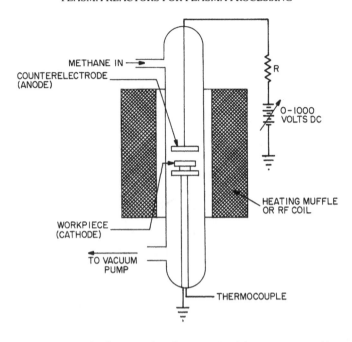

Figure 17.29. A schematic diagram of a plasma carbonizing apparatus with an external heater to maintain the workpiece temperature during treatment.

17.2.2   Reactors for Plasma Ion Implantation

Plasma ion implantation is a recently introduced process that can improve the hardness, wear resistance, and corrosion resistance of metals by implanting energetic ions a few tenths of a micron below their surface. The process consists of inserting the workpiece to be implanted into a glow discharge plasma, as shown in figure 17.30. The workpiece is pulsed to a high negative voltage, V_0, such that it acts like a Langmuir probe in ion saturation, and attracts ions from the plasma. The ions acquire enough energy from the potential drop across the sheath that they are implanted below the surface of the workpiece. Further information on this process may be found in Conrad (1988), Keebler (1990) and Matossian (1994).

Plasma ion implantation is accomplished in apparatus such as that shown schematically in figure 17.31. Characteristically, a large-volume (at least several hundred liters), glow discharge plasma is generated by microwave power, or a DC or RF glow discharge. Recombination losses on the walls may be reduced by magnetizing the entire plasma to keep electrons from reaching the walls, or by surface magnetic confinement with permanent magnets in an array of cusps (see section 3.4 of Volume 1). This improved plasma confinement increases the plasma source efficiency and/or increases its electron number density. Electron

PLASMA ION IMPLANTATION

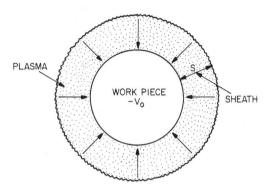

Figure 17.30. Plasma ion implantation of a workpiece by negative biasing and the attraction of ions from the plasma.

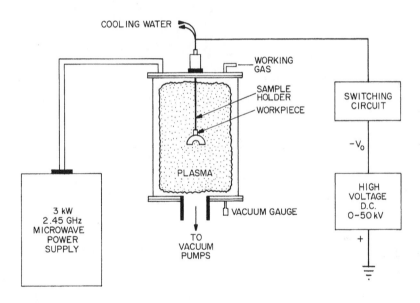

Figure 17.31. A plasma ion implantation apparatus in which the workpiece to be implanted is immersed in a large-volume uniform plasma.

number densities of an implantation plasma characteristically range from 10^{14} to 10^{17} electrons/m^3.

A workpiece is inserted in this plasma, and a pulsed, high negative DC voltage is applied by a fast switching circuit. DC voltages as low as 10 kV have been used to impart corrosion resistance; voltages above 200 kV have been used

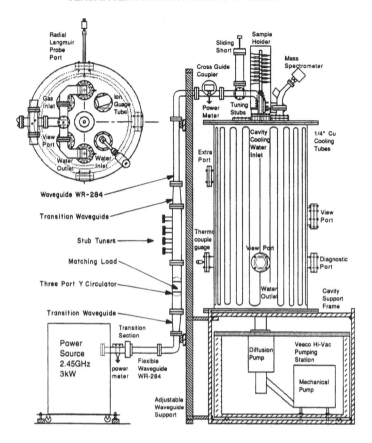

Figure 17.32. An elevation view of the vacuum tank and microwave hardware of the UT Microwave Plasma Facility (MPF).

for implantation and wear applications where deep implantation (up to 1 μm) is desired. Depending on the plasma density and the size of the workpiece, the power supply and the switching circuit must be capable of providing from one to several hundred amperes during an implantation pulse (see Keebler *et al* 1989).

The switching circuit must be capable of limiting the duration of the current pulse, as well as adjusting the pulse repetition frequency. This is necessary in order to control the total dose, to avoid quenching the plasma, and to avoid overheating the surface. If the surface is overheated, the implanted species may thermally diffuse to a depth so great that they would have no useful effect. Pulse rise times are characteristically a microsecond, their durations are 10–30 μs, and their repetition rates are usually below 500 Hz, with 200 Hz being typical.

One example of a plasma ion-implantation system is the UT Microwave Plasma Facility (MPF), illustrated in figure 17.32, and discussed previously in

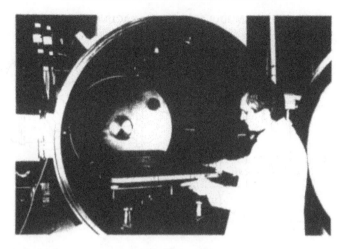

Figure 17.33. Photograph of the Hughes Research Laboratory Plasma Ion Implantation Facility. The aluminum mounting plate for workpieces is 1 m^2, and can be pulsed to voltages of more than 100 kV (Matossian 1994).

section 13.5.3. The MPF is a large-volume, high-density, steady-state, non-resonant microwave plasma reactor intended for plasma ion implantation and other industrial plasma-processing studies (Spence *et al* 1991, Keebler 1990). The MPF consists of a cylindrical stainless steel vacuum tank with its axis vertical, approximately 50 cm in diameter and 1 m high. The outer surface of its cylindrical side wall is fitted with strips of permanent magnets that produce a multipolar cusp at the plasma boundary. The magnetic induction is no more than 30 mT on the inner surface of the vacuum vessel. The multipolar magnetic fields improve the confinement and electron number density of the plasma by a factor of 1.7, but are too weak to lead to non-uniform ECR plasma generation.

The plasma is maintained in the 200 liter volume by 2.45 GHz non-resonant microwave power at levels that typically range between 1 and 2 kW. The vacuum tank is illustrated in elevation in figure 17.32, along with the components of the microwave power supply subsystem. The working gas in this facility is commercial dry nitrogen at pressures ranging from 2.6 to 13.3 Pa (20 to 100 mTorr). A high negative DC voltage is applied to the workpiece in pulses from 5 to 50 μs in length, with a repetition rate of 100 to 2000 Hz (Keebler *et al* 1989). Ions are delivered to the workpiece at each pulse, and the process repeated until the required dose is built up on the workpiece.

A second example is the Hughes Research Center's Plasma Ion Implantation Facility, a photograph of which is shown in figure 17.33 (Matossian 1994). This large vacuum tank, approximately 1.8 m in diameter and 3 m long, is used for studies of the hardness and wear characteristics of plasma ion implanted workpieces having dimensions up to 1 m. The plasma source used in this facility

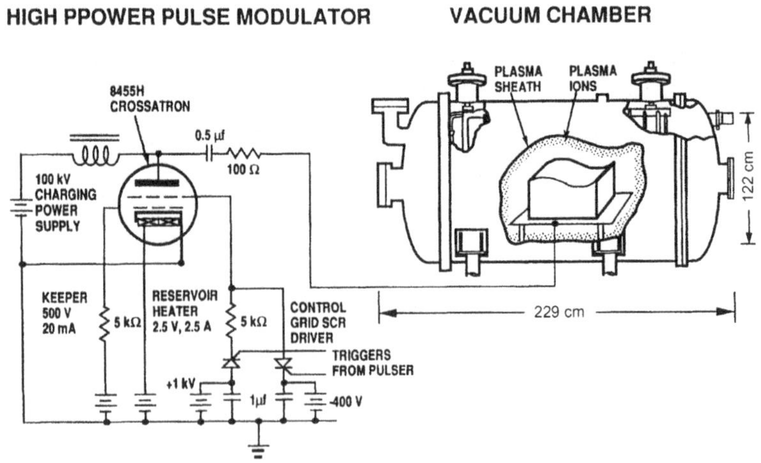

HIGH PPOWER PULSE MODULATOR   **VACUUM CHAMBER**

Figure 17.34. Schematic diagram of the Hughes Research Laboratory's Plasma Ion Implantation Facility (Matossian 1994).

is a large-scale electron bombardment (Kaufman) source. The major components of this facility are shown in figure 17.34. Peak currents up to 100 A can be supplied at voltages up to 200 kV, in pulses with rise times on the order of a microsecond, durations of tens of microseconds, and at repetition rates up to 1 kHz.

17.3   REACTORS FOR ION-BEAM-INDUCED SPUTTER DEPOSITION

Because of the different technologies involved, the sputter deposition of thin films has been divided into two major groups, illustrated schematically in figure 17.35. The first is *ion-beam-induced sputter deposition*, shown in figure 17.35(*a*), in which a beam of energetic ions from an external ion source sputter atoms from a target, which are then deposited on the workpiece. An ambient plasma in the vicinity of the target or workpiece may or may not be present.

The second major group is termed *plasma/cathode sputter deposition*, illustrated in figure 17.35(*b*). In this form of deposition, an energetic ion source and target are not present. Instead, the deposited atoms are produced by sputtering a sacrificial cathode with ions accelerated across the sheath of a glow discharge plasma. The workpiece may either be mounted on the anode, immersed in the plasma, or located external to the plasma opposite the cathode.

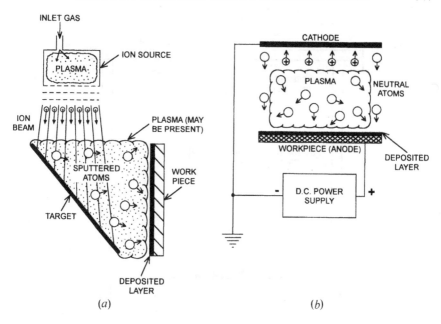

(a)   (b)

Figure 17.35. The two major approaches to the sputter deposition of thin films: (a) *ion-beam-induced sputter deposition*, in which energetic ions from an external ion source sputter atoms from a target, which are then deposited on the workpiece; (b) *plasma/cathode sputter deposition*, in which the deposited atoms originate from cathode sputtering.

17.3.1 Ion-Beam-Induced Sputter Deposition Reactors

In this section, we consider ion-beam-induced sputter deposition. Ion-beam-induced sputtering technologies have many variants, some of which do not use plasma. These technologies have many subsystems in common, however, and they find widespread industrial applications. For the sake of completeness, we discuss both the ion-beam-induced sputter deposition technologies that do not incorporate plasmas, as well as those that do.

17.3.1.1 Primary-Ion-Beam Deposition

In *primary-ion-beam deposition*, illustrated in figure 17.36, the 'primary' ion beam is generated at high vacuum by a space-charge-limited source. The beam consists of ions of the material that is to be deposited on the surface, having energies below the self-sputtering yield of 1.0, which occurs at the energy E_c. At energies below E_c, net deposition of metallic and other materials that are solid at room temperature may occur. In this concept, a source of metal or insulator ions produces a beam that is focused to dimensions smaller than the design rule of the circuit being fabricated. The ion beam is then electrostatically deflected

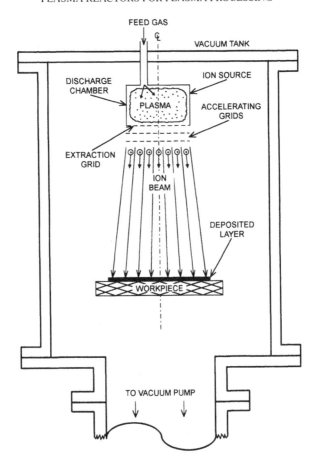

Figure 17.36. Illustration of *primary-ion-beam deposition*, in which the 'primary' ion beam is generated in high vacuum by a space-charge-limited source. Ions of the material to be deposited with energies below the self-sputtering threshold coat the substrate.

(by means discussed in section 5.7.3 of Volume 1) to deposit an electrically conducting circuit or insulating coating on the substrate without masks or etching. This process is illustrated in figure 17.37, and discussed further in section 23.4.5 of this volume.

17.3.1.2 Secondary-Ion-Beam–Target Deposition

The process of *secondary-ion-beam–target deposition* is illustrated in figure 17.38, in which a Child law-limited ion beam operating in the high-vacuum regime (below 0.13 Pa or 1 mTorr) sputters atoms from a target. These sputtered atoms, in a secondary process, deposit a thin film on the workpiece. The essen-

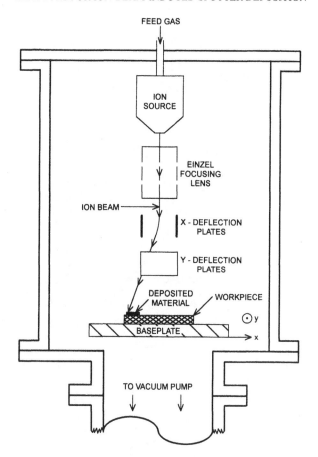

Figure 17.37. Schematic diagram of apparatus for maskless circuit deposition using rastered primary-ion-beam deposition.

tials of this technology are illustrated in figure 17.38, and discussed further in section 23.4.3.

When simple sputtering on unconditioned surfaces like that just described is not adequate to produce a high quality layer, features added to secondary-ion-beam–target deposition chambers to improve film quality may include a rotating workpiece holder and a shutter, illustrated in figure 17.39. This form of conditioning the workpiece surface is discussed further in section 23.4.3

Sometimes it is desired to create a coating of oxides, nitrides, or other chemical compounds with secondary-ion-beam deposition alone, and without the complication of an ambient plasma at the target or workpiece. If the desired heterogeneous chemical reactions take place, this can be done by *ion-beam reactive sputter deposition*, three versions of which are illustrated in figure 17.40.

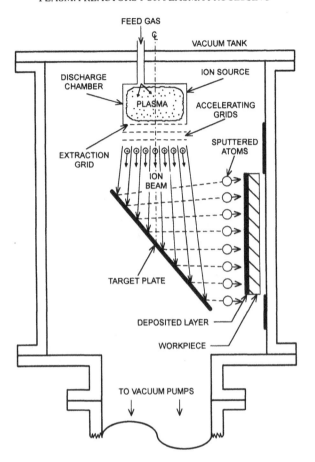

Figure 17.38. A schematic illustration of *secondary-ion-beam–target deposition*, in which a Child law-limited ion beam sputters atoms from a target. These sputtered atoms, in a secondary process, impinge on the workpiece and deposit a thin film.

Figure 17.40(*a*) illustrates the deposition of a chemical compound by reactions between the reactive (with the target) ions and the target material. The ion beam may have an admixture of inert gas ions (for example, argon) to ensure a high sputtering rate of the material ejected from the target surface. Figure 17.40(*b*) illustrates passive target sputtering by an inert ion beam, followed by chemical reactions on the workpiece surface induced by reactive feed gas flowing over its surface. Finally, the two-beam configuration illustrated in figure 17.40(*c*) can be used to induce chemical reactions on a surface by sputtering a layer of target atoms on the workpiece with an inert ion beam, and causing these target atoms to react with reactive ions reaching the workpiece from a second ion source.

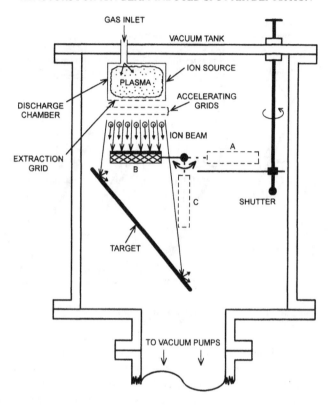

Figure 17.39. An illustration of features added to secondary-ion-beam–target deposition chambers to improve film quality. These may include a rotating workpiece holder and a shutter, to allow decontamination of the surface by ion-beam impingement.

17.3.1.3   Secondary-Ion-Beam Plasma Deposition

In *secondary-ion-beam plasma deposition*, illustrated schematically in figure 17.41, the target and/or the workpiece are surrounded by an ambient glow discharge plasma, characteristically operated at neutral gas pressures of about 0.5 Pa (a few mTorr). The plasma is available to condition and decontaminate the surfaces of the target and workpiece, increase the surface energy of the workpiece to produce a more adherent film, and provide highly oxidizing or reducing plasma active species to react chemically with the target and/or deposited film. The operational characteristics of this configuration are discussed further in section 23.5.1.

17.3.1.4   Operational Issues

Ion-beam-induced sputter deposition makes possible independent control of most inputs to the sputtering process. The high monoenergetic ion energy, ion current

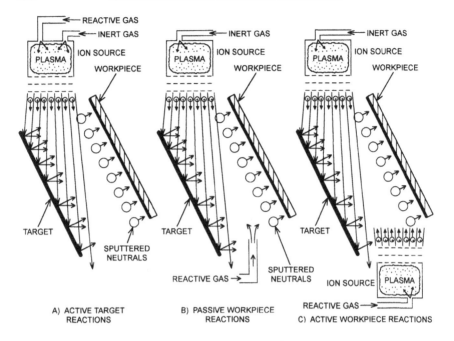

Figure 17.40. A schematic illustration of three versions of *ion-beam reactive sputter deposition*: (*a*) the deposition of a chemical compound by reactions between the reactive (with the target) ions and the target material; (*b*) passive target sputtering by an inert ion beam, followed by chemical reactions on the workpiece surface induced by reactive feed gas flowing over it; and (*c*) induction of active chemical reactions on the workpiece by sputtering a layer of target atoms on the workpiece, and reacting these atoms with reactive ions deposited on the workpiece by a second ion source.

density, angle of incidence on the target, and angle of incidence on the workpiece are all independent variables. The primary ion mass and energy can be adjusted independently of the beam current density (within the Child law current density limit), and hence of deposition rate.

Since deposition rates usually must be maximized for economic reasons, Child-law-limited ion sources of the kinds discussed in chapter 6 of Volume 1 are usually used. These sources operate at ion energies from 500 eV to several keV, above the energies readily attained with plasma cathode sputter deposition, discussed later in section 17.4. Ion current densities on the target may be on the order of 1.0 mA/cm^2, with power densities on the sputtering target ranging from a few tenths to tens of W/cm^2. This heat load on the target must be removed by cooling, which may require an additional subsystem. However, this heat load does not appear on the workpiece, which can be maintained near room temperature. The diameter of the widely used *Kaufman ion sources* may exceed several tens

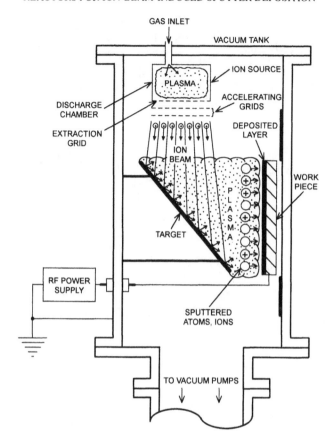

Figure 17.41. A schematic illustration of a *secondary-ion-beam plasma deposition* configuration, in which the target and/or workpiece are surrounded by an ambient glow discharge plasma.

of centimeters; that of some other sources used for the production of ions of non-volatile materials may be restricted to beams only a few millimeters in diameter.

17.3.2 Reactor System for Ion-Beam-Induced Sputtering

An example of a reactor system for ion-beam-induced sputtering is illustrated in figure 17.42. This system can be operated with or without a background plasma in contact with the sputtering target and the workpiece. A background plasma may decontaminate or activate the surface of the workpiece in a way that improves the quality or adhesion of the deposited film.

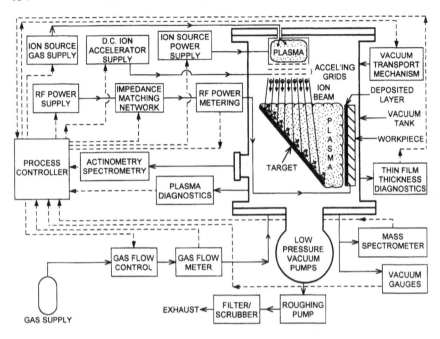

Figure 17.42. A schematic drawing of the subsystems of a plasma-assisted ion-beam sputtering deposition system.

17.3.2.1   Plasma-Assisted Ion-Beam Sputtering

A complete reactor system for plasma-assisted sputter deposition is illustrated in figure 17.42, with its subsystems and the vacuum tank in which deposition takes place. In plasma-assisted ion-beam sputtering, an ion source, shown at the upper left of the vacuum tank in figure 17.42, produces an energetic, unidirectional, space-charge-limited beam of ions, that strikes a *target* or *sputter plate* immersed in a background plasma. The target produces sputtered atoms that accumulate as a deposited layer on the workpiece at the center right of figure 17.42. The target is placed at a small angle to the direction of the ion beam. This practice enhances the sputtering yield, which increases up to about 70° off-normal as the target assumes progressively smaller angles with respect to the direction of the ion beam.

In plasma-assisted ion-beam sputtering, the background plasma serves at least two functions: one of these is to remove adsorbed monolayers and decontaminate the surface of the workpiece. This surface cleaning will increase the surface energy and improve related characteristics that affect the deposited layer. A second function is to allow energetic electrons in the plasma to dissociate or excite the sputtered species while in transit in a way that improves the quality or adhesion of the deposited layer.

17.3.2.2   Vacuum Subsystem

The vacuum subsystem of plasma-assisted sputter deposition systems must be capable of high-vacuum operation. Pressures below 1.3 Pa (10 mTorr) ensure that the mean free path of sputtered atoms is comparable to or longer than the distance from the target to the workpiece on which the sputtered atoms are deposited. Such pressures in the region between the target/cathode and the workpiece are achieved with commercial forepumps in line with turbomolecular pumps and, sometimes, with cryopumps.

The requirement for long mean free paths at low pressures can conflict with the desirability of high neutral atom densities to generate a sufficiently dense background plasma. The latter may require pressures above 1.3 Pa (10 mTorr), especially with DC, obstructed abnormal glow discharges and some RF parallel-plate reactors. At pressures above 1.3 Pa (10 mTorr), the deposited layer can become less adherent or spongy in texture.

The vacuum subsystem of a plasma sputter deposition reactor may, in addition to high vacuum pumps, also include a mass spectrometer, vacuum gauges, an exhaust line scrubber to remove toxic or other unwanted byproducts of the deposition process, and a robotic transport mechanism to insert wafers or workpieces into and remove them from the vacuum system.

17.3.2.3   Gas Supply Subsystem

The plasma-assisted ion-beam sputtering system illustrated in figure 17.42 and the plasma/cathode ion sputtering system illustrated in figure 17.43, to be discussed in section 17.4 of this chapter, have most of their subsystems in common. Both systems normally require only a single input gas to create the background plasma and/or to supply the ion source. Usually argon is used, because it is chemically inert, has high sputtering yields at low energies when it strikes common target materials, and is a low cost, widely available industrial gas. The gas supply subsystem characteristically consists of a supply bottle, a gas flow control valve, a gas flow meter, and a filter to remove particulate contamination from the flowing gas.

In deposition systems having a microprocessor-based process controller, data from a vacuum gauge, the gas flow meter, other sensors, and the programmed operating protocol are input to a microprocessor. The microprocessor sends an effector signal to the gas flow control valve to maintain a desired set point. Inside the vacuum chamber, there may be a sophisticated gas flow distribution system surrounding the workpiece. However, details of the gas flow within the vacuum tank tend to be less an issue with plasma-assisted sputtering deposition systems than they are with etching or PCVD systems.

17.3.2.4   Plasma Reactor Subsystem

Both plasma-assisted ion-beam sputtering systems and the plasma/cathode ion sputtering systems to be discussed in section 17.4 require the production of a background plasma. This is done in several ways, of which the obstructed abnormal DC glow discharge and the RF parallel-plate capacitive plasma reactor are the most widely used. The plasma reactor subsystem consists of an RF or DC power supply, an impedance matching network for an RF power supply, and either RF or DC input power metering. The plasma input power meter characteristically sends a sensor signal to the system controller, which adjusts the power supply settings to achieve an effect specified by the system programming.

17.3.2.5   Ion-Beam Subsystem

In plasma-assisted ion-beam sputtering systems, an ion-beam subsystem is required. The most widely used ion source is based on the Kaufman ion source discussed in section 6.5 of Volume 1. This, or a functionally equivalent ion source, is designed to produce a (usually) space-charge-limited beam of ions with energies of several keV. The ion energy selected is usually near the maximum of the sputtering yield curve for the material of the target or sputter plate.

The ability to adjust the ion energy over a wide range and up to values of tens of keV is a distinct advantage of the plasma-assisted ion-beam sputtering configuration of figure 17.42 over the plasma/cathode ion sputtering system of figure 17.43. In the latter system, ion energies are limited by sheath potential drops to no more than a few hundred eV.

In addition to a Kaufman ion source or its equivalent, an ion-beam subsystem must also include two DC power supplies to operate the ion source and its discharge chamber, and a separate gas supply for the ion source. These additional requirements make a plasma-assisted ion-beam sputtering system more complex than plasma/cathode ion sputtering systems. The latter do not require the additional power supplies, an additional gas supply subsystem, or the extra process control capability to manage parameters associated with the ion-beam subsystem. The target or sputter plate, which serves as a source of the sputtered species which form the deposited layer, is also a part of the ion-beam subsystem. This plate may have to be cooled and must be periodically replaced.

17.3.2.6   Process Controller

A complete plasma-assisted sputter deposition system will have a microprocessor-based process controller, which receives sensor signals from plasma diagnostic instruments, thin-film diagnostic instruments, vacuum gauges, a mass spectrometer, and possibly actinometry or ion energy analyzers. Data from such sensors are input to software designed for endpoint detection and to maximize thin-film deposition rates, while maintaining conditions that will ensure adequate quality in the deposited layer. To accomplish this end, effector signals are sent to other subsystems until the deposition process is complete, as well as to such post-endpoint effectors as robotic manipulators.

17.4   PLASMA/CATHODE SPUTTER DEPOSITION REACTORS

The ion-beam-assisted sputtering technologies discussed in the previous section have a requirement for an ion-source subsystem with its own gas supply, high voltage and discharge chamber power supplies, and high vacuum operation (below about 1.3 Pa, or 10 mTorr). These requirements can be a disadvantage in large-scale industrial applications such as the coating of architectural and automotive glass and brightwork.

The *plasma/cathode sputter deposition* system illustrated in figure 17.43 has been developed for these large-scale applications, and contains the plasma-related elements shown in figure 17.35(*b*). The major subsystems are essentially the same as those of the ion-beam-assisted sputtering system illustrated in figure 17.42, except that the ion-beam subsystem is not present. The ion-beam subsystem is replaced by ions accelerated across the cathode sheath between a glow discharge plasma and a target-cathode. In addition to producing the ions responsible for sputtering, the negative glow plasma in the plasma/cathode configuration of figure 17.43 can serve at least two additional functions. One function is to condition or decontaminate the surface of the anode/workpiece that receives the deposited layer. A second function can be to excite sputtered neutral atoms by inelastic collisions as they transit the negative glow plasma on their way from the cathode to the anode/workpiece.

Except for a few research and minor industrial applications, plasma/cathode sputter deposition is accomplished with either one of two plasma sources; the *parallel-plate* (see section 16.1.1), or the *magnetron* (see section 16.1.2) glow discharge. Applications of these sources to industrial sputtering are discussed in Thornton and Penfold (1978), Vossen and Kern (1978), Vossen and Cuomo (1978), Smith (1995), and Waits (1978, 1997).

17.4.1   Parallel-Plate Configurations

A geometry used for plasma/cathode sputtering is the *parallel-plate configuration*, the DC-energized version of which is illustrated in figure 17.44. The plasma in the DC version is normally operated as an obstructed, abnormal glow discharge. Such an operation ensures that the plasma covers the cathode and operates at a high current density (characteristics of the *abnormal glow discharge*). The workpiece is as close as possible to the cathode-target, while maintaining a high cathode fall voltage that gives ions enough energy to produce a useful sputtering yield (*obstructed normal glow discharge* operation).

In this device, ions created in the *negative glow plasma* are accelerated through the cathode fall voltage and impact on the cathode. The sputtered atoms

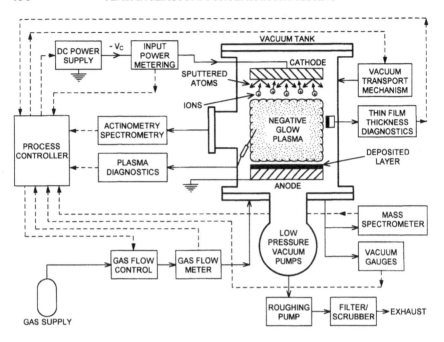

Figure 17.43. A PCVD system based on plasma-assisted sputtering of atoms from the cathode of a DC glow discharge.

of cathode material are available to be deposited on a workpiece located at the anode or at the periphery of the plasma. A significant limitation of this device as a sputtering source arises from two competing requirements. One of these is low operating pressures, below 1.3 Pa (10 mTorr) to ensure long enough mean free paths for the sputtered material atoms to reach the workpiece. The second requirement is dense enough plasmas to ensure the high ion and sputtering fluxes required for commercially acceptable deposition rates.

A related parallel-plate sputtering configuration utilizes the RF glow discharge illustrated in figure 17.45. For sputter deposition of workpieces, the upper electrode is energized by RF. A sufficiently high sheath voltage to cause sputtering of this electrode can be maintained in two ways. One method is by a negative DC bias on the electrode imposed by a separate power supply. A second method is by maintaining a large ratio between the area of the grounded electrode on which the workpieces are mounted, and the energized electrode area (see the discussion in section 16.1.1.5 of this volume). In this configuration, sputtered atoms transit the RF glow discharge plasma and are deposited on workpieces mounted on the opposite grounded electrode. Compared to the DC obstructed abnormal glow discharge, this configuration has the advantage of being able to

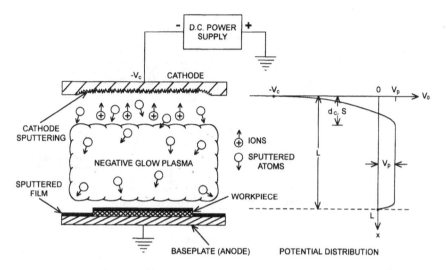

Figure 17.44. The DC-energized version of the *parallel-plate configuration* used for plasma/cathode sputtering. The plasma is normally operated as an obstructed, abnormal glow discharge.

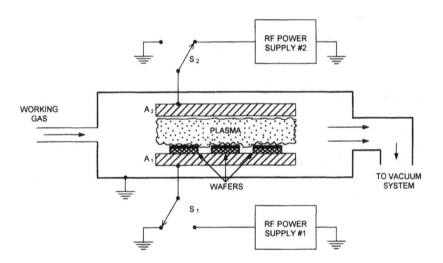

Figure 17.45. Illustration of a parallel-plate RF glow discharge plasma reactor used for plasma deposition. For deposition, the workpieces are usually on the grounded electrode.

operate at higher plasma densities, but the disadvantage of greater difficulty in achieving sheath potentials high enough to produce useful sputtering rates.

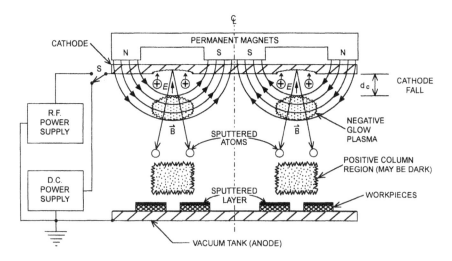

Figure 17.46. Drawing of a parallel-plate magnetron configuration for the deposition of sputtered cathode material on workpieces located on the anode.

17.4.2   Planar Magnetron Configurations

Among the most widely used plasma/cathode sputtering sources are two closely related *planar magnetron sources* previously discussed in section 16.1.2: the *parallel electrode magnetron* source, illustrated in figure 17.46, and the *co-planar magnetron* source, illustrated in figure 17.47. These sources were discussed by Waits in chapter II-4 of Vossen and Kern (1978), and were described in an early patent by Corbani (1975). Here we will use the co-planar magnetron of figure 17.47 as an example of this configuration. On the left are indicated the direction of the electric and magnetic fields that determine the plasma drift motion, and the trapped orbit of an electron associated with the negative glow plasma. The electric field is associated with the cathode fall voltage, and accelerates ions across the cathode dark space to energies high enough to sputter the cathode.

As is illustrated on the right-hand side of figure 17.47, sputtered atoms transit the plasma and are deposited on the workpiece. The working gas (and its ions) is typically argon at pressures from 0.13 to 1.3 Pa (1 to 10 mTorr), low enough to ensure a mean free path comparable to or larger than the cathode–workpiece distance. Characteristic biasing voltages used for architectural glass are illustrated in figure 17.47.

A schematic cross section of a co-planar magnetron used for the sputter deposition of architectural and automotive glass is illustrated in figure 17.48. This dual-track geometry may extend for several meters in the direction normal to the diagram. As was true of the parallel-plate configuration discussed in section 16.1.1, the glow discharge plasma in the planar magnetron can be

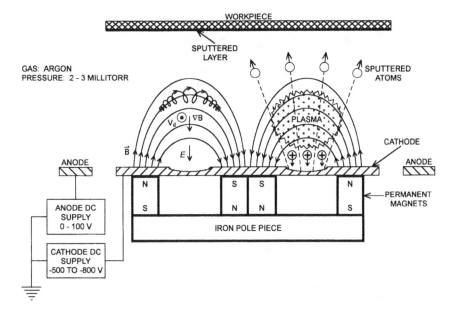

Figure 17.47. Schematic drawing of the co-planar magnetron, in which sputtered material from the cathode is deposited on a workpiece above the magnetron cathode. The workpiece need not be an electrode of the magnetron configuration.

generated by a DC power supply or by RF. The use of RF may require either additional DC bias voltage on the workpiece or large RF electrode area ratios to ensure sufficiently energetic ions to sputter the cathode-target.

Careful quantitative studies of the performance characteristics of industrial magnetron sputtering configurations, such as those described in Vossen and Kern (1978), are relatively few. An investigation of co-planar magnetron sputtering in a geometry similar to that of figure 17.48 was made by Leahy and Kaganowicz (1987). These authors measured the magnetic induction profile perpendicular to the cathode fall electric field, and the deposition rate on a static anode/workpiece parallel to the cathode. Their results are plotted in figure 17.49, in which the approximate position of the cathode, the negative glow plasma, and the permanent magnets are indicated. The magnetic induction in the negative glow plasma is about 10 mT, sufficient to magnetize the electrons, but not the ions. The sputter-deposition rate is highest, about 2.3 nm/s (1380 Å/min), immediately below the negative glow plasma. At such a deposition rate, it would take less than 8 min to deposit a 1 μm thick metallic layer on a workpiece immediately under the sputtering source.

Such a non-uniform deposition profile as that shown in figure 17.49 has little utility, so uniformity in the horizontal direction is ensured by *motional averaging*, discussed in section 18.4 of this volume. It is also necessary to average out

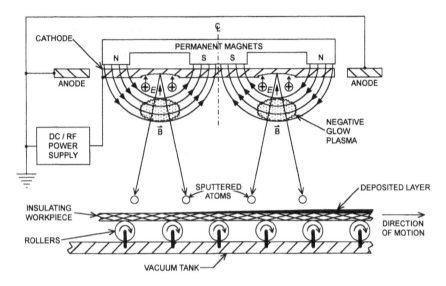

Figure 17.48. A schematic cross section of a co-planar magnetron used for the sputter deposition of architectural and automotive glass.

non-uniformities arising from small geometrical or plasma-related effects in the direction normal to the plane of the diagrams of figures 17.48 and 17.49. Such averaging is accomplished by taking advantage of E/B drift in this direction, and curving the ends of these dual-track geometries into the *racetrack configuration* illustrated in figure 17.50. The crossed electric and magnetic fields cause the magnetized electrons to drift around the racetrack, dragging the ions with them, and ensuring plasma uniformity. The E/B drift velocity discussed in section 5.3 will be at least 100 km/s for the electrons. This velocity leads to a lapping of a 3 m long racetrack plasma in less than 0.1 ms, a time short compared with the dwell time of the workpiece in the vicinity of the deposition plasma.

17.4.3 Axisymmetric Magnetron Configurations

The *axisymmetric magnetron configuration* is variously known as the *S-Gun* (a trademark of Varian Associates), *Sputter-Gun* (a trademark of Sloan Technology), and by other trade names. The S-Gun configuration is illustrated in figure 17.51, and the Sputter-Gun configuration in figure 17.52. Commercial application of this configuration is based on the original patent of Clark (1971). A survey of the early history of this configuration is given by Fraser in chapter II-3 of Vossen and Kern (1978). A third configuration, also proprietary (to Stanford University), is an axisymmetric version of the planar magnetron configuration illustrated in figure 17.48. All of these configurations produce an approximate point source of sputtered atoms that deposit a coating on a static workpiece up to about 5 cm in

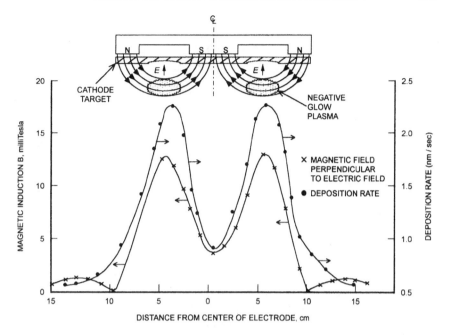

Figure 17.49. Results of an investigation of co-planar magnetron sputtering in a geometry similar to that of figure 17.48 by Leahy and Kaganowicz (1987). These authors measured the magnetic induction profile perpendicular to the cathode fall electric field, and the deposition rate on a static anode/workpiece.

diameter, and with a uniformity of thickness of a few percent over that diameter.

On figure 17.53 are data on the uniformity of a copper layer deposited by a commercial axisymmetric planar magnetron source (US Inc. 1989). The relative thickness of the layer deposited on an extended workpiece by a 5 cm diameter copper cathode target is shown for three spacings between the target and the workpiece. The uniformity quoted was based on the ratio of the layer thickness on the axis and at the edge of the target. As the workpiece approaches the target, the deposition rate increases, but the coverage becomes less uniform. On figure 17.54, the deposition rate on the axis is plotted as a function of the target/workpiece separation, d. In spite of the relatively complex geometry of this axisymmetric magnetron, an inverse square dependence emerges, as though the extended target were a point source of sputtered atoms. The commercially significant deposition rate of 1.67 nm/s (1000 Å/min) is indicated on the ordinate.

Like most glow discharge sputtering sources, axisymmetric magnetrons have a deposition rate that depends on background pressure. This rate is the result of a competition between low plasma density and the resulting low ion fluxes at low pressures, and the scattering at high pressures of sputtered atoms before they reach the workpiece. This effect is illustrated by data from Fraser (1978) taken from a

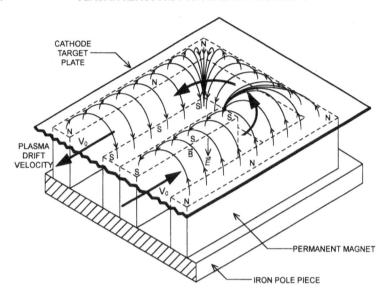

Figure 17.50. Schematic illustration of how averaging potential plasma non-uniformities is accomplished by taking advantage of E/B drift, and curving the ends of these dual-track geometries into a racetrack configuration. The crossed electric and magnetic fields cause the magnetized electrons to drift around the racetrack, ensuring plasma uniformity.

2.5 cm diameter S-Gun configuration with an aluminum cathode, that is plotted in figure 17.55. The peaking of the deposition rate at 0.4 Pa (3 mTorr) represents the outcome of these two competing effects.

17.4.4 Cylindrical Magnetron Configurations

Another magnetron sputtering technology developed since the early 1970s is the cylindrical magnetron described by Penfold and Thornton (1975), and discussed further in chapter II-2 of Vossen and Kern (1978). The *cylindrical magnetron* has two principal configurations, each with two variants. Figure 17.56 illustrates the *cylindrical post magnetron*, with the *normal* variant on the left that is used to deposit a layer on the inside of an axisymmetric tube (or on an array of workpieces arrayed circumferentially around it). The *inverted* variant on the right is useful for depositing a layer on the outside of an axisymmetric tube or workpiece. The cylindrical post magnetron may generate the axial magnetic field with a permanent magnet, or an electromagnet which allows the magnetic field to be adjusted.

Another widely used cylindrical magnetron configuration is the *cylindrical multipolar magnetron* illustrated in figure 17.57, with the *normal* variant on the left and the *inverted* variant on the right. These treat the same kinds of workpiece

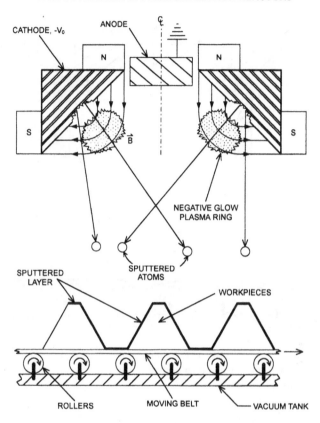

Figure 17.51. The S-Gun^{TM} configuration, an axisymmetric commercial variation of the sputtering magnetron for coating workpieces with sputtered cathode material.

as their corresponding variant of the cylindrical post magnetron. The multipolar cusp magnetic fields trap the negative glow plasma locally, like a smaller version of the planar magnetron, and their magnetic induction may be provided by permanent magnets. Axisymmetric deposition is promoted by E/B drift in the perpendicular radial electric and axial magnetic fields. Uniformity of deposition along the axis is achieved by movement of the cylindrical magnetron structure along the axis of the workpiece.

Cylindrical magnetrons have been extensively studied, and several interesting and puzzling relationships have been observed that require theoretical interpretation and identification of the physical processes responsible. One example is the current–voltage curve reported by Thornton and Penfold (1978) for cylindrical post magnetrons, discussed in section 9.6.3 of Volume 1, and reproduced in figure 17.58. These data show a power law dependence of the current density on voltage, over four orders of magnitude in current density. This

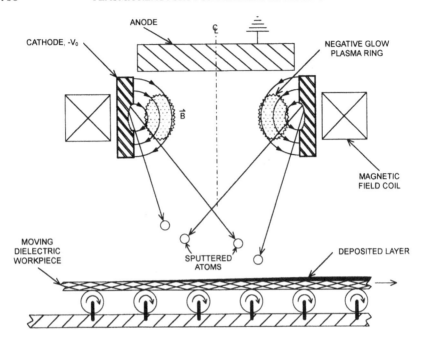

Figure 17.52. Schematic illustration of the Sputter Gun^{TM} configuration. A negative glow plasma ring produces ions that sputter cathode material on a workpiece moving by the sputter gun magnetron.

relationship holds for a wide range of sizes of normal and inverted geometries and is more characteristic of normal glow than abnormal glow operation.

Another significant observation reported for cylindrical post magnetrons, which may be of wider application, is the narrowing of the angular distribution of sputtered atoms with increasing pressure. If the sputtered atoms arrive at an angle θ measured from the normal to the surface, the flux of atoms as a function of the arrival angle θ from a cathode-target at low pressures can be written as

$$\Gamma = \Gamma_0 \cos\theta \qquad \text{atoms/m}^2\text{-s.} \qquad (17.1)$$

Equation (17.1) represents an isotropic distribution, and applies to pressures low enough that the sputtered atoms are collisionless between the cathode-target and the workpiece. However, as the pressure increases above about 0.4 Pa (3 mTorr) of argon (the same pressure at which the deposition rate of figure 17.55 reaches a maximum before scattering effects dominate), Eisenmenger-Sittner *et al* (1995) find that the angular distribution of copper atoms sputtered by argon ions has the form

$$\Gamma = \Gamma_0 \cos^n\theta \qquad \text{atoms/m}^2\text{-s.} \qquad (17.2)$$

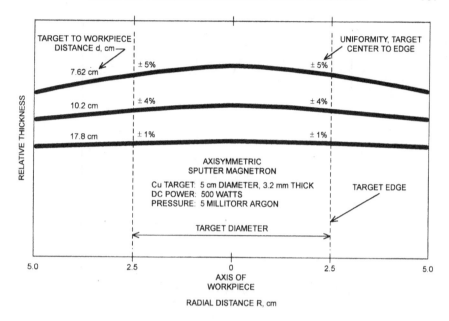

Figure 17.53. Data on the uniformity of a copper layer deposited by a commercial axisymmetric planar magnetron source (US Inc. 1989). The relative thickness of the layer is shown for three spacings between the target and the workpiece.

The exponent n is the phenomenologically determined function of pressure plotted in figure 17.59, and rises with argon pressure in the atom scattering regime. The narrowing of the atom emission to a more beamlike character with increased pressure is illustrated in figure 17.60, which shows polar plots of equation (17.2) for $n = 1.0$ (isotropic emission), 1.5, 2.0, and 3.0. This behavior is qualitatively consistent with the progressive loss at higher pressures of atoms emitted with shallow angles (that is, $\theta \approx 90°$). Such atoms undergo multiple scattering collisions, while the atoms emitted normal to the surface travel directly across the cathode–workpiece gap and are less likely to be lost by scattering.

17.5 REACTORS FOR PLASMA CHEMICAL VAPOR DEPOSITION

In the plasma-assisted sputtering deposition systems discussed in the previous two sections, the deposited layer is built up by the accumulation of atoms sputtered from a target by energetic ions. Another important thin-film deposition method is plasma chemical vapor deposition (PCVD), in which a film is deposited by polymerization of monomers produced in the plasma, or by other heterogeneous chemical reactions on the surface being coated.

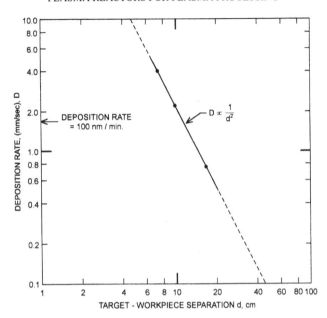

Figure 17.54. The deposition rate on the axis from figure 17.53 is plotted as a function of the target/workpiece separation, d.

17.5.1   Non-Polymeric PCVD Reactor Systems

PCVD systems can be divided into two broad areas of application. The first of these is the deposition of non-polymeric thin films that result from surface chemical reactions that do not result from polymerization. These non-polymeric PCVD coatings are frequently oxides that are used as insulating layers, optical coatings, and oxygen barriers for food packaging.

The second and largest area of application is the deposition of polymeric thin films formed by the polymerization of monomers generated as active species in the plasma. Such films are useful as insulating layers in microelectronic chips, as protective coatings, and as reflective or anti-reflective optical coatings. Plasma reactors for this area of application are discussed further in Hollahan and Rosler (1978), Yasuda (1978, 1985), Mort and Jansen (1986), Claude *et al* (1987), d'Agostino (1990), and Danilich and Marchant (1994). The non-polymeric PCVD reactor systems will be considered first.

17.5.1.1   *Workpiece Exposure Configurations*

PCVD systems have two principal reactor configurations and a variety of minor variants. The first configuration is the *external plasma-activated PCVD system*, illustrated in figure 17.61. In such reactors, the feed gases are provided by a gas

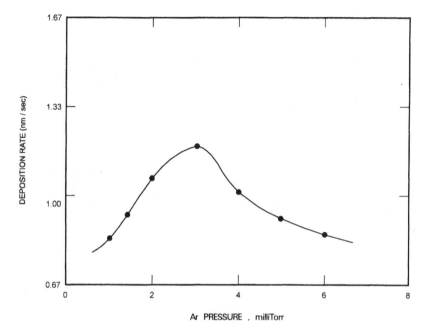

Figure 17.55. Data from Fraser (1978) on the deposition rate of aluminum as a function of argon pressure taken with a 2.5 cm diameter S-Gun configuration with an aluminum cathode.

supply subsystem in which the flow of each feed gas is separately metered and controlled. The feed gases flow through an external plasma, shown schematically in figure 17.61, generated in the delivery tube between the gas flow meter and the reactor in which deposition takes place. The external plasma may, for example, generate active species that convect with the gas flow to deposit protective coatings on the surface of workpieces located within the reactor. As illustrated in figure 17.61, such workpieces may be located on a rotating turntable as a form of motional averaging to promote uniformity of deposition.

A second principal PCVD configuration is the *plasma immersion PCVD system* (or *in situ system*) illustrated in figure 17.62. In this configuration, the feed gas that forms active species/precursors for the deposited layer flows from the gas supply subsystem into the vacuum vessel, and undergoes plasma–chemical reactions in a glow discharge. The glow discharge is most frequently generated either by an electron cyclotron resonance (ECR) source, or by a 13.56 MHz RF capacitive plasma source. In plasma immersion PCVD systems, the plasma is in direct contact with the workpiece undergoing deposition.

The subsystems required for PCVD systems are very similar to those required for plasma-assisted sputter deposition. A significant difference, however, is that PCVD systems have fewer components (no target plate or sacrificial

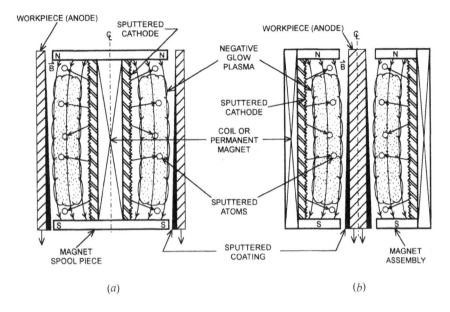

Figure 17.56. The cylindrical post magnetron with axial magnetic field generated by current-carrying coils or permanent magnets: (*a*) normal and (*b*) inverted cylindrical post magnetrons used to deposit a thin film on (*a*) the inside or (*b*) the outside of a workpiece.

cathode, for example), and offer greater ease of operation and control than the plasma/cathode sputter deposition systems discussed in section 17.4.

17.5.1.2 Vacuum Subsystem

The *vacuum subsystem* of PCVD reactors operates at higher pressure than that of plasma-assisted sputter deposition systems, normally in the range from 13.3 Pa (100 mTorr) to 133 Pa (1 Torr). These pressures usually do not require the free molecular flow, high vacuum pumping technology used in the plasma-assisted sputter deposition systems discussed in sections 17.3 and 17.4. When appropriate, the vacuum pumps are backed up by a scrubber to remove toxic or other chemically active species from the gas flow before venting to the atmosphere. The vacuum subsystem of PCVD reactors may include a robotic transport mechanism to install the workpieces before deposition, and remove them afterward.

17.5.1.3 Gas Supply Subsystem

The *gas supply subsystem* required by PCVD reactors is more complex than that of plasma-assisted sputter deposition reactors, since more than one feed gas is normally required to produce precursors for the deposited layer. The input gases (three are shown in figures 17.61 and 17.62) each require their own provision for

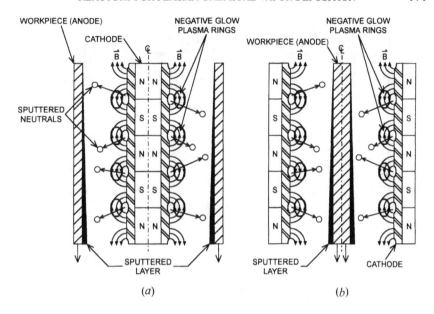

Figure 17.57. The multipolar cylindrical magnetron with axial multipolar cusp magnetic field generated by permanent magnets: (*a*) normal and (*b*) inverted multipolar cylindrical magnetrons, used to deposit a thin film on (*a*) the inside or (*b*) the outside of a workpiece.

flow control and flow monitoring. In some cases, sophisticated programming of the system controller is needed to maximize the deposition rate, while maintaining the quality of the deposited layer throughout the deposition process. In addition to control and metering the gas flow, the gas supply subsystem may have a filter to remove particulates from the input gas flow. Once the feed gases enter the vacuum chamber, they may require a carefully engineered flow distribution system to produce uniform deposition of a high quality thin film on the workpiece, while minimizing the formation of dust or oils.

17.5.1.4   Plasma Reactor Subsystem

PCVD systems require a *plasma reactor subsystem* to produce the glow discharge plasma that converts the feed gases into precursors of a deposited layer on the workpiece. Although figure 17.61 shows an inductive plasma source in the input gas duct of the system, and figure 17.62 shows a parallel-plate RF capacitive plasma reactor, other plasma sources can be and are widely used.

In RF plasma reactors, an impedance matching network is normally used to couple the maximum possible amount of RF power to the variable plasma load, while minimizing the circulating, reactive power. The impedance matching network includes bi-directional power metering to monitor both the total power flowing to the plasma, and the reflected power caused by an impedance mismatch.

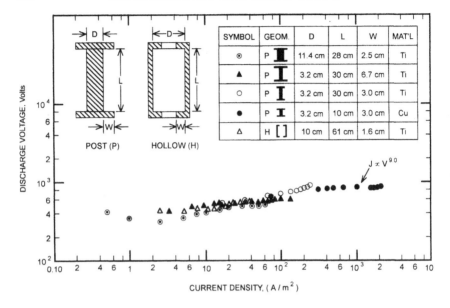

Figure 17.58. A current–voltage curve reported by Thornton and Penfold (1978) for cylindrical post magnetrons. These data show a striking power law dependence of the current density on voltage.

The process controller is usually programmed to maximize the net power flowing to the plasma by adjusting the impedance matching network and/or the independent variables that determine the plasma parameters.

17.5.1.5   Workpiece Mounting Subsystem

The *workpiece mounting subsystem* is usually more sophisticated in PCVD reactors than in plasma-assisted sputter deposition systems. In order to produce uniformity of effect over the surface of single workpieces and collectively over the surfaces of multiple workpieces exposed simultaneously, the gas flow distribution over the workpieces has to be carefully designed. Some approaches to handling this gas flow will be discussed in section 18.5 of this volume.

In some deposition processes the workpiece must be maintained at several hundred degrees centigrade to achieve the desired results. Such deposition applications require a heater in the baseplate (or '*chuck*') on which the workpiece is mounted. This heater may be operated by the process controller, and the workpiece temperature measured by a thermocouple or other sensor. Frequently, motional averaging is used in order to produce a uniform effect on all workpieces in a reactor. This is achieved, for example, by rotating the workpieces on turntables within the vacuum system in the manner illustrated in figure 17.61.

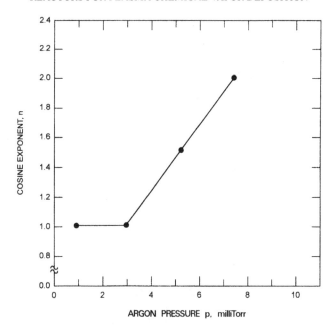

Figure 17.59. Data reported by Eisenmenger-Sittner *et al* (1995) for the angular distribution of copper atoms sputtered by argon ions. The exponent *n* from equation (17.2) is a function of the argon gas pressure.

17.5.1.6 Process Controller

Finally, the *process controller* performs most of the same functions as its counterpart in plasma-assisted sputter deposition systems. The major differences are that in PCVD reactors, no ion source is required, but the monitoring and control of more than one input gas, and measurement and control of the workpiece temperature are additional parameters to be monitored.

17.5.2 Polymeric PCVD Reactor Systems

The characteristics of intermediate-pressure reactors specialized for non-polymeric PCVD have just been discussed in section 17.5.1. These reactors have many features in common with reactors used to deposit polymeric films. These common features normally include operation at intermediate pressures below 1.33 kPa (10 Torr), batch processing of workpieces, RF glow discharge plasma sources (usually at 13.56 MHz), a gas flow control subsystem capable of metering several gases, a vacuum system designed for quick turnaround between atmospheric and operating pressure, a robotic handling subsystem to insert and remove workpieces, sufficient plasma and plasma–chemical diagnostics for process monitoring, and a process controller.

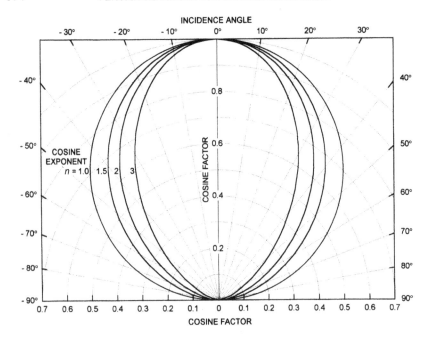

Figure 17.60. The narrowing of the atom emission described by equation (17.2) to a more beamlike character with increased pressure is illustrated by polar plots of equation (17.2) for $n = 1.0$ (isotropic emission), 1.5, 2.0, and 3.0.

Polymeric PCVD reactors are most likely to differ from other intermediate-pressure plasma reactors in their mode of workpiece exposure, their mode of workpiece handling, and whether and what type of motional averaging is used. In the next section, we will not discuss those features in common any further, but concentrate on those that differ for PCVD reactors.

17.5.2.1 Workpiece Exposure Configurations

As in the non-polymeric PCVD reactors previously discussed, two principal modes of workpiece exposure are widely used in polymeric PCVD reactors. The first of these is the *remote exposure* configuration, illustrated in figure 17.61, with the workpiece exposed to active species at a distance from the plasma-generating region. The second mode of exposure is the *in situ* configuration, illustrated in figure 17.62, with the workpiece 'in place', and in direct contact with the plasma.

The choice between these configurations can have important effects on the chemical and morphological nature of the deposited polymers. Films deposited by remote exposure are more likely to consist of conventional linear polymers, while polymeric films deposited *in situ* exhibit much more cross-linking. Such cross-linking leads to greater strength and to other bulk physical properties useful in

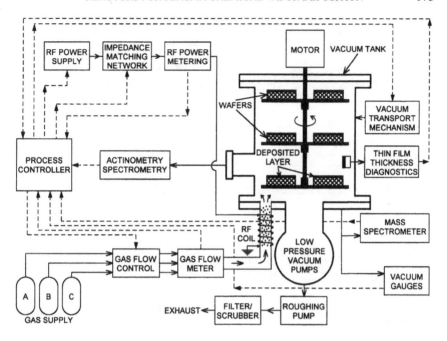

Figure 17.61. A PCVD system that utilizes monomers or active species generated outside the deposition chamber.

many applications, including membranes. Key issues in determining the mode of workpiece exposure are whether the application envisioned requires cross-linking, and whether the polymer-forming species have a long enough lifetime to convect from the plasma and arrive at the workpiece.

17.5.2.2 Mode of Workpiece Handling

One factor that determines the mode of workpiece handling is that PCVD takes place in vacuum systems that enforce batch processing. In such systems, it is necessary to quickly and efficiently put (usually numerous) workpieces in place, and remove them after PCVD for further processing. Other factors include the value and quantity of the workpieces and the competitiveness of the market, with higher throughputs and more competitive markets justifying a high degree of automation.

The least sophisticated mode of batch processing of workpieces for PCVD is that in which numerous workpieces are installed or removed by hand from a deposition reactor such as those illustrated in figures 17.61 and 17.62. High value workpieces or those with high throughputs, such as microelectronic wafers or circuits, are normally stacked in cassettes and handled robotically. Some PCVD reactors come close to realizing the functional ideal of continuous processing.

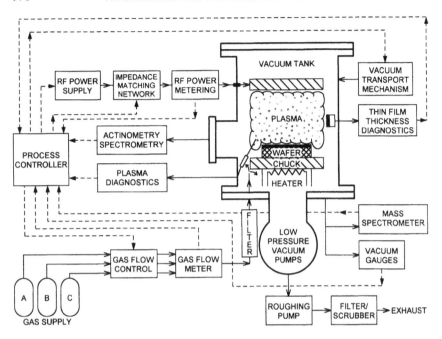

Figure 17.62. Illustration of a PCVD reactor based on immersion of the workpiece in a glow discharge plasma, that provides the required active species.

This may be done either by insertion and removal of workpieces through airlocks or sliding seals while PCVD occurs continuously under vacuum, or by putting large rolls of webbing or film (some up to 30 km in length) into a vacuum system for roll-to-roll PCVD vacuum coating.

17.5.2.3 Mode of Motional Averaging

Maintaining uniformity of effect over the entire surface of a web or workpiece subject to polymeric PCVD can be difficult. For some applications (such as barrier coatings or coatings designed to change the surface energy), uniformity of thickness is not essential to the function of a thin film. In other applications, such as microelectronic circuits or optical applications, uniformity is essential. When uniformity is a secondary consideration, the workpiece can remain static with respect to the plasma for *in situ* exposure, or with respect to the convected active species for remote exposure.

When uniformity is essential, the simplest way to achieve it is *motional averaging*. The workpiece may be moved through the (non-uniform) plasma during *in situ* deposition, or through the flow of active species downstream of a plasma during remote exposure. Various methods for motional averaging involving a moving workpiece are discussed in section 18.4 of this volume. If a workpiece is exposed *in situ*, uniformity of effect can be enhanced by moving the plasma with respect to the workpiece. Such movement can result from drift motions induced by crossed electric and magnetic fields, by magnetic field gradients, or by moving magnetic fields.

17.5.2.4   Exposure Environment

The deposition rate for PCVD can be affected by several factors. These factors include the workpiece location relative to the working gas inlet and plasma, the gas flow pattern through the plasma and around the workpiece, and whether or not the workpiece is heated.

The effects of electrode and gas flow geometry have been extensively investigated by Yasuda (1978, 1985 section 8.3). He investigated the polymer deposition rate in a tubular reactor as a function of axial location with respect to the monomer inlet, starting material flow rate, concentration of inert carrier gas, and other factors. Yasuda found that the lower the operating pressure was, the longer in the direction of flow is the axial region of polymeric deposition. He also found that the higher the chemical reactivity for polymer formation of the feed gases, the narrower is the axial region of deposition. He further found that the addition of an inert gas such as argon will narrow the axial region of deposition.

Perhaps surprisingly, in the polymerization of ethylene the deposition rate downstream of the inlet is higher when the polymerizable starting material is injected from a sidearm of the tubular reactor, than when it flows directly through the plasma. Data on this phenomenon from Yasuda (1978, p 389) are reproduced in figure 17.63, and indicate that exposure of monomers to energetic electrons in the plasma is not a requirement for deposition in this case.

Another factor affecting the deposition rate is the workpiece temperature, which may be varied by resistive heaters or a liquid from a temperature-controlled bath in the baseplate. The deposition rate of many polymers is an exponentially decreasing function of temperature. Temperatures above 100 °C tend to volatilize many deposited polymers and are therefore of limited interest during deposition. However, the stability of the dielectric characteristics of some polymer coatings can be improved by annealing the workpiece after deposition at temperatures above 100 °C.

17.5.2.5   Plasma Sources for Polymeric PCVD

The plasma sources used for polymeric deposition are among those discussed in section 16.1 of this volume. Note, however, that not all plasma sources discussed in that section are suitable for or are used in industrial polymeric PCVD. Some sources are not suitable because the objective of depositing insulating coatings by PCVD is inconsistent with the use of DC glow discharge plasma sources that require the flow of real currents. Such coatings may cover not only workpieces, but also other surfaces in the vacuum system, including electrodes that must draw

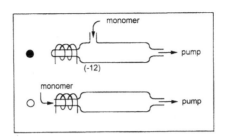

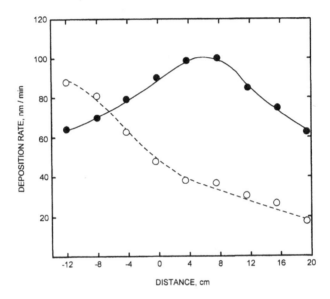

Figure 17.63. Data from Yasuda (1978, p 389) on the effect of feed gas injection location on the polymerization of ethylene. The deposition rate downstream of the inlet is plotted as a function of axial position, and is higher when the polymerizable starting material is injected from a sidearm of the tubular reactor than when it flows directly through the plasma.

real currents from the plasma. The tendency of polymeric PCVD operations to produce coatings that prevent the flow of real currents has led to the widespread use for this application of so-called *'electrodeless' RF reactors* that operate on displacement rather than real currents.

When DC glow discharges *are* used for polymeric PCVD, the electrodes, but not the workpieces, may be kept free of an insulating coating by selectively heating either the workpiece or electrode, by suitably arranging the flow of feed gases, or by arranging for remote exposure, rather than *in situ* deposition. When *in situ* polymeric PCVD is conducted in a DC glow discharge, the workpieces may be mounted on the cathodes to promote cross-linking and other unconventional forms of polymer formation induced by bombardment of ions and other active species.

17.6   PLASMA ETCHING REACTORS

In this section we discuss the reactor systems required for plasma etching of microelectronic circuits. These systems have recently undergone a major evolutionary advance in their technology, as competitive pressures have reduced the design rule of microelectronic circuits below 0.5 μm. Older plasma etching systems intended for operation with design rules larger than 0.5 μm characteristically use RF parallel-plate glow discharge plasma sources operating at neutral gas pressures ranging from 6.7 Pa (50 mTorr) to 133 Pa (1 Torr). Etching design rules below 0.5 μm, however, require plasma sources capable of operating at pressures below 1.3 Pa (10 mTorr), while producing plasmas with higher electron number densities. These plasma sources include ECR microwave sources at 2.45 GHz, and planar helical coil inductive plasma reactors operating at 13.56 MHz. Further information on reactors for plasma etching may be found in Manos and Flamm (1989), d'Agostino (1990), Lieberman and Lichtenberg (1994), and Roth (1995).

17.6.1   Plasma Etching Reactor Systems ('Cluster Tools')

The fabrication of a monolithic microelectronic chip requires four fundamental operations, which are carried out on a single or a few *'cluster' tools*. These four operations include: (1) *deposition of thin films*, covered in chapters 23 and 24 of this volume; (2) *masking* of the layer to be etched, which is done by microlithographic techniques beyond the scope of this text; (3) *etching* of the pattern on the mask into the layer to be etched, covered in this section and in chapter 25 of this volume; and (4) *stripping* away the mask material after pattern transfer.

All but masking technology utilize or require plasmas or plasma-related methods. All four operations can be done in a single vacuum system (the *'cluster tool'*), in which wafers are robotically transferred from one station to the next for sequential operations necessary to build up the layers needed for a microelectronic chip. The etching systems discussed in this section characteristically are a major portion of such a cluster tool.

17.6.1.1   Clean Room Environment

The cluster tools in which microelectronic circuit fabrication is carried out are normally operated in a *clean room*, to reduce the likelihood of contamination of the wafers by the dust and chemical contaminants found in a typical industrial environment. The facilities required to remove submicron dust and contaminants in clean rooms used for microelectronic circuit fabrication are expensive. As a result, floor space in clean rooms is some of the most expensive manufacturing floor space in all of industry. In order to reduce the exposure of partially completed wafers to the atmosphere, the cluster tools are compact, and each tool carries individual wafers through as many sequential fabrication steps under vacuum as possible.

17.6.1.2 Types of Etching Operation

At least four kinds of etching operation may be conducted during the course of microelectronic circuit fabrication. These operations include a *metal etch*, required to produce electrically conducting connections between components; an *oxide etch*, which is required to produce holes for electrical connections through insulating oxide layers; a *silicon trench etch*, for capacitors or other components, or isolation of components; and finally, etching is required for the *etchback of intermetallic dielectric connections* by way of high aspect ratio holes (or *vias*) between conductors on opposite sides of a dielectric. The connections between electrical conductors on parallel, separated layers are built up by a series of deposition and etchback steps that also require plasma etching.

17.6.1.3 Plasma Pressure Regimes

Some characteristics of these four types of microelectronic etching are summarized in table 17.1. These etching operations may be conducted at either intermediate or low pressure, in order to meet the requirements of design rules greater than 0.5 μm (intermediate pressure) or the more demanding requirements of design rules below 0.5 μm (low pressure). Table 17.1 also lists some examples of the material etched, etching gases used for those materials, characteristic values of the flow rate of the etching gases used, and the base pressure commonly achieved in the vacuum systems used.

17.6.1.4 Wafer Handling and Size

The basic unit of microelectronic circuit fabrication is the individual circular wafer, cut from a single crystal silicon ingot. This circumstance, with the requirement for vacuum processing, prevents the microelectronic industry from fabricating its product in a continuous production process, and enforces batch processing of individual wafers. In the early history of microelectronic technology, it was common to process many wafers 7.5 or 10 cm in diameter in a single reactor.

The evolution of the microelectronic industry, however, has been in the direction of processing single, large diameter wafers containing more individual microelectronic chips than can be fitted onto multiple smaller wafers subject to simultaneous processing. The processing of a large number of individual chips

Table 17.1. Characteristics of four types of etching which are important in microelectronic circuit fabrication.

Type of etch	Pressure regime	Size scale (μm)	Typical material etched	Typical etching gas	Flow rate (SCCM)	Typical operating pressure (mTorr)
Metal etch	Intermediate	> 0.5	Al, Cu	$BCl_3 + Cl_2$	~100	100
	Low	< 0.5	Al, Cu	$BCl_3 + Cl_2$	~100	≤ 10
Oxide etch	Intermediate	> 0.5	SiO_2	$F_2 + C_x, O_x$	10–100	100–500
	Low	< 0.5	SiO_2	$F_2 + C_x, O_x$	10–100	5–10
Silicon trench etch	Intermediate	> 0.5	Si, poly-Si	$Cl_2 + HBr$	~100	500
	Low	< 0.5	Si, poly-Si	$Cl_2 + HBr$	~100	10
Etchback of intermetallic dielectric	Intermediate	—	Si, poly-Si	Ar^+ sputter, CF_4	100–1000	100–500

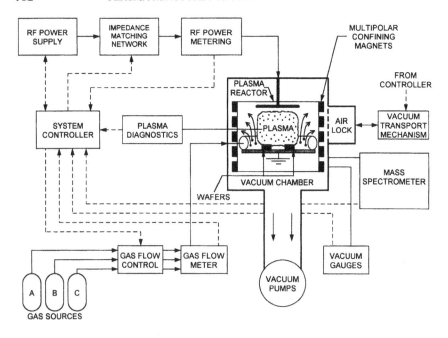

Figure 17.64. The subsystems of a plasma etching reactor.

on a single wafer facilitates its robotic handling in cassettes between individual fabrication steps on a cluster tool. Evolutionary trends within the microelectronic industry have been in the direction of substantially increasing wafer diameters, from 7.5 or 10 cm several decades ago to 20 and 30 cm in diameter at the time of writing.

17.6.2   Plasma Etching Reactor Subsystems

Plasma etching systems are normally part of the cluster tools in which the four fundamental operations of microelectronic circuit fabrication are carried out (deposition, masking, etching, and stripping). However, the *plasma etching system* can be considered separately, and has the components illustrated in the block diagram of figure 17.64.

The plasma etching reactor shown within the heavy boundary will characteristically have four major subsystems. These include the *gas supply subsystem*, shown to the lower left; the *vacuum subsystem*, shown below and to the lower right of the etching reactor; and the *plasma reactor subsystem*, shown to the upper left and in the interior of the plasma etching reactor. The etching reactor will normally also have a *process control subsystem*, shown to the middle left, with its sensors and effectors linked as indicated by the dotted lines.

Plasma etching reactors like those illustrated in figure 17.64 may cost between several hundred thousand dollars and several million dollars each, depending on the application. The high cost of such units results in part because of a requirement for chemically inert materials in the vacuum chamber and vacuum system to resist common etching gases. Such units also are costly because of a requirement for robotic manipulation of wafers, and on-line process control of the etching process.

17.6.2.1   Gas Supply Subsystem

The *gas supply subsystem* shown at the lower left-hand side of figure 17.64 provides the gas or mixture of gases needed to effect a given etching operation, and controls the pressure of the working gas in the reactor. The *background pressure* is a control variable for the electron collision frequency and mean free paths of the ions, electrons, and active species responsible for etching.

The gas supply subsystem must be carefully designed and operated, since many gases used in microelectronic circuit fabrication are flammable, explosive, toxic to humans, or otherwise hazardous. A common practice is to keep gas bottles, shown as A, B, and C at the lower left-hand side of figure 17.64, all supply tubing, and flow control and metering instrumentation in continuously vented enclosures or ductwork to prevent the accumulation of dangerous gases. The gas sources A, B, and C, may include such inert gases as argon or helium for flushing the system before or after an etching operation; etching gases such as chlorine or fluorine; or other gases such as silane (SiH_4).

Gases from pressurized supply bottles are usually controlled by motorized high-precision vacuum valves, and the resulting flow rates monitored by flowmeters for each gas. After the control and metering of each gas used in the etching operation, the gas flow is mixed, and allowed to flow into the vacuum system. The flow distribution of the etching gas around the wafers may have an important effect on the ultimate outcome of the etching process. If the flow of etching gas is not uniform, significant variations in the flux of active species can occur across the diameter of a wafer, or among several wafers etched at the same time.

The geometry of the gas flow over the wafers can affect the formation of unwanted dust particles in the plasma and their deposition on the surface of the wafers. The formation, and especially the deposition, of such dust particles is to be avoided. The patent and journal literature contain a great many approaches to the problem of distributing etching gas over the surface of wafers in order to produce a uniform effect. In the microelectronic industry, a uniformity of four to six percent in the etching rates and etching depth across the diameter of a wafer is normally considered acceptable.

17.6.2.2   Vacuum Subsystem

The *vacuum subsystem* shown at the bottom and to the lower right of figure 17.64 must, as previously noted, be made of inert metals or ceramics highly resistant to the corrosive effects of etching gases. Such materials are expensive. The vacuum subsystem consists of turbomolecular vacuum pumps for pressures below 1.3 Pa (10 mTorr) used in the microelectronic industry, backed up by pumps which minimize contamination of the vacuum chamber with oil or any other volatile fluid.

The vacuum subsystem also may contain one or more airlocks that allow a computer-controlled robotic manipulator to insert and withdraw wafers without breaking the vacuum in the etching chamber. The airlock and vacuum transport mechanism may also be used to move wafers from one station dedicated to etching operations, to another station, either in the same chamber, or in an external vacuum system, where other deposition, masking, or stripping operations are conducted.

An important part of the vacuum subsystem are the *vacuum gauges* that measure the total pressure in the vacuum chamber, and a mass spectrometer. The *mass spectrometer* yields information about the relative concentration of etching gases (A, B, and C in figure 17.64) fed into the system; about the concentrations of reaction products that result from etching; and about the concentration of any reaction products from attack by the etching gases of the substrate or masking material.

17.6.2.3   Plasma Source Subsystem

The *plasma source subsystem*, shown to the upper left and in the center of figure 17.64, produces the plasma which yields active species for etching. In the US microelectronic industry, the *RF power supply* characteristically provides power at 13.56 MHz to a plane-parallel capacitive RF glow discharge reactor of the type discussed in chapter 12 of Volume 1, and illustrated in figure 17.65. Such plasma source subsystems have operated at background pressures above 6.7 Pa (50 mTorr), when the design rule of the microelectronic wafers processed is above 0.5 μm. More recent practice in the US microelectronic industry has been to use flat-coil inductive plasma reactors at pressures below 1.3 Pa (10 mTorr) for the more demanding design rules below 0.5 μm. A characteristic RF power supply might be rated from 100 to 500 W at 13.56 MHz for the RF parallel-plate glow discharge plasmas, and from 500 W to 2 kW for the ECR and flat-coil inductive plasma reactors.

An *impedance matching network* is desirable in all RF plasma source subsystems, since the plasma impedance varies with input power and other operating conditions. This variable load can lead to the dissipation or reflection back to the source of a large fraction of the input power, and inefficient coupling of RF power to the plasma. The impedance matching network is usually followed by

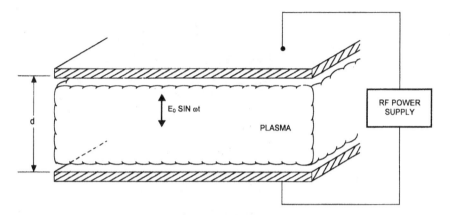

Figure 17.65. Essential components of an RF parallel-plate plasma source.

RF power metering, capable of measuring both the *forward power* flowing to the plasma, and the *reflected power* returned to the power supply by an impedance mismatch. The system controller may adjust the impedance matching network to minimize the reflected power, while at the same time maximizing the power actually coupled to the plasma.

The configuration of the plasma source subsystem depends on the RF frequency and the technology chosen to couple the RF power to the plasma. The resulting glow discharge plasma is operated by the plasma source subsystem to provide an as uniform as possible flux of active species onto the wafer surface for plasma etching. The radial uniformity and the efficiency of generating the plasma are improved by minimizing plasma losses. One tactic for doing this is to place permanent magnets in a multipolar confining configuration at the boundary of a plasma reactor, as discussed in section 3.4 of Volume 1.

Another important component of the plasma source subsystem is the plasma diagnostic instruments. These are important in research and development, but are much less used in production reactors. Probably the most common plasma diagnostic instrument in microelectronic circuit fabrication is the *retarding potential energy analyzer*, placed below or adjacent to a wafer. Its usual function is to measure the energy distribution function of ions that reach the surface of a wafer. Sometimes an ordinary, emissive, or double *Langmuir probe* is used to monitor the plasma parameters in research and development reactors. Chapter 20 contains a further discussion of plasma diagnostic instruments.

17.6.2.4   *Process Control Subsystem*

Finally, plasma etching reactors intended for production usually have a *process control subsystem* consisting of *sensors* and *effectors* for important process control variables, and a *system controller*, typically based on a microprocessor or

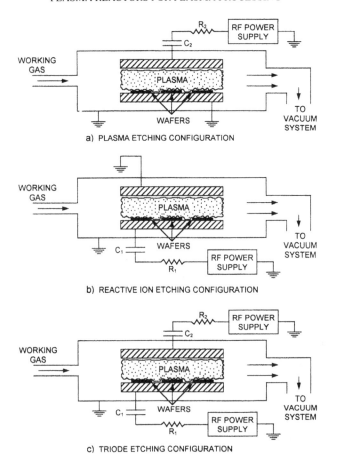

Figure 17.66. Configurations of the RF parallel-plate plasma reactor: (*a*) the plasma etching configuration; (*b*) the reactive ion etching (RIE) configuration; and (*c*) the triode configuration.

personal computer. In a production sequence, the system controller might install the wafer(s) in place; adjust the mixture ratio of the gases and total pressure to preset values; optimize and stabilize the RF power flow to the plasma source; monitor the etching process, by, for example, monitoring the level of etching reaction product gases measured with the mass spectrometer; use some form of *endpoint detection* to indicate when to turn the plasma and gas flow off; and move the wafer(s) on to the next deposition, etching, or stripping operation called for by the manufacturing protocol.

17.6.3 Reactor Configurations for Plasma Etching

In the previous section we discussed the plasma reactor systems used for plasma etching. In this section, we will discuss various geometrical configurations and electrical and RF arrangements used to generate plasmas for microelectronic etching.

17.6.3.1 Parallel-Plate Reactor Configurations

Some configurations of the *DC parallel-plate reactor* already have been discussed in sections 9.2 and 9.5.2 of Volume 1 and will not be repeated here. The *parallel-plate RF reactor* is illustrated in figure 17.65. It consists of a plane slab plasma between two parallel plates separated by a distance *d*. The RF electric field between the plates accelerates electrons that ionize the neutral working gas, in the manner discussed in sections 12.1 and 12.2 of Volume 1.

When parallel-plate RF plasma reactors are used for microelectronic etching, one of the three configurations shown in figure 17.66 is used. The three configurations are distinguished by the manner in which the RF power supply is connected to the electrodes. Figure 17.66(*a*) shows the *plasma etching* configuration. The *reactive ion etching* (RIE) configuration is shown in figure 17.66(*b*), and the *triode etching* configuration in figure 17.66(*c*). The phenomenology of these configurations has been discussed in section 16.1.1.2. A point of nomenclature is that the RIE configuration shown in figure 17.66(*b*) is sometimes referred to in the older literature as *reactive sputter etching*.

17.6.3.2 Multiple-Wafer Reactors

Two related RF plasma etching reactor configurations are the *hexagonal* and *barrel reactors*, discussed in section 12.4.1 of Volume 1. These configurations were widely used during the early history of microelectronic circuit fabrication to simultaneously process many individual wafers up to 10 cm in diameter. These and related multiple-wafer configurations are now less used, as trends in the microelectronic industry move to single wafers with large diameters.

17.6.3.3 Downstream Etching Configurations

There are at least three *downstream etching configurations* in which the etching plasma is generated in a resonant or non-resonant source, and then flows downstream to the workpiece or wafer being etched. One such configuration, which can operate at pressures below 1.3 Pa (10 mTorr), is the electron bombardment plasma source (also known as the *Kaufman source*). This source was discussed in section 16.2.1, and is located upstream of workpieces or wafers as shown in figure 17.67. In etching, deposition, and other applications, the Kaufman source produces a plasma in a diverging magnetic field of a few tens of millitesla, which is sufficient to magnetize the electron population, and assists in

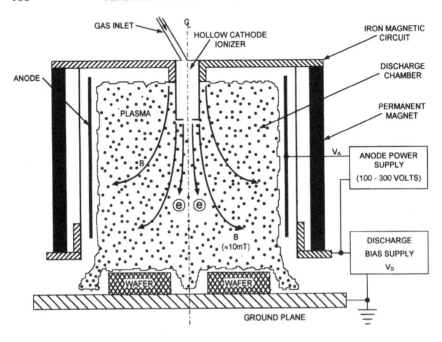

Figure 17.67. The Kaufman or electron bombardment plasma source as a plasma etching reactor.

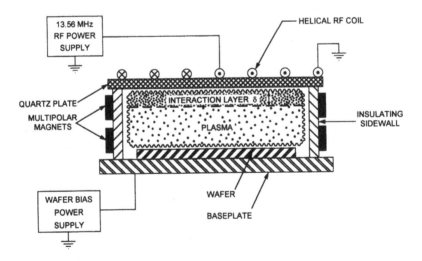

Figure 17.68. Configuration of a planar helical coil RF inductive plasma reactor.

confining the plasma. This plasma flows to the wafers or workpieces undergoing plasma treatment through a cylindrical volume along the magnetic field.

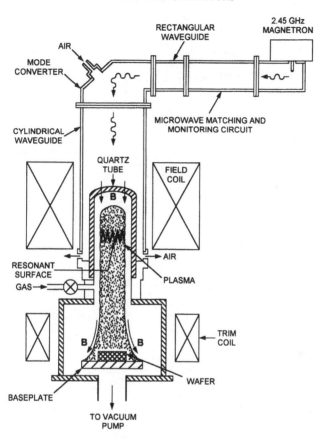

Figure 17.69. Configuration of an ECR resonant microwave plasma reactor for plasma etching.

The *flat-coil inductive plasma source* discussed in chapter 11 of Volume 1 can be used to treat large diameter wafers in the configuration shown in figure 17.68. In this figure, the flat coil is separated from the vacuum system by a quartz plate, chosen for its high resistance to thermal shock. In addition, there is an insulating (usually quartz) sidewall between the upper plate and the baseplate on which a wafer or other workpiece is mounted. The plasma may be confined by multipolar permanent magnets around its cylindrical sidewalls. The plasma is generated in the interaction layer, then flows downstream to interact with the wafer below. The energy of the ions reaching the wafer and their directionality can be adjusted by biasing the wafer with respect to the plasma, using a DC bias power supply as indicated in figure 17.68.

The *ECR downstream etching configuration* most frequently used is based on the ECR plasma reactor illustrated in figure 17.69. In this reactor configuration,

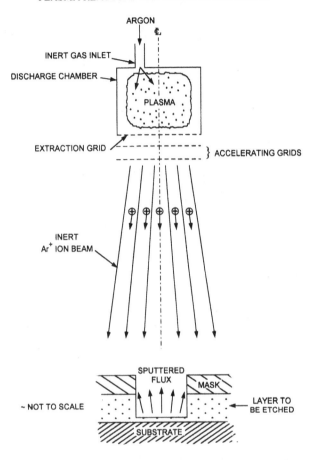

Figure 17.70. Configuration of an inert ion-beam milling configuration for plasma etching.

the ECR plasma is generated in the resonant region (where $\omega = \omega_c$) and flows downstream along diverging magnetic surfaces to a workpiece or wafer located on a baseplate. The baseplate may be connected to a DC power supply, and the ion energy reaching the surface of the wafer changed by adjusting the DC bias. The angle of arrival of the ions can be made more nearly vertical with respect to the wafer by accelerating them across the sheath above it. The magnetic induction at the wafer is typically high enough to magnetize the electron population, but not strong enough to magnetize the ion population and provide a significant magnetically induced improvement in directionality of the ion motion.

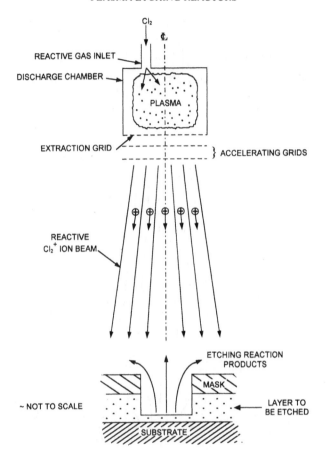

Figure 17.71. A reactive ion-beam etching (RIBE) configuration for plasma etching.

17.6.3.4 Etching with Ion Beams

Finally, several etching processes rely on separate ion source subsystems to produce the energetic ion beams responsible for etching. The first of these is *ion-beam milling*, or *sputter etching*, illustrated in figure 17.70. In this configuration, an inert gas such as argon is used as the working gas in a space-charge-limited ion source like those discussed in chapter 6 of Volume 1. After the working gas is ionized in the discharge chamber, and the ions extracted and accelerated through grids, the ion beam impinges on the workpiece or wafer to be etched. Since argon ions are chemically inert, and the ion-beam milling process is conducted in a high vacuum, the only etching mechanism operating is the sputtering of material from the layer to be etched. The sputtered flux is not normally volatile enough to be carried away by the vacuum system, and can redeposit on the sidewalls of a trench

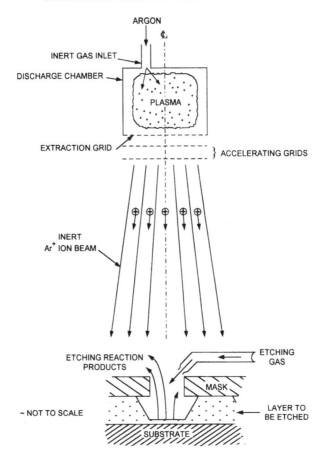

Figure 17.72. Configuration of a reactive etching configuration in which the etching reaction is promoted by energetic beam ions.

and the upper surface of the mask. This redeposition can severely limit the utility of ion-beam milling.

A related process is *reactive ion-beam etching* (RIBE), illustrated on figure 17.71. In this process, an energetic beam of chemically reactive ions from a separate, space-charge-limited ion source impinges on the wafer to be etched. In the example shown in figure 17.71, the reactive ions are chlorine, which react chemically with the layer to be etched. This process is most useful if a volatile reaction product is formed that can be pumped away by the vacuum system, and does not leave a residue behind as does the inert ion-beam milling process discussed previously.

The final ion-beam-etching process is *chemically assisted ion-beam etching* (CAIBE), a version of which is illustrated in figure 17.72. In CAIBE, an inert gas such as argon is used to produce a space-charge-limited ion beam. The ion beam impinges on the wafer to be etched, while an etching gas is fed from small nozzles in the vicinity of the etching site. The ions promote a chemical reaction between the etching gas and the layer to be etched. This process is most useful if the reaction products are volatile and can be pumped away. If the etching gas removes material on the sidewall more slowly than the ion-promoted reaction rate at the bottom of the trench, a directional etch is produced.

REFERENCES

Asmussen J 1989 Electron cyclotron resonance microwave discharges for etching and thin-film deposition *J. Vac. Sci. Technol.* A **7** 883–93

Batenin V M, Klimovskii I I, Lysov G V and Troitskii V N 1992 *Superhigh Frequency Generators of Plasma* (Boca Raton, FL: Chemical Rubber Company) ISBN 0-8493-93 05-1.

Ben Gadri R 1997 Numerical simulation of an atmospheric pressure and dielectric barrier controlled glow discharge *PhD Thesis* Order No 2644, University Paul Sabatier, Toulouse III, France

——1999 One atmosphere glow discharge structure revealed by computer modeling *IEEE Trans. Plasma Sci.* **27** 36–7

Boenig H V 1988 *Fundamentals of Plasma Chemistry and Technology* (Lancaster, PA: Technomic) ISBN 87762-538-7

Claude R, Moisan M, Wertheimer M R and Zakrzewski Z 1987 Comparison of microwave and lower frequency discharges for plasma polymerization *Plasma Chem. Plasma Process.* **7** 451–64

Clark P J 1971 Sputtering apparatus *US Patent* 3,616,450

Coburn J W 1991 Surface processing with partially ionized plasma *IEEE Trans. Plasma Sci.* **19** 1048–62

Conrad J R 1988 Method and apparatus for plasma source ion implantation *US Patent* 4,764,394

Corbani J F 1975 Cathode sputtering apparatus *US Patent* 3,878,085

d'Agostino R (ed) 1990 *Plasma Deposition, Treatment, and Etching of Polymers* (New York: Academic) ISBN 0-12-200430-2

Danilich M J and Marchant R E 1994 Plasma deposition of polymeric thin films *J. Appl. Polym. Sci.: Applied Polymer Symp.* vol 54 (New York: Wiley) CCC 0570-4898/94/010001-02

Deeds W E 1992 Charging apparatus for meltblown webs *US Patent* 5,122,048

Dusek V and Musil J 1990 Microwave plasmas in surface treatment technologies *Czech. J. Phys.* **40** 1185–204

Eisenmenger-Sittner C, Beyerknecht R, Bergauer A, Bauer W and Betz G 1995 Angular distribution of sputtered neutrals in a post magnetron geometry: measurement and Monte Carlo simulation *J. Vac. Sci. Technol.* A **13** 2435–43

Flinn J E (ed) 1971 *Engineering, Chemistry, and Use of Plasma Reactors* (New York: American Institute of Chemical Engineers) LCCCN 74-164050

Fraser D B 1978 The Sputter and S-Gun magnetrons *Thin Film Processes* ed J L Vossen and W Kern (New York: Academic) ISBN 0-12-728250-5, ch II-3, pp 115–29

Gross B, Grycz B and Miklossy K 1969 *Plasma Technology* (New York: Elsevier) LCCCN 68-27535

Grube W L and Gay J G 1978 High rate carburizing in a glow discharge methane plasma *Metall. Trans.* A **9** 1421–9

Harper J M E 1978 Ion beam deposition *Thin Film Processes* ed J L Vossen and W Kern (New York: Academic) ch II-5, pp 175–206 ISBN 0-12-728250-5

Hollahan J R and Bell A T 1974 *Techniques and Applications of Plasma Chemistry* (New York: Wiley) ISBN 0-471-40628-7

Hollahan J R and Rosler R S 1978 Plasma deposition of organic thin films *Thin Film Processes* ed J L Vossen and W Kern (New York: Academic) ch IV-1 ISBN 0-12-728250-5

Kanazawa S, Kogoma M, Moriwaki T and Okazaki S 1988 Stable glow plasma at atmospheric pressure *J. Phys. D: Appl. Phys.* **21** 838–40

Kaufman H R 1963 *The Electron Bombardment Ion Rocket (Advanced Propulsion Concepts, 1) (Proc. 3rd Symp., October, 1962)* (New York: Gordon and Breach)

——1965 Performance correlation for electron bombardment ion sources *NASA Technical Note* TND-3041

Keebler P F 1990 A large volume microwave plasma facility for plasma ion implantation studies *MS Thesis* University of Tennessee, Knoxville

Keebler P F, Crowley J E and Roth J R 1989 A high-voltage switching circuit for rapid plasma ion implantation *Proc. 1989 IEEE Int. Conf. on Plasma Science (May 22–24, Buffalo, NY)* IEEE Catalog No 89 CH2760-7, p 149

Kohl W H 1995 *Handbook of Materials and Techniques for Vacuum Devices* (Woodbury, NY: American Institute of Physics) ISBN 1-56396-387-6

Leahy M F and Kaganowicz G 1987 Magnetically enhanced plasma deposition and etching *Solid State Technol.* **30** 99–104

Lieberman M A and Lichtenberg A J 1994 *Principles of Plasma Discharges and Materials Processing* (New York: Wiley) ISBN 0-471-00577-0

Lieberman M A, Selwyn G S and Tuszewski M 1996 Plasma generation for materials processing *MRS Bull.* **21** 32–7

Loeb L B 1965 *Electrical Coronas* (Berkeley, CA: University of California Press) LCCCN 64-18642

Madou M 1995 *Fundamentals of Microfabrication* (Boca Raton, FL: Chemical Rubber Company) ISBN 0-8493-9451-1

Manos D M and Flamm D L (ed) 1989 *Plasma Etching* (New York: Academic) ISBN 0-12-469370-9

Matossian J N 1994 Plasma ion implantation technology at Hughes Research Laboratories *J. Vac. Sci. Technol.* B **12** 850–3

Montie T C, Kelly-Wintenberg K and Roth J R 2000 An overview of research using a one atmosphere uniform glow discharge plasma (OAUGDP) for sterilization of surfaces and materials *IEEE Trans. Plasma Sci.* **28** 41–50

Mort J and Jansen F (ed) 1986 *Plasma Deposited Thin Films* (Boca Raton, FL: Chemical Rubber Company) ISBN 0-8493-5119-7

Okazaki S 1992 Private Communication

Penfold A S and Thornton J A 1975 Electrode type glow discharge apparatus *US Patent* 3,884,793

Raizer Y P, Schneider M N and Yatsenko N A 1995 *Radio-Frequency Capacitive Discharges* (Boca Raton, FL: Chemical Rubber Company) ISBN 0-8493-8644-6

Rakowski W 1989 Plasma modification of wool under industrial conditions *Melliand Textilberichte* **70** 780–5

Roth J R 1995 *Industrial Plasma Engineering: Volume I, Principles* (Bristol: Institute of Physics Publishing) ISBN 0-7503-0318-2

——1997 Method and apparatus for covering bodies with a uniform glow discharge plasma and applications thereof *US Patent* 5,669,583

——1999 Method and apparatus for cleaning surfaces with a glow discharge plasma at one atmosphere of pressure *US Patent* 5,938,854

Roth J R, Sherman D M, Gadri R B, Karakaya F, Chen Z, Montie T C, Kelly-Wintenberg K and Tsai P P-Y 2000 A remote exposure reactor (RER) for plasma processing and sterilization by plasma active species at one atmosphere *IEEE Trans. Plasma Sci.* **28** 56–63

Roth J R, Tsai P P and Liu C 1995a Steady-state, glow discharge plasma *US Patent* 5,387,842

Roth J R, Tsai P P, Liu C, Laroussi M and Spence P D 1995b One atmosphere, uniform glow discharge plasma *US Patent* 5,414,324

Smith D L 1995 *Thin-Film Deposition: Principles and Practice* (New York: McGraw-Hill) ISBN 0-07-058502-4

Spence P D, Keebler P F, Freeland M S and Roth J R 1991 A large-volume, uniform unmagnetized microwave plasma facility (MPF) for industrial plasma-processing applications *Proc. Workshop on Industrial Plasma Applications and Engineering Problems, 10th Int. Symp. on Plasma Chemistry (Bochum, August 10)*

Tsai P P-Y and Wadsworth L C 1995 Electro-static charging of meltblown webs for high-efficiency air filters *Adv. Filtration Separation Technol.* **9** 473–91

Thornton J A and Penfold A S 1978 Cylindrical magnetron sputtering *Thin Film Processes* ed J L Vossen and W Kern (New York: Academic) ch II-2, pp 75–113 ISBN 0-12-728250-5

US Inc. 1989 *Planar Magnetron Sputter Sources* Sales literature on the US Gun II

Von Engle A, Seeliger R and Steenback M 1933 On the glow discharge at high pressure *Z. Phys.* **85** 144–60

Vossen J L and Cuomo J J 1978 Glow discharge sputter deposition *Thin Film Processes* ed J L Vossen and W Kern (New York: Academic) ch II-1, pp 11–73 ISBN 0-12-728250-5

Vossen J L and Kern W (ed) 1978 *Thin Film Processes* (New York: Academic) ISBN 0-12-728250-5

Waits R K 1978 Planer magnetron sputtering *Thin Film Processes* ed J L Vossen and W Kern (New York: Academic) ch II-4, pp 131–73 ISBN 0-12-728250-5

——1997 Edison's vacuum coating patents *AVS Newsletter* May/June, pp 18–19

Yasuda H 1978 Glow discharge polymerization *Thin Film Processes* ed J L Vossen and W Kern (New York: Academic) ch IV-2 ISBN 0-12-728250-5

——1985 *Plasma Polymerization* (Orlando, FL: Academic) ISBN 0-12-768760-2

18

Specialized Techniques and Devices for Plasma Processing

Plasma reactors intended for surface treatment, etching, thin-film deposition, or other forms of plasma processing use a variety of specialized techniques and devices designed to facilitate processing or to improve the quality and uniformity of the result. The following discussion is not intended to be exhaustive; many minor variants of the devices, and alternate methods for achieving a given result, may be found in the patent literature or in industrial use. It is the objective of this chapter to present some of the more widely used of these specialized techniques and devices.

18.1 VACUUM SYSTEM OPERATION

When implementing a control strategy for plasma processing, it is necessary to measure and control the flow rate, composition, and total pressure of the working gas. A characteristic *vacuum subsystem* of a plasma reactor used for plasma processing is illustrated in figure 18.1. This subsystem consists of a gas supply, inlet flow control; inlet flowmeters; a vacuum tank with airlock; vacuum pumps; a mass spectrometer to measure gas composition; and a vacuum gauge to measure the total gas pressure. A recent handbook on vacuum measurements and techniques has been prepared by Kohl (1995).

18.1.1 Measurement of Vacuum System Parameters

In a characteristic plasma-processing application, the vacuum-related parameters that need to be known include the *gas flow rate*, the working gas pressure, the vacuum system volume, the neutral gas *dwell time*, and the vacuum *pumping*

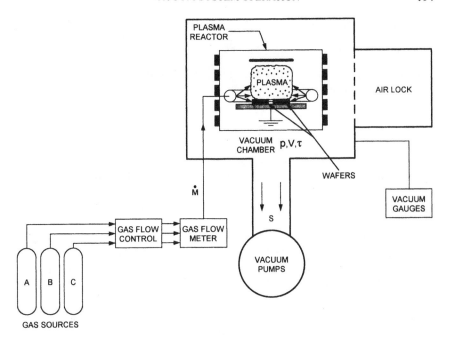

Figure 18.1. Components of the vacuum subsystem of a plasma reactor.

speed. The working gas flows into the vacuum system in a way normally intended to promote uniformity of effect over the workpiece surface.

The input *gas flow rate,* $\dot{M}$, in liters/s maintains a constant pressure, p, in the vacuum system. This pressure is properly measured in Pascal in SI units, but is often measured in the older units of Torr. The vacuum system has a total volume V, measured in cubic meters in the SI system, but conventionally expressed in liters (1 m^3 is equal to 1000 liters). A particular combination of flow rate, pressure, and system volume will result in a *residence time* (or *dwell time*) τ of the individual atoms or molecules in the vacuum system, measured in seconds. In dealing with these conventional and non-SI units, recall from section 2.1 of Volume 1 that at 300 K, the number density of a gas at 1 Torr is equal to 3.22×10^{22} atoms/m^3, and 1 Pa corresponds to 2.42×10^{20} atoms/m^3. Further formulae for conversion among these non-SI units may be found in appendix D.

The equation of continuity for the pumping system shown on figure 18.1 may be written

$$V\frac{\mathrm{d}p}{\mathrm{d}t} = \dot{M} - Sp \qquad (18.1)$$

where S is the *pumping speed* of the vacuum pump, conventionally measured in liters per second. If the vacuum pump removes S liters/s from a vacuum system with a total volume of V liters, the *dwell time* of an average molecule in the vacuum system is given by

$$\tau \equiv \frac{V}{S} \quad \text{s.} \tag{18.2}$$

The solution to the differential equation (18.1) for the vacuum pressure as a function of time is given by

$$p(t) = \frac{\dot{M}}{S} + \left(p_0 - \frac{\dot{M}}{S} \right) e^{-t/\tau}. \tag{18.3}$$

This equation describes both steady-state operation and the transient response of the vacuum system to changes in pressure or other operating parameters.

18.1.2 Gas Handling Tactics

It may be necessary to measure the actual (as opposed to the manufacturer's rated) speed of the vacuum pump, S. This pumping speed may be measured by waiting until the system equilibrates to a steady state for which $dp/dt = 0$ (or t goes to infinity). In the steady state, equation (18.1) yields

$$\dot{M} = Sp_0 \quad \text{Torr liters/s.} \tag{18.4}$$

One usually knows the mass flow rate, $\dot{M}$, and steady-state working gas pressure p_0 in Torr. Therefore, the steady-state condition of equation (18.4) will yield a measure of the *effective pumping speed* S of the system. The quantity $\dot{M}$ appearing in equation (18.4) is conventionally known as the *throughput* of the vacuum system, and has the units of Torr liters/s.

   Some vacuum systems may not be metered in such a way that the throughput, $\dot{M}$, is available from calibrated instruments, and one must determine it by other means. If one can measure the working gas pressure of the vacuum system as a function of time, and if one knows its volume, V, one can measure the throughput $\dot{M}$ using the procedure illustrated in figure 18.2(a). In this procedure, one closes a valve leading to the vacuum pump at time $t = t_0$ when the pressure is p_0, and then monitors the increase of pressure with time, which will be a linear function. Once the vacuum pump is isolated with a gate valve, $S = 0$, and equation (18.1) yields

$$\dot{M} = V \frac{dp}{dt} \quad \text{Torr liters/s.} \tag{18.5}$$

The pressure increase dp/dt is found from the slope of the pressure rise on figure 18.2(a), and since the volume of the vacuum system is known, the mass flow rate $\dot{M}$ follows directly.

   Finally, one may be able to measure the total pressure as a function of time, but not know the total volume of the vacuum system and/or the pumping speed S. In this case, the dwell time τ can be determined by shutting off the input gas flow at t_1, after which the pressure as a function of time is given by equation (18.3) as

$$p(t) = p_0 e^{-t/\tau} \quad \text{Torr.} \tag{18.6}$$

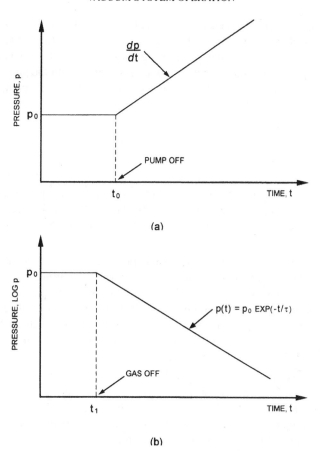

Figure 18.2. Strategies for measuring the gas handling characteristics of a vacuum subsystem for a plasma reactor by monitoring the pressure as a function of time: (*a*) turning off the vacuum pump; and (*b*) turning off the feed gas.

If one plots the pressure after shutoff as a function of time on a semilogarithmic graph, as shown in figure 18.2(*b*), the pressure decrease will be linear. The slope of the linear portion of the graph determines the dwell time τ, which is also the ratio given by equation (18.2). From equation (18.2), one can calculate the pumping speed S if the vacuum system volume V is known.

18.1.3 Pressure Adjustment

During plasma processing, and particularly during its initial and final phases, the evolution of reaction products and the gas load on the pumping system may vary. This circumstance has given rise to several approaches to controlling the gas handling system, which ensure a constant pressure of working gases and/or a constant flux of active species on the workpiece. Since it is difficult to adjust the pumping speed of most vacuum systems, the most common strategy is to operate the vacuum pumps at their maximum pumping speed S. With the pumping speed fixed, the steady-state relationship between the throughput $\dot{M}$, the pumping speed S, and the steady-state gas pressure p_0 is given by equation (18.4), which may be written

$$p_0 = \frac{\dot{M}}{S} = \frac{\dot{M}\tau}{V}. \qquad (18.7)$$

Equation (18.7) indicates that with a fixed pumping speed, the pressure p_0 can be kept constant by adjusting the throughput $\dot{M}$. This is the usual mode of pressure control in vacuum systems, and it is normally effected with a servo-controlled inlet valve.

Alternatively, one can keep the throughput $\dot{M}$ constant, and let the pressure vary. This latter strategy is sometimes followed when a significant amount of the working gas is consumed in processing, in order to minimize pressure or density gradients in the working gas and ensure uniformity of effect on the workpiece. In vacuum systems capable of precise control of the pumping speed S, one can set the throughput $\dot{M}$ at a constant value, and adjust the pressure p, or keep it constant, by adjusting the pumping speed.

One of two strategies may be used to vary the setpoint pressure p_0, based on equation (18.7). One can vary the dwell time τ by adjusting the pumping speed S, and keeping the throughput $\dot{M}$ constant. Since most vacuum systems do not have a variable pumping speed, the preferred strategy varies the throughput $\dot{M}$, while keeping the dwell time τ constant by maintaining a constant pumping speed in a fixed vacuum system volume.

18.2   WORKPIECE CURRENT COLLECTION

Workpieces undergoing plasma processing can either be part of an *active electrode*, through which a real DC current is drawn from the plasma; they can be part of a *passive electrode* that draws only displacement currents; they can be part of the wall surrounding the plasma, where they do not serve as an electrode; or they can be remotely exposed to plasma active species. The bombardment of a workpiece surface by plasma ions—associated with the flow of real currents— can lead to contamination or erosion of its surface. As a result, the so-called '*electrodeless discharges*' produced by microwave or RF power are preferred for most plasma-processing applications. The issues surrounding workpiece current collection are most pressing in deposition applications, particularly the deposition of insulating thin films.

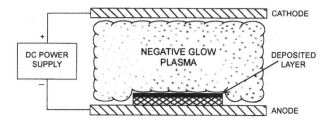

A) ELECTRICALLY CONDUCTING WORKPIECE

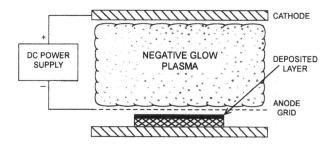

B) ELECTRICALLY INSULATING WORKPIECE

Figure 18.3. Two forms of active electrode PCVD: (*a*) using an electrically conducting workpiece on which an electrically conducting thin film is deposited; and (*b*) deposition on an electrically insulating workpiece, with a grid anode serving as the active electrode.

18.2.1 Active Electrode Processing

The term '*electrode*' is derived from the Greek words for 'electrical road', since an electrode is the origin or terminus of real or displacement currents that energize a plasma. It can make a significant difference to the quality of a plasma-processing operation whether the workpiece is an electrode or not. Deposition or etching of an active electrode/workpiece to which real currents flow can be effected in either of the two ways illustrated in figure 18.3, depending on whether the workpiece is electrically conducting or electrically insulating.

Figure 18.3(*a*) illustrates an electrically conducting workpiece undergoing deposition of an electrically conducting layer, for example by plasma ion sputtering of metal from a cathode. In this situation the (real) DC electrical current to the power supply flows through the deposited layer, which serves as an anode for the negative glow plasma. When a thin film is deposited on an electrically insulating workpiece, as in figure 18.3(*b*), the DC electrical current to the power supply can be collected on a grid with sufficient open area not to interfere with the deposition process.

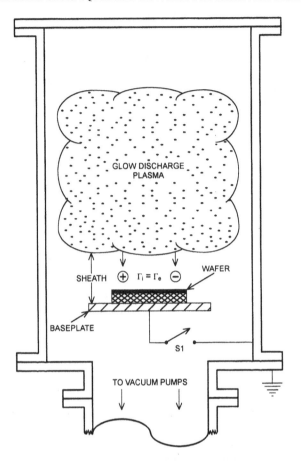

Figure 18.4. Schematic illustration of PCVD using a ambipolar flux of charged particles on an electrically isolated workpiece, with switch S1 open.

18.2.2   Passive Workpiece

Processing an insulating, floating, or electrically isolated workpiece is accomplished in the manner illustrated in figure 18.4. The workpiece with its chuck and baseplate are electrically isolated from the plasma. The workpiece receives an *ambipolar current* from the plasma, for which the ion and electron fluxes reaching the workpiece are equal.

Another form of passive workpiece processing occurs in *electrodeless reactors*, three examples of which are illustrated in figure 18.5. In electrodeless reactors, the plasma is generated either capacitively, inductively, or with microwave power, in such a way that the workpiece need not collect a real current from the plasma. The workpiece, however, may receive an ambipolar current

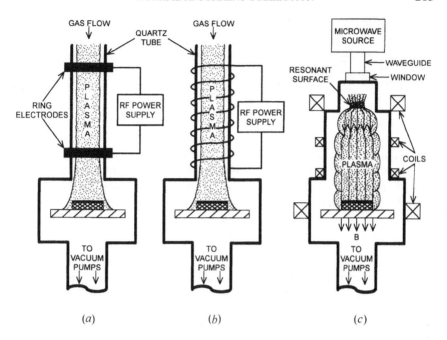

Figure 18.5. Three types of electrodeless plasma source suitable for PCVD reactors: (*a*) capacitively coupled; (*b*) inductively coupled; and (*c*) microwave. In such reactors, active species flow with the gas or along magnetic field lines before deposition on the workpiece.

with an equal flux of ions and electrons. Such electrodeless RF reactors are sometimes preferred, since the absence of current-carrying electrodes in direct contact with the plasma can reduce contamination of the deposited film (due to cathode sputtering), and improve the quality of the deposited layer.

18.2.3   Relative Processing Rates

An issue in choosing between active electrode and passive workpiece plasma processing is the magnitude of the surface particle flux, which must be sufficiently large to reduce the processing time to acceptable levels, typically a few minutes. Below, the potential processing rate of active and passive particle fluxes on a workpiece is compared.

18.2.3.1   Active Electrode Processing

On an active electrode or workpiece that receives a real ion current from the plasma, the flux of ions is related to the ion current density J_i by

$$J_i = e\Gamma_i \qquad A/m^3. \tag{18.8}$$

If ions etch the surface, or if neutralized ions accumulate to produce a deposited layer, the ion flux will determine the rate at which the plasma processing of the workpiece proceeds. This ion flux is obtained from equation (18.8) and is

$$\Gamma_i = \frac{J_i}{e} \qquad \text{ions/m}^2\text{-s}. \qquad (18.9)$$

An electrode current density at the high end of the range encountered in industrial DC glow discharge plasmas is $J = 0.1$ A/cm^2 or 1000 A/m^2. Substituting this value into equation (18.9) yields an ion flux on the workpiece of 6.24×10^{21} ions/m^2-s. This is approximately the highest active ion flux (and hence plasma-processing rate) that can reasonably be expected from a DC abnormal glow discharge plasma with an active electrode (cathode) as workpiece.

18.2.3.2   Passive Workpiece Processing

Consider a plasma-processing operation that depends on the passive arrival of neutral active species or precursors from the plasma. The flux of such species on the workpiece can be written in terms of the thermal neutral flux given by equation (2.28) of Volume 1,

$$\Gamma_n = \frac{F}{4}n\bar{v} \qquad \text{atoms/m}^2\text{-s} \qquad (18.10)$$

where F is the fraction of the neutral gas density which consists of active species that effect the plasma-processing operation of interest. The mean thermal velocity is given by equation (2.9) of Volume 1,

$$\bar{v} = \sqrt{\frac{8kT}{\pi M}} \qquad \text{m/s}. \qquad (18.11)$$

At 300 K, the relationship between the neutral gas density and the pressure in Torr is given by equation (2.3) from Volume 1,

$$n = 3.220 \times 10^{22} p \text{ (Torr)} \qquad \text{particles/m}^3. \qquad (18.12)$$

At 300 K, a processing plasma with argon as the dominant neutral component would have a mean thermal velocity of 397 m/s. If such a reactor were operated at a working pressure of 13.3 Pa (0.1 Torr), equation (18.10) yields a neutral flux on the workpiece

$$\Gamma_n = 3.2 \times 10^{23} F \qquad \text{precursors/m}^2\text{-s}. \qquad (18.13)$$

If the gas flow parameters of a passive workpiece reactor are adjusted to produce a long particle residence time τ, the precursor fraction in the gas may build up to 10%. In such a case, passive workpiece processing would proceed at a rate at least five times faster than the active electrode deposition process estimated by equation (18.8).

18.3 REMOTE EXPOSURE CONFIGURATIONS

There exist some plasma-processing applications, including sputtering and PCVD, for which it is useful to place the workpiece at a location remote from the target that produces sputtered atoms, or from the plasma–chemical reactor that produces the precursors responsible for PCVD. To illustrate, we consider two *remote exposure reactors* used industrially, the remote sputtering and remote PCVD configurations.

18.3.1 Sputtering Applications

Sputtering applications in which the workpiece is remote from the plasma/ion source will be discussed first as an illustration of this configuration.

18.3.1.1 Ion-Beam–Target Configuration

An *ion-beam–target configuration* in which a source of ions with keV energies is operated in a high vacuum chamber and impinges on a sputtering target is shown in figure 18.6. The ion energies may be chosen to lie close to the maximum of the sputtering yield curve of the target. The target is inserted at an angle to the ion beam, in order to take advantage of the increased sputtering yield associated with oblique incidence of the ions on the target, as discussed in section 14.5.2 of this volume. The sputtered atoms leave the target over a wide range of angles, and pass through a glow discharge plasma, generated in the example of figure 18.6 by an RF capacitive reactor. The background glow discharge can be generated by any of a variety of plasma sources described in chapter 16.

The glow discharge plasma in figure 18.6 can increase the surface energy of the workpiece shown at the right, and prepare it for better adhesion of the deposited layer by removing such surface contamination as monolayers of hydrocarbon machining oils. Sputtered atoms from the target that transit the plasma may become excited by inelastic collisions, and this may assist the deposition or etching process. The remote sputtering configuration illustrated in figure 18.6 has the disadvantage that it must be operated at low pressures, usually below 1.3 Pa (10 mTorr), thus increasing the pump-down and turn-around time for batch processing of workpieces.

18.3.1.2 Plasma–Target Configuration

A simpler and more widely used deposition tool is the *plasma–target remote sputtering configuration* shown in figure 18.7. In this configuration, ions from the negative glow plasma of a DC *abnormal glow discharge* are accelerated across the cathode sheath, and sputter neutral atoms from the cathode surface. These neutral atoms travel through the negative glow plasma and impinge on the anode/workpiece. This configuration requires no energetic ion source, and the

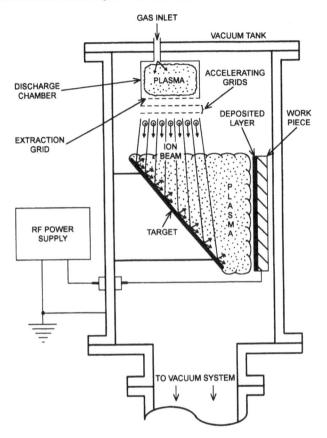

Figure 18.6. Schematic diagram of a plasma-activated ion-beam sputtering reactor, in which atoms sputtered off the target pass through the plasma, and are deposited as a thin layer on the workpiece.

negative glow plasma that produces the ions responsible for sputtering can be generated at relatively high pressures, up to 6.7 Pa (50 mTorr).

In figure 18.7, a negative glow plasma is formed by a parallel-plate DC obstructed abnormal glow discharge in which a large sheath voltage drop (shown to the right) exists. This cathode fall voltage can range from tens of volts to more than 100 V. Ions that reach the cathode surface therefore are not as energetic as those produced by a separate space-charge-limited ion source, like that illustrated in figure 18.6. Nonetheless, the sputtering yield of many combinations of ions on materials (for example, argon on copper or aluminum) is sufficiently high at these energies that the system illustrated in figure 18.7 is satisfactory for industrial sputter deposition. An important variant of the system illustrated in figure 18.7 is the widely used *magnetron reactors* discussed in section 17.4 of this volume.

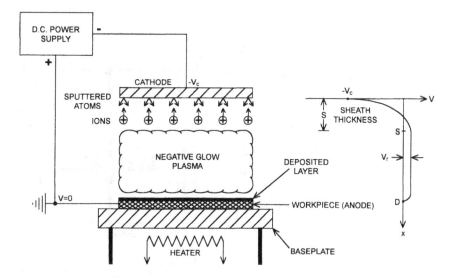

Figure 18.7. Illustration of a parallel-plate DC abnormal glow discharge plasma reactor in which the workpiece acts as an active anode.

18.3.2 PCVD Applications

A variety of remote exposure configurations have been developed for plasma chemical vapor deposition (PCVD). Remote PCVD exposure may be desirable for several reasons. The production of suitable monomers may require a long residence time in the plasma. Convection of the active species to a remote location may purge them of oils or dust which would contaminate the workpiece. In polymeric PCVD, remote deposition may be needed to promote the formation of linear polymers or to discourage cross-linking in polymeric thin films. Two of the more widely used remote deposition reactors are discussed here. These are the continuous-flow PCVD configuration, and a PCVD reactor in a common vacuum.

18.3.2.1 Continuous-Flow PCVD Configuration

A *continuous-flow PCVD configuration* is shown in figure 18.8, in which a glow discharge plasma is generated by a pair of RF capacitive ring electrodes. This remote plasma acts as a plasma–chemical reactor to convert the feed gas (or gases) into monomers or other precursors. When these precursors reach the workpieces, they deposit a thin layer of polymeric or other material. In such reactors the working gas pressure is usually below 133 Pa (1 Torr), although seldom below 6.7 Pa (50 mTorr). Relatively high working gas pressures—in the upper range of the intermediate-pressure regime—are required to maintain industrially acceptable deposition rates on workpieces.

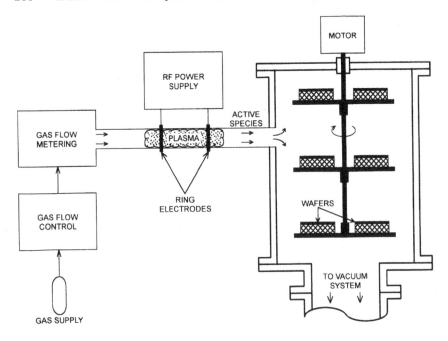

Figure 18.8. A remote exposure plasma PCVD reactor system in which active species are generated externally and flow into a chamber where deposition on workpieces occurs.

In remote exposure PCVD reactors, it is difficult to ensure uniformity of the deposited thin film both on individual workpieces, and on all workpieces simultaneously exposed. Such non-uniformity is usually dealt with by *motional averaging*, discussed in section 18.4 of this chapter. Figure 18.8 illustrates a remote exposure PCVD reactor in which the workpieces are mounted on rotating turntables in order to motionally average their deposition rates and produce a more nearly uniform thickness. Systems similar to that illustrated in figure 18.8 are widely used in PCVD applications because of their simplicity. They can be employed if the active species have a lifetime long enough to travel from the remote plasma source to the workpieces in the deposition chamber.

18.3.2.2   PCVD in a Common Vacuum

A second remote exposure configuration is the remote workpiece *PCVD in a common vacuum*, in which the plasma source and the workpiece are in a common chamber, illustrated in figure 18.9. In this example, an ECR source provides a relatively high density PCVD plasma in a magnetic induction strong enough to magnetize the electron population. Magnetization of the electrons is maintained as they are transported to the remote location in the lower part of the chamber by a uniform or slightly diverging magnetic field. A combination

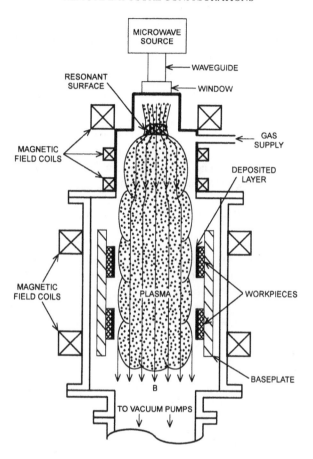

Figure 18.9. A remote exposure plasma ECR microwave plasma source for PCVD. In the example shown, the plasma is channeled from the ECR source to the deposition chamber by a magnetic nozzle, followed by a uniform magnetic field.

of magnetic channeling and neutral gas flow carries the plasma, which contains precursors for PCVD deposition, to workpieces where deposition occurs. The ECR microwave plasma source illustrated in figure 18.9 is not employed in all PCVD remote exposure configurations. It is, however, useful for applications in which a relatively high density of active precursor species must be generated at low working gas pressures.

18.3.3 Relative Deposition Rates

When planning industrial thin-film deposition, a choice between plasma-aided sputtering and PCVD methods may have to be made. If other factors are equal, the process that takes the least time and uses an expensive deposition tool for the shortest period is preferable. A few general considerations affecting such a choice can be illustrated as follows.

18.3.3.1 Deposition by Sputtering

In the sputter deposition systems illustrated in figure 18.6, the deposition rate on the workpiece cannot be any faster than the rate at which atoms are sputtered from the target. One therefore can estimate the maximum possible deposition rate on the workpiece by setting it equal to the rate of target (cathode) sputtering. This maximum *deposition rate* (DR) may be written

$$DR = \varepsilon \Gamma_i = 6.24 \times 10^{21} \text{ ions/m}^2\text{-s.} \tag{18.14}$$

The ion arrival flux Γ_i is the same as that previously calculated in connection with equation (18.8), and can be taken to be 6.24×10^{21} ions/m^2-s. The sputtering yield for many materials is about $\varepsilon = 0.1$ or less at ion energies of a few hundred eV, implying a maximum deposition rate

$$DR \leq 6.24 \times 10^{20} \text{ atoms/m}^2\text{-s.} \tag{18.15}$$

18.3.3.2 PCVD Deposition

For PCVD, however, the deposition rate due to polymerization of monomers on the surface of a thin film may be written

$$DR = \frac{F}{4} n \bar{v} \qquad \text{atoms/m}^2\text{-s} \tag{18.16}$$

where the deposition rate is set equal to the arrival rate of neutral monomers on the surface. The neutral monomers constitute a fraction F of the working gas density. If we take as an example CH_2 monomers with a mass of 14 amu, at 300 K they will have a mean thermal velocity $v = 671$ m/s. If all such monomers polymerize when they hit the surface, if the working gas pressure is $p = 13.3$ Pa (0.1 Torr), and if the monomer fraction is $F = 0.1$, the deposition rate of equation (18.16) may be as high as

$$DR \leq 5.4 \times 10^{22} \text{ atoms/m}^2\text{-s.} \tag{18.17}$$

A comparison of equations (18.15) and (18.17) indicates that a plasma-assisted sputtering deposition tool may have a deposition rate approximately a factor of 100 less than a PCVD reactor. Very often one does not have a choice between a PCVD reactor, which generally produces insulating films, and a sputter deposition reactor, which usually produces metallic conducting films. In those cases where a choice is possible for ornamental applications such as brightwork or protective coatings, the PCVD process is likely to be faster.

18.4 MOTIONAL AVERAGING TO ACHIEVE UNIFORMITY OF EFFECT

A variation in thickness or uniformity of effect no more than about 6% across a workpiece diameter is normally required for microelectronic deposition and etching. Other applications, such as coating architectural glass for office buildings, also have stringent requirements on the uniformity of sputtered or deposited coatings. Purely decorative coatings, such as brightwork, or oxygen barrier coatings, have less demanding requirements on the uniformity of thickness.

A number of methods have been developed to achieve uniform plasma-processing effects. One of the most widely used is *motional averaging*, which compensates for non-uniformities in the plasma. Motional averaging can be accomplished by workpiece motion; by moving magnetic fields, which drag a magnetized plasma across the workpiece; or by plasma drift or convection induced by crossed electric and magnetic fields, or magnetic field gradients.

18.4.1 Workpiece Motion

The mechanical motion of workpieces can be accomplished by translation, rotation, or compound epicyclic motion of a workpiece through the plasma in a plasma reactor. A widely used industrial example of motional averaging by workpiece translation is the co-planar magnetron sputtering source illustrated in figure 18.10, in which sputtered atoms are produced by a DC abnormal glow discharge at the top of the diagram. These atoms transit the negative glow plasma and are deposited below it on a flat workpiece. The workpiece could be a large sheet of architectural glass or plastic film that is to be coated with a thin layer of aluminum or other metal.

Uniformity of the coating in the horizontal direction of figure 18.10 is ensured by moving the workpiece at constant velocity from left to right past the co-planar magnetron source. To facilitate this motion, the workpiece may be mounted on rollers in a large rectangular vacuum system. Uniformity of the glow discharge plasma and of the sputtering process perpendicular to the plane of figure 18.10 is ensured by E/B drift, in a manner that will be discussed later.

Uniformity of a deposited or etched layer may also be improved by rotation of the workpiece. A system for accomplishing this is illustrated in figures 18.8 and 18.11, in which wafers are placed on turntables and rotated around the axis of a vacuum chamber.

Another method to improve the uniformity of deposition or etching is illustrated in figure 18.12, in which wafers are robotically removed from external cassettes, and inserted through an airlock onto a rotating turntable inside the vacuum tank on the left. Plasma processing is initiated after the gate valve is closed, and the vacuum chamber pumped down to its operating pressure. The turntable rotates to average out any non-uniformities associated with the flow of

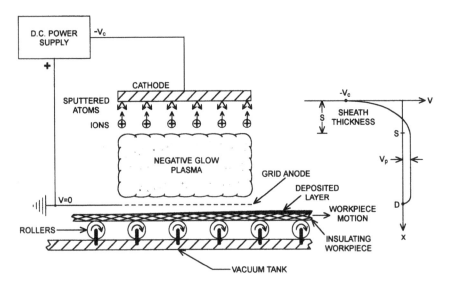

Figure 18.10. Illustration of motional averaging by moving a flat workpiece past a DC abnormal glow discharge reactor that generates sputtered atoms from the cathode.

precursor gas or the plasma in the reactor vessel.

Motional averaging also can be accomplished by compound motion. In some applications, such as the deposition of brightwork on three-dimensional objects, or the etching or deposition of uniform layers on microelectronic wafers, motional averaging in two or three dimensions is required. Such compound motion is achieved by systems of gears which provide compound epicyclic rotation through the region in which plasma processing takes place.

18.4.2 Moving Magnetic Fields

Uniformity of effect can be improved in etching or PCVD reactors by convecting the plasma past a workpiece during processing. This plasma motion can be achieved by moving magnetic fields, several variants of which have been developed in industrial practice (Leahy and Kaganowicz 1987).

Figure 18.13 illustrates a *peristaltic polyphase driver* arrangement in which the plasma above a workpiece is made to rotate about the vertical axis by energizing with polyphase RF power a set of coils spaced around the periphery of the plasma. These coils can be energized in such a way that their magnetic field rotates relatively slowly in a plane parallel to the workpiece surface. If the electron population is magnetized, the plasma will follow the magnetic field, to produce azimuthal (but not necessarily radial) averaging of the processing effect. When motional averaging is achieved by a rotating magnetic field, its rotational frequency can be adjusted over a wide range. Relatively slow rates of rotation,

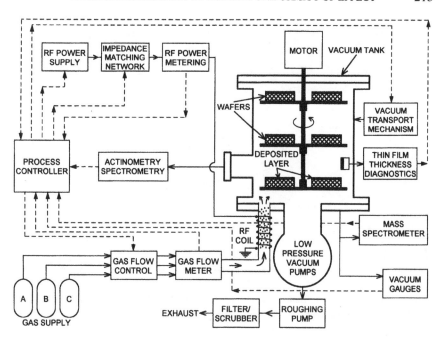

Figure 18.11. Motional averaging in a plasma reactor system that achieves uniformity of effect by axially rotating turntables on which workpieces are mounted.

a hertz or less, is usually adequate for the processing times used for industrial applications.

Another form of peristaltic polyphase averaging can be applied to such linear reactor configurations as a barrel reactor, around which individual axisymmetric coils are wrapped, and connected to a polyphase power supply. The phasing of adjacent coils can be adjusted such that the plasma flows along the axis of the reactor shown in figure 18.14. In this configuration, the workpieces are near the axis of the plasma (generated by means not shown). The plasma and workpieces are surrounded by a dielectric chamber wall, on the outside of which are located the individual polyphase RF coils. The phase of adjacent coils is adjusted to move the plasma axially past the workpieces, thus ensuring motional averaging along the axis of the reactor.

It characteristically requires ten or more millitesla to magnetize the electron population of industrial glow discharge plasmas. In many industrial contexts, the additional polyphase RF power supplies required for the configurations of figures 18.13 and 18.14 are not acceptable because of their additional complexity, impact on reliability, and capital cost. For this reason, motional averaging is sometimes accomplished by moving permanent magnets close to the processing plasma and the workpiece. One such *moving magnet configuration* is illustrated

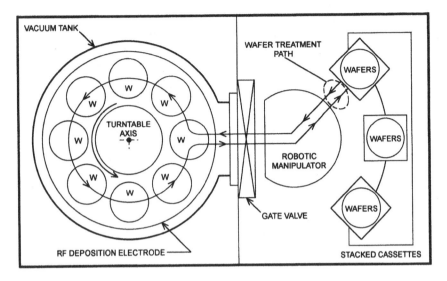

Figure 18.12. A robotic manipulator system for removing microelectronic wafers from stacked cassettes on the right, and feeding them through an airlock for placement on a rotating turntable in a vacuum tank.

in figure 18.15, in which a deposition or etching reactor is fitted with permanent magnets mounted on a turntable. When the turntable is rotated mechanically around the vertical axis, the plasma is convected along with it, producing azimuthal (but not radial) averaging of the plasma effects.

Perhaps because of patent or proprietary restrictions, some moving magnetic field configurations designed to produce motional averaging are elaborate. One such configuration is illustrated in figure 18.16, in which the plasma is made to flow past workpieces by moving permanent magnets mounted on a conveyer belt under the workpieces. This arrangement ensures motional averaging of effect in the horizontal direction. By rotating the belt assembly 90°, the direction of motion of the magnetic field moves in and out of the plane of the diagram of figure 18.16, ensuring motional averaging in a second orthogonal direction.

18.4.3 Plasma Drift/Convection

Finally, if the plasma electron population is magnetized, motional averaging can be achieved by causing the plasma to flow over the workpiece as the result of E/B or magnetic field gradient drifts. A characteristic configuration, found in magnetrons and related deposition reactors, is illustrated in figure 18.17. This rectangular slab plasma contains an electric field E and a perpendicular magnetic induction B. The drift of ions and electrons in the crossed fields of figure 18.17

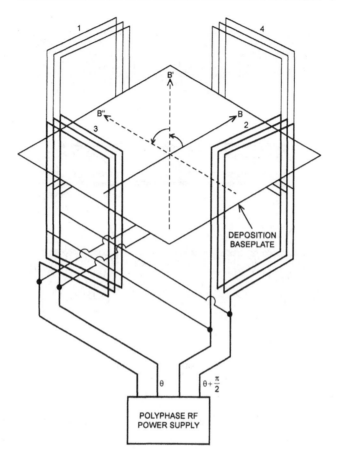

Figure 18.13. A four-coil polyphase RF system for rotating the magnetic induction in a plasma volume which contains workpieces at its boundary.

is given by

$$v_d = \left| \frac{E \times B}{B^2} \right| = \frac{E}{B} \qquad \text{m/s}. \tag{18.18}$$

This drift is shown to the lower left-hand side of figure 18.17 for ions and electrons with the indicated magnetic induction and electric field vectors. If only the electron population is magnetized (the usual situation in industrial glow discharges), the high ion collision rate with the neutral gas will prevent the ion population from drifting as fast as the electrons.

In a typical industrial processing plasma, the magnetic induction required to ensure magnetization of the electron population is approximately 10 mT, which can be generated by inexpensive and maintenance-free permanent magnets. The electric field is approximately 10 V/cm, or 1000 V/m, in an industrial magnetron

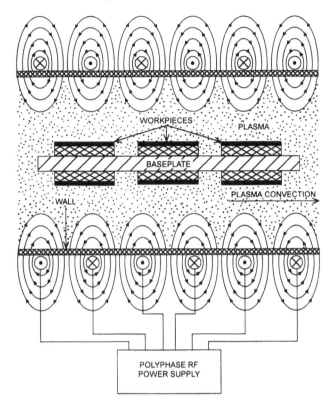

Figure 18.14. A polyphase system of coils designed to move plasma uniformly past wafers on which PCVD is taking place.

configuration. Substituting these values of electric and magnetic field into the right-hand side of equation (18.18) yields an electron drift velocity of 100 km/s. This drift velocity is much less than the electron thermal velocity of 5 eV electrons, which is approximately 1500 km/s. Since the ion population is not magnetized, the drifting electrons will drag the ions behind them, but at less than the electron drift velocity. The E/B drift produces a convective motion of the plasma as a whole, which averages out any non-uniformity in the E/B direction.

The permanent magnets used for magnetizing industrial glow discharge plasmas have significant magnetic field gradients, which cause an additional plasma drift. The magnetic gradient drift velocity is given by equation (3.27) of Roth (1986) as

$$v_{dm} = \frac{mv_{\perp}^2}{4qB^2} \frac{\boldsymbol{B} \times \nabla B^2}{B^2}. \qquad (18.19)$$

The electric-field-free magnetized plasma illustrated in figure 18.18 is a characteristic magnetic gradient drift geometry. Strong magnetic field gradients

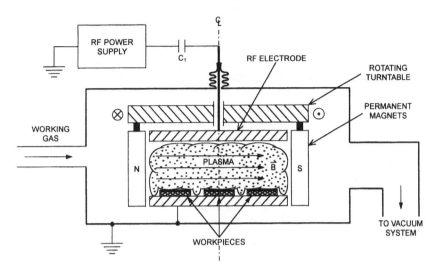

Figure 18.15. Illustration of motional averaging of plasma effects by rotating the magnetic induction generated by permanent magnets mounted on a turntable.

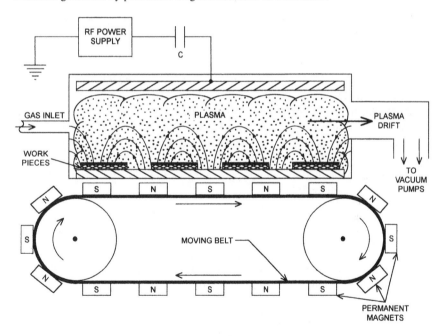

Figure 18.16. Motional averaging of plasma effects by moving magnetic fields generated by permanent magnets on a rotating belt under the PCVD reactor.

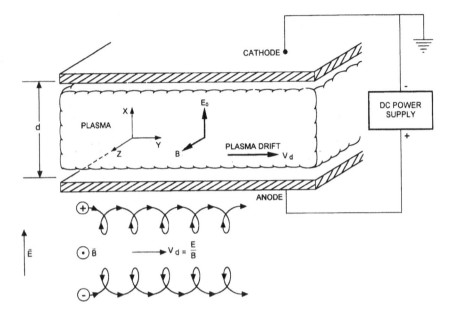

Figure 18.17. A parallel-plate reactor in which crossed electric and magnetic fields cause plasma drift past workpieces undergoing PCVD.

exist in the vertical direction, at right angles to the magnetic induction generated by coils or permanent magnets on the outside of the reactor vessel.   The perpendicular magnetic induction and magnetic field gradients produce the drift indicated at the lower right of figure 18.18. In industrial glow discharge plasmas the ions are not magnetized, and so will not drift in the manner indicated by equation (18.19).

If the magnetic field gradient is perpendicular to the magnetic induction, the electron drift velocity is

$$v_{dm} = \frac{mv_{\perp}^2}{2qB^2}\nabla B = \frac{E_{\perp}}{qB^2}\nabla B \qquad \text{m/s.} \qquad (18.20)$$

The thermal electron energy perpendicular to the magnetic field indicated by $E_{\perp}$ in equation (18.20) may be written in terms of the electron kinetic temperature T' in eV as

$$E_{\perp} = \tfrac{2}{3}\bar{E} = \tfrac{2}{3}(\tfrac{3}{2}kT) = eT' \qquad \text{J.} \qquad (18.21)$$

Substituting equation (18.21) into equation (18.20) yields an expression for the magnetic gradient drift velocity,

$$v_{dm} = \frac{T'}{B^2}\nabla B \qquad \text{m/s.} \qquad (18.22)$$

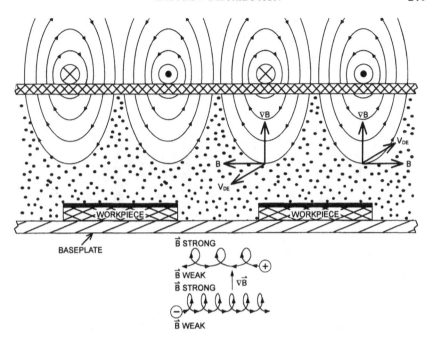

Figure 18.18. Schematic drawing of magnetic field gradients near workpieces undergoing PCVD deposition, and the direction of the magnetic gradient induced plasma drift.

Characteristic values for the parameters of equation (18.22) in an industrial glow discharge plasma are $T' = 5$ eV, $B = 10$ mT, and a magnetic field gradient $\nabla B = 0.2$ T/m. Substituting these values into equation (18.22) yields a magnetic gradient drift velocity of 10 km/s for electrons, about 10% of the drift velocity associated with the crossed electric and magnetic field (magnetron) drifts discussed previously. It should be noted, however, that the magnetic gradient drifts described by equation (18.19) can occur in the absence of an electric field. An electric field must be present to produce E/B drift.

18.5   GAS FLOW DISTRIBUTION

In plasma reactors intended for deposition, etching, or surface treatment, the active species responsible for these forms of plasma processing are convected along with the neutral working gas.   This has led to a variety of gas flow distribution configurations designed to improve the uniformity of plasma processing by controlling the flow of active species past the workpiece. Gas flow control has little effect on sputter deposition reactors, since such reactors are operated at low pressures for which sputtered atoms have long mean free paths. Such sputtered atoms interact only weakly if at all with the neutral gas

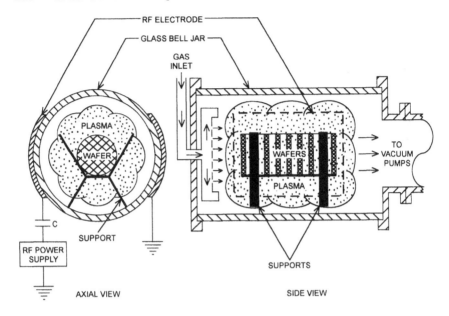

Figure 18.19. Plasma processing of microelectronic wafers in a barrel reactor with showerhead gas injection and axial gas flow.

between the site of their sputtering and the workpiece. In this section, we restrict the discussion to conventional fluid dynamic flow control methods, leaving a discussion of electrohydrodynamic (EHD) flow control for section 18.6.

18.5.1   Axial Flow Reactors

Achieving uniformity of plasma-processing effect by gas flow manipulation is usually accomplished by axial or radial flow of the working gas.   An axial flow reactor widely used during the early development of microelectronic deposition and etching technology was the cylindrical *barrel reactor* illustrated in figure 18.19. The working gas is injected at one end of the reactor, the walls of which are quartz or a dielectric. An RF capacitive glow discharge is created by two hemi-cylindrical electrodes on the outer surface of the vacuum vessel, illustrated on the left-hand side of figure 18.19. Active species generated by the RF glow discharge plasma flow axially, to ensure axial uniformity of effect on wafers lined up on a rack on the axis of the reactor.

Another variation of the axial flow reactor is the *hexagonal reactor* illustrated in figure 18.20. In this reactor, the working gas flows in at the top of a grounded cylindrical metal bell jar, in which an RF glow discharge plasma is maintained. This discharge is energized by connecting an RF power supply between the inner hexagonal electrode on which wafers are mounted, and the grounded outer wall of

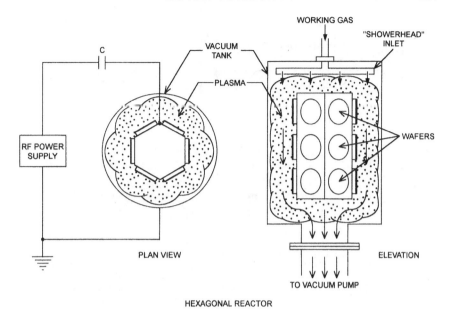

Figure 18.20. A hexagonal reactor with axisymmetric axial gas flow past microelectronic wafers mounted on the central hexagonal electrode.

the reactor. The axial flow of working gas vertically downward past the wafers is intended to ensure axial and azimuthal uniformity of effect in plasma-processing operations.

In both the barrel and hexagonal reactors, the use of many wafers (as small as 7.5 cm in diameter during early phases of microelectronic etching technology) produced difficulties with *loading effects* during etching operations, whereby depletion of the active species as they flow past the workpieces results in an axial non-uniformity of effect.

In barrel and hexagonal reactors, radial and azimuthal uniformity in the flow of etching species or deposition precursors is promoted by *showerhead injectors*, illustrated on the right-hand side of the vacuum tank in figure 18.19. Such injectors consist of a large plenum behind a circular plate containing many small holes through which the gas exits in a radially and azimuthally symmetric manner.

18.5.2 Radial Flow Reactors

In many plasma-processing reactors, azimuthal and, to a lesser extent, radial uniformity of effect is achieved by radial flow of the working gas. Two of many possible *radial flow reactor* configurations are shown in figures 18.21 and 18.22. In figure 18.21, the input gas is admitted symmetrically from a circumferential plenum at the edge of a circular electrode plate. This circular

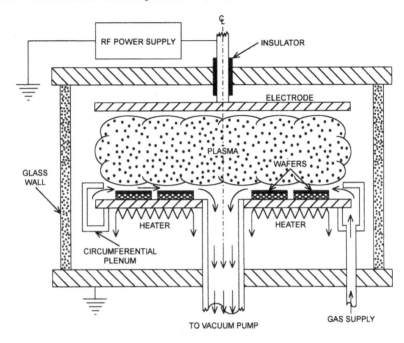

Figure 18.21. A parallel-plate RF glow discharge reactor, with radial gas flow past microelectronic wafers undergoing processing.

electrode serves as a *platform* or *chuck* for wafers or other workpieces, and the plasma is characteristically generated by an RF parallel-plate glow discharge plasma source. In the arrangement shown in figure 18.21, the gas may undergo plasma chemistry with the glow discharge above the workpieces, and the spent gas exits through a central exhaust line on the axis of the circular plate. This arrangement will promote azimuthal uniformity of effect for wafers on the lower electrode plate. In some applications, the working gas may enter on the axis, and be pumped out around the circumference of the electrode.

Loading effects are non-uniformities in processing outcomes that result from the depletion of active species as the working gas flows over the workpiece. Such loading effects can be minimized if, as a radial gas flow converges toward the center of the baseplate in figure 18.21, the number density of precursors or active species entrained in the gas flow remains independent of radius. The loading effect that removes active species/precursors from the flow as it moves radially inward across the workpieces can be compensated. This may be done by adjusting the radial gas flow rate to balance the geometrical effect that concentrates those same species as they flow inward to a smaller radius.

A similar arrangement, more suitable for deposition reactors than that of figure 18.21, is shown in figure 18.22. Here, the supply plenum is located under

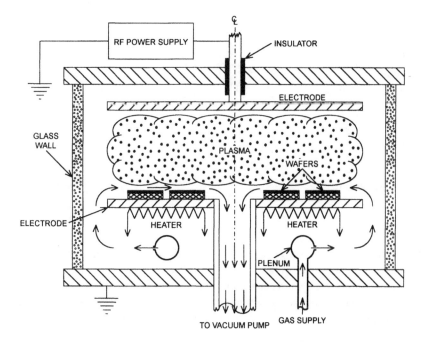

Figure 18.22. A variant of the radial gas flow reactor shown in figure 18.21, in which the working gas enters from an axisymmetric plenum below a circular electrode outfitted with a central exhaust port.

the lower electrode on which the wafers or workpieces are mounted. Azimuthal uniformity of the gas flow is ensured by outward flow of gas from equally spaced fine holes around the outer circumference of the plenum. This sheet of gas establishes an azimuthal symmetry in the gas flow around the edges of the electrode plate. This symmetry should continue into the glow discharge where the working gas may undergo plasma chemistry and generate active species necessary for deposition or etching. The gas finally exits through the central exhaust port located on the axis of the reactor. As in the previous discussion, the gas flow described here can be reversed. If the gas flow is inward, however, loading effects that reduce the active species concentration as the gas flows inward may be compensated by the geometrically induced radially increasing concentration of active species as they flow to smaller radii.

18.5.3 Dust Control

An important issue in gas flow management is control of spurious coatings, dust, and oils that may form in a reactor intended for plasma deposition, along with desirable active species (Barnes *et al* 1992, I *et al* 1996). PCVD is the result of

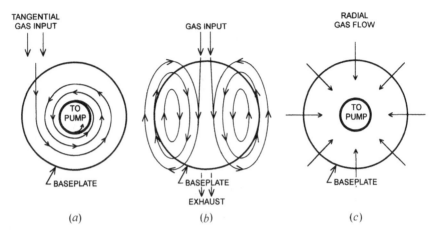

Figure 18.23. Three patterns of gas flow in PCVD reactors: (*a*) tangential injection with a central pumping port on the workpiece axis; (*b*) recirculating gyre vortex flow, in which the input gas is fed across a diameter and forms recirculating vortices over the wafer which trap dust particles for extended periods of time in the vicinity of the workpieces; and (*c*) laminar radial flow configuration, in which the gas moves in minimum time radially from the edge to the center of the electrode past the workpieces undergoing deposition or etching.

precursors that may form polymeric compounds on surfaces other than that of the workpiece to be coated. These spurious coatings can be deposited on viewports, diagnostic access ports, the interior of the vacuum system, and other surfaces where they are not wanted.

In addition to coating surfaces other than the workpiece, precursors in deposition plasmas can form particles of dust or droplets of oils in the plasma or in regions adjacent to the glow discharge plasma. These oil or dust particles can grow to dimensions of several microns or larger, and are of sufficient size to ruin microelectronic circuit elements if they fall on a wafer during the deposition or etching process. Such dust particles can become electrically charged, and remain levitated above the surface of a workpiece for relatively long periods. They can fall on the surface of a workpiece when the plasma and its accompanying sheath electric field are turned off at the end of a particular step in plasma processing.

In order to minimize the number and impact of oil and dust particles on plasma processing, several precautions are taken with respect to gas flow control. In processing reactors in which levitation of dust particles above the workpiece surface occurs, the concentration and size of the dust particles is determined by their dwell time above the surface. Two undesirable situations with respect to dust control are shown in figure 18.23.

One method of admitting the input gas which may cause dust control problems is that in which a gas jet is directed tangentially to a circular electrode on which the workpieces are mounted. Such a tangential gas flow can result in an injection vortex illustrated in figure 18.23(*a*). The longer dwell time of the gas in the vortex flow allows more and larger dust particles to build up, which have the potential to accumulate on the wafer. Such vortex flow may be associated with a vacuum pump outlet located at the center of a circular array of workpieces like those shown in figures 18.21 and 18.22.

Another type of gas input that also results in long dwell times and can lead to the buildup of dust or oils is the *recirculation gyre* vortex flow illustrated in figure 18.23(*b*). In this case, there is no exhaust port on the electrode that holds the workpieces, and the gas is injected across the diameter. If precautions are not taken, it is possible for the injected gas to form recirculating gyres in the manner indicated in figure 18.23(*b*).

The best and most effective way to minimize the dwell time of active species above the workpieces is to use a laminar radial flow configuration like those illustrated in figures 18.21 and 18.22. The radial gas flow in such reactors is illustrated in figure 18.23(*b*), where the precursors and the gas flow are convected radially from the outside to the inside of the baseplate, across the workpieces. This configuration minimizes the particle dwell time above the workpiece, as well as the likelihood and amount of dust or oil buildup.

18.6   ELECTROHYDRODYNAMIC (EHD) FLOW CONTROL

Industrial processing plasmas and the neutral working gas in which they are imbedded can be manipulated by magnetic or electric fields through the respective application of magnetohydrodynamic (MHD) or electrohydrodynamic (EHD) body forces.

MHD forces are sometimes useful in manipulating arcs, plasma torches, or other thermal plasmas with high current densities and electrical conductivity (Clauser and Meyer 1964, Tsinober 1989). The attractiveness of MHD flow control in industrial settings is compromised, however, by the requirement of a strong magnetic field. The magnetic field coils may raise reliability issues and have high electrical power and cooling water requirements. MHD forces are rarely useful in manipulating the plasma or neutral gas flow in industrial glow discharge processing plasmas, because the electrical conductivity and current density associated with such plasmas are too low.

The electric field can be used to manipulate industrial glow discharge plasmas and the neutral working gas in which they are imbedded, even at 1 atm. Creation of electric fields requires only relatively robust sheet metal electrodes, which do not pose the power or cooling requirements associated with the magnetic field coils required for MHD flow control. The electric fields so generated can be used to manipulate the flow of glow discharge plasmas by EHD body forces. EHD is the study of the behavior of electrically charged fluids in electric fields, and it

STATIC POLARIZED PLASMA

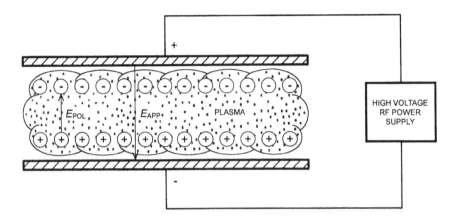

Figure 18.24. Parallel-plate plasma reactor with a polarized slab plasma in stable equilibrium.

finds application in electrostatic paint spraying and electrostatic precipitators, to be discussed in Volume 3, as well as other industrial processes.

18.6.1 EHD Flow Induced by Paraelectric Effects

The first of three mechanisms we will discuss for electrohydrodynamically manipulating glow discharge plasmas and their neutral working gas is the *paraelectric EHD body force*. This force arises when the applied electric field acts on the net charge density of the plasma, to provide a body force capable of accelerating the neutral gas to velocities up to about 10 m/s (Roth 2001).

18.6.1.1 Conceptual Basis of the Paraelectric Body Force

A conceptual aid to understanding EHD phenomena is to utilize the fact that electric field lines terminate on charges (either free charges or charged conductors). These electric field lines behave like rubber bands in tension to pull charges of opposite sign together. In glow discharge plasmas, the polarization electric field causes the charges, the plasma, and the working gas to move toward regions with shorter electric field lines and stronger electric fields. During this behavior, the plasma will move *paraelectrically* toward increasing electric field gradients, and drag the neutral gas along with it as the result of frequent ion–neutral and electron–neutral collisions.

The physical processes responsible for paraelectric gas flow can be visualized with the aid of figure 18.24, a two-dimensional slab plasma confined

PARAELECTRIC PLASMA ACCELERATOR

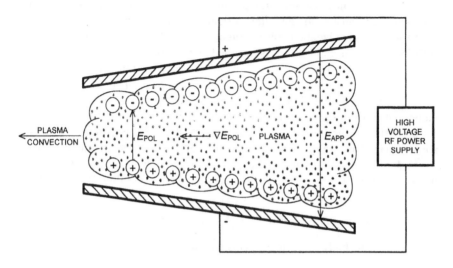

Figure 18.25. Tilted-plate plasma reactor with polarized plasma wedge in unstable equilibrium due to paraelectric body forces.

between parallel plates. This slab plasma will polarize in the manner indicated, resulting in a *polarization electric field* in the plasma volume, the electric field lines of which terminate on charges at the plasma boundary. These electric field lines act like rubber bands in tension and draw the two sides of the plasma together, but the plasma will remain in equilibrium as long as the external electric field remains in place. The electrostatic pressure will manifest itself by a force per unit area tending to draw the two electrodes together, as a result of the force between the charges at the plasma boundary and the charges on the electrodes. There are no electric field gradients in the bulk of the plasma, so it will remain in a stable equilibrium. The fringing electric fields at the sides act to pull the polarized slab back to its equilibrium position, with no tendency to move to the right or left.

If the geometry of figure 18.24 is changed by tilting the two electrodes as shown in figure 18.25, an electric field gradient will exist horizontally. The plasma will be accelerated toward the left, in the direction of increasing electric field gradient, by the tendency of the electric field lines of the polarization electric field to contract. This can be understood as an imbalance in electrostatic pressure that provides a net paraelectric body force. Lorentzian collisions of ions and electrons with the neutral gas will drag the latter to the left in figure 18.25, along with the charged species. This electrostatic body force toward the left is independent of the direction of the electric field, and thus is as strong in RF electric fields as in DC fields of the same magnitude. Further, this force is independent of the sign of the charge species being accelerated, and both species move in the same direction.

EHD forces in glow discharge plasmas are best studied with an individual particle rather than a continuous-fluid formalism, and the *Lorentzian formalism* is frequently a productive theoretical approach. In the Lorentzian formalism, each collision of the ions or electrons is assumed to give up to the neutral working gas all the momentum and energy that each charge gained, on the average, since its last collision. As will be shown later, in glow discharge plasmas the large ratio of neutrals to ions does not 'dilute' the momentum lost by the ions, because the large number of collisions per second compensates for the small ionization fraction. In atmospheric air, the ion collision frequency is about 7 GHz; that of electrons about 5 THz. These high collision frequencies are why the electric fields are well coupled to the neutral gas through the ion/electron populations, and are why the induced neutral gas velocities are comparable to the ion mobility drift velocity.

18.6.1.2   Theory of Paraelectric Flow Acceleration

The electrostatic body force on a plasma with a net charge density ρ_c is given by

$$F_E = \rho_c E \qquad N/m^3. \tag{18.23}$$

The net charge density, ρ_c, is given by

$$\rho_c = e(Zn_i - n_e) \qquad C/m^3. \tag{18.24}$$

Note that equation (18.24) is the *difference* between the ionic and electron charge densities. It is a term usually ignored in quasineutral theoretical formulations. This net charge density is related to the electric field in the plasma through *Poisson's equation*,

$$\nabla \cdot E = \frac{\rho_c}{\varepsilon_0}. \tag{18.25}$$

If one substitutes equation (18.25) into equation (18.23), the electrostatic body force is

$$F_E = \varepsilon_0 E \nabla \cdot E = \frac{1}{2}\varepsilon_0 \nabla \cdot E^2 => \frac{d}{dx}\left(\frac{1}{2}\varepsilon_0 E^2\right). \tag{18.26}$$

The last two terms in equation (18.26) are equal for the one-dimensional geometry of interest here. The expression in the parentheses in the last term of equation (18.26) is the electrostatic pressure p_E,

$$p_E = \tfrac{1}{2}\varepsilon_0 E^2 \qquad N/m^2 \tag{18.27}$$

that is numerically and dimensionally the energy density in J/m^3, as well having the units of N/m^2, or pressure. In the present context, it is useful to regard p_E as a pressure, because of its influence on the neutral gas flow. Using equation (18.27), equation (18.26) may be written as

$$F_E = \frac{d}{dx}(p_E) \qquad N/m^3. \tag{18.28}$$

The body force given by equation (18.28) results because the electrostatic pressure is transmitted to the ions and electrons by acceleration in the electric field, and the momentum acquired by the ions/electrons is then transmitted in turn to the neutral gas by Lorentzian collisions.

The ordinary gasdynamic pressure of the neutral gas is given by

$$p_g = nkT. \tag{18.29}$$

If viscosity forces, centrifugal forces, etc are neglected, the body forces due to gasdynamic and electrostatic gradients will be approximately in equilibrium, as described by

$$\nabla p_g + \nabla p_E = \frac{d}{dx}(p_g + p_E) = 0. \tag{18.30}$$

As a result, the sum of the gasdynamic and electrostatic pressures are approximately constant,

$$p_g + p_E = constant. \tag{18.31}$$

It is important to bear in mind that, with reference to figure 18.24, the gasdynamic pressure on the parallel plates is repulsive (the gas pushes the plates apart), while the electrostatic pressure is attractive, drawing the plates together. The sign conventions for these pressures, and for the associated body forces, must be treated cautiously in partially ionized gases.

Substituting equations (18.27) and (18.29) into equation (18.31), one obtains an approximate relation between the gasdynamic parameters and the electric field,

$$nkT + \frac{\varepsilon_0}{2}E^2 = constant. \tag{18.32}$$

Equation (18.32) predicts that in regions of high electric field (p_E large), the neutral gas pressure p_g is lower than that of the surroundings. This low pressure causes an inflow of surrounding, higher pressure gas. This pumping action is a *paraelectric* effect by which the plasma ions and electrons, and the neutral gas to which they are coupled by collisions, are accelerated toward regions of high electric field gradient. Global operation of the neutral gas continuity equation will draw neutral gas toward the plasma, in regions outside the volume in which Lorentzian momentum transfer occurs. A potential advantage of the paraelectric EHD flow acceleration mechanism described by equation (18.32) is that the required electric fields can be set up with a very simple, robust, and lightweight system of fixed electrodes.

The neutral gas velocity due to paraelectric gas flow acceleration effects can be estimated in the following way, if the neutral mean free paths are much smaller than the flow dimensions. The electrostatic pressure is given by equation (18.27) above, and this accelerates the neutral gas to a velocity v_0. The kinetic energy of this flow creates a *stagnation pressure*, p_s, equal to the electrostatic pressure that drives the gas flow,

$$p_s = \tfrac{1}{2}\rho v_0^2 = \tfrac{1}{2}\varepsilon_0 E^2 \qquad \text{N/m}^2. \tag{18.33}$$

In equation (18.33), it is assumed that the electrostatic pressure compresses the gas to a *stagnation* (or *dynamic*) *pressure* given by the middle term of the equation. When the gas is accelerated, a time-reversed version of stagnated gas flow occurs. Solving equation (18.33) for the induced neutral gas flow velocity v_0, one obtains

$$v_0 = E\sqrt{\frac{\varepsilon_0}{\rho}} \qquad \text{m/s.} \tag{18.34}$$

The paraelectrically induced gas flow velocity in a glow discharge therefore depends on the electric field (or in a fixed geometry, the applied voltage), and the mass density of the neutral working gas.

18.6.1.3   Paraelectric Gas Flow at 1 atm

In figure 18.26 a digital image of a smoke flow test involving an OAUGDP illustrates the paraelectric effects just discussed (Roth *et al* 1998, 2000, Sherman 1998). In the upper image (figure 18.26(*a*)), a low velocity (about 1 m/s) jet of smoke flows horizontally 1.5 cm above a flat panel. The panel has located on it a single, asymmetric, unenergized electrode like that shown schematically in figure 18.27(*a*). The electrode is energized in the lower image (figure 18.26(*b*)). The geometry of the electrode is asymmetric in such a way that the paraelectrically induced neutral gas flow on the surface of the panel flows to the left, with a velocity of a few meters per second.

The descent of the smoke jet to the surface occurs because a low-pressure region is generated in the plasma by paraelectric pumping of the neutral gas from its vicinity. In the digital image shown in figure 18.26(*b*), the blue glow of the plasma is overwhelmed by the strong illumination needed to observe the smoke, and is not visible. The plasma itself is confined to within about 1–2 mm of the panel surface, and extends no more than about 5 mm from the asymmetric electrode shown in figure 18.27(*a*).

The geometry of the OAUGDP flat panel used in the flow visualization experiments shown in figure 18.26 is fixed, and hence the electric field (and the flow velocity) is directly proportional to the applied voltage. The electric field E is approximately 10^6 V/m, and the mass density of air at STP is about 1.3 kg/m^3. When these values are substituted into equation (18.34), the predicted neutral gas flow velocity is

$$v_0 = 10^6\sqrt{\frac{8.854 \times 10^{-12}}{1.3}} = 2.6 \text{ m/s.} \tag{18.35}$$

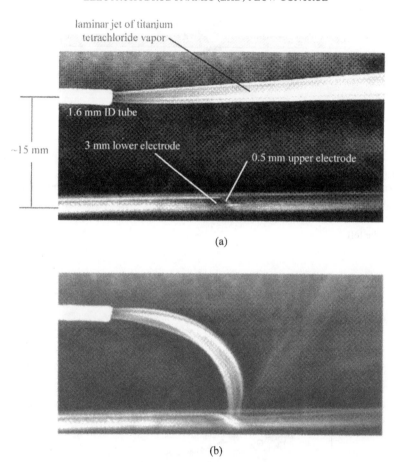

Figure 18.26. Demonstration of the paraelectric force in an OAUGD air plasma due to a single asymmetric electrode like that shown in figure 18.27(a). A single 0.5 mm wide electrode is embedded on the upper surface of the panel with a 3 mm wide lower electrode offset to the left on the lower surface. The jet is 1.5 cm above the surface, and its exit velocity is estimated to be in the range 1–2 m/s. (a) Plasma off; (b) plasma on, $V_{max} \sim 4.5$ kV rms, $\nu_0 = 3$ kHz.

The blowing velocity above the surface of a panel covered with asymmetric electrodes similar to those in figure 18.27(b) was measured as a function of height above the surface with a pitot tube, and is plotted for three excitation voltages in figure 18.28 (Roth *et al* 1998). The maximum velocities observed in figure 18.28 are plotted in figure 18.29 as a function of the excitation voltage. These data demonstrate induced neutral gas velocities of several meters per second, consistent with equation (18.35). These data also demonstrate that,

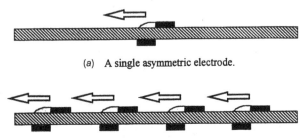

(a)   A single asymmetric electrode.

(b)   An array of asymmetric electrode sets.

Figure 18.27. Schematic diagram of the electrode configurations used to generate a layer of OAUGD plasma on a flat panel: (*a*) a panel with a single asymmetric electrode strip to induce EHD flow to the left; and (*b*) an array of asymmetric electrode strips to induce EHD flow to the left.

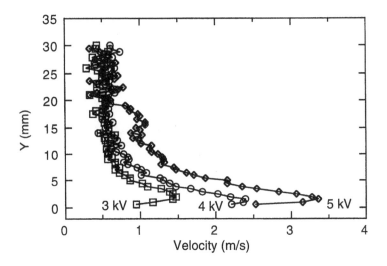

Figure 18.28. Blowing velocity profiles measured with a pitot tube above the surface of a flat panel with multiple asymmetric strip electrodes like those shown in figure 18.27(*b*) (but with the asymmetry reversed to induce flow to the right). The three profiles were taken for rms electrode excitation voltages of 3, 4, and 5 kV.

above a threshold voltage at which the plasma initiates, the neutral gas velocity is linearly proportional to the excitation voltage, also consistent with the dependence on electric field predicted by equation (18.34).

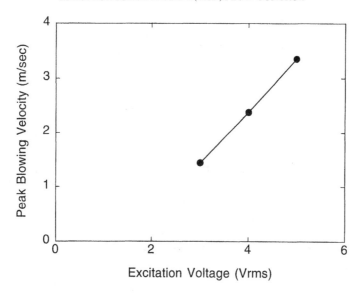

Figure 18.29. The maximum blowing velocities from figure 18.28 plotted as a function of rms electrode excitation voltage in kilovolts.

18.6.2   EHD Flow Induced by DC Mobility Drift

The second EHD-related neutral gas flow acceleration mechanism is induced by Lorentzian collisions of either electrons or ions as they drift in a DC electric field. This can be implemented with the geometry illustrated in figure 18.30, a flat panel containing strip electrodes, which is covered by OAUGDP plasma. The bottom surface of the panel is covered by a solid copper ground plane, and the top surface contains parallel strip electrodes. These electrodes are connected to an RF power supply that generates the OAUGDP, and a high voltage DC power supply which, through a string of voltage dropping resistors, maintains an electric field along the panel perpendicular to the axis of the individual electrodes. This DC electric field causes ions and electrons to drift in opposite directions. As they drift, they transfer momentum to the neutral gas as the result of Lorentzian collisions.

18.6.2.1   *Flow Induced by Electron Mobility Drift*

The neutral gas flow induced by momentum transferred to it by electron collisions will now be derived. The Lorentzian term for the loss of momentum by individual electrons in a DC electric field is

$$m v_{de} v_{en} \qquad \text{kg m/s}^2. \tag{18.36}$$

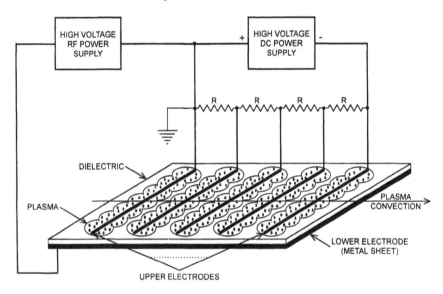

Figure 18.30. A flat panel of strip electrodes, each with a progressively higher DC voltage to impose a DC electric field along the surface of the panel, and thereby induce DC mobility drift of the charged species of the plasma.

The volumetric Lorentzian momentum loss is given by multiplying equation (18.36) by the electron number density,

$$\dot{M}_e = mn_e v_{de} \nu_{en} \qquad \text{kg m/s}^2\text{-m}^3 \tag{18.37}$$

where mv_{de} is the electron mobility drift momentum acquired from the electric field, and

$$\nu_{en} = n_0 \langle \sigma v \rangle_{ne} \qquad \text{electron collisions/s} \tag{18.38}$$

is the collision frequency of an individual electron with the neutral working gas. By combining equations (18.37) and (18.38), one obtains the total momentum transfer rate, per unit volume, of the electrons to the neutral working gas,

$$\dot{M}_e = mv_{de}n_0 n_e \langle \sigma v \rangle_{ne} = mv_{de} R_{ne}. \tag{18.39}$$

In this equation, R_{ne} is the volumetric reaction rate, the total number of electron–neutral collisions per second and per cubic meter.

In a Lorentzian gas, equation (18.39) must be exactly equal to the momentum received by the neutral population as the result of electron collisions,

$$\dot{M}_0 = M_0 n_0 v_{oe} \nu_{ne} \tag{18.40}$$

where $M_0 v_{oe}$ is the neutral drift momentum acquired from the electron population. The collision frequency of an individual neutral atom with the electrons is given by

$$\nu_{ne} = n_e \langle \sigma v \rangle_{ne}. \tag{18.41}$$

Combining equations (18.40) and (18.41), one obtains the total momentum received per unit volume and per second by the neutral gas from the Lorentzian collisions with the electrons,

$$\dot{M}_0 = M_0 \nu_{oe} n_o n_e \langle \sigma v \rangle_{ne} = M_0 \nu_{oe} R_{ne} \tag{18.42}$$

where M_0 is the neutral mass, and ν_{oe} is the electron-drift-induced neutral convection velocity.

Equations (18.39) and (18.42) must be equal, so that

$$\dot{M}_0 = M_0 \nu_{oe} R_{ne} = \dot{M}_e = m \nu_{de} R_{ne} \tag{18.43}$$

from which it follows that the induced neutral gas drift velocity due to Lorentzian electron momentum transfer to the neutral gas is given by

$$\nu_{oe} = \frac{m \nu_{de}}{M_0}. \tag{18.44}$$

Equation (18.44) states that the electron-induced neutral gas drift velocity is equal to the electron mobility drift velocity, modified by the ratio of the electron mass to the average neutral atom mass. The electron mobility drift velocity can be written in terms of the electric field, the electron mobility, and the electron–neutral collision frequency as

$$\nu_{de} = \mu_e E = \frac{eE}{m \nu_{ec}} \qquad \text{m/s}. \tag{18.45}$$

Further substituting equation (18.45) into equation (18.44), one obtains an expression for the induced neutral gas velocity due to electron–neutral momentum transfer,

$$\nu_{oe} = \frac{eE}{M_0 \nu_{ec}} \qquad \text{m/s}. \tag{18.46}$$

The Lorentzian-induced neutral gas drift velocity due to electron–neutral collisions in a characteristic OAUGDP in air is approximately 0.5 m/s, and is less than that induced by the paraelectric effects previously analyzed. This low velocity is a consequence of the very small electron/neutral atom mass ratio appearing in equation (18.44).

18.6.2.2   Flow Induced by Ion Mobility Drift

The neutral gas flow velocity induced by the DC mobility drift of *ions* will now be determined. By analogy with equation (18.37) for electrons, the Lorentzian volumetric momentum loss term for ions is given by

$$\dot{M}_i = M_i n_i \nu_{di} \nu_{in} = M_i \nu_{di} R_{in} \tag{18.47}$$

with v_{di} the ion mobility drift velocity induced by the electric field. If heating, viscous, and centrifugal effects are ignored, this momentum is transferred from the ions to the neutral population to produce an ion-induced neutral gas convection velocity v_{oi}. The momentum acquired by the neutral gas from Lorentzian collisions with ions is given by

$$\dot{M}_0 = M_0 v_{oi} n_0 v_{ni} = M_0 v_{oi} R_{ni}. \tag{18.48}$$

The ion mobility drift velocity is given by

$$v_{di} = \mu_i E = \frac{eE}{M_i v_{in}} \qquad \text{m/s}. \tag{18.49}$$

Equating equations (18.47) and (18.48), and incorporating equation (18.49), one obtains for the neutral gas drift velocity induced by ion–neutral Lorentzian momentum transfer,

$$v_{oi} = \frac{M_i v_{di}}{M_0} = \frac{eE}{M_0 v_{in}}. \tag{18.50}$$

In this case, the induced neutral gas drift velocity is essentially equal to the ion mobility drift velocity, since the ion mass and the neutral atom mass are the same if the ions are formed from the neutral working gas. For a characteristic electric field of 10 kV/cm in air, the induced gas flow velocity calculated from equation (18.50) may be as high as several hundred meters per second. However, heating and viscous effects will reduce this to lower values. This method of imparting a net velocity to air was suggested in the patent literature of the 1960s (Hill 1963). Hill's method is not likely to be practical because the necessary electric fields, on the order of 10^4 V/cm, require impracticably high voltages across an extended surface.

The analytical theory of DC mobility drifts implies that the Lorentzian momentum transfer from the electrons to the neutral gas is much smaller than that of the ions. The theory also indicates that the neutral gas drift velocity is comparable to the ion drift velocity in an applied DC electric field. Such drifts may be responsible for the 'ion wind' or 'corona wind' reported in connection with DC corona discharges (Malik *et al* 1983, El-Khabiry and Colver 1997).

18.6.3   EHD Flow Induced by Peristaltic Excitation of Strip Electrodes

For DC mobility drift acceleration, high voltages are required across an extended surface in order to impose a useful DC voltage between a series of strip electrodes. These high voltages can be avoided by using phased voltages on adjacent electrode strips in the *peristaltic plasma accelerator* (Roth 1997), shown in figure 18.31. This device is based on an OAUGDP surface plasma on a flat panel, and has a series of electrically isolated electrode strips similar to those shown in figure 18.30 or figure 18.27(*b*). In this device, the radiofrequency (RF) voltages that energize the strip electrodes and generate the OAUGDP are provided

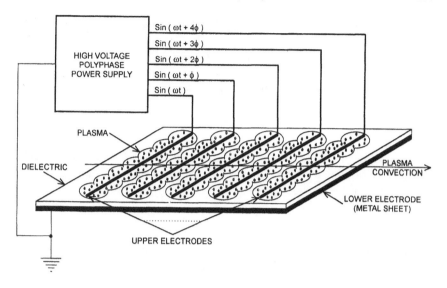

Figure 18.31. A flat panel of strip electrodes, each excited by progressively more advanced phases of a moderate rms excitation voltage provided by a polyphase power supply. The resulting peristaltic wave of voltage produces an electric field along the surface of the panel, and thereby induces a directional mobility drift of the charged species of the plasma.

by a polyphase RF power supply. The phased RF voltages energize successive electrode strips at progressively increasing phase angles, like the individual lights on a theater marquee intended to produce the illusion of motion. The individual electrodes are energized with a phase angle that increases (in figure 18.31) from left to right, giving the effect of a traveling wave from left to right. On a flat panel covered with plasma, the traveling wave will create an electric field along the surface of the panel that transports the plasma and the neutral gas in the direction perpendicular to the strip electrodes.

18.6.3.1 Peristaltic Phase Velocity

Consider the strip electrodes in figure 18.31, each spaced L m apart and each energized by a single sequenced phase of the polyphase power supply output. Each electrode therefore has a phase angle $360°/N$ with respect to adjacent electrodes, where N is the number of electrodes per cycle. A traveling (peristaltic) wave of voltage propagates to the right, given by

$$V = V_0 \sin(\omega t - kx) \qquad \text{V} \qquad (18.52)$$

where the wavenumber, k, of the electrostatic wave is

$$k = \frac{2\pi}{\lambda} = \frac{2\pi}{NL} \qquad \text{m}^{-1}. \qquad (18.53)$$

In equation (18.53), the phase angle ϕ between individual electrode pairs is given by

$$\phi = \frac{360°}{N}. \tag{18.54}$$

The phase velocity that acts on the net charge density is given by

$$v_p = \frac{\omega}{k} = \frac{2\pi v_0 N L}{2\pi} = v_0 N L \tag{18.55}$$

where v_0 is the driving frequency in Hz. In order for the ions to be able to keep up with this traveling wave, the ion mobility drift velocity of equation (18.50) has to be of the same magnitude.

18.6.3.2 Induced Peristaltic Flow Velocity

The peristaltic electric field that drags the ions and the neutral gas along the surface is given by

$$E = -\nabla V = -\frac{dV}{dx} = +V_0 k \cos(\omega t - kx). \tag{18.56}$$

The ions will 'snowplow' to the maximum electric field by a modification of the diocotron instability (see Roth 1986, ch 4), in which drifting charges move toward charge concentrations and away from charge rarefactions as the result of the operation of Poisson's equation. The maximum electric field of the traveling potential wave is, from equation (18.56),

$$E_{max} = \frac{2\pi V_0}{NL} \quad \text{V/m}. \tag{18.57}$$

The previous derivation of equation (18.50) has shown that the ion drift velocity will equal the ion-induced neutral convection velocity or

$$v_{oi} = v_{di} = \mu_i E_{max} = \frac{e}{M_i v_{in}} \times \frac{2\pi V_0}{NL} \quad \text{m/s}. \tag{18.58}$$

In this expression, equation (18.57) has been used for the driving electric field. For best effect, this ion-induced convection velocity should be approximately equal to the peristaltic phase velocity of equation (18.55). The velocity of equation (18.58) also ignores viscosity, heating, and other real gas effects.

The EHD flow acceleration effects discussed previously might be used to pump input gases over the workpieces of an OAUGDP or other plasma deposition or etching reactor. Such flow effects might control active species dwell time, uniformity of effect, uniformity of the plasma itself, formation of dust and oils, deposition of dust or oils, and maximize the utilization of rare or expensive feed gases. Such pumping of feed gases could be done externally to the plasma reactor, or on the surface of the chuck or baseplate on which workpieces are located.

This EHD effect should be useful at all working gas pressures for which the mean free paths of the ions are smaller than the dimensions of the flow, including pressures greater than 1 atm. These effects can also be used to pump feed and effluent gases in plasma-assisted CVD, and in ordinary CVD reactors, at all pressures from 1 Pa (a few mTorr) to 10 atm. This would be particularly useful in those chemical reactors in which vortical mixing is not desired, and the reaction must proceed in a laminar flow.

REFERENCES

Barnes M S, Keller J H, Forster J C, O'Nell J A and Coultas D K 1992 Transport of dust particles in glow-discharge plasmas *Phys. Rev. Lett.* **68** 313–16

Clauser M U and Meyer R X 1964 Magnetohydrodynamic control systems *US Patent* 3,162,398

El-Khabiry S and Colver G M 1997 Drag reduction by DC corona discharge along an electrically conductive flat plate for small Reynolds number flow *Phys. Fluids* **9** 587–99

Hill G A 1963 Ionized boundary layer fluid pumping system *US Patent* 3,095,163

I L, Juan W-T, Chiang C-H and Chu J H 1996 Microscopic particle motions in strongly coupled dusty plasmas *Science* **272** 1626–8

Kohl W H 1995 *Handbook of Materials and Techniques for Vacuum Devices* (Woodbury, NY: American Institute of Physics) ISBN 1-56396-387-6

Leahy M F and Kaganowicz G 1987 Magnetically enhanced plasma deposition and etching *Solid State Technol.* **30** 99–104

Malik M R, Weinstein L M and Hussani M Y 1983 Ion wind drag reduction *AIAA Paper* 83-0231

Roth J R 1986 *Introduction to Fusion Energy* (Charlottesville, VA: Ibis) ISBN 0-935005-07-2

——1995 *Industrial Plasma Engineering: Vol I—Principles* (Bristol: Institute of Physics Publishing) ISBN 0-7503-0318-2

——1997 Method and apparatus for covering bodies with a uniform glow discharge plasma and applications thereof *US Patent* 5,669,583

——2001 Neutral gas flow induced by Lorentzian collisions at plasma boundaries and the possibility of operating fusion reactors at one atmosphere *Current Trends in International Fusion Research—Proc. 3rd Symp.* ed E Panarella (Ottawa: NRC Press) at press

Roth J R, Sherman D M and Wilkinson S P 1998 Boundary layer flow control with a one atmosphere uniform glow discharge *AIAA Paper* 98-0328, *36th AIAA Aerospace Sciences Meeting (Reno, NV)*

——2000 Electrodynamic flow control with a glow-discharge plasma *AIAA J.* **38** 1166–72

Sherman D M 1998 Manipulating aerodynamic boundary layers using an electrohydrodynamic effect generated by a one atmosphere uniform glow discharge plasma *MS Thesis* Department of Physics, University of Tennessee, Knoxville, TN

Tsinober A 1989 MHD flow drag reduction *Viscous Drag Reduction in Boundary Layers (AIAA Progress in Astronautics and Aeronautics, 123)* ed A R Seabass, pp 327–49 ISBN 0-930403-66-5

19

Parametric Plasma Effects On Plasma Processing

This chapter is concerned with the effects of the plasma characteristics on plasma processing. The relevant characteristics include the mean free paths of ions, neutrals, and electrons; the collision frequencies of these species; the effect of neutral gas pressure on RF power coupling to the plasma; the thickness of the ion-accelerating sheath and the resulting effect on etching directionality; the sheath voltage drop; the production of active species in the plasma by electron–neutral collisions; and the effect of a magnetic field on the relative composition of the active species.

19.1 THE ROLE OF THE PLASMA

The *glow discharge* plasmas used for plasma processing perform several essential functions, which include supplying the surface to be processed with active species that are more energetic and/or more chemically reactive than can be obtained by purely chemical means. In commercially important etching and sputtering operations, the plasma serves as a source of energetic ions. As these ions are accelerated through the sheath potential, they acquire enough energy to promote sputtering or directional etching (Coburn and Winters 1979, 1983, Coburn 1982, Winters *et al* 1983).

19.1.1 Source of Energetic Ions

The energetic ions produced by sheath acceleration can be chemically inert, and their only function is sputtering the surface; they can be chemically reactive with the surface to be processed; or they can be *'catalytic' ions*, which are themselves chemically inert, but promote chemical reactions between a working gas and the material of the surface (Gerlach-Meyer *et al* 1981). Such 'catalytic' ions do not combine chemically to produce etching or other processing effects.

19.1.2 Source of Active Neutral Species

The plasma may also provide *active neutral species* that participate in plasma processing. Active neutral species may include such *chemically reactive species* as atomic oxygen, atomic chlorine, or monomers that undergo chemical reactions in the absence of ion bombardment to produce such processing effects as isotropic etching or the deposition of polymers. Active neutral species may also include *ion-induced reactive species*, which undergo energetic ion-induced chemical reactions to produce such processing effects as a directional etch.

19.1.3 Pressure Regimes for Processing with Energetic Ions

Industrial plasma processing may involve surface treatment, deposition, sputtering, etching, or other operations, with magnetron sputtering and microelectronic etching and deposition having the largest economic impacts. Microelectronic etching has been extensively studied because of its economic importance, and will be discussed as an example of plasma processing with energetic ions. Most microelectronic etching operations involve *reactive ion etching* (RIE) that utilizes reactive neutral species or ion-induced reactive species. In RIE, ions accelerated across the sheath above the wafer are responsible for catalyzing or inducing the etching reaction.

Plasma-processing operations that use energetic ions, including RIE, can be divided into the two pressure regimes illustrated in figure 19.1. On the right in figure 19.1(*b*), is the *intermediate-pressure RIE* regime, in which the working neutral gas pressure is characteristically between 6.7 and 133 Pa (50 and 1000 mTorr). In this regime, the ion mean free paths are shorter than the sheath thickness, S, between the plasma and the surface of the wafer, and ions approach the surface with significant parallel velocity components that result from scattering collisions. These parallel components allow ions to impinge on the sidewalls of an etched trench, to produce a directional (not a vertical) etch, as indicated at the bottom of figure 19.1(*b*).

The *low-pressure RIE* regime is illustrated in figure 19.1(*a*). In this regime, the neutral working gas pressure is below 1.3 Pa (10 mTorr), and ions have a sufficiently long mean free path that they are accelerated without collisions between the plasma and the surface. Such ions arrive with a nearly vertical velocity, producing a vertical trench if neither the etching gas nor active species from the plasma react with the sidewalls in the absence of ion bombardment. In this regime, the ion–neutral mean free path is greater than the sheath thickness S, and the only horizontal ion velocity components are associated with the ion thermal velocity in the plasma. The ion kinetic temperature in most plasmas is close to room temperature (0.025 eV). For a 20 eV argon ion arriving from such a

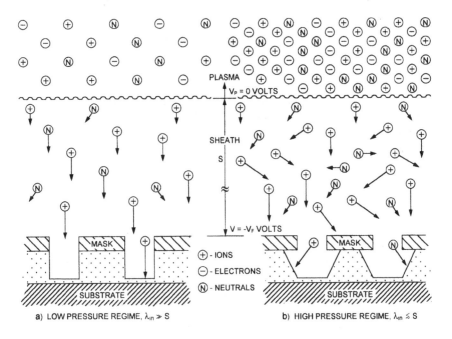

Figure 19.1. Ion collisions in a sheath between the plasma and a workpiece. (*a*) The 'low-pressure' regime, for which the ion–neutral mean free path is greater than the sheath thickness S; and (*b*) the 'high-pressure' regime, for which the ion–neutral mean free path is shorter than the sheath thickness, resulting in scattered ions reaching the etched layer with velocities parallel to the mask.

plasma, the horizontal to vertical velocity ratio would be 0.035, and the sidewall angle of the resulting etch would be 2° from the vertical.

19.1.4 Optimization of Processing Effects

Data have been taken by Mutsukura *et al* (1994) that illustrate the optimization of plasma processing as a function of operating pressure. These data describe the etching of silicon by a CF_4 plasma in the intermediate-pressure regime over the range 2.6–106 Pa (20–800 mTorr), and are shown in figure 19.2. At the low end of this range, a vertical etch was produced below a pressure of approximately 6.7 Pa (50 mTorr) in the low-pressure regime. The decrease in etching rate at low pressure is a result of the decreasing concentration and flux of fluorine radicals, the ion-induced reactive species. Above approximately 13 Pa (100 mTorr) in the intermediate-pressure regime, the etching rate decreased with increasing pressure as a result of scattering collisions in transit that prevented energetic ions from reaching the bottom of the trench. Scattering also kept ions that reached the surface from attaining the full energy available across the sheath voltage drop.

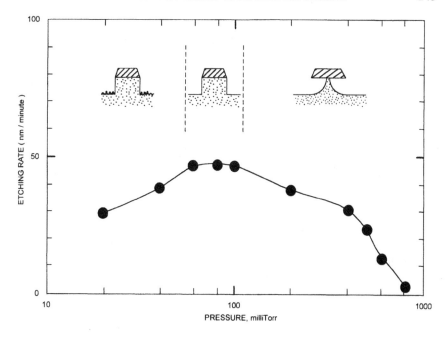

Figure 19.2. The etching rates and etching profiles of silicon in CF$_4$ plasmas as functions of gas pressure (Mutsukura *et al* 1994).

The dependent variables of reactive ion etching can be optimized by varying such plasma parameters as the pressure, as the data in figure 19.2 from Mutsukura *et al* (1994) indicate. In this example, the etching rate was an optimum at pressures between approximately 6.7 and 13 Pa (50 and 100 mTorr), where the maximum etching rate also produced a highly desirable vertical etch. Many etching and other processing reactions behave in a qualitatively similar manner to figure 19.2. The maximum in the etching rate–pressure curve, as well as the directionality of the etch, are both functions of the neutral gas pressure.

In addition to maximizing the etching rate or flux of etching species, one also should maximize the *directionality* of the etch, as illustrated in figure 19.2, so that the mask will be undercut as little as possible. The most desirable vertical etch is associated with low pressure and long mean free paths. This can result in a tradeoff against maximizing the etching rate, since at low pressure there may not be a high enough flux of the parent etching species to produce a rapid etch.

19.2 KINETIC PARAMETERS OF PLASMA PROCESSING

The mean free paths and collision frequency of the neutrals, ions, and electrons play an essential role in determining the nature and intensity of the flux of active species on a workpiece, the directionality of an etching process, and the functional dependence of the input power to the plasma on working gas pressure. In this section, we discuss the magnitude of the mean free paths and collision frequencies for neutrals, ions and electrons in several selected gases. These gases span a range of characteristics encountered in plasma processing, and those for which reliable data on cross sections and reaction rate coefficients are available.

19.2.1   Mean Free Paths

As was discussed in section 2.4 of Volume 1, the *mean free path* for a hard sphere atom or molecule of *cross section* σ and *number density n* is given by

$$\lambda = \frac{1}{n\sigma} \quad \text{m.} \tag{19.1}$$

At 300 K, the perfect gas law yields a relationship between the number density and pressure measured in Torr that is given by equation (2.3) of Volume 1 as

$$n = 3.220 \times 10^{22} p \,(\text{Torr}) \quad /\text{m}^3. \tag{19.2}$$

where the coefficient of equation (19.2) is dependent on the gas temperature, assumed to be 300 K.

19.2.1.1   Neutral–Neutral Mean Free Path

In table 19.1 are shown elastic sphere cross sections for atom–atom and ion–atom collisions, in units of 10^{-19} m^2, for eight common gases and their ions (McDaniel 1964). These cross sections were used in conjunction with equations (19.1) and (19.2) to plot the corresponding neutral mean free paths as a function of pressure in Torr (1 Torr = 133 Pa) for helium, argon, and nitrogen in figure 19.3. It is notable on this graph that the intermediate-pressure regime from 6.7 to 133 Pa (50 to 1000 mTorr) is associated with neutral–neutral mean free paths between 100 μm and 1 mm in length, while in the low-pressure regime below 1.3 Pa (10 mTorr) the neutral mean free paths are 1 cm or longer.

19.2.1.2   Ion–Neutral Mean Free Path

The ion–neutral mean free paths are plotted in figure 19.4 in the same way, from the ion cross sections in table 19.1. The intermediate-pressure regime, in which parallel-plate RF glow discharge reactors typically operate, is associated with ion mean free paths that range from 50 μm to approximately 1 mm in length. Inductive and ECR reactors, which usually operate in the low-pressure regime below 1.3 Pa (10 mTorr), have ion mean free paths that range from approximately half a centimeter to many centimeters.

Table 19.1. Elastic sphere cross sections for atom–atom and ion–atom collisions (from McDaniel 1964).

Gas	Neutral atoms Cross section σ_{in} $(\times 10^{-19} \text{ m}^2)$	Ion	Ions Cross section σ_{nn} $(\times 10^{-19} \text{ m}^2)$
He	1.5	He^+	3.8
Ne	2.1	Ne^+	4.5
Ar	4.2	Ar^+	9.3
Kr	5.5	Kr^+	10.9
Xe	7.6	Xe^+	13.4
O_2	4.1	O_2^+	5.5
N_2	4.4	N_2^+	8.5
CO	4.5	CO^+	10.1

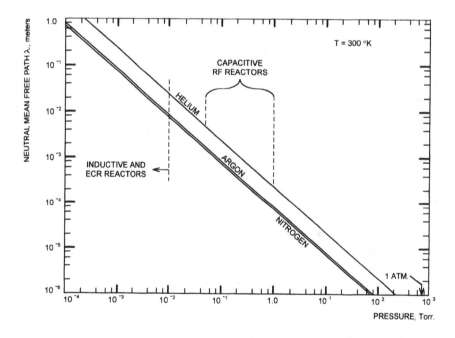

Figure 19.3. Neutral mean free paths at a gas temperature $T = 300$ K, as a function of background pressure for helium, argon, and nitrogen.

19.2.1.3 Electron–Neutral Mean Free Path

The cross sections and reaction rate coefficients required to calculate the electron–neutral mean free paths are functions of the electron kinetic temperature, as

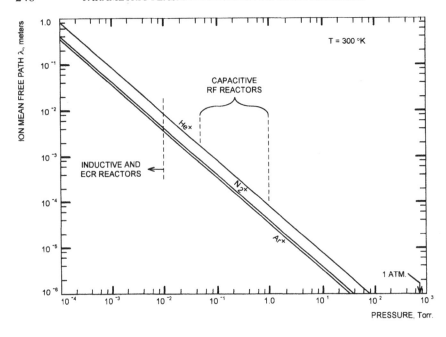

Figure 19.4. Ion–neutral mean free paths at a neutral gas temperature of 300 K, as a function of background pressure for argon, nitrogen, and helium ions.

discussed in section 4.8 of Volume 1 (see also Colgan *et al* 1991). The electron–neutral mean free paths may be written

$$\lambda_{en} = \frac{\bar{v}_e}{n \langle \sigma v \rangle_s} \qquad m \qquad (19.3)$$

where the parameter $\langle \sigma v \rangle_s$ is the electron-energy-dependent reaction rate coefficient for elastic electron–neutral scattering. The denominator of equation (19.3) is the electron–neutral collision frequency, and the electron thermal velocity in the numerator of equation (19.3) is given by equation (2.9) of Volume 1 as

$$\bar{v}_e = \sqrt{\frac{8kT}{\pi m}} \qquad m/s. \qquad (19.4)$$

The electron–neutral collision mean free path is plotted on figure 19.5, using the electron–neutral reaction rate coefficients plotted in figure 4.9 of Volume 1. Figure 19.5 shows that electron–neutral mean free paths in the intermediate-pressure regime between 6.7 and 133 Pa (50 and 1000 mTorr) range from approximately 1 mm to 1 cm. This is the pressure regime of parallel-plate RF glow discharge plasma reactors. The inductive and ECR reactors widely used in the low-pressure regime below 1.3 Pa (10 mTorr) have electron–neutral mean free paths several centimeters or longer.

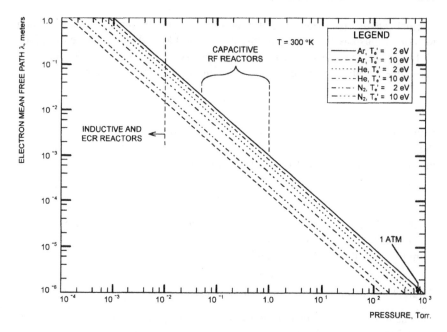

Figure 19.5. The electron–neutral mean free path as a function of pressure in Torr for argon, nitrogen, and helium background gas at the electron kinetic temperatures shown.

Table 19.2. Kinetic parameters of common gases and their ions at $T = 300$ K and $p = 1$ Torr.

Gas	Molecular weight	Mean thermal velocity (m/s)	Neutrals		Ions	
			λ_{nn} (μm)	ν_{nn} (MHz)	λ_{ni} (μm)	ν_{ni} (MHz)
He	4.00	1255	207	6.04	82	15.3
Ne	20.2	559	148	3.77	69	8.07
Ar	39.9	398	74	5.38	33	11.9
Kr	82.9	276	56	4.89	28	9.70
Xe	130.2	220	41	5.38	23	9.48
O_2	32.0	444	76	5.87	56	7.84
N_2	28.0	474	71	6.71	37	13.0
CO	28.0	474	69	6.88	31	15.5

19.2.2 Collision Frequency

The *collision frequency* of the electrons, ions, and neutrals determines the power transferred from the RF electric field to these species. In table 19.2 the mean thermal velocity, the mean free paths, and the collision frequency for neutrals and ions of various gases at a temperature of $T = 300$ K, and at a pressure of 133 Pa (1 Torr). The cross sections for these parameters are taken from table 19.1. The hard-sphere collision frequency of these neutrals and ions is given by

$$\nu_s = n\sigma_s \bar{v} \quad \text{Hz} \tag{19.5}$$

where σ_s is the hard-sphere cross section shown on table 19.1, and is the mean thermal velocity given by equation (19.4). For the electron population, the electron–neutral collision frequency is given by

$$\nu_{ec} = n\langle \sigma v \rangle_s \quad \text{Hz} \tag{19.6}$$

where n is the neutral number density and $\langle \sigma v \rangle_s$ is the electron kinetic temperature-dependent reaction rate coefficient for elastic electron scattering from the neutral population, given in figure 4.9 of Volume 1.

19.2.2.1 Neutral–Neutral Collision Frequency

The neutral–neutral collision frequency for nitrogen, helium, and argon are shown in figure 19.6. In the low-pressure regime, in which inductive and ECR reactors operate, the neutral collision frequencies are well below the standard RF frequency of 13.56 MHz. The collision frequencies, however, are comparable to this standard frequency in the intermediate-pressure regime from 6.7 to 133 Pa (50–1000 mTorr), where most parallel-plate RF glow discharge reactors operate.

19.2.2.2 Ion–Neutral Collision Frequency

The ion–neutral collision frequency associated with table 19.2 is plotted in figure 19.7 for the same gases. These collision frequencies are below 13.56 MHz in the intermediate-pressure regime in which most parallel-plate RF glow discharge reactors operate, and also in the low-pressure regime used for inductive flat-coil and ECR reactors.

19.2.2.3 Electron–Neutral Collision Frequency

The electron–neutral collision frequency, ν_c, is plotted in figure 19.8 as a function of pressure for argon, nitrogen, and helium over a range of electron kinetic temperatures characteristic of RF glow discharge processing plasmas. This collision frequency appears in all the transport coefficients for electrons in partially ionized glow discharges, whether DC or RF. For the intermediate-pressure regime from 6.7 to 133 Pa (50 to 1000 mTorr), in which most parallel-plate RF glow discharge reactors operate, the electron–neutral collision frequency is much greater than 13.56 MHz. However, the collision frequency ν_c is generally below 2.45 GHz, used for ECR and non-resonant microwave glow discharge

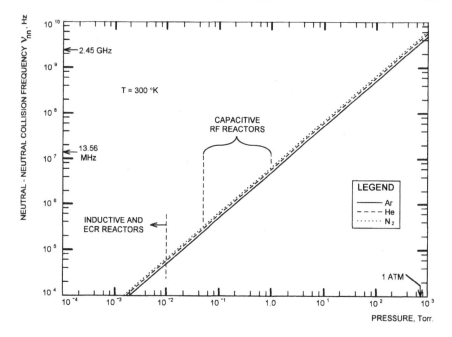

Figure 19.6. The neutral–neutral collision frequency of nitrogen, helium and argon as functions of pressure at a gas temperature of 300 K.

plasma production. In the low-pressure regime below 1.3 Pa (10 mTorr), the electron–neutral collision frequencies are far below 2.45 GHz, characteristic of resonant ECR microwave plasma reactors, but are comparable to or lower than the 13.56 MHz operating frequency of inductive low-pressure plasma reactors.

The operating regimes for low- and intermediate-pressure plasma-processing reactors are shown in table 19.3. The intermediate-pressure regime is associated with parallel-plate RF glow discharge plasma reactors, while the values listed in table 19.3 for the low-pressure regime apply to ECR and inductive plasma sources, except for the resonant parameters in line 10 of this table. The numerical values of the parameters on table 19.3 are only approximate; certain gases or plasma operating conditions that involve unusually high or low electron kinetic temperatures can produce values outside the range of those shown.

19.3   RF POWER COUPLING

The time required to complete a plasma-processing operation is directly proportional to the flux of active species on the workpiece, and this flux is, in turn, proportional to the electron number density in the glow discharge plasma which produced the active species. As a consequence of the linkage between the

Table 19.3. Characteristic plasma parameters for low- and intermediate-pressure operating regimes.

Plasma parameters	Source technology			
	Low-pressure regime			High-pressure regime
	ECR	Inductive	Helicon	Capacitive
1. Chip design rule		$\leq 0.5\ \mu m$		$\geq 0.5\ \mu m$
2. Pressure		$p < 10$ mTorr		$50 \leq p \leq 1000$ mTorr
3. Operating frequency, ν	2.45 GHz	13.56 MHz	13.56 MHz	13.56 MHz
4. Ion–neutral mean free path (λ_{in})		$4\ mm < \lambda_{in} < 1\ m$		$50 < \lambda_{in} < 1700\ \mu m$
5. Electron–neutral mean free path (λ_{en})		$2\ cm < \lambda_{en} < 1\ m$		$0.2\ mm < \lambda_{en} < 2\ cm$
6. Electron–neutral collision frequency		$1.0 < \nu_{ec} < 50$ MHz		$50 < \nu_{ec} < 5000$ MHz
7. Electron Debye length (λ_{de})		$6 < \lambda_{de} < 150\ \mu m$		$50 < \lambda_{de} < 700\ \mu m$
8. RF collisionality	$\omega \gg \nu_{ec}$	$\omega \gtrsim \nu_{ec}$	$\omega \gg \nu_{ec}$	$\nu_{ec} \gg \omega$
9. Sheath collisionality		$\lambda_{en} > \lambda_{in} \gg \lambda_{de} \ll S$		$\lambda_{en} > \lambda_{in} \approx \lambda_{de} \ll S$
10. RF resonance	$\omega = \omega_{ce}$	$|\omega| \approx \omega_{LH}$		

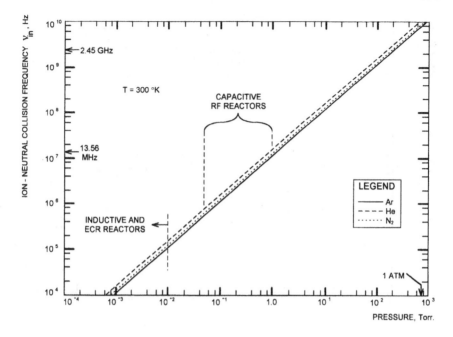

Figure 19.7. The ion–neutral collision frequency of helium, nitrogen and argon as a function of pressure at a background gas temperature of $T = 300$ K.

electron number density and the input RF power density, it is interesting to use the information just developed to examine the regimes of RF power coupling, and thereby obtain guidance as to what forms of power coupling are most useful for plasma processing in the intermediate-pressure regime, and also in the rapidly developing low-pressure regime required for microelectronic etching.

19.3.1 Operating Regimes

The operating regimes for plasma processing can be characterized by the working gas pressure and the magnetic induction in the plasma. These regimes each have their characteristic dependence of power input to the plasma on pressure, and hence active-species fluxes and processing rates. This dependence, in turn, determines whether a plasma source will have a favorable or unfavorable RF power scaling in the pressure regime in which it is required to operate.

19.3.1.1 Pressure Regimes

It is useful to divide the pressure range into *intermediate-* and *low-pressure regimes*, as discussed in section 19.2, because of the effect of pressure on the scaling of the processing effects discussed in connection with figure 19.1.

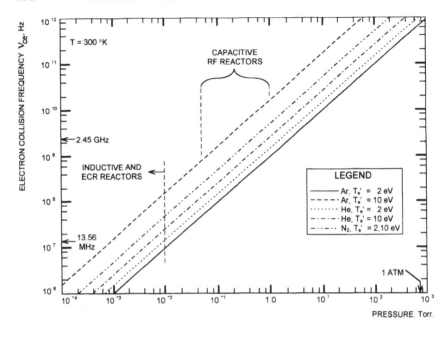

Figure 19.8. The electron–neutral collision frequency as a function of neutral gas pressure for argon, nitrogen, and helium at the electron kinetic temperatures shown.

19.3.1.2   Regimes of Magnetic Induction

In addition to gas pressure, the operating regime of plasma-processing reactors can be characterized by whether and what strength of magnetic induction is used. A plasma is *unmagnetized* if the magnetic induction is zero *or* if it is so weak that the gyroradius is larger than the plasma dimensions, and the product $\omega_c \tau_c < 1$. *Magnetized* plasmas have these relations reversed, as discussed in section 3.4.3 of Volume 1. The ion population of industrial plasmas is rarely magnetized; the electron population can be magnetized in the low-pressure regime by a few tens of millitesla.

For an unmagnetized RF plasma, the power density absorbed by the plasma is given by equation (12.15) from Volume 1,

$$P = \frac{e^2 n_e E_0^2}{2m} \frac{v_c}{\omega^2 + v_c^2} \qquad \text{W/m}^3 \qquad B = 0 \text{ or } \omega_c \ll v_c. \qquad (19.7)$$

For a magnetized RF plasma, the power density absorbed by the plasma is given by equation (12.41) from Volume 1,

$$P = \frac{e^2 n_e E_0^2}{4m} \left[\frac{v_c}{(\omega + \omega_c)^2 + v_c^2} + \frac{v_c}{(\omega - \omega_c)^2 + v_c^2} \right] \qquad \text{W/m}^3 \qquad (19.8)$$

which is appropriate when $\omega_c \tau_c \gg 1$, and the charged particle gyrates several times between collisions. The magnetized plasmas described by equation (19.8) may be at a disadvantage for many industrial applications because of a requirement for electromagnets to maintain the magnetic induction. Electromagnets, with their significant power consumption and other operational and reliability problems, are obviously not required for unmagnetized plasmas.

In addition to unmagnetized and magnetized plasmas, one can also distinguish a *resonant plasma*, for which the magnetic induction is adjusted to one of the resonant frequencies described by the *Appleton equation*. The most commonly used is the electron cyclotron resonance (ECR) frequency which, for the widely used microwave frequency of 2.45 GHz, occurs at a magnetic induction of 87.5 mT.

19.3.2 Intermediate-Pressure Unmagnetized RF Plasmas

In the atmospheric- and intermediate-pressure regimes above 6.7 Pa (50 mTorr), magnetized plasmas are rarely used for industrial applications, so the power coupling formula of equation (19.7) is appropriate. For operation in this regime at the standard frequency of 13.56 MHz, figure 19.8 indicates that the electron–neutral collision frequency for most gases is well above this value. In this limit, the power coupling can be approximated as

$$P \approx \frac{e^2 n_e E_0^2}{2m\nu_c} \quad \text{W/m}^3 \propto \frac{1}{p} \qquad \nu_c \gg \omega. \tag{19.9}$$

The power density coupled to the plasma is inversely proportional to the collision frequency, and thus to the pressure. This inverse dependence of power density on pressure is favorable. It implies that as one goes to the lower pressures necessary for directional etching or collisionless magnetron sputtering, the plasma power density, and hence the electron number density and active-species fluxes, should tend to increase with decreasing pressure.

If, however, one were to operate an intermediate-pressure plasma reactor at the microwave frequency of 2.45 GHz, then, as indicated on figure 19.8, the driving frequency, ω, would be much larger than the electron collision frequency, ν_{ec}, yielding

$$P \approx \frac{e^2 n_e E_0^2}{2m\omega^2} \nu_c \quad \text{W/m}^3 \propto p \qquad \omega \gg \nu_c. \tag{19.10}$$

In this case, the power density coupled to the plasma is directly proportional to the pressure. This is an undesirable functional dependence, since it implies that if one were to lower the pressure in order to improve directionality, then the power density, the electron number density and the active-species fluxes would decrease with the pressure. In such a situation, one is forced into an undesirable tradeoff, in which the low pressures required for high directionality and long mean free path sputtering are inconsistent with higher active-species fluxes and processing rates.

19.3.3   Low-Pressure Resonant ECR Plasmas

Figure 19.8 indicates that for electron cyclotron resonant plasmas operating in the low-pressure regime, the driving frequency ω, which is operated at the electron gyroresonance frequency $\omega = \omega_c$, is much greater than the electron–neutral collision frequency, ν_c. For such plasmas, equation (19.8) becomes

$$P \approx \frac{e^2 n_e E_0^2}{4m} \left[\frac{1}{\omega_c^2 + \nu_c^2} + \frac{1}{\nu_c^2} \right] \approx \frac{e^2 n_e E_0^2}{4m\nu_c} \propto \frac{1}{p} \qquad \omega_c = \omega \gg \nu_c. \qquad (19.11)$$

Under these resonant low-pressure conditions, the power density absorbed by the plasma depends inversely on the working neutral gas pressure. This is the same favorable situation described by equation (19.9) for unmagnetized parallel-plate reactors, in which the electron–neutral collision frequency is greater than the driving frequency. The inverse dependence of power density (and also active-species flux) on pressure is desirable, since operation at progressively lower pressures will simultaneously increase the mean free paths, improve etching directionality, and increase the flux of active species on the workpiece.

19.3.4   Low-Pressure Non-Resonant Plasmas

At low pressure, below 1.3 Pa (10 mTorr), RF plasma reactors operating at frequencies at or above the standard 13.56 MHz have driving frequencies much greater than the electron–neutral collision frequency. For this limit of low-pressure and low-frequency excitation, the power density coupled to the plasma is given by

$$P \approx \frac{e^2 n_e E_0^2}{2m} \frac{\nu_c}{(\omega^2 + \nu_c^2)} \approx \frac{e^2 n_e E_0^2 \nu_c}{2m\omega^2} \propto p \qquad \omega \gg \nu_c. \qquad (19.12)$$

In this case, the power density coupled to the plasma is directly proportional to the neutral gas pressure. This is an unfavorable scaling, which reduces the power density coupled to the plasma (and hence the active-species fluxes and processing times) at the same time the pressure is lowered to increase the mean free paths for sputtering or to improve directionality of an etching process.

Finally, if one operates, at low pressure, a magnetized, non-resonant RF plasma such that the electron cyclotron frequency is much greater than the RF driving frequency ($\omega_c \gg \omega$), a situation characteristic of 13.56 MHz RF reactors operated with a magnetized electron population, the RF power coupled to the plasma is given by equation (19.8), which is approximately

$$P = \frac{e^2 n_e E_0^2}{2m} \left(\frac{\nu_c}{\omega_c^2 + \nu_c^2} \right) \approx \frac{e^2 n_e E_0^2}{2m\omega_c^2} \propto p \qquad \omega \gg \omega \approx \nu_c. \qquad (19.13)$$

This relationship also shows an undesirable scaling of power density, proportional to the neutral gas pressure.

Those conditions that produce a linear dependence of the power density on neutral gas pressure place the designer of processing reactors in the difficult position of trading off high active-species fluxes (and hence short processing times) against the directionality and longer mean free paths associated with lower pressure. For this reason, the RF reactors that yield a direct proportionality between power density and pressure are less widely used and less desirable for most forms of plasma processing. Reactors such as the intermediate-pressure parallel-plate 13.56 MHz reactor, and the low-pressure 2.45 GHz ECR reactors, which yield an inverse dependence of power density on pressure, are more desirable for plasma-processing applications, because they allow the designer simultaneously to achieve high coupled RF power densities (and therefore higher active-species fluxes) and longer mean free paths (with their improved directionality and collisionless sputtering).

19.3.5   Comparison with Experimental Data

It is interesting to compare the theoretical functional dependence just discussed with experimental data taken in a 13.56 MHz parallel-plate reactor by Mutsukura *et al* (1994) on the etching rate and trench profiles of silicon in unmagnetized CF_4 plasmas. These data have been presented in figure 19.2, which shows that at pressures from 13.3 to 133 Pa (100 to 1000 mTorr), where figure 19.8 indicates that $\nu_c \gg \omega$, the etching rate has an inverse dependence on the neutral gas pressure in the reactor, consistent with equation (19.9). Figure 19.2 also indicates that at pressures below approximately 6.7 Pa (50 mTorr), where the driving frequency is comparable to or greater than the electron neutral collision frequency ($\omega > \nu_c$), the etching rate is proportional to the neutral gas pressure, consistent with equation (19.10). While the data of Mutsukura *et al* do not exhibit a clear-cut direct or inverse dependence on pressure, the general trend of rising etching rates in the low-pressure regime, and falling etching rates in the intermediate-pressure regime, are consistent with the general arguments presented here.

19.4   SHEATH THICKNESS ABOVE WORKPIECE

In plasma reactors intended for sputtering or etching, it is desirable that the ion mean free paths illustrated in figure 19.4 be greater than the thickness of the sheath across which the ions are accelerated before impinging on the workpiece. To assess whether this requirement is satisfied, one must be able to estimate or predict analytically the sheath thickness in a particular application. As was pointed out in sections 9.4 and 12.3 of Volume 1, the theory of DC and RF sheaths is not well understood. As a result, the discussion here depends on experimental observation and phenomenological correlation of sheath properties to provide guidance concerning this difficult subject.

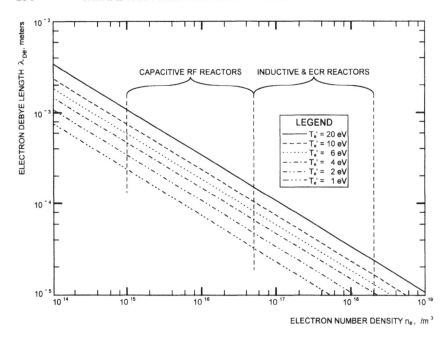

Figure 19.9. The electron Debye length as a function of electron number density for selected electron kinetic temperatures.

19.4.1 The Low-Voltage DC Sheath Model

The simple low-voltage DC sheath approximation discussed in section 9.4.1 of Volume 1 predicts a lower bound on the sheath thickness, S, of a few Debye lengths. The *electron Debye length*, derived in section 4.4 and given by equation (4.57) of Volume 1, is

$$\lambda_{de} = \left(\frac{kT_e\varepsilon_0}{n_e e^2}\right)^{1/2} = \left(\frac{T'_e\varepsilon_0}{n_e e}\right)^{1/2} = 7434\left(\frac{T'_e(ev)}{n_e}\right)^{1/2} \qquad \text{m.} \qquad (19.14)$$

The electron Debye length is plotted as a function of the electron number density in figure 19.9 for electron kinetic temperatures spanning the range likely to be found in RF glow discharge reactors. The parallel-plate RF reactors operated in the intermediate-pressure regime have Debye lengths that range from approximately 50 μm to perhaps 500 μm, while low-pressure reactors operating below 6.7 Pa (50 mTorr) tend to have electron Debye lengths ranging from 10 to 100 μm.

It is evident from figure 19.9 that the Debye sheath thickness will be far greater than the size of the smallest element of most microelectronic circuits, than the size of micro-organisms to be treated during sterilization, or the thickness of layers to be deposited or removed by plasma processing. Even if the sheath were no thicker than the electron Debye length, this dimension is still sufficiently great that one must be concerned about ion mean free path effects, especially in the intermediate pressure regime of parallel plate RF reactors.

Beyond this concern, however, is the frequent observation that the sheaths around actual workpieces are much thicker than the values indicated in figure 19.9 for the Debye sheath thickness. This observation prompted the formulation of the *Bohm presheath model*, characterized by a thick ($\approx$10–100 Debye lengths) presheath (Zheng *et al* 1995). Even if the sheath were no thicker than the electron Debye length, this dimension is still sufficiently great that one must be concerned about ion mean-free-path effects, especially in the intermediate-pressure regime of parallel-plate RF reactors.

19.4.2 Experimentally Determined Sheath Thickness

Because of the poor state of RF sheath theory (see section 12.6 of Volume 1), it is necessary to appeal to experiment for an estimate of the sheath thickness to be expected in industrial glow discharge processing reactors. Such an experimental investigation has been reported by Mutsukura *et al* (1994), who used optical emission spectrometry to measure the plasma sheath thickness in a parallel-plate RF glow discharge plasma reactor operated over the pressure range from 1.3 to 133 Pa (10 to 1000 mTorr). The sheath thickness was measured as a function of pressure and input power for a variety of gases that included the monatomic inert gases helium and argon, and the polyatomic reactive gases oxygen, methane, and carbon tetrafluoride.

Mutsukura *et al* determined the sheath thickness by scanning the light intensity profile above the workpiece, integrated across the discharge diameter. Their results for three reactive polyatomic gases are shown in figure 19.10(a), in which the sheath thickness is plotted on log–log paper as a function of pressure. Their results for the inert monatomic gases helium and argon are plotted in figure 19.10(b). Many of their data fall along straight lines, indicating a power law relation between the observed sheath thickness, and the working gas pressure.

The straight line portions of the curves in figures 19.10(a) and (b) each have approximately the same slope, thus making it possible to identify a power law relationship between the neutral gas pressure and the sheath thickness in centimeters,

$$p^n S = \text{constant} \quad \text{or} \quad S = S_0 p^{-n} \quad \text{m.} \tag{19.15}$$

Equations (19.15) provide an approximation to the sheath thickness over most of the pressure range of interest in parallel-plate RF glow discharge plasma reactors operated at 13.56 MHz.

In addition to identifying the power law of equation (19.15), Mutsukura *et al* have also found that the exponent n is approximately equal to one-third for the reactive polyatomic gases, and is approximately equal to 0.5 for the inert

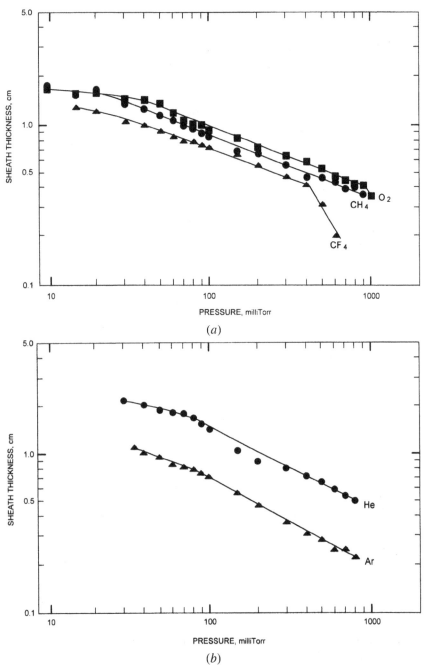

Figure 19.10. The sheath thickness measured by Mutsukura *et al* (1994) as a function of pressure: (*a*) for plasmas generated with the polyatomic gases O_2, CH_4, and CF_4 and (*b*) for plasmas generated with the monatomic gases He and Ar.

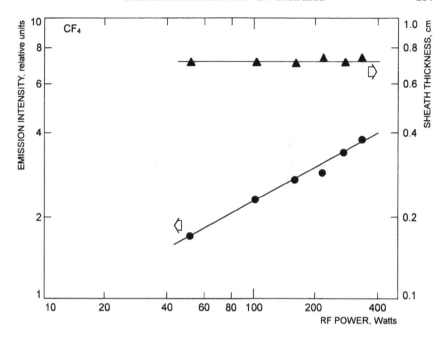

Figure 19.11. The sheath thickness and actinometric maximum emission intensities in CF_4 plasmas at a gas pressure of $p = 53$ Pa (0.4 Torr) (from Mutsukura *et al* 1994).

monatomic gases. The approximate range of validity of equation (19.15) is over pressures from 6.7 to 133 Pa (50 mTorr to 1 Torr).

When the data of figure 19.10 are compared with the electron Debye distance of figure 19.9, it is clear that the observed sheath thickness is between a factor of 10 and a factor of 100 times larger than the electron Debye length, a significant result, consistent with the common observation that the plasma sheaths in glow discharges are much thicker than a few electron Debye lengths.

19.4.3 Dependence of Sheath Thickness on RF Power

In addition to measuring the sheath thickness as a function of pressure, Mutsukura *et al* (1994) also measured the sheath thickness as a function of total RF power delivered to their 13.56 MHz parallel-plate reactor. Their results are shown in figure 19.11, and indicate that the sheath thickness is independent of the total RF power delivered to the plasma over nearly a factor of 10 in the latter quantity. This range spans most of the intermediate-pressure operating regime of parallel-plate RF reactors.

This result can be justified theoretically in the following way. The power density delivered to a intermediate-pressure ($p > 6.7$ Pa (50 mTorr)) unmagnetized RF plasma is given by equation (19.7) in the limit for which the

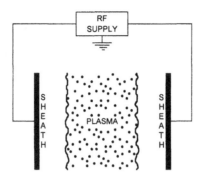

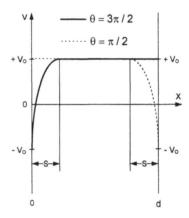

Figure 19.12. Sheath characteristics of a parallel-plate RF reactor with sheath thickness S and maximum RF voltage V_0.

driving frequency $\omega = 13.56$ MHz is much less than the collision frequency $(\nu_c \gg \omega)$,

$$P_{(x)} = \frac{e^2 n_e E^2}{2m\nu_c} = \frac{e\mu_e n_e E^2}{2} \qquad \text{W/m}^3 \qquad (19.16)$$

where the parameter μ_e is the electron mobility.

We adopt the sheath model for a symmetric, parallel-plate plasma reactor shown in figure 19.12. In this model, the sheath potential profile near the instantaneous cathode is the same as that in the cathode fall region of a DC normal glow discharge. The RF electric field is confined to the instantaneous cathode sheath of thickness S, which corresponds to the *cathode fall distance* in a DC normal glow discharge. The sheath profiles are linear in electric field (*Aston's law*), parabolic in potential, and alternate between the two electrodes when they become the instantaneous cathode during each negative half cycle. For the sake of simplicity, we assume that the potential of the bulk plasma (the *positive column*) remains near that of the instantaneous anode, with a negligible *anode fall voltage*.

The RF voltage drop across the instantaneous cathode sheath varies between $+V_0$ and $-V_0$, to produce the potential profiles diagrammed at the bottom of figure 19.12. In this model, the total power delivered to the plasma is dissipated in the sheath of cross-sectional area A m^2 and axial thickness S. The total power input to the sheath is given by

$$P = 2AS\bar{p} = eAS\mu_e \times \frac{1}{S} \int_0^S n_e(x) E^2(x) \, dx \qquad \text{W} \qquad (19.17)$$

where equation (19.16) has been substituted for the sheath power density.

It is further assumed that both the electron number density and the electric field in the sheath are functions of distance from the electrode. The instantaneous current I is constant along the axis of the plasma, and is given by

$$I = J_1 A_1 = A_1 e n_e v_{de} = A_1 e n_e \mu_e E \qquad (19.18)$$

where J_1 is the current density flowing in the sheath normal to the electrode, and A_1 is the cross-sectional area of the sheath. The electron drift velocity in the sheath has been written in terms of the electron mobility and the electric field in the last term of equation (19.18). Solving for the product of electron number density and electric field yields the relation

$$n_e E = \frac{I}{A_1 e \mu_e} \approx \text{constant.} \qquad (19.19)$$

Thus, the product of the electron number density and electric field is a constant independent of the coordinate x, since all the parameters on the right-hand side of equation (19.19) are approximately constant in the sheath. Substituting equation (19.19) into equation (19.17) yields the power delivered to the plasma,

$$P = I \int_0^S E(x) \, dx. \qquad (19.20)$$

To estimate the functional dependence of the electric field on the distance x, we now adopt the *quasi-static Aston model* for the left-hand RF sheath, in which the potential on the instantaneous cathode and at the sheath-negative glow plasma boundary is taken to be

$$V = V_c = V_0 \sin \omega t \qquad (19.21a)$$

at $x = 0$, and

$$V = V_0 \qquad (19.21b)$$

at $x = S$, respectively. The RF electric field is assumed to obey *Aston's law* of DC glow discharges, discussed in connection with equation (9.4) in Volume 1.

Aston's law states that in the cathode fall region, the electric field at any point in time decreases linearly from the cathode to the plasma boundary, and can be written

$$E = \frac{dV}{dx} = C(S - x).$$ (19.22)

Integrating the potential from the middle expression of equation (19.22) yields

$$\int_{V_c}^{V} dV = V - V_c$$ (19.23)

whereas integration of the right-hand expression of equation (19.22) yields

$$\int_0^x C(S - x)\,dx = Cx\left(S - \frac{x}{2}\right).$$ (19.24)

Equating equation (19.23) and equation (19.24) yields

$$V(x) = V_c + Cx\left(S - \frac{x}{2}\right).$$ (19.25)

Substituting the boundary conditions of equations (19.21a) and (19.21b) into equation (19.25) allows the constant C to be determined,

$$C = -\frac{2V_c}{S^2}.$$ (19.26)

Substituting this constant into equation (19.25) yields an expression for the sheath potential profile as a function of the distance x,

$$V(x) = V_c\left(1 - \frac{x}{S}\right)^2.$$ (19.27)

Taking the derivative of equation (19.27) to find the electric field in the sheath yields

$$E(x) = \frac{dV}{dx} = -\frac{2V_c}{S}\left(1 - \frac{x}{S}\right).$$ (19.28)

Substituting this expression for the sheath electric field profile into equation (19.20) and performing the indicated average yields

$$P = -I\int_0^S \frac{2V_c}{S}\left(1 - \frac{x}{S}\right)dx = -IV_c \qquad \text{W}.$$ (19.29)

Equation (19.29) yields the total power input into the parallel-plate RF glow discharge, which is independent of the sheath thickness S, a result consistent with the data of Mutsukura et al shown in figure 19.11. Their data and the previous physical arguments indicate that the RF sheath thickness is not a function of the total power dissipated in the discharge, a result analogous to that from Townsend theory for the cathode region of a DC normal glow discharge, in which the cathode fall distance, d_c, is determined by the Stoltow point, and not by the power dissipated in the discharge.

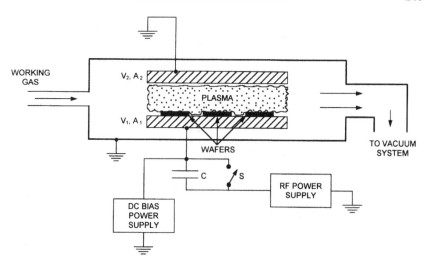

Figure 19.13. DC biasing modes of a parallel-plate RF reactive ion etching plasma reactor: (*a*) in the floating mode with the switch S open; (*b*) in the grounded mode with the switch S closed; and (*c*) in the biased mode, with the switch S open and a DC bias voltage imposed by the bias power supply.

19.5   RF SHEATH PHENOMENOLOGY

To produce a highly directional etch for microelectronic circuit fabrication, or a high sputtering yield for plasma sputter deposition, it is necessary that the ion mean free path for elastic collisions be greater than the thickness of the sheath between the plasma and the workpiece. In addition, if the ions are to be useful as energetic agents for plasma processing, they must have energies of at least several electronvolts to promote etching reactions, and several tens of electronvolts to induce sputtering. In this section, we discuss aspects of RF sheath phenomenology which have a bearing on the potential drop across the sheath, and hence on the energy of the ions which arrive at the surface of a workpiece.

19.5.1   Wafer Biasing

In order to ensure that ions have sufficient energy to promote etching or other plasma-processing reactions, the surface of the workpiece must have an electrostatic potential at least several electronvolts below that of the plasma in which the ions originate. Ions that fall across this *sheath potential* without colliding arrive at the workpiece surface with an energy equal to the potential drop. If collisions occur, the ions lose energy, and acquire velocity components parallel to the workpiece surface that make it impossible to perform a vertical etch.

Workpieces can be biased with respect to the plasma in at least three ways, indicated schematically on figure 19.13 for a parallel-plate RF plasma reactor. One method is to leave the switch S closed, and directly power the electrode with the RF power supply, leaving the workpieces on an electrode having a low impedance to ground. Alternatively, one can open the switch S, and float the powered electrode using the blocking capacitor C. The workpieces assume a DC *floating potential* that results when ions and electrons from the plasma impinge on the powered electrode at equal rates. Other effects of an applied RF voltage also may come into play, as will be discussed later. A third alternative is to leave switch S open, and bias the electrode with a DC power supply. Such negative DC bias voltages will increase the energy of ions impinging on the workpiece. Finally, the applied DC bias applied to a workpiece can be modulated periodically to manipulate the chemistry of the etching or deposition process (Bounasri *et al* 1993).

19.5.2   Floating DC Potential

The floating DC potential expected if the switch S on figure 19.13 is open, and no DC bias is applied to the electrode containing the wafers can be estimated, based upon section 9.4 of Volume 1. For the *low-voltage DC sheath approximation*, the simplest of the DC sheath theories, the *floating potential* of an electrically isolated electrode brought into contact with a plasma is given by equation (9.132) of Volume 1,

$$V_F = -\frac{1}{2} T_e' \ln \left(\frac{M T_e'}{m T_i'} \right). \qquad (19.30)$$

In this expression, the electron kinetic temperature T_e', the ion kinetic temperature T_i', the ion mass M, and the electron mass m all come into play.

The floating potential given by equation (19.30) is plotted (the upper curves) as a function of the electron kinetic temperature in figure 19.14 for helium, nitrogen, and carbon tetrafluoride ions. Most RF parallel-plate glow discharges used for plasma processing have electron kinetic temperatures between 2 and 10 eV, thus yielding floating potentials in equation (19.30) ranging from 15 to 80 V. These energies are consistent with those measured with retarding potential energy analyzers in parallel-plate RF glow discharge reactors. The simple sheath theory that yields equation (19.30) also predicts a sheath only a few Debye lengths thick. This prediction is inconsistent with the much thicker sheaths experimentally observed, and discussed earlier in connection with the data of Mutsukura *et al* (1994).

The *Bohm sheath model* discussed in section 9.4.2 of Volume 1 is more consistent with the observations of Mutsukura *et al*, since it postulates a thick, quasi-neutral presheath across which a profile of decreasing electron number density and potential exists. The Bohm sheath model predicts a floating potential

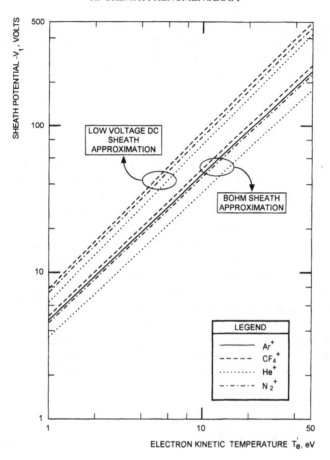

Figure 19.14. The floating sheath potential with the switch S open in figure 19.13, predicted by the low-voltage DC sheath approximation given by equation (19.30) (upper curves), and given by the Bohm sheath approximation of equation (19.31) (lower curves).

on an insulated electrode in contact with a glow discharge plasma given by

$$V_F = -\frac{T'_e}{2} \ln \left(\frac{M}{2\pi m} \right). \tag{19.31}$$

This floating potential is negative, and directly proportional to the electron kinetic temperature in the plasma. The logarithmic term contains only the ratio of the ion to electron mass. Equation (19.31) is plotted as the four lower curves in figure 19.14.

The Bohm sheath model predicts a floating potential for RF parallel-plate glow discharge plasmas approximately a factor of two lower than that of the low-voltage DC sheath model discussed previously. It predicts, for electron kinetic temperatures between 2 and 10 eV, floating potentials ranging from about 10 to 40 V. This range is also consistent with measurements of the energy of ions reaching the surface of electrodes of parallel-plate reactors, so the data are not useful in distinguishing between these two sheath models.

19.5.3   Sheath Transit Time Effects for Ions

The DC floating potential discussed in connection with figure 19.14 is modified by the application of an RF, time-varying voltage and the effects of the ion current drawn across the sheath (Godyak *et al* 1991). The sheath between the plasma and the surface of the workpiece can be approximated by the equivalent electrical circuit shown in figure 19.15. With respect to DC currents, the sheath acts like a diode that allows ions to flow from the plasma to the workpiece surface, but not in the reverse direction. The equivalent circuit of the sheath also contains a capacitance, C, associated with the plane-parallel geometry of the sheath, and a resistance R, to account for the energy dissipated in the sheath.

Whether the sheath behaves capacitively or resistively depends on the *critical sheath frequency*, ν_{sh}. If the critical sheath frequency is less than the applied RF frequency, ions can transit the sheath in less than one period of the RF oscillation; the sheath will behave resistively with respect to ions. If the RF frequency is greater than the critical sheath frequency, ions are not able to transit the sheath during one period of RF oscillation (although electrons may), and the sheath behaves capacitively. In the capacitive regime, ions arrive at the workpiece surface as the result of a DC component of the sheath electric field; not as a direct result of RF electric fields.

The critical sheath frequency may be obtained by considering the force on an ion of mass M in the sheath,

$$F = Ma = eE_s \qquad N \qquad (19.32)$$

where a is the acceleration of the ion, and E_s is the sheath electric field. The sheath electric field may be approximated by the average of equation (19.28) over the sheath thickness S, and is

$$E_s = \frac{V_1}{S} \qquad V/m. \qquad (19.33)$$

Using equation (19.33), the acceleration of an ion in the sheath may be found from equation (19.32), and is

$$a = \frac{eV_1}{MS} \qquad m/s^2. \qquad (19.34)$$

The *critical sheath frequency* is the inverse of the *transit time* T of an ion across the sheath thickness,

$$\nu_{sh} = \frac{1}{T}. \qquad (19.35)$$

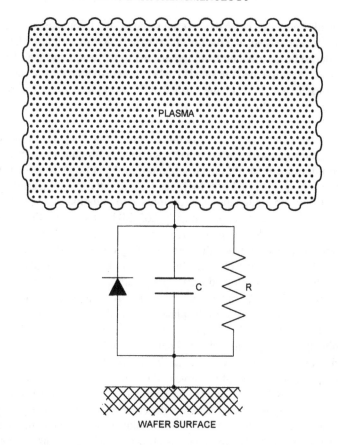

Figure 19.15. An equivalent circuit for the sheath between a plasma and the surface of a wafer.

The relationship between the sheath thickness S, the ion acceleration a, and the transit time T, is given by

$$S = \frac{1}{2}aT^2 = \frac{a}{2v_{sh}^2} = \frac{eV_1}{2MSv_{sh}^2} \qquad \text{m} \qquad (19.36)$$

where the transit time T has been replaced by the sheath frequency of equation (19.35). Equation (19.36) can be solved for the critical sheath frequency,

$$v_{sh} = \frac{1}{S}\sqrt{\frac{eV_1}{2M}} \qquad \text{Hz.} \qquad (19.37)$$

The data of Mutsukura *et al* (1994) in figure 19.10 indicate that the sheath thickness in their reactor is approximately $S = 0.5$ cm, with a sheath voltage drop $V_1 = 20$ V. For an argon ion transiting a sheath with these characteristics, the critical sheath frequency is 1 MHz. At the standard RF frequency of 13.56 MHz, the sheaths are capacitive, and the ions must experience an additional DC electric field in order to be accelerated from the plasma to the wafer surface. The sheath in the previous example becomes resistive below approximately 1 MHz. At about 100 kHz, ions are able to reach the surface of the workpiece directly during a half cycle of the RF. The effects of resistive or capacitive sheath operation on the ion energy distribution function are discussed later.

Since processing plasmas are in a quasi-steady state, they must lose electrons and ions at the same average rate. The electrons are highly mobile, can travel across the interior dimension of most vacuum systems in a small fraction of an RF cycle at a frequency of 13.56 MHz, and are usually lost to nearby grounded surfaces. If one substitutes the electron mass into equation (19.37), the critical frequency *for electrons* is approximately 265 MHz, far above the frequencies normally used in RF glow discharge reactors. Above this frequency, neither the ions nor the electrons can reach the surrounding grounded surfaces during an RF cycle. The plasma becomes electrically isolated, in that real currents cannot flow across the sheath as a result of the RF voltage applied to the electrodes. It is in this (microwave) regime, above the critical sheath frequency for electrons, that the interaction of RF energy with the plasma enters the *collective mode* discussed in chapter 13. In this mode, the plasma behaves more like a dielectric material than a collection of individual charged particles, and the plasma is energized by displacement rather than real currents (Batenin *et al* 1992, Raizer *et al* 1995).

The equivalent circuits of resistive and parallel-plate RF plasmas as a whole are shown in figures 19.16(*a*) and (*b*), respectively. Figure 19.16(*a*) is appropriate when the RF driving frequency is below the critical sheath frequency. Since ions and electrons can reach the workpiece surface during an RF cycle, the sheath acts like a resistor. When the RF frequency is above the critical sheath frequency, the model illustrated in figure 19.16(*b*) is appropriate, in which the sheath behaves as a capacitor. The diode in each case is present to account for the DC current of ions driven by the positive DC floating potential of the plasma.

19.5.4   Ion Energy by Sheath Acceleration

The energy distribution function of ions reaching the surface of a workpiece during plasma processing depends upon whether the driving RF frequency is below or above the critical sheath frequency just discussed (see Hope *et al* 1993, Olthoff *et al* 1994). In figure 19.17 the characteristic data cited by Coburn and Winters (1979) on the energy distribution function of ions reaching the surface of a plane parallel RF glow discharge reactor using argon at a pressure of 6.7 Pa (50 mTorr) are shown. The 100 kHz data, below the approximately 1 MHz critical sheath frequency, were recorded using the Ar^+ ion, whereas the 13.56 MHz data, well above the critical sheath frequency, were recorded using the Ar_2^+ ion. At 100 kHz, low-energy ions from the plasma are able to travel across the sheath and reach the energy analyzer during most of the RF cycle. The workpiece

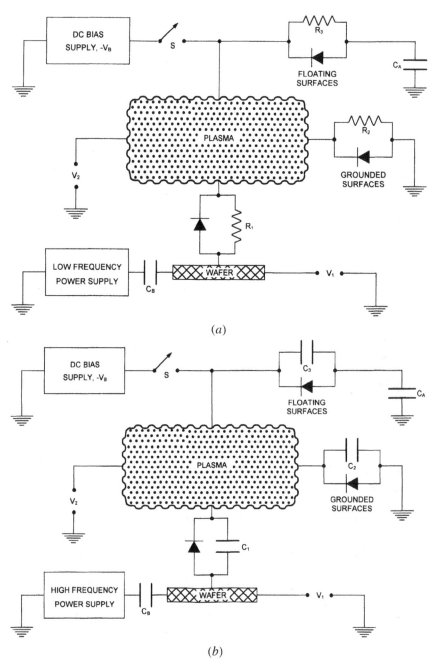

Figure 19.16. The equivalent circuits for an RF etching plasma in two frequency regimes: (*a*) the low-frequency resistive sheath regime below 1 MHz; and (*b*) the high-frequency capacitive sheath regime above 1 MHz.

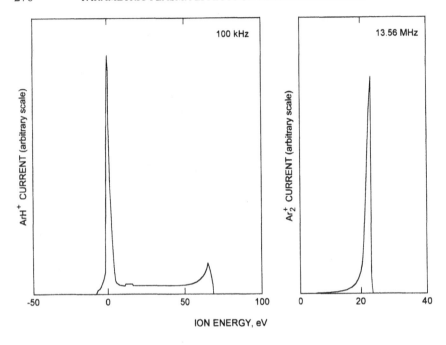

Figure 19.17. Ion energy spectra at the surface of a wafer exposed to an RF parallel-plate glow discharge plasma operating at frequencies of 100 kHz and 13.56 MHz (Coburn and Winters 1979).

assumes the instantaneous plasma potential, creating no electric field between the plasma and the electrode to accelerate the ions. The distribution of ions, which extends with a low intensity up to approximately 65 eV, is the result of ions accelerated across the sheath during the maximum of the RF waveform. When the same discharge was operated at 13.56 MHz, a DC potential of approximately 20 V appeared between the plasma and electrode, yielding a nearly delta function distribution of Ar_2^+ ions at that energy.

As was discussed in section 12.3 of Volume 1, the analytical theory of RF sheaths is not in a satisfactory state, although rapid progress appears to be occurring in the computer modeling of simple RF discharges with particle-in-cell computational techniques. In section 12.3.3, it was shown that if equal ion currents flow to both electrodes in an asymmetric parallel-plate discharge like that shown in figure 19.13, the following relationship holds,

$$\frac{V_1}{V_2} = \left(\frac{A_2}{A_1} \right)^2 \tag{19.38}$$

where V_1 and A_1 are, respectively, the sheath voltage drop and area of the powered electrode, and V_2 and A_2 are the voltage drop and surface area of the grounded electrode.

The scaling relation of equation (19.38) is consistent with a phenomenological scaling law obtained in the microelectronics industry that relates the sheath voltage drops to the RF electrode areas by the power law relation:

$$\frac{V_1}{V_2} = \left(\frac{A_2}{A_1}\right)^q \qquad 1 \leq q \leq 2.5 \qquad (19.39)$$

where the parameter q depends on operating conditions, and can vary between 1 and 2.5. This range includes the equal ion current model of equation (19.38) at $q = 2.0$.

It often is not possible to bias the wafers being etched, or workpieces undergoing other forms of plasma processing, to the desired voltages by adjusting the electrode area ratios. One approach is to use the arrangement of figure 19.13, in which the desired negative bias on the workpiece is achieved by a DC power supply connected to the powered electrode, with the switch S open. An example is illustrated in figure 19.18 from Reinke *et al* (1992), in which the electrode containing the workpiece was biased at the negative voltages indicated. The resulting ion energy distribution functions were measured in a methane plasma in the low-pressure regime at 0.055 Pa (0.41 mTorr). The biasing of the wafer clearly increased the energy of the ion population, although at the expense of a somewhat reduced ion flux.

19.6 FORMATION OF ACTIVE SPECIES

The primary function of a glow discharge plasma in processing applications is to generate the ions and chemically active species required. These species result from inelastic electron–neutral collisions by an electron population with a characteristic electron kinetic temperature ranging from 2 to 10 eV. These relatively high kinetic temperatures (as compared with room temperature neutral species) provide enough energetic electrons to produce such inelastic events as ionization, excitation, and molecular breakup when they collide with the plasma constituents.

Relatively few inelastic collisions result in an ionization event, and the reaction rates for the production of excited states, molecular fragments, and other neutral active species are much greater than that for ionization. The lifetime of most molecular fragments and neutral active species is at least as long as that of the ions, with the result that the number density of neutral active species is usually greater (by one or more orders of magnitude) than the ion/electron number density in the plasma. Because the plasma is quasi-neutral and nearly all ions are singly charged in industrial glow discharge plasmas, the ion and electron number densities are approximately equal.

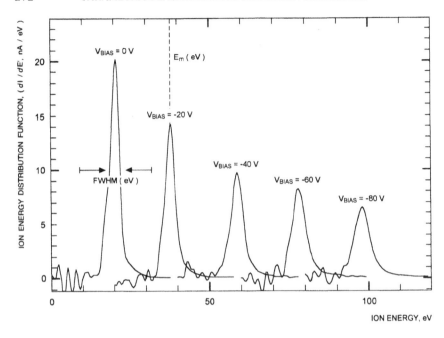

Figure 19.18. Ion energy spectra measured *in situ* at a wafer surface as a function of bias voltage on the wafer. From Reinke *et al* (1992).

19.6.1   Electron-Based Reactions

In industrial glow discharge plasmas, the characteristic energy of the ion population in the plasma is low, not far above that of the working neutral gas at a kinetic temperature of 0.025 eV. This low energy is the result of frequent ion–neutral collisions associated with a small ionization fraction, and the fact that the masses of the ion and neutral background species are comparable, allowing the ions to lose a large fraction of any energy they may have gained from electric fields in a single collision. The low ion energy greatly reduces the probability of their contributing to the formation of active species through inelastic collisions. The inelastic collisional processes of electrons are therefore of primary interest in industrial glow discharges.

19.6.1.1   Ionization

The energetic electrons in an RF parallel-plate glow discharge plasma reactor can produce *ionization reactions*, of which a few characteristic examples are listed in equation (19.40):

$$e + Ar \rightarrow Ar^+ + 2e \qquad (19.40a)$$

$$e + O_2 \rightarrow O_2^+ + 2e \qquad (19.40b)$$

$$e + O \rightarrow O^+ + 2e. \qquad (19.40c)$$

These reactions produce ion–electron pairs directly, and are necessary to maintain the plasma ionization fraction in a steady state.

19.6.1.2 Dissociation

Equations (19.41) illustrate *dissociation reactions* in which energetic electrons fragment the molecules with which they collide, knocking off active species such as atomic oxygen or fluorine. Such atomic species are highly reactive chemically and can participate in etching reactions.

$$e + O_2 \rightarrow O + O + e \qquad (19.41a)$$
$$e + CF_4 \rightarrow CF_3 + F + e \qquad (19.41b)$$
$$e + CF_3 \rightarrow CF_2 + F + e \qquad (19.41c)$$
$$e + CF_2 \rightarrow CF + F + e \qquad (19.41d)$$
$$e + CF \rightarrow C + F + e. \qquad (19.41e)$$

If the principal product of an initial dissociation reaction has a long enough lifetime, it can undergo further dissociation reactions, such as the sequence of dissociations illustrated for carbon tetrafluoride in equations (19.41b)–(19.41e).

19.6.1.3 Dissociative Ionization

Three examples of electron-induced *dissociative ionization reactions* are illustrated in equations (19.42):

$$e + O_2 \rightarrow O^+ + O + 2e \qquad (19.42a)$$
$$e + CF_4 \rightarrow CF_3^+ + F + 2e \qquad (19.42b)$$
$$e + CF_3 \rightarrow CF_2^+ + F + 2e. \qquad (19.42c)$$

In these reactions, the electrons are sufficiently energetic not only to dissociate the target molecule, but also to ionize the heaviest fragment of the reaction. This process produces two highly energetic active species, as well as an ion–electron pair.

19.6.1.4 Dissociative Attachment

Finally, some *electrophilic* (electron attaching) gases can undergo *dissociative attachment*, an example of which is

$$e + SF_6 \rightarrow SF_6^- \rightarrow SF_5^- + F. \qquad (19.43)$$

In this reaction, an energetic electron undergoes attachment to sulfur hexafluoride, which retains enough internal energy to dissociate into a sulfur pentafluoride ion and atomic fluorine. The atomic fluorine is highly chemically reactive.

19.6.2   Species Concentrations in Processing Plasmas

The inelastic collisions of electrons may result in ionization and the production of ion–electron pairs; the production of neutral and ionized active species which participate directly in the plasma-processing reactions of interest; and the production of other neutral and ionized molecular fragments which do not participate in processing reactions. Some of these intermediate reaction products may, after further electron collisions, produce the active species required for specific processing applications.

19.6.2.1   Neutral Active Species

The ionization fraction in RF glow discharges used for plasma processing under vacuum appears in figure 19.19 as the parameter on the sloping lines. At the upper left of this graph is a limit line which represents the fully ionized state of a working neutral gas, the number density of which is related to pressure by the perfect gas law. The number density is given in terms of the pressure in Torr for a gas with a temperature $T = 300$ K by

$$n = 3.22 \times 10^{22} p \,(\text{Torr}) \quad /\text{m}^3. \tag{19.44}$$

If a glow discharge plasma contains only singly charged ions, the 'fully ionized' line at the upper left of figure 19.19 represents the maximum electron number density possible in singly ionized industrial glow discharge plasmas.

If the ordinate of figure 19.19 is identified with the electron number density, the *intermediate-pressure regime* frequently used in parallel-plate RF glow discharge reactor technology, is shown to the lower right. Such plasma reactors characteristically have small ionization fractions, between 10^{-4} and 10^{-6}. In such intermediate-pressure plasmas, the number density of neutral molecular fragments and other neutral active species may be many orders of magnitude higher than the electron or ion number densities. In the low-pressure regime below 1.3 Pa (10 mTorr), the ionization fractions are significantly greater, ranging from approximately 10^{-4} to as high as 10% in some ECR reactors operating at pressures below 0.13 Pa (1 mTorr). In the low-pressure regime, the number density of active species may be comparable to or greater than the electron number density, but by a smaller margin than in the intermediate-pressure regime.

One reason for a lower concentration of active species relative to the electron number density in the low-pressure regime is the longer mean free paths of active species in that regime. These species may collide more frequently with the walls than with electrons or other active species in the plasma itself. Heterogeneous wall reactions lead to de-excitation, recombination, and polymerization, all of which lower the number density of active species and molecular fragments relative to the electron number density in the plasma.

A characteristic example of the relative concentration of molecular fragments measured in an RF glow discharge used for etching is listed in

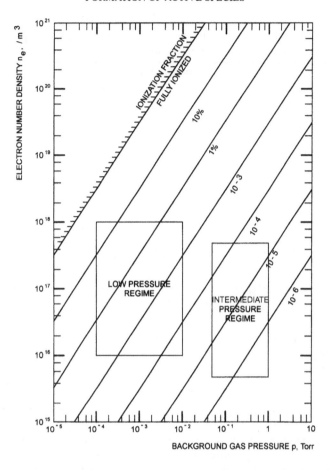

Figure 19.19. Plasma electron and active-species number densities as functions of the background gas pressure in Torr. A fully ionized background gas is shown as the last curve to the upper left. The low- and intermediate-pressure regimes for microelectronic etching are indicated schematically.

table 19.4, taken from Vossen and Kern (1978). This table shows the principal fragments inferred from the relative intensity of mass spectrometric peaks, when the etching gas was bombarded by 70 eV electrons. Several combinations of working gas and materials to be etched were investigated. The intensity of the carbon trifluoride (CF_3) mass spectrometric peak was taken as a normalizing value of 100 units, and the intensities of other species given relative to this value. In some instances, molecular fragments stripped all the way down to atomic carbon and fluorine are evident above the noise level of the mass spectrometer.

Table 19.4. Some halocarbon gases for plasma etching and their principal fragments from mass spectrometer data (Vossen and Kern 1978).

Etching gas	Etchable materials	Principal fragments and relative intensities from mass spectra					
CF_4	Si, SiO_2, Si_3N_4, Ti, Mo, Ta, W	CF_3 100	CF_2 12	F 7	CF 5		
C_2F_6	SiO_2, Si_3N_4	CF_3 100	C_2F_5 41	CF 10	CF_2 10	C 2	F 1
$CClF_3$	Au, Ti	CF_3 100	$CClF_2$ 21	CF_2 10	Cl 7	CF 4	
$CBrF_3$	Ti, Pt	CF_3 100	$CBrF_3$ 15	$CBrF_2$ 14	Br 5		
C_3F_8	SiO_2, Si_3N_4	CF_3 100	CF 29	C_3F_7 25	CF_2 9	C_2F_5 9	C_2F_4 7
CCl_4	Al, Cr_2O_3, Cr	CCl_3 100	Cl 41	CCl 41	CCl_2 29		

19.6.2.2 Ionized Active Species

Some RF glow discharge plasmas, especially those with *electrophilic* (electron attaching) species such as nitrogen (N_2) and sulfur hexafluoride (SF_6), produce significant numbers of negative ions. Because of the requirement for quasi-neutrality, the number density of positive and negative charges in the plasma must be equal, as shown in equation (19.45) for the ionic species S,

$$n(S^+) = n(e) + n(S^-). \tag{19.45}$$

Only the most electrophilic species (such as sulfur hexafluoride, SF_6) will have a higher concentration of negative ions than of electrons. In RF glow discharge processing plasmas, the number density of positive ions will normally be much greater than the number density of negative ions,

$$n(S^+) > n(S^-). \tag{19.46}$$

19.6.3 Parametric Dependence of Active-Species Production

The length of time required to process a workpiece in a glow discharge is a function of the active-species flux, the energy and power fluxes of ions which may promote the processing reaction, and the chemistry of the processing reaction itself. The *random ion flux* from a plasma, Γ_i, is given by

$$\Gamma_i = \frac{1}{4}n_i\bar{v}_i = \frac{1}{4}n_i\sqrt{\frac{8eT_i'}{\pi M}} \qquad \text{ions/m}^2\text{-s.} \qquad (19.47)$$

The ion and active-species fluxes will be roughly proportional to one another, since the same reactions (discussed earlier) which produce ions also yield other active species by dissociation, dissociative ionization, or dissociative attachment. The exact ratio of ions to active species depends on such factors as the details of the electron energy distribution function and the pressure of the working gas.

The ion flux given by equation (19.47) is proportional to the ion (and hence to the electron) number density in the plasma. The electron number density is, in turn, approximately proportional to the RF power density coupled to the plasma, and to the neutral gas pressure, as discussed in section 19.3. Therefore, the most direct way to increase the flux of ions or active species on a wafer, and to decrease the processing time, is to increase the electron number density by either increasing the RF power density or operating in a pressure regime for which the RF power coupling is increased. The ion flux is also proportional to the mean *ion thermal velocity*, but this parameter has very little scope for variation, since the ion kinetic temperature T_i' is approximately room temperature, and the ion mass is determined by the requirements of the processing operation.

A global model of high density plasma discharges in argon, oxygen, chlorine, and other gases has been published by Lee and Lieberman (1995). They find that the electron kinetic temperature in the plasmas they modeled decreases with increasing pressure over the range from below 0.13 to nearly 133 Pa (1 mTorr to 1 Torr), a result consistent with higher electron collision frequency and greater electron energy loss at higher pressures. They also find that the positive ion density increases with pressure over the same range for argon and chlorine, whereas the ion density for an oxygen plasma reaches a maximum at approximately 2.6 Pa (20 mTorr), and then declines at higher pressure.

19.7   THE EFFECT OF ELECTRON MAGNETIZATION ON ACTIVE-SPECIES CONCENTRATION

A charged species in a plasma is said to be *magnetized* if the gyroradius of individual particles in an applied magnetic field is smaller than the plasma radius, *and* they gyrate more than one full turn between collisions. The ion population is rarely magnetized in industrial glow discharge plasmas, but in the low-pressure regime below 6.7 Pa (50 mTorr), it is possible to magnetize the electron population in magnetic inductions of a few tens of millitesla, values that may be achieved inexpensively with simple coils or permanent magnets. There is evidence in the literature (Mayer and Barker 1982) that the degree of magnetization of the electron population in a glow discharge plasma can strongly affect the relative concentration of active species in glow discharge plasmas of interest in plasma processing, and particularly in etching.

19.7.1   Experimental Study of Electron Magnetization

The magnetic confinement and *magnetization* of electrons in a glow discharge plasma can have significant effects on the relative concentrations of molecular fragments among its active species. In general, the better confined and more highly magnetized are the electrons, the more fragmented are the molecules and molecular fragments remaining in the plasma. This effect was measured by Mayer and Barker (1982), who used the apparatus shown in figure 19.20. They operated the electron bombardment discharge chamber of a *Kaufman ion source* (see section 6.5 of Volume 1) on the right, using carbon tetrafluoride (CF_4) gas at a pressure of approximately 0.1 Pa (0.8 mTorr) in the discharge chamber. The active species produced in the discharge chamber flowed into a high vacuum region in which a workpiece was undergoing etching. The ions and neutral species in this efflux were measured with the quadrupole mass spectrometer shown on the left of figure 19.20. The magnetic induction in the ion source was generated by electromagnets and was varied from 0 to 16 mT.

19.7.2   Electron Magnetization

The magnetization of electrons in the discharge chamber is characterized by the product $\omega\tau_c$, which is the ratio of the electron gyrofrequency to the electron collision frequency with the carbon tetrafluoride working gas in the discharge chamber. This ratio, which must be greater than one to ensure magnetization of the electrons, is given by

$$\omega\tau_c \approx \frac{\nu_{ge}}{\nu_{ce}} \approx \frac{eB}{m_e n_0 \langle \sigma v \rangle_{ec}} \gg 1. \tag{19.48}$$

The electron collision frequency in the discharge chamber, for primary electrons of energy 30 eV, and a neutral working pressure of 0.1 Pa (0.8 mTorr) of carbon tetrafluoride, is approximately

$$\nu_{ec} \approx 5.1 \times 10^7 \ \text{Hz}. \tag{19.49}$$

Under these conditions the magnetization parameter of equation (19.48) is given by

$$\frac{\nu_{ge}}{\nu_{ce}} \approx 0.55B \ (\text{mT}) \gg 1 \tag{19.50}$$

where the magnetic induction B is expressed in millitesla.

The second of the magnetization conditions discussed in section 3.4.3 of Volume 1 requires that the gyroradius of the electrons be smaller than the

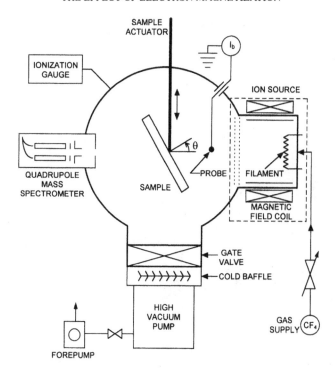

Figure 19.20. Schematic diagram of an ion-beam etching apparatus used to measure the flux of ion and neutral molecular fragments produced by an electron bombardment (Kaufman) discharge chamber, taken from Mayer and Barker (1982).

dimensions of the plasma. In the experiment of Mayer and Barker shown in figure 19.20, the radius of the electron bombardment discharge chamber was 5 cm, which implies that the electrons can be considered confined only if their gyroradius is less than this value,

$$\frac{R_g}{R_p} = \frac{m\bar{v}_e}{Re\,B} = \frac{0.417}{B(\text{mT})} \ll 1. \tag{19.51}$$

For electrons of 30 eV, the ratio of the gyroradius to the plasma radius has the numerical value given by the right-hand term of equation (19.51), with the magnetic induction B expressed in millitesla.

The magnetization criteria of equations (19.50) and (19.51) are plotted in figure 19.21. This indicates that in the experiment of Mayer and Barker, electrons are confined (their gyroradius is smaller than the plasma radius) at magnetic inductions above 0.4 mT, and they are fully magnetized at magnetic inductions above 2 mT, at which point the parameter $\omega\tau$ is greater than unity.

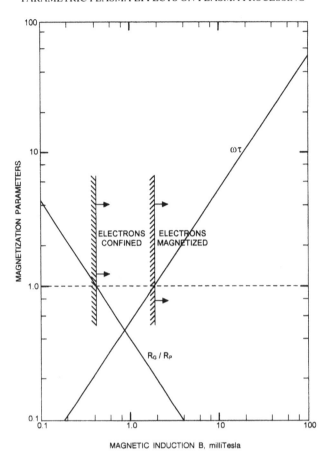

Figure 19.21. The magnetization parameters given by equations (19.50) and (19.51) as functions of the magnetic induction B in millitesla in the electron bombardment discharge chamber of figure 19.20.

19.7.3　Variation of Species Concentration with Magnetic Induction

Mayer and Barker (1982) measured the relative concentrations of both ionic and neutral active species produced by the plasma source as a function of the applied magnetic field, with the results plotted in figures 19.22(a) and (b). These data indicate a significant change in active-species concentration as the magnetic induction is varied from an electron unmagnetized to a magnetized state.

The relative concentration of ionic molecular fragments observed with a mass spectrometer is shown in figure 19.22(a) as a function of the magnetic induction in millitesla. As electrons in the plasma make a transition from unmagnetized to magnetized confinement in the region between 0.4 and more

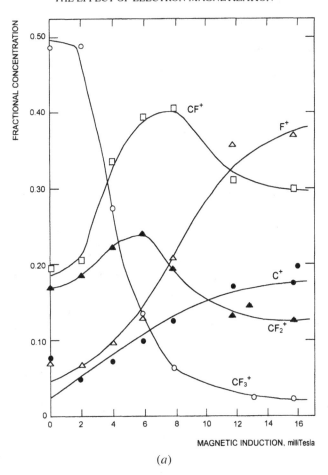

(a)

Figure 19.22. The relative concentration of molecular fragments resulting from changing the magnetic induction in the discharge chamber illustrated in figure 19.20, taken from Mayer and Barker (1982): (a) singly ionized molecular fragments in a carbon tetrafluoride plasma as a function of magnetic induction; and (b) the neutral species produced as a function of magnetic induction.

than 2 mT, the species balance of the ions exiting the discharge chamber changes significantly, from higher molecular weight with more complex molecular fragments when the electrons are poorly confined in low magnetic fields, to simpler and lower molecular weight ions as the electrons become better confined and more highly magnetized.

The same general trend is evident in figure 19.22(b), which shows the relative concentration of neutral molecular fragments leaving the discharge chamber as a function of the magnetic induction. Again, as the electrons become increasingly

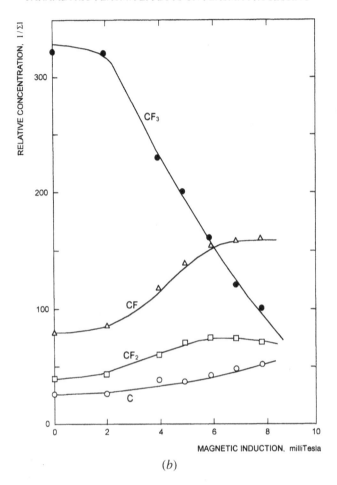

Figure 19.22. (Continued)

magnetized and better confined above 2 mT, the proportion of higher molecular weight, more complicated molecules declines, while the relative concentration of simpler and lower molecular weight fragments increases.

This investigation by Mayer and Barker exhibits a strong dependence of the species balance in a glow discharge plasma on its degree of electron magnetization. This dependence allows the operator of glow discharge processing reactors to control the active species balance by adjusting the magnetic induction. It should be noted that, unlike most industrial glow discharge plasma reactors that use a magnetic field for confinement, the apparatus in figure 19.20 uses an electromagnet, which produces an externally adjustable, relatively uniform magnetic field throughout the plasma volume. This contrasts with the multipolar permanent magnets used for wall confinement, discussed in section 3.4.3 of

Volume 1, which do not affect the entire volume of the plasma, and the magnetic field strength of which cannot be independently adjusted.

REFERENCES

Batenin V M, Klimovskii I I, Lysov G V and Troitskii V N 1992 *Superhigh Frequency Generators of Plasma* (Boca Raton, FL: Chemical Rubber Company) ISBN 0-8493-93 05-1

Bounasri F, Moisan M, Sauve G and Pelletier J 1993 Influence of the frequency of a periodic biasing voltage upon the etching of polymers *J. Vac. Sci. Technol.* B **11** 1859–67

Coburn J W 1982 Plasma-assisted etching *Plasma Chem. Plasma Process.* **2** 1–41

Coburn J W and Winters H F 1979 Plasma etching—a discussion of mechanisms *J. Vac. Sci. Technol.* **16** 391–403

——1983 Plasma-assisted etching in microfabrication *Annu. Rev. Mater. Sci.* **13** 91–116

Colgan M J, Kwon N, Li Y and Murnick D E 1991 Time- and space-resolved electron-impact excitation rates in an RF glow discharge *Phys. Rev. Lett.* **66** 1858–61

Gerlach-Meyer U, Coburn J W and Key E 1981 Ion-enhanced gas–surface chemistry: the influence of the mass of the incident ion *Surf. Sci.* **103** 177–88

Godyak V A, Piejak R B and Alexandrovich B M 1991 Electrical characteristics of parallel-plate RF discharges in argon *IEEE Trans. Plasma Sci.* **19** 660–71

Hope D A D, Monnington G J, Gill S S, Borsing N, Smith J A and Rees J A 1993 Ion energy distributions in $SiCl_4$ and Ar/O_2 dry etching discharges *Vacuum* **44** 245–8

Lee C and Lieberman M A 1995 Global model of Ar, O_2, Cl, and Ar/O_2 high-density plasma discharges *J. Vac. Sci. Technol.* A **13** 368–80

Mayer T M and Barker R A 1982 Reactive ion beam etching with CF_4: characterization of a Kaufman ion source and details of SiO_2 etching *J. Electrochem. Soc.* **129** 585–91

McDaniel E W 1964 *Collision Phenomena in Ionized Gases* (New York: Wiley) table 9.9.2, p 439 LCCCN 64-13219

Mutsukura N, Fukasawa Y, Machi Y and Kubota T 1994 Diagnostics and control of radio-frequency glow discharge *J. Vac. Sci. Technol.* A **12** 3126–30

Olthoff J K, Van Brunt R J, Radovanov S B, Rees J A and Surowiec R 1994 Kinetic-energy distributions of ions sampled from argon plasma in a parallel-plate, radio frequency reference cell *J. Appl. Phys.* **75** 115–25

Raizer Y P, Schneider M N and Yatsenko N A 1995 *Radio-Frequency Capacitive Discharges* (Boca Raton, FL: Chemical Rubber Company) ISBN 0-8493-8644-19

Reinke P, Schelz S, Jacob W and Moller W 1992 Influence of a direct current bias on the energy of ions from an electron cyclotron resonance plasma *J. Vac. Sci. Technol.* A **10** 434–8

Roth J R 1995 *Industrial Plasma Engineering: Vol. I—Principles* (Bristol: Institute of Physics Publishing) ISBN 0-7503-0318-2

Vossen J L and Kern W (ed) (1978) *Thin Film Processes* (New York: Academic) ISBN 0-12-728250-5

Winters H F, Coburn J W and Chuang T J 1983 Surface processes in plasma-assisted etching environment *J. Vac. Sci. Technol.* B **1** 469–80

Zheng J, Brinkmann R P and McVittie J P 1995 The effect of the presheath on the ion angular distribution at the wafer surface *J. Vac. Sci. Technol.* A **13** 859–64

20

Diagnostics for Plasma Processing

This chapter describes selected diagnostic methods used in plasma processing, with an emphasis on each method roughly proportional to its utility or frequency of use. Little coverage is given to diagnostic methods used in isolated research contexts. The diagnostic methods selected include those used for the measurement of plasma parameters, active-species concentrations, effects on the surface treated, and the performance of the treated surface. Those wishing for more information about the general subject of plasma-parameter diagnostics are referred to the books by Swift and Schwar (1970) and Hutchinson (1987). Information on diagnostic methods developed specifically for plasma processing outcomes may be found in Winters (1980), Manos and Flamm (1989), and in Smith (1995).

20.1 EXPERIMENTAL PARAMETERS

The plasma reactors or '*tools*' used for plasma processing have been discussed previously in chapter 17. The reactors used for plasma surface treatment were discussed in section 17.1, for ion implantation in section 17.2, for thin-film deposition in sections 17.3 to 17.5, and for plasma etching in section 17.6. The plasma sources in these reactors are very similar, with similar glow discharges operating with the same plasma parameters and in the same pressure regime. Only the working gases, active species, and effects differ substantially.

As illustrated in figure 20.1, the pertinent variables of plasma-processing systems can be divided into *independent input (or control) parameters*, *plasma parameters*, and *dependent output parameters*. Each of these sets of variables has its own diagnostic methods and devices. Many of those relevant to the microelectronic industry are available off the shelf, because of the large market provided by that industry. Other industrial sectors are not as well provided for, and it is sometimes necessary to use diagnostic methods for which the instrumentation

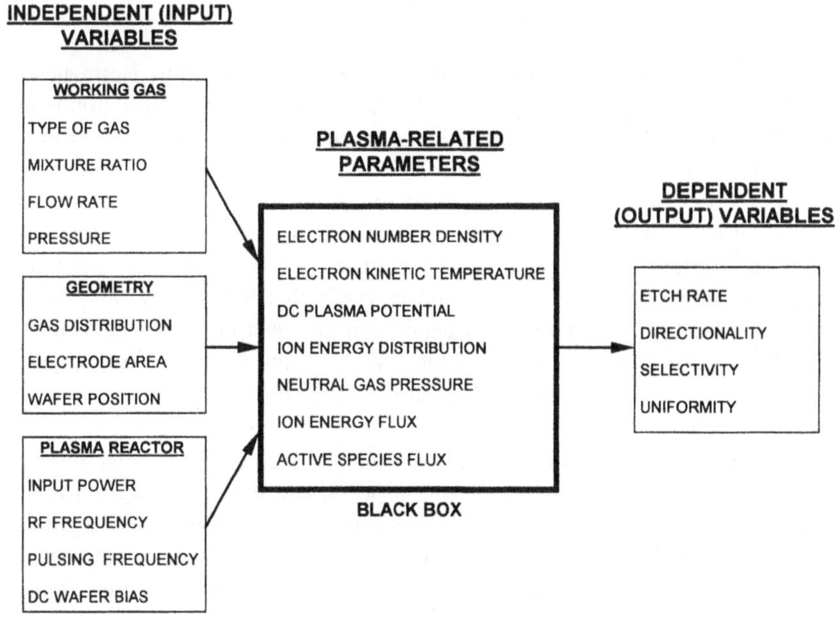

INDEPENDENT (INPUT)
     VARIABLES

```
WORKING GAS
TYPE OF GAS
MIXTURE RATIO                    PLASMA-RELATED
FLOW RATE                          PARAMETERS
PRESSURE                                                            DEPENDENT
                          ELECTRON NUMBER DENSITY             (OUTPUT) VARIABLES
                          ELECTRON KINETIC TEMPERATURE
GEOMETRY                  DC PLASMA POTENTIAL              ETCH RATE
GAS DISTRIBUTION          ION ENERGY DISTRIBUTION         DIRECTIONALITY
ELECTRODE AREA                                            SELECTIVITY
                          NEUTRAL GAS PRESSURE
WAFER POSITION                                            UNIFORMITY
                          ION ENERGY FLUX
PLASMA REACTOR            ACTIVE SPECIES FLUX
INPUT POWER
RF FREQUENCY                        BLACK BOX
PULSING FREQUENCY
DC WAFER BIAS
```

Figure 20.1. Independent input variables, plasma-related parameters, and dependent output variables of a plasma-processing reactor.

must be built in its entirety for a specific purpose. The independent input diagnostics selected for discussion are the gas-phase diagnostics discussed in sections 20.2 and 20.9; the plasma-parameter diagnostic devices discussed in section 20.3; and the dependent output parameters (what has been done to the surface during plasma processing) are covered in sections 20.4–20.8.

20.1.1   Independent Input Parameters

Information about the magnitude and time variation of the independent input parameters of the plasma-processing system illustrated in figure 20.1 is usually supplied by sensors built into the commercially available gas supply and plasma reactor subsystems. The diagnostic information available may include information on the gas feed pressures, from pressure gauges; gas flow rates, from flow meters; the net RF input power to the plasma reactor, from the RF power metering system; and the geometry of the electrodes and plasma reactor, established by measurement before production runs.

20.1.2   Dependent Output Parameters

Diagnostic methods are required to measure the dependent output parameters of the plasma-processing system illustrated in figure 20.1. This information is not normally obtained from each individual tool during production runs, but from an identical, specially instrumented tool before committing a process to production. One then relies on the identical behavior of production reactors to achieve the desired results in durations, and under conditions, employed in the test tool. The variables of interest include the deposition or etching rate, which is required to predict the endpoint of the process and to verify that it is not too lengthy for production applications. Etching or deposition rate measurements are available from knowledge of the thickness of the layer, and the length of time required to reach the endpoint. These rates can also be measured directly using a quartz crystal microbalance.

Additional dependent output parameters include the *directionality*, *selectivity*, and *uniformity* of an etching process, and the *uniformity*, *thickness*, and *adhesion* of a deposition process. These characteristics will be defined and discussed further in chapters 23–25. These characteristics can be assessed by tomographic cross sections of a processed workpiece subject to scanning electron microscopy, by one of several interference-based diagnostics outlined here, and by other methods. Information about the relative rate of etching a layer and its mask or substrate (the *selectivity*) may be provided by monitoring the reaction products of etching the mask and of the substrate during an etch. The uniformity of an etching or deposition process over the surface of a workpiece can be assessed by processing a uniform layer on a special workpiece, or by examining the detailed structure of the surface by scanning electron microscopy.

20.2   GAS-PHASE PROCESS MONITORING

This section covers *gas-phase process monitoring*, and describes instruments frequently used to measure the characteristics of the neutral working gas in a plasma reactor. The gas-phase species monitored usually include the *background neutral gas* resulting from leakage and outgassing of the vacuum system; the (input) *neutral working gas*; the active species produced by interaction of the background and working gas with the plasma; and the chemical reaction products of plasma processing. The gaseous species monitored during processing also may include ionic active species or the ionic chemical reaction products of processing.

20.2.1   Neutral Gas Pressure

The principle of operation of three widely used types of vacuum gauge are now described, in order of decreasing pressure range of operation. These gauges are used to monitor the total pressure (and hence number density) of the neutral gas

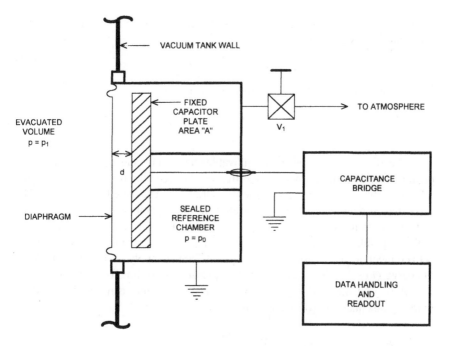

Figure 20.2. A schematic drawing of a capacitance manometer, in which the distance (and hence capacitance) between the movable diaphragm and fixed capacitor plate varies as the pressure in the evacuated volume, p_1, changes with respect to the pressure p_0 in the reference chamber.

in intermediate- and low-pressure plasma reactors, a major independent variable in any form of plasma processing conducted in a vacuum system.

20.2.1.1 Capacitance Manometer

The first of these vacuum gauges is the *capacitance manometer* illustrated in figure 20.2. This instrument consists of a thin diaphragm exposed to the vacuum to be measured, and located on the surface of a sealed reference chamber. The diaphragm is separated by a small distance d from a fixed capacitor plate of area A. The operation of the instrument is based on the capacitance changes that occur as pressure variations in the vacuum system cause the diaphragm to move with respect to the fixed capacitor plate. The relationship between the charge Q, the capacitance C, and the voltage V, applied between the diaphragm and the fixed capacitor plate is given by

$$Q = CV \qquad \text{C.} \qquad (20.1)$$

For a plane-parallel capacitor with plates of area A separated by a distance d, the capacitance C is given by

$$C = \frac{\varepsilon_0 A}{d} \quad \text{F.} \tag{20.2}$$

Substituting equation (20.2) into equation (20.1) yields

$$Q = \frac{\varepsilon_0 A V}{d} \quad \text{C.} \tag{20.3}$$

The distance d between the movable diaphragm and the fixed capacitor plate is linearly proportional to the difference between the pressure in the sealed reference chamber, $p = p_0$, and the pressure in the evacuated volume, p_1. The capacitance manometer uses the variation in the capacitance that occurs as the distance d changes with pressure, to convert the pressure-induced displacement of the diaphragm into an electrical signal. This may be done in two ways. The first is to connect the fixed capacitor plate to a DC power supply, and measure the flow of charge Q as the distance d varies with pressure according to equation (20.3).

A second method of producing an electrical signal can be illustrated by rearranging equation (20.3) to obtain

$$V = \frac{Qd}{\varepsilon_0 A} \quad \text{V.} \tag{20.4}$$

In this version of the capacitance manometer, the charge Q is kept constant, and the voltage on the fixed plate is monitored as the distance d changes with pressure. The constant charge mode of operation has the advantage of producing a linear voltage signal that is directly proportional to the distance d between the diaphragm and the fixed capacitor plate, and thus to the net pressure on the diaphragm.

The capacitance manometer can be operated with a sealed reference chamber, in which the fixed capacitor plate is in contact with a known gas trapped at a specified pressure. A second mode of operation is to open the valve V_1 in figure 20.2 to set the reference pressure equal to that of the ambient atmosphere. If the gases involved obey the perfect gas law, the pressure readings are not sensitive to the molecular weight or atomic characteristics of the gases used, only to their pressure. Commercially available capacitance manometers can be used over the pressure range from 1 atm down to approximately 0.13 Pa (1 mTorr).

20.2.1.2   Thermocouple Gauge

The components of the *thermocouple gauge*, a second widely used vacuum gauge, are shown schematically in figure 20.3. The operation of the thermocouple gauge is based upon the fact that the thermal conductivity of a gas depends on its pressure. Hence, the temperature of a heated resistor is a function of the pressure of the surrounding gas.

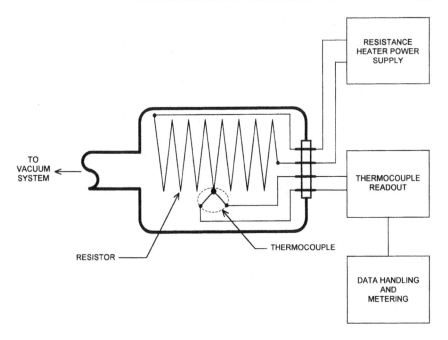

Figure 20.3. A schematic drawing of a thermocouple gauge, in which the temperature of the resistor varies with pressure in the gauge tube as the result of convection by the background gas.

The thermocouple gauge illustrated in figure 20.3 contains a resistor heated by a power supply that provides a constant power input. The temperature of this resistor, which is nonlinearly proportional to the pressure of the surrounding gas, is monitored by the attached thermocouple shown in figure 20.3. The output voltage of the thermocouple is therefore a nonlinear but monotone function of the pressure of the gas surrounding the resistor to which it is attached.

At a fixed pressure, the temperature of the resistor and the output of the thermocouple may differ in different gases, since the thermal conductivity may be different for each gas. The response of a thermocouple gauge is highly nonlinear because in addition to the gas cooling, the resistor is also being cooled by blackbody radiation and conduction through its electrical leads. The useful range of thermocouple gauges for most gases is between about 0.67 and 67 Pa (5 and 500 mTorr).

20.2.1.3 Bayard–Alpert (Ionization) Gauge

The most widely used vacuum gauge at low pressures (below 0.13 Pa, 1 mTorr) is the *Bayard–Alpert ionization gauge*, a schematic diagram of which is shown in figure 20.4. The operation of this gauge relies on the fact that the electron–neutral

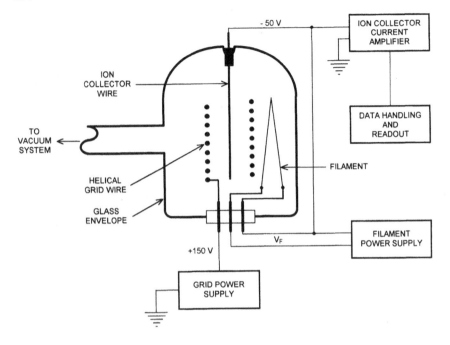

Figure 20.4. A schematic drawing of a Bayard–Alpert ionization gauge, in which the current flowing to the ion collector wire is directly proportional to ion–electron pair production, and hence to the background neutral gas pressure.

ionization rate is directly proportional to the density of neutral atoms available to be ionized, provided that the gauge contains a fixed number of electrons with energies above the ionization potential of the gas.

In the ionization gauge of figure 20.4, electrons are emitted from the heated filament, and accelerated toward a helical grid wire that is maintained at a relatively high positive potential. These energetic electrons collide with and ionize the neutral gas that is in pressure equilibrium with the vacuum system to the left. The positive ions produced by ionization events are collected on a fine wire located on the axis of the helical grid. The collected ion current is amplified, and read out as a signal proportional to the neutral gas pressure in the vacuum system.

Because of its requirement for a hot filament and long electron mean free paths, the Bayard–Alpert ionization gauge cannot be used with oxidizing gases, or at pressures above 0.13 Pa (1 mTorr). The relationship between the ion current and the pressure is linear down to pressures below $\sim 10^{-11}$ Pa (10^{-10} Torr), but the constant of proportionality between the current and pressure differs for each gas, because of their different ionization potentials. Like the other vacuum gauges discussed earlier, the Bayard–Alpert ionization gauge and its associated readout circuits are widely available commercially.

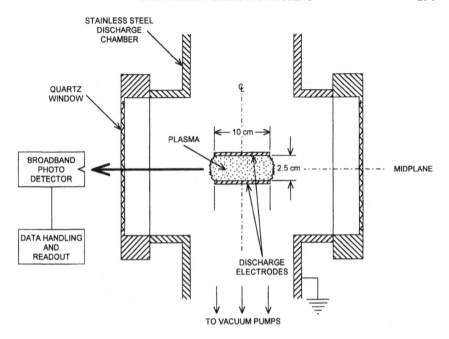

Figure 20.5. A schematic drawing of plasma actinometry, performed with a broadband photodetector.

20.2.2 Actinometry

A global gas phase diagnostic is *actinometry*, which consists of the gross, broadband, but often spatially resolved measurement of optical (particularly ultraviolet) emissions from the plasma, illustrated in figure 20.5. In this arrangement, a broadband photodetector monitors the total collimated light output from a chord across the plasma. Such a measurement is sometimes capable of revealing the presence of contamination, failure in the reactor subsystems, or failure of the process control system itself. In addition, the gross light output of some etching reactions may serve as a means of endpoint detection. The intensity of the emitted light may depend on the chemical reaction products of etching, the production of which ceases when the etching process is complete.

In some applications, the broadband photodetector shown in figure 20.5 is replaced by an optical spectrometer set on a particular emission line characteristic of the process to be monitored. The intensity of such a line can indicate progress of a deposition or etching process, and serve as a form of endpoint detection when applied to emission lines from reaction products.

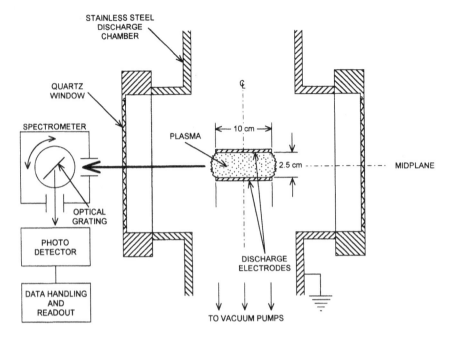

Figure 20.6. An experimental setup for plasma spectroscopy, using a plasma spectrometer with an optical grating.

20.2.3 Optical Spectrometry

A more sophisticated, non-perturbing, and passive diagnostic of glow discharge plasmas can be provided by either *optical, ultraviolet,* or *infrared spectrometry.* Most spectrometry is done in the visible part of the spectrum (*optical spectrometry*), because of the experimental inconvenience and additional expense of working at infrared or ultraviolet wavelengths. Optical spectrometers are relatively inexpensive and widely available commercially. A characteristic plasma spectroscopy system is illustrated in figure 20.6. Radiation from the plasma is focused on a spectrometer, in which a prism or a grating disperses the light to one or more photodetectors. As the prism or grating is rotated, the photodetector produces a record of the emission intensity as a function of wavelength over the range of the spectrometer.

The usual result of this process is a spectrum of dozens or hundreds of emission lines from both neutral and ionized species in the plasma. These emission lines can originate from active species formed by the plasma, and from chemical reaction products of plasma processing. The intensity of the active-species emissions can be used to indicate whether the plasma is producing the optimum conditions required for a particular process. The intensity of reaction product emissions can be used as a means of endpoint detection. The presence or absence of such emissions indicates whether an etching or *stripping* process is still underway, or has been completed as the process reached the bottom of a layer. (*Stripping* is the use of oxidizing reactions to remove an etching mask.)

More quantitative applications of plasma spectroscopy than process monitoring by emission line intensity are usually reserved for research investigations rather than industrial process control. In principle, the relative intensity of optical emission lines can be used to determine the electron kinetic temperature, and the relative and absolute number densities of the emitting species in the plasma. Such quantitative applications are often infeasible if the electron energy distribution function is not known, if there are insufficient data on relevant excitation cross sections, or if the electron population and excited species fail to be in a state of local classical kinetic equilibrium.

20.2.4 Neutral Mass Spectrometry

In addition to the optical emission lines discussed earlier, gas-phase process monitoring also can be performed by *neutral mass spectrometry*. In this technique, the mass of neutral species in a glow discharge plasma is measured, and their relative concentrations monitored as a function of time. Like the optical spectrometers discussed before, neutral mass spectrometers are available commercially from several instrument suppliers, and are almost universally based on variants of the *quadrupole mass filter*, illustrated in figure 20.7. In this instrument, four cylindrical rods are equally spaced at the corners of a square, to form an electrostatic quadrupole configuration that acts as a mass filter for ionized species moving along its axis.

In a quadrupole mass analyzer, neutral species flow into an ionizing chamber (not shown in figure 20.7). The resulting ions travel along a trajectory near the axis of the four rods of the quadrupole mass filter. The rods are energized by a combination of a variable frequency RF voltage and a DC bias, which allow only ions of a particular mass to exit through the upper orifice of the filter shown in figure 20.7. The mass of ions reaching the collector electrode is a function of the RF frequency applied to the quadrupole filter rods. As the frequency of the RF is swept over its range, a corresponding spectrum of mass is covered. Commercially available instruments usually offer ranges of 0–50 and 0–200 atomic mass units (amu).

Accurate quantitative measurement of the concentration of each species with the quadrupole mass analyzer is difficult, in part because of the requirement to ionize neutral species before their mass can be determined in the quadrupole filter. In addition, the mean free paths of the species analyzed must be greater than the dimensions of the quadrupole mass filter, restricting its operation to low pressures, below 1.3 Pa (10 mTorr). This restriction is sometimes dealt with by operating the mass spectrometer itself in a high vacuum, but bleeding in a flow of gas to be measured from a higher pressure system through a control valve.

A quadrupole mass analyzer can provide the approximate relative

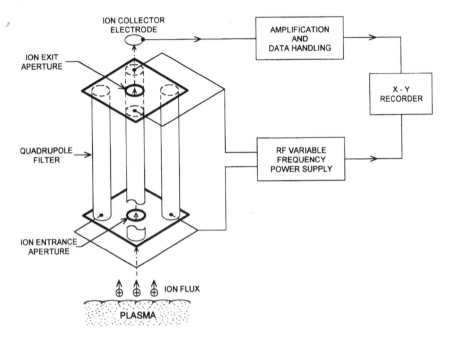

Figure 20.7. Illustration of the principle of the quadrupole mass analyzer. The four rods that make up the quadrupole filter are shown on the center left. The ion flux from the plasma passes along the common axis of the four quadruple rods.

concentration of the neutral species of working or feed gases, of active species, and of the chemical reaction products of processing. The relative intensity of reaction products can be used for endpoint detection of a stripping or etching process, since the relative intensity of etching reaction products will normally drop significantly when the process has reached its substrate.

20.2.5   Ion Mass Spectrometry

Ion mass spectrometry may also be accomplished more directly by the quadrupole mass analyzer illustrated in figure 20.7. When it is desired to monitor the mass of ionic active species or reaction products, the ions can be input directly into the quadrupole filter. The filter output yields a spectrum of the relative intensity of the ionic species as a function of their mass.

A characteristic arrangement for process monitoring by neutral and ion mass spectrometry is shown schematically in figure 20.8. A small aperture adjacent to a workpiece allows ions from the sheath to exit the plasma reactor. This aperture is intended to sample characteristic ions that reach the surface of the workpiece being processed. The ions that pass through this aperture enter a *curved-plate energy analyzer* (to be discussed in section 20.3.4.2). As a result, ions of a known

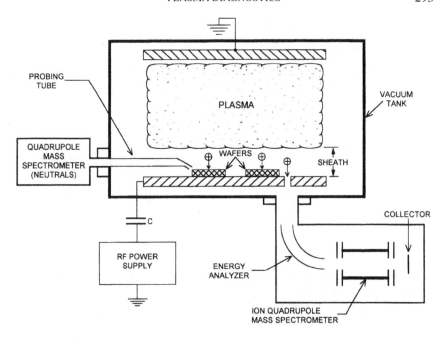

Figure 20.8. A schematic diagram of an experimental setup for gas-phase mass spectrometry, using a quadrupole mass spectrometer to sample the neutral flux above the wafer surface, and an opening in the base plate for determination of the ion energy distribution function with an ion quadrupole mass spectrometer.

energy impinge on the quadrupole mass filter, to yield a final collector current that is both energy and mass resolved. In this way, one can diagnose the energy distribution function of specific ions or isotopes important to a plasma-processing application.

In addition to energy and mass analysis of ions impinging on a workpiece, it is sometimes useful to monitor the characteristics of the neutral gas immediately above its surface. This can be done by connecting a quadrupole mass spectrometer to a probing tube that samples gas from above the workpiece, as indicated on the left-hand side of figure 20.8. The mass spectrum so measured can be used to indicate whether a normal or a contaminated process is underway, and it may also detect reaction products useful for endpoint detection.

20.3 PLASMA DIAGNOSTICS

Diagnostic instruments are used in plasma processing to measure the electron number density, the ion and electron energy distribution functions (or kinetic temperature, if Maxwellian), the plasma potential, and the floating potential

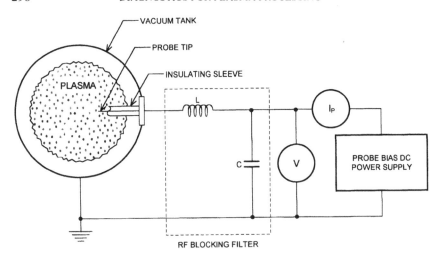

Figure 20.9. A schematic drawing of an RF Langmuir probe circuit for plasma diagnostics, including an RF blocking filter.

associated with the bulk plasma. The instruments selected for discussion here include the *Langmuir probe*, the *double probe*, the *microwave interferometer*, the *retarding potential energy analyzer*, and the *curved-plate energy analyzer*. These instruments are available commercially, and are widely used, particularly in the microelectronic industry. These instruments may also be used for endpoint detection and process control applications (Mutsukura *et al* 1994).

20.3.1 Electrostatic (Langmuir) Probes

The *Langmuir probe*, also called the *electric probe* or *electrostatic probe*, is a widely used diagnostic for DC and RF glow discharge plasmas at intermediate and low pressures. The Langmuir probe typically consists of a bare wire inserted in a plasma, as indicated in figure 20.9. In order to obtain an unmodified sample of the plasma electron population, it is important that electrons do not suffer collisions in the sheath between the plasma and the probe surface. This implies that the electron mean free path must be greater than the sheath thickness, of which the Debye length is characteristic. Part of the Langmuir probing circuit for RF glow discharge plasmas is an RF blocking filter to prevent leakage of RF power into the diagnostic instrumentation.

As indicated in figure 20.9, the Langmuir probe is connected to a DC power supply that biases the probe with respect to the plasma (see, for example, Cali *et al* 1995). The DC bias voltage is varied from negative to positive values while the DC current flowing to the probe is measured. Commercially available versions of Langmuir probing systems characteristically have microprocessor controlled

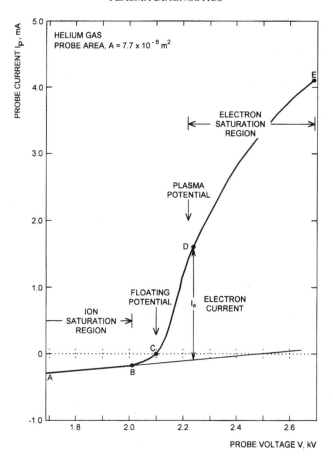

Figure 20.10. A characteristic Langmuir probe curve, taken from a low-pressure turbulent Penning discharge plasma.

power supplies that sweep the DC voltage; digital data logging of the voltage and current; and data reduction that calculates the desired plasma parameters.

When the probe bias voltage, V, is swept over a range that includes the *plasma potential* V_p, the current flowing to the probe, I_p, produces a *Langmuir probe curve*, an example of which is shown in figure 20.10. For these data, the area of the probe wire was 7.7×10^{-6} m^2. The probe curve was taken in a highly turbulent, high vacuum (6.7×10^{-6} Pa, 0.05 mTorr) *Penning discharge* helium plasma of the type described in section 9.5.5 of Volume 1. The plasma is subject to *magnetoelectric heating*, which results in very high ion energies and lower, but still significant, electron kinetic temperatures.

20.3.1.1   Ion Saturation Current

In the region from A to B on the probe curve of figure 20.10, the potential is sufficiently negative with respect to the plasma potential that only ions are collected by the probe. This is called the *ion saturation regime*, and the total current drawn is

$$I_{is} = A_0 J_{is} = e A_0 \Gamma_{is} = \tfrac{1}{4} e A_0 n_i \bar{v}_i \qquad \text{A} \qquad (20.5)$$

where A_0 is the area of the probe, n_i is the ion number density, and $\bar{v}_i$ is the mean thermal velocity of ions that leave the plasma. Substituting the ion thermal velocity into equation (20.5) yields

$$I_{is} = \frac{e A_0 n_i}{4} \sqrt{\frac{8 e T_i'}{\pi M}} \qquad \text{A} \qquad (20.6)$$

where T_i' is the ion kinetic temperature in eV, and M the ion mass.

The ion current curve between A and B is not horizontal, because the sheath boundary moves away from the probe surface at progressively more negative potentials, thus increasing the effective surface area, A_0, of the probe. If the electron population did not come into play, the ion current would continue to rise along the extrapolated straight line indicated. Since the sheath boundary is closest to the probe at the most positive potential on the ion saturation region at point B, the current at the point B is identified as the *ion saturation current*, 0.15 mA in this case.

20.3.1.2   Floating Potential

For potentials more positive than the point B on the Langmuir probe curve, the probe is no longer sufficiently negative with respect to the plasma potential to repel all electrons back to the plasma. As the potential on the probe rises from B to C to D and finally to E, more electrons are collected by the probe as it becomes more positive. Exactly as many electrons as ions are collected by the probe at the point C on the curve, yielding no net current. The potential at which this net current is zero is known as the *floating potential* of the plasma. This is the potential assumed by any electrically isolated surface in contact with the plasma. Immediately to the right of point B, only the most energetic electrons in the tail of the electron energy distribution are able to reach the probe. As the probe potential becomes progressively more positive, the electron current increases exponentially according to the *electrostatic Boltzmann equation* discussed in section 4.3 of Volume 1. In the notation used here, the electrostatic Boltzmann equation may be written

$$I_e = I_{es} \exp\left[-\frac{V_p - V}{T_e'}\right] \qquad \text{A} \qquad V \le V_p \qquad (20.7)$$

where I_{es} is the electron saturation current, reached at the plasma potential, $V = V_p$.

20.3.1.3 Electron Kinetic Temperature

The electron current flowing to the probe in figure 20.10 is identified with the distance between the extrapolated portion of the ion saturation curve and the probe curve itself. If the electron energy distribution is Maxwellian, its energy can be characterized by the electron kinetic temperature, T_e'. The value of T_e' can be determined by plotting the electron current I_e from figure 20.10 as a function of the probe voltage on a semi-logarithmic graph, as illustrated in figure 20.11. When the electron energy distribution function is Maxwellian, the electron current will be a straight line on this semi-logarithmic plot, as predicted by the electrostatic Boltzmann equation (20.7). The electron kinetic temperature can be obtained from the slope of the straight-line portion of the data, based on taking the logarithms of both sides of equation (20.7),

$$T_e' = \frac{V - V_p}{\ln(I_e/I_{es})} \qquad eV. \qquad (20.8)$$

For the data of figure 20.11, the electron kinetic temperature is $T_e' = 42$ eV.

20.3.1.4 Plasma Potential

When the probe voltage reaches the plasma potential, essentially all electrons available from the plasma are collected by the probe, and the probe is in *electron saturation*. The *electron saturation regime* of the Langmuir probe curve, from D to E, is characterized by a straight line on the semi-logarithmic plot of figure 20.11. Like the ion saturation portion of the curve, the electron saturation line has a positive slope. This positive slope is the result of the increasing effective area of the probe as the sheath boundary moves away from the probe surface at voltages above the plasma potential. It is customary in reducing Langmuir probe data to take the point of intersection of the electron saturation line and of the electron current portion on the semi-logarithmic plot as the defining point for the plasma potential. In figure 20.11 these values are $V_p = 2220$ V for the plasma potential, and $I_{se} = 2.35$ mA for the electron saturation current.

20.3.1.5 Electron Number Density

Since the Langmuir probe in electron saturation collects all the electrons incident on the plasma–sheath boundary, one may write the total electron current collected by the probe as

$$I_{es} = A_0 J_{es} = \frac{1}{4} e A_0 n_e \bar{v}_e = \frac{e A_0 n_e}{4} \sqrt{\frac{8 e T_e'}{\pi m}} \qquad (20.9)$$

where A_0 is the surface area of the probe, n_e is the electron number density, and m the electron mass. In the example of figure 20.11, one can substitute the electron kinetic temperature, 42 eV, and the electron saturation current of 2.35 mA

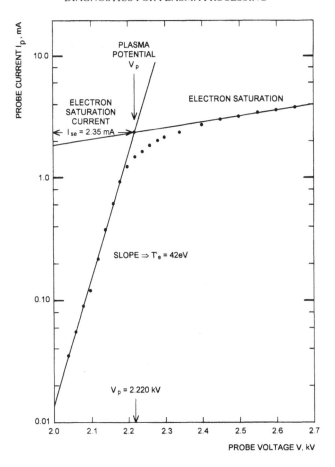

Figure 20.11. The electron current of the Langmuir probe curve in figure 20.10, plotted on semi-logarithmic coordinates.

into equation (20.9) to obtain the electron number density. For this example, $n_e = 1.75 \times 10^{15}$ electrons/m^3.

This procedure allows one to obtain the floating potential, plasma potential, electron kinetic temperature, and electron number density from a Langmuir probe curve. It is also possible to obtain an estimate of the ion kinetic temperature from the ion saturation current of equation (20.6), based on the assumptions that the ion energy distribution function is Maxwellian, that quasi-neutrality as expressed by

$$n_e = n_i \qquad (20.10)$$

holds, and that no other magnetic field or collisional effects are important. One can substitute the electron number density just obtained and the ion saturation

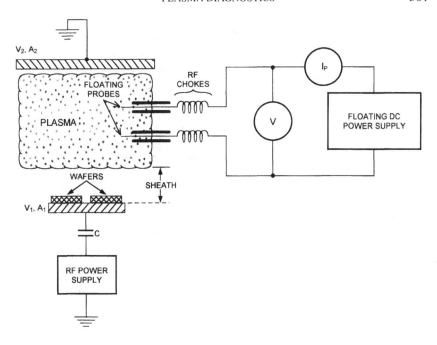

Figure 20.12. A schematic drawing of an RF double-probe circuit, intended for applications in which a ground reference cannot easily be established.

current (0.15 mA in this example) into equation (20.6), to obtain an effective ion kinetic temperature for this singly ionized helium plasma of 1270 eV. Such a high ratio of ion-to-electron kinetic temperature is expected in this magnetoelectrically heated Penning discharge plasma, as discussed in section 3.3.3 of Volume 1. Further references on Langmuir probes and Langmuir probe theory may be found in Swift and Schwar (1970), Chatterton *et al* (1991), and Cali *et al* (1995).

20.3.2 Double Probe

A variation of the Langmuir probe that is sometimes useful in situations where a reference electrode is not available or the plasma potential is poorly defined, is the so-called *double-probe* circuit shown in figure 20.12. In this circuit, the need for a reference electrode is eliminated by electrically floating both probes and their power supply with respect to the plasma. A current–voltage curve is traced out that is symmetric, if the probes and their contact potentials are identical. The double-probe arrangement allows one to determine the electron kinetic temperature and number density, but does not yield a measurement of the plasma potential.

When RF glow discharge plasmas are being diagnosed with a double-probe circuit, inductors are used to prevent RF from leaking from the plasma reactor

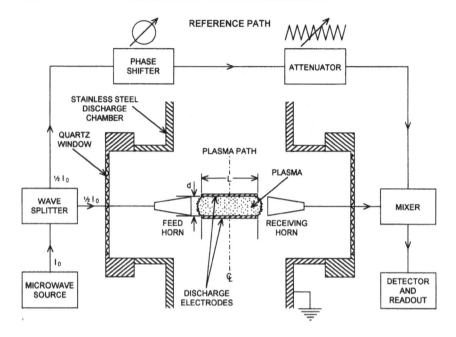

Figure 20.13. A schematic drawing of a Mach–Zehnder microwave interferometer. The *reference path* is at the top of the diagram. The microwave radiation propagates through the plasma volume in the lower, *plasma path* of the interferometer.

to the probe circuit. A characteristic value for the inductors used at 13.56 MHz might be 1 mHenry. Further information on the theory and technique of double probes may be found in Lochte-Holtgreven, chapter 11, section 3.

20.3.3   Microwave Interferometry

Microwave interferometry is widely used to measure the electron number density of DC and RF glow discharge plasmas. It has the advantage of being non-perturbing, with negligible effects on the plasma being measured. Microwave interferometry makes use of the fact that microwave radiation is slowed down by its passage through a plasma, the index of refraction of which is greater than unity. This slowing is measured with a Mach–Zehnder interferometer, illustrated in figure 20.13. The phase of the slowed radiation passing through the plasma is compared with that propagating over the same equivalent distance, but without plasma present.

In the Mach–Zehnder interferometer of figure 20.13, a microwave source produces an intensity I_0 that is divided into two equal signals, each of intensity $I_0/2$. One signal propagates through the *plasma path*, and is slowed down by the increased index of refraction of the plasma. The microwave radiation propagating in the second *reference path* goes through a phase shifter and attenuator, adjusted such that when the signals are recombined, the relative phase angle can be measured by comparing the signals propagating through the two paths. Once the attenuator is adjusted so that the signals reaching the detector are of equal intensity, the plasma can be turned off, and the phase shifter adjusted until the signals null out after being recombined. If the plasma is turned on, the phase shifter can be readjusted to again null out the recombined signal, and the change in the phase shifter reading will equal the phase angle change through a length L of the plasma.

The relationship between the measured phase angle in radians and the real part of the index of refraction is given by (Roth 1995, ch 13)

$$\Delta\phi = \frac{2\pi}{\lambda} \int_0^L [1 - \mu(x)]\,dx \qquad \text{rad} \qquad (20.11)$$

where the quantity in the square brackets represents the difference in the indices of refraction of the reference arm of the interferometer (which is unity) and the plasma arm, in which the index of refraction is $\mu(x)$. When the difference in the index of refraction is integrated over the length, L, of the plasma, one obtains the change in phase angle shown. If the collision frequency ν_c is much less than the microwave frequency ω, equation (13.7) of section 13.2.3 of Volume 1 gives for the real part of the index of refraction in an unmagnetized plasma,

$$\mu(x) = \sqrt{1 - \frac{\omega_{pe}^2}{\omega^2}} = \sqrt{1 - \frac{n(x)}{n_c}} \qquad (20.12)$$

where the *critical* or *cutoff density* n_c is given by

$$n_c \equiv \frac{\varepsilon_0 m_e \omega^2}{e^2} \qquad \text{electrons/m}^3. \qquad (20.13)$$

In most glow discharge plasmas, the electron number density is far below the cutoff density, so the right-hand term of equation (20.12) may be approximated

$$\mu(x) \approx 1 - \frac{n(x)}{2n_c}. \qquad (20.14)$$

Substituting equation (20.14) into equation (20.11), one can write the phase angle in terms of the integral of the electron number density,

$$\Delta\phi = \frac{\pi}{\lambda n_c} \int_0^L n(x)\,dx = \frac{\pi\nu}{cn_c} \int_0^L n(x)\,dx \qquad \text{rad} \qquad (20.15)$$

where ν is the frequency in hertz of the microwave radiation, and L is the distance through which the microwave radiation propagates in the plasma. An average electron number density in the plasma can be defined as

$$\bar{n}_e \equiv \frac{1}{L} \int_0^L n(x)\,dx \qquad \text{electrons/m}^3 \qquad (20.16)$$

and the phase angle can be written

$$\Delta\phi = \frac{\pi v L \bar{n}_e}{c n_c} \qquad \text{rad.} \qquad (20.17)$$

Since the phase angle $\Delta\phi$ is obtained experimentally from the Mach–Zehnder microwave interferometer, one can write the average electron number density in terms of this phase angle in the form

$$\bar{n}_e = \frac{c n_c}{\pi L v}\Delta\phi = \frac{4\pi c \varepsilon_0 m v}{L e^2}\Delta\phi \qquad \text{electrons/m}^3. \qquad (20.18)$$

Thus, when the electron number density in the plasma is much smaller than the cutoff density, the average electron number density is a linear function of the phase angle measured by microwave interferometry.

One of the limitations of microwave interferometry for measuring the electron number density of processing plasmas is that the plasma dimensions are often comparable to or smaller than the free space wavelength of readily available microwave power. This can lead to spurious reflections and interference effects involving the microwave radiation. To avoid this problem in processing plasma reactors, one needs to use microwave frequencies of 60 GHz or higher, a regime in which operation is sometimes technically difficult. Nonetheless, complete microwave interferometer systems are available commercially. These units can be furnished with the necessary hardware and computer-based controllers capable of automating the data-taking process and following the changes in plasma density in real time. Such changes of density are often associated with the endpoint of an etching process, in which the changing composition of reaction products at the endpoint will significantly change the average electron number density of the plasma.

20.3.4 Ion Energy Analysis

Probably the most widely used plasma diagnostic technique is ion energy analysis, which can be accomplished by either a *retarding potential energy analyzer*, or a *curved-plate energy analyzer*. Both of these instruments are commercially available as off-the-shelf units specifically designed for microelectronic plasma deposition and etching applications. References on these and related ion energy measurement techniques can be found in Roth and Clark (1969), Hope *et al* (1993), Bohm and Peril (1993), and Olthoff *et al* (1994).

20.3.4.1 Retarding Potential Energy Analyzer

A schematic diagram of a *retarding potential energy analyzer* for *in situ* monitoring of the ion flux on a surface is shown in figure 20.14. In this arrangement, a small hole in the surface samples the ion flux accelerated across

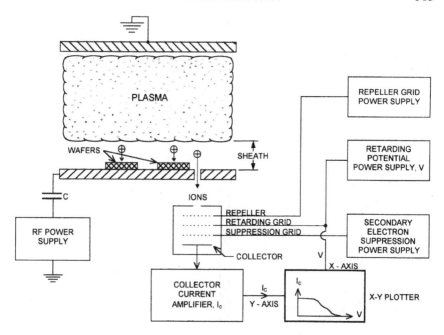

Figure 20.14. A schematic drawing of a retarding potential energy analyzer, used for determining the ion energy distribution function of ions arriving at the surface of a workpiece.

the sheath. The ion flux that enters the retarding potential energy analyzer is assumed to be the same as that on nearby workpieces.

Upon entering the analyzer, the ion flux first encounters a *repeller grid*, that is biased to a potential that repels the species having the sign of charge that is not of interest. For example, if one were interested in monitoring positive reactive ions responsible for etching, one would bias the repeller grid negatively to prevent electrons and negative ions from reaching the collector. The second grid is the *retarding grid*, the voltage of which is increased to repel positive ions traveling to the collector. There is usually a third *electron suppression grid*, which is sufficiently negatively biased to repel secondary electrons emitted from the collector as the result of bombardment by the positive ion flux. The suppression of secondary electron emission from the collector is necessary because the current metering system cannot distinguish between a flux of positive ions arriving at the collector, and a flux of secondary electrons leaving it.

The collector current is plotted on the y-axis of a graph, and the retarding potential on the x-axis, to produce an integrated ion energy distribution function for ions entering the analyzer. The ions may not arrive at the entrance to the analyzer over a full hemisphere in velocity space. In some situations, they are

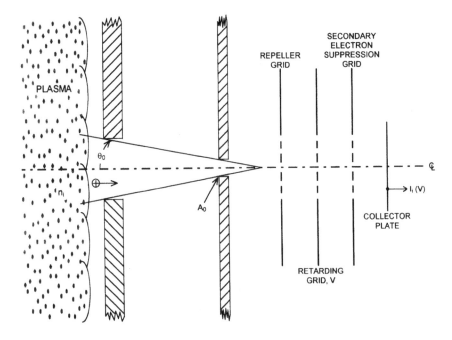

Figure 20.15. The geometrical configuration of a retarding potential energy analyzer with acceptance cone angle θ.

constrained to an angle θ_0 by aperture plates through which the ion flux must pass, as shown in figure 20.15.

When ions arriving at the entrance to a retarding potential energy analyzer have a velocity distribution that is Maxwellian and isotropic in velocity space, the total ion current passing through an entrance aperture of area A_0 is given by

$$I_{max} = -\frac{e A_0 n_i \bar{v}_i}{4} \qquad A \qquad V \le V_p. \tag{20.19}$$

It has been shown by Roth and Clark (1969) that a steady-state ion current reaching the collector plate of a retarding potential energy analyzer is given by

$$I_i(V) = I_{max} \left\{ \exp\left[-\frac{(v - V_p)}{T_i'} \right] - \cos^2\theta_0 \exp\left[-\frac{(V - V_p)\sec^2\theta_0}{T_i'} \right] \right\} \qquad A$$

$$V \ge V_p, \ 0 \le \theta \le \theta_0 \tag{20.20}$$

where I_{max} is given by equation (20.19), V is the voltage on the retarding grid, T_i' is the ion kinetic temperature, and V_p is the plasma potential. The parameter θ_0 is the maximum entrance angle of the ions into the retarding potential energy analyzer for anisotropic velocity distributions, illustrated in figure 20.15. In plasma-processing applications, the ion distribution can be cut off in an angle in the velocity space because of columnation due to passage through apertures or motion through a thick plasma presheath (Zheng *et al* 1995).

Equation (20.20) was derived on the assumption that the ions reaching the analyzer are uniformly spread through the angles $0-\theta_0$ in velocity space, that the ion energy distribution function in the plasma is Maxwellian, and that the plasma is steady state. For an ideal analyzer geometry, in which the ions approach the entrance aperture with velocity vectors distributed over a hemisphere, the integrated ion current impinging on the collector for a Maxwellian distribution of ions is given by setting $\theta_0 = \pi/2$, to yield

$$I_i = I_{max} \exp\left[-\frac{(V - V_p)}{T_i'}\right] \quad \text{A.} \tag{20.21}$$

A retarding potential curve from a Maxwellian, hot-ion, steady-state Penning discharge plasma of the kind described in section 9.5 of Volume 1 is shown in figure 20.16. In this case, deuterium ions had a kinetic temperature just over 1900 eV, and the plasma potential, V_p was 1623 V. This integrated energy distribution function illustrates the ability of the Penning discharge, discussed in section 9.5.5 of Volume 1, to produce hot, Maxwellian ions. This plasma was operated under conditions for which the ions were magnetized, which allowed magnetoelectric heating to take place. The ion energies characteristic of plasma-processing applications, however, are acquired by acceleration across the sheath above the workpiece. The ions are unmagnetized, and rarely exceed two or three hundred electronvolts in energy.

In plasma-processing applications, the ions rarely have Maxwellian distributions, and enter the analyzer normal to the surface of the workpiece. The integrated ion energy distribution functions for such applications may be written as

$$I_i(V) = I_{max}\left[1 - \int_{V_p}^{V} f(V)\,dV\right] \quad V_p \leq V \leq \infty \tag{20.22}$$

where I_{max} is given by

$$I_{max} = n_i e \tilde{v}_i \quad \text{A} \tag{20.23}$$

n_i is the ion number density in the sheath above the analyzer opening, and $\tilde{v}_i$ is the most probable ion velocity given by

$$\tilde{v}_i = \sqrt{\frac{2e E_i'}{M}} \quad \text{m/s} \tag{20.24}$$

in which M is the ion mass, and E_i' is the dominant ion energy of the distribution function in electronvolts.

In equation (20.22), the ion distribution function, $f(V)$, is integrated from the plasma potential V_p that is the most positive potential at which the collector

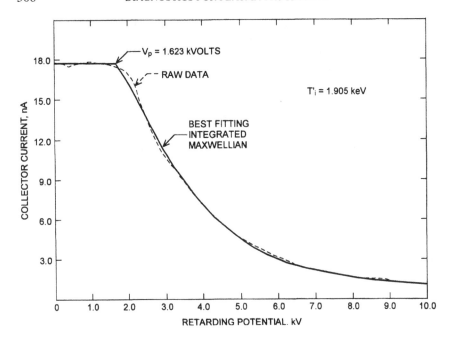

Figure 20.16. Characteristic examples of a best-fitting Maxwellian and raw data taken from a turbulent low pressure Penning discharge with a plasma potential $V_p = 1623$ V, and an ion kinetic temperature $T_i' = 1905$ eV.

current of the retarding potential analyzer remains unchanged, to the potential, V, on the retarding potential grid. If integrated over all energies, the ion distribution function $f(V)$ is normalized such that

$$\int_{V_p}^{\infty} f(V)\,dV = 1.0. \tag{20.25}$$

The retarding potential curve given by equation (20.25) is an integrated ion energy distribution function, and the distribution function of the ions themselves, $f(V)$, can be obtained by differentiating equation (20.25), to yield

$$f(V) = -\frac{1}{I_{max}}\frac{dI_i(V)}{dV}. \tag{20.26}$$

Commercially available retarding potential energy analyzers are usually equipped with software to perform the differential operation implied by equation (20.26), and plot the resulting ion energy distribution function.

As an aid in interpreting the integrated ion energy distribution functions encountered in plasma-processing applications, four characteristic examples are illustrated on figure 20.17. The integrated ion energy distribution function is

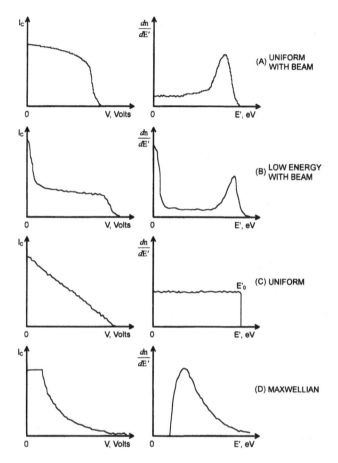

Figure 20.17. Illustrative examples of integrated ion energy distribution functions resulting from retarding potential energy analyzer traces in the left-hand column, and the ion energy distribution functions themselves, corresponding to these integrated distributions, in the right-hand column.

shown in the left-hand column. The corresponding distribution function, the derivative of the curve in the first column, is shown in the right-hand column.

Illustration 20.17(a) is that of an approximately monoenergetic ion energy distribution function, of a kind that can be observed at the workpiece of plane-parallel RF plasma etching reactors operating at 13.56 GHz. Illustration 20.17(b) is a bimodal ion energy distribution function, such as may result from operating a plane-parallel RF plasma reactor at less than 1 MHz. Illustration 20.17(c) illustrates the appearance of a uniform ion energy distribution function, in which the numbers of ions are equally distributed over all energies from $0 \le E' \le E'_0$. Finally, illustration 20.17(d) shows the integrated and differential ion energy

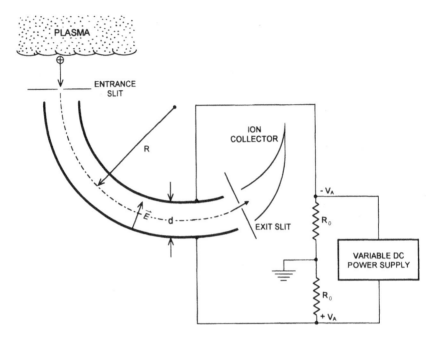

Figure 20.18. Principle of the curved-plate electrostatic energy analyzer, used to measure the energy of ions accelerated across a sheath to a surface on which workpieces are mounted.

distribution functions for an approximately Maxwellian distribution of energies similar to that of figure 20.16.

20.3.4.2 Curved-Plate Energy Analyzer

The retarding potential energy analyzer discussed before is a simple passive, non-perturbing diagnostic instrument for the energy of ions passing through its entrance aperture, but it cannot yield information about the mass of those ions. When several positive ionic species are produced in a processing plasma, it is usually desired to know their mass as well as their energy. In energy-analyzed mass spectrometers, the usual method to determine the ion energy is the *curved-plate energy analyzer* shown schematically in figure 20.18.

In such an analyzer ions from the plasma are accelerated across the sheath, and pass through a small aperture adjacent to the workpiece. The ions travel along the mid-chord of a pair of curved, parallel plates, across which an electric field is maintained. Only those ions reach the detector that travel along the mid-chord of the curved plates, and through the exit slit. This occurs when the centrifugal forces due to the ion motion on a curved trajectory are exactly balanced by the centripetal electrostatic forces generated by the electric field.

The balance of the centripetal electrostatic force and the centrifugal force may be written

$$F_r = e E_r = \frac{M v_i^2}{R} \qquad \text{N.} \qquad (20.27)$$

The radial electric field E_r between the two parallel plates is approximately that of plane-parallel capacitor plates having a median radius of curvature R with a separation d and a total potential $2V_A$ across them, given by

$$E_r = \frac{2V_A}{d} \qquad (20.28)$$

as shown in figure 20.18. The electric field across the plane parallel plates is maintained by a variable DC power supply with a center-tapped ground. This eliminates a fringing electric field at the entrance and exit, so ions will not be deflected as they enter and leave the curved plates. Also, the total voltage applied between the plates and surrounding grounded surfaces is reduced to the value V_A with the center-tapped ground arrangement, from the value $2V_A$ that would be required if one plate were grounded.

The relationship between the energy in Joules of ions of mass M and the ion velocity v_i is given by

$$E_i = \frac{M v_i^2}{2} = e E_i'. \qquad (20.29)$$

By substituting equations (20.28) and (20.29) into equation (20.27), one obtains

$$\frac{2e V_A}{d} = \frac{2E_i}{R} = \frac{2e E_i'}{R}. \qquad (20.30)$$

From this, one can obtain the ion energy E_i' in terms of the applied DC voltage to the curved plates, V_A,

$$E_i' = \frac{R}{d} V_A \qquad \text{eV.} \qquad (20.31)$$

The energy of the ions passing through the exit slit is related to the voltage applied between the plates, V_A, and the analyzer geometry. When only the ion energy distribution function is required, one can collect the ions passing through the exit slit and produce the ion energy distribution function directly.

An apparatus to measure the ion energy distribution function is shown schematically in figure 20.19(a). In most plasma-processing applications, however, it is desired to have *both* energy *and* mass analyzed ion energy distribution functions. This is accomplished by the energy analyzed mass spectrometer illustrated in figure 20.19(b), in which ions first pass through a curved plate analyzer, where their energy is determined. These ions of known energy then pass through a quadrupole mass filter where their mass spectrum is analyzed. Alternatively, the quadrupole mass filter can be set to pass ions only of a specific mass, and their ion energy distribution function swept out by changing the voltage on the curved-plate energy analyzer. Such studies of ion energy

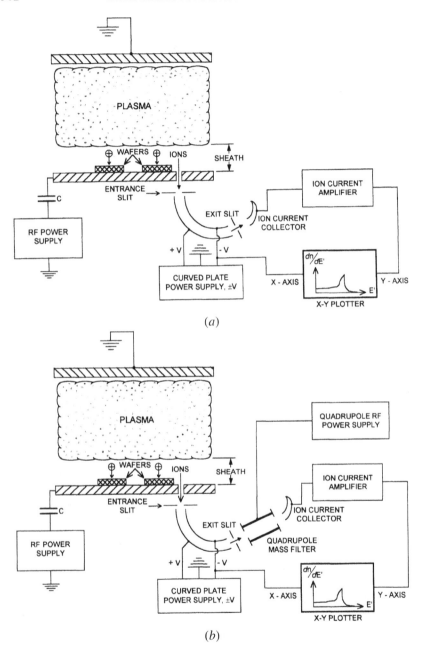

Figure 20.19. Two configurations of the curved-plate energy analyzer for the analysis of ion energy distribution functions: (*a*) direct measurement of the ion energy distribution function without mass analysis; (*b*) measurement of the ion energy distribution function with subsequent mass analysis by a quadrupole mass analyzer.

distribution functions in an etching plasma have been reported, for example, by Olthoff *et al* (1994).

20.4 MEASUREMENT OF SURFACE TOPOGRAPHY

The literature of plasma processing contains a wide variety of diagnostic methods designed to assess the physical effects of plasma exposure of surfaces, a selected group of which are surveyed below. These methods are used in the fields of plasma ion implantation, in the plasma-assisted deposition and etching of thin films, and in documenting the effects of plasma surface treatment (see, for example, Iwata and Asakawa 1991). Surface diagnostic methods can be divided into methods that visualize the topography of a surface, discussed in this section; methods that measure the chemical composition of superficial layers at and immediately below the surface, discussed in section 20.5; and methods that measure properties such as the surface energy, discussed in sections 20.6 and 20.7.

There is no diagnostic instrument more powerful than the human eye operating in conjunction with a trained human mind. The function of *surface microscopy* is to bring this combination to bear on surface features too small for the unaided human eye to resolve. To better understand the nature of surfaces and the physical processes that take place on them, various forms of surface microscopy are used to aid in visualizing the surface before and after treatment. Microscopy is generally *non-perturbing* of the surface under study, and is *active*, because of the need for electromagnetic radiation or energetic electrons as probing media.

20.4.1 Resolution

The *resolving power* or *resolution* of a microscope is measured by the smallest distance between two points on a surface that can be detectably separated. The greater the resolving power is, the smaller this distance will be. At the lower limit of resolution, images formed by electromagnetic radiation or electrons are limited by diffractive effects. Consequently, the resolution of two points is rigorously defined in terms of *Rayleigh's criterion* (John W Strutt, Lord Rayleigh 1842–1919). This criterion states that the edge of the inner dark ring of a diffraction pattern of one point should pass through the center of the diffraction pattern of a second point. The *resolution distance*, l, is then given by

$$l \equiv \frac{0.61\lambda}{n \sin i} \quad \text{m} \qquad (20.32)$$

where λ is the wavelength of the probing radiation, n is the index of refraction of the ambient medium, and 'i' is one-half the angle subtended at the surface under observation by the edges of the last lens. The quantity ($n \sin i$) is known as the *numerical aperture*, and may be as much as 1.60 for optical oil-immersion microscopes.

Based on experimental measurements, Ernst Abbe (1840–1905) suggested a slightly less conservative resolution distance,

$$l = \frac{0.50\lambda}{n \sin i} \qquad \text{m.} \qquad (20.33)$$

In materials science work, oil immersion techniques are rarely used ($n = 1.00$, air). The numerical aperture is no more than ($n \sin i$) < 1.00, so the resolution distance is no less than the estimate of the resolution length known as the *Abbe criterion*,

$$l \approx 0.50\lambda \qquad \text{m.} \qquad (20.34)$$

20.4.2 Optical Microscopy

One of the most widely used diagnostics to characterize surface topology is *optical microscopy*, which is capable of detecting gross failures in plasma processing, as well as yielding abundant information about the quality of the process. *Optical microscopy* of surfaces is done in air with monocular or binocular microscopes, using visual inspection by eye, or photography of the surface image. At the blue end of the spectrum, $\lambda \approx 400$ nm, and the resolution predicted by equation (20.34) is $l \approx 200$ nm $\approx 0.2\ \mu$m. This is comparable to or greater than current microelectronic design rules (scale sizes), so other microscopic techniques are required to visualize microelectronic circuits in full detail.

20.4.3 Ultraviolet Microscopy

Ultraviolet microscopy (UVM) uses UV radiation at $\lambda < 200$ nm in an instrument with UV-transparent lenses to produce a photographic image of the object under study. UVM is rarely used in surface studies, and offers an improved resolution little better than a factor of two, to $l \approx 100$ nm $\approx 0.10\ \mu$m.

20.4.4 Scanning Electron Microscopy

In *scanning electron microscopy* (SEM) electrons rather than photons are used as a probe to scan the surface topography and produce a photographic image. In this application, the electrons behave as a wave phenomenon, with the equivalent energy-dependent (*Compton*) *wavelength* given by

$$\lambda_C = \frac{hc}{e E'} \qquad \text{m} \qquad (20.35)$$

where E' is the electron energy in electronvolts. Substituting equation (20.35) into the criterion in equation (20.33) yields the resolution

$$l = \frac{0.5\lambda_C}{n \sin i} = \frac{0.5hc}{e E' n \sin i}$$

$$= \frac{6.2 \times 10^{-7}}{E'n \sin i} \quad \text{m.} \tag{20.36}$$

If the numerical aperture is $(n \sin i) \approx 1.0$ then equation (20.36) suggests that at $E' = 200$ keV, a characteristic SEM electron energy, $l \approx 3.1 \times 10^{-12}$ m $= 0.0031$ nm, about 1% of an atomic diameter. In practice, it is not possible to form and focus electron beams to such small dimensions. The numerical aperture is smaller than one, and the best SEM resolutions are therefore in the range of an atomic diameter, $0.2 \leq l \leq 0.4$ nm, or $2 \leq l \leq 4$ Å.

20.4.5 Atomic Force Microscopy

Atomic force microscopy (AFM) is the most powerful and highly resolved method for visualizing the topological features of a surface, and is capable of resolving details on an atomic scale ($\approx 10^{-10}$ m). It is accomplished by scanning an electrically charged fine point, with a tip radius approaching atomic dimensions, over a surface in a rastered, parallel-line scan. The distance between the tip and the surface is sensed capacitively or by other means. A single scan will give a profile of the surface, and a series of such parallel scans can be used to build up a three-dimensional image of the surface topography. Resolutions of at least $0.03 \leq l \leq 0.05$ nm $= 0.3 \leq l \leq 0.5$ Å are possible.

20.5 MEASUREMENT OF SURFACE COMPOSITION

When plasma processing is complete, it may be necessary to characterize the changes in surface composition. A large number of particle- and photon-based methods have been developed to measure the composition of surfaces. These methods tend to be active, as they require bombardment by energetic beams of photons or particles, and they are perturbing, at least on the atomic scale, since they excite and/or knock atoms off the surface. The depth of penetration of these methods can vary from a few atomic layers to several tens of nanometers into the bulk of the material. For a survey of methods used in microelectronic circuit fabrication, see Smith (1995). The characteristics of some of the more widely used of these surface diagnostic techniques are now briefly discussed.

20.5.1 Photoelectron Spectroscopy

Photoelectron spectroscopy uses an intense beam of photons to knock photoelectrons off the surface of a sample under study, by the *photoelectric effect* discussed in section 5.2 of Volume 1. The energy of the emitted photoelectrons is concentrated at peaks characteristic of the energy levels of the surface material (Matieno *et al* 1989).

If $E_0 = h\nu$ is the energy of the probing photon beam, peaks will appear in the energy spectrum of the photoelectrons at an energy ΔE_k given by

$$\Delta E_k = h\nu - E_k \qquad \text{J} \qquad (20.37)$$

where E_k is the kth ionization potential of the surface material. Having measured ΔE_k and knowing $E_0 = h\nu$, one may calculate E_k, and thereby identify the atomic species responsible.

20.5.1.1   Ultraviolet Photoelectron Spectroscopy

Ultraviolet photoelectron spectroscopy (UVPS) is a version of photoelectron spectroscopy that utilizes the ultraviolet portion of the spectrum. The photon beam usually consists of photons with an energy $E_0 = 21.21$ eV, provided by a helium discharge tube. UVPS is limited by this relatively low photon energy to excitation of the outermost electron shells of high-Z materials, and to the inner shells of low-Z materials.

20.5.1.2   X-Ray Photoelectron Spectroscopy

In *x-ray photoelectron spectroscopy* (XPS), also known as *electron spectroscopy for chemical analysis* (ESCA), the photon beam characteristically consists of 1.2 to 1.5 keV monoenergetic x-rays. These photon energies permit the excitation of, and emission from, strongly bound inner shell electrons of high-Z materials. The binding energies E_k of such inner shell electrons depend not only on the atomic number of the atom excited, but also on the chemical compound of which the atom is a part. Hence, XPS can yield information about the chemical composition or nature of the surface (hence, ESCA). The depth below the surface sampled by XPS is typically 3–5 nm, or roughly 20 atomic layers.

20.5.2   Auger Electron Spectroscopy

In *Auger electron spectroscopy* (AES), *Auger electrons* are emitted from atoms by a process first studied in the mid-1920s by the French physicist Pierre Victor Auger (1899–). If a vacancy exists in the innermost atomic K-shell of an atom, it can be filled in one of two principal ways. It can be filled by x-ray emission as an outer shell electron falls in to fill the vacancy. It can be filled also by the *Auger process*, in which the K-shell vacancy is filled by an L-shell electron (for example), while at the same time a second L-shell electron (for example) is emitted by the atom. The emitted electron is an *Auger electron*. The process is written K–$L_1 L_2$, and the energy of the Auger electron, $E(A)$, is

$$E(A) = E(K) - E(L_1) - E(L_2) \qquad \text{eV} \qquad (20.38)$$

where $E(K)$ is the energy of the K-shell electron, and $E(L_1)$, $E(L_2)$ are, respectively, the energy of the Auger electron and of the upper-shell electron that filled the K-shell. Auger electron emission is much more probable than x-ray emission from low-Z materials; for high-Z materials, x-ray emission is more likely.

As equation (20.38) implies, the Auger electron has an energy characteristic of the atom that emitted it. *Auger electron spectroscopy* (AES) consists of bombarding a solid material with a beam of x-rays or electrons energetic enough (typically, keV to tens of keV) to knock electrons out of the K-shell of the atoms under study. The energy spectrum of the Auger electrons is measured, and the peaks evident in this spectrum identify the atoms near the surface. The useful depth of AES depends on the Auger electron's mean free path in the material; it ranges from 0.5 nm (a few atomic diameters) to 100 nm. AES can be used to identify the composition of the surface material, and depth profiling is possible by measuring the Auger energy spectrum while the surface is eroded away by sputtering or etching.

20.5.3   Rutherford Backscattering Spectroscopy

Rutherford backscattering spectroscopy (RBS), also called *Rutherford ion backscattering spectroscopy*, is based on the nuclear scattering of energetic (usually helium) ions in the range from several hundred keV to several MeV, a process first investigated by Ernest Rutherford (1871–1937). The intensity, angular deflection, and energy of the scattered ions depends on the atomic number and mass of the scattering nuclei and allows them to be identified. Data on the surface composition can be obtained from depths of 20 nm to 10 μm, depending on the atomic number Z of the surface species. Depth profiles are possible by varying the energy of the probing ions, or by sputtering or etching away the surface as measurements proceed.

20.5.4   Secondary Ion Mass Spectrometry

Secondary ion mass spectrometry (SIMS), and two variants, *static secondary ion mass spectrometry* (SSIMS), and *secondary neutral mass spectrometry* (SNMS) are based on sputtering the surface of interest with energetic ions (1–10 keV) of known mass. The sputtered (secondary) ions or neutrals are mass-analyzed, yielding knowledge of the surface composition and/or the nature and atomic weight of sputtered molecular fragments. SIMS is capable of very superficial measurements, typically yielding information from the upper 0.5 nm (~2 atomic diameters) of the surface. The surface can be eroded away by prolonged sputtering with the incident ions, to yield a depth profile.

20.5.5   Reflection–Absorption Fourier Transform Infrared Spectroscopy

Reflection–absorption Fourier transform infrared spectroscopy (RA-FTIRS, FTIR) is an optical spectroscopic technique that uses the reflection of infrared radiation from a surface (Green *et al* 1991). The incident frequency is chosen to correspond to characteristic resonant chemical bond vibrations from the surface under study, which allow the chemical nature of the material to be determined. Polarization of the infrared radiation allows anisotropy of the surface material (i.e. aligned hydrocarbon chains) to be identified. Depending on the opacity of the surface material, the probing depth can range from 1 nm, to the interior, bulk material.

20.5.6   Photoelastic Modulation in Infrared Reflection–Absorption Spectroscopy

This method, related to FTIR, is useful for measuring the composition of species adsorbed on a surface that have a thickness of a few monolayers. The method is particularly valuable on metal surfaces. In this technique a polarized beam of infrared radiation, spanning wavelengths that include chemical bond resonance peaks of the surface monolayers, is allowed to fall on the sample surface at a shallow angle, 60–80° from the vertical (Barren *et al* 1991). The reflected infrared radiation propagates to a detector where its intensity is measured as a function of wavelength. The characteristic resonance peaks allow the surface composition to be determined, an approach particularly useful for polymer surface films. This method is fast, and can monitor the build-up of surface layers on metals in real time.

20.5.7   Laser Desorption–Fourier Transform Mass Spectrometry

In laser desorption–Fourier transform mass spectrometry (LD-FTMS), a laser is used to desorb a sample of surface material, or its molecular fragments, by thermal ablation. A typical probing depth for this technique is about 10–50 nm, depending on the laser energy and surface material. Mass spectroscopy yields the atomic mass and hence the composition of the surface materials and/or its ablated fragments.

20.6   SURFACE ENERGY RELATED DIAGNOSTICS

Measurement of the *surface energy* is necessary for a wide range of industrial plasma surface treatment applications, including the printability, bondability, adhesion, and wettability of papers, films, and fabrics. The molecules in the interior of a liquid are attracted isotropically by atomic forces from nearby molecules, but those on the surface are only attracted inward or horizontally. This *surface tension* force (or *surface energy*) is manifested by the tendency of the surface of a liquid to contract to form a sphere as a freely floating droplet, and by the tendency of a thin sheet of liquid to pull inward at its boundaries.

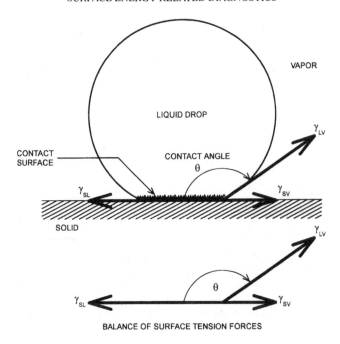

VAPOR

LIQUID DROP

γ_{LV}

CONTACT
SURFACE

CONTACT ANGLE
θ

γ_{SL} γ_{SV}

SOLID

γ_{LV}

θ

γ_{SL} γ_{SV}

BALANCE OF SURFACE TENSION FORCES

Figure 20.20. Characteristic sessile water drop on an unwettable surface, showing the balance of surface tension forces.

20.6.1 Measurement of Surface Tension/Energy

The *surface tension* of a liquid, γ, is the force per unit length required to hold in equilibrium the straight boundary of a thin, plane liquid surface. The surface tension is measured in N/m of boundary length in the SI system of units, and in dynes/cm in CGS units. These units are dimensionally (and happen to be numerically) equal to the *free surface energy*, expressed in J/m^2, or $ergs/cm^2$, in the respective units. The *free surface energy* may be understood as the work done against surface tension forces in creating a unit area of surface on the liquid at constant temperature.

The *contact angle*, θ, is the tangent angle between a surface and a liquid droplet at their contact point, as shown for an unwettable liquid–surface combination in figure 20.20. The *contact angle*, θ, is a measure of the relative amounts of *adhesive* (liquid-to-solid) and *cohesive* (liquid-to-liquid) force acting on a liquid, and varies over the range $0 \leq \theta \leq 180°$. As illustrated in figure 20.21, if $\theta = 0°$, the adhesion is as strong as the cohesion. If $\theta > 90°$, the solid is said to be *unwettable* by the liquid; if $\theta \ll 90°$, it is *wettable*. In industrial practice, a contact angle $\theta = 20°$ may be used as the threshold of wettability.

Rigorously, the relation between the contact angle and the surface tension is defined by an equilibrium at constant temperature among the surface tension

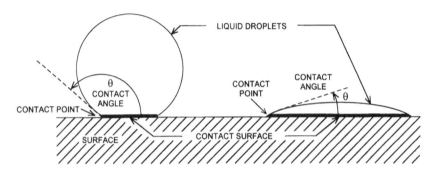

Figure 20.21. Two extremes of adsorptive wettability as revealed by the sessile water drop test.

forces,

- γ_{LV}, surface tension between liquid and vapor;
- γ_{SL}, surface tension between solid and liquid; and
- γ_{SV}, surface tension between solid and vapor.

These add vectorially, as indicated in figure 20.20, to yield

$$\gamma_{SV} - \gamma_{SL} = \gamma_{LV} \cos\theta. \tag{20.39}$$

For a material to be *wettable*, $(\gamma_{SV} - \gamma_{SL}) > 0$. For *unwettable* materials, $(\gamma_{SV} - \gamma_{SL}) < 0$. The *spreading coefficient*, S, is defined as

$$S \equiv \gamma_{SV} - \gamma_{SL} - \gamma_{LV}. \tag{20.40}$$

If $S > 0$, the solid is *wettable* by the liquid; if $S < 0$, the solid is *non-wettable* or *repellent* to the liquid. To increase the wettability of a solid, one must increase γ_{SV}. This is often best done by removing dirt, adsorbed films, or monolayers of unwettable contaminants from the surface.

An equivalent definition of the *contact angle* is that angle between the normal to a liquid surface at the contact point, pointing into the body of the liquid. It is also the normal to a solid, pointing into the body of the solid. This is equivalent to the tangent contact angle shown in figure 20.21. In this illustration, the larger the *contact surface* of a drop of fixed volume (the *footprint* of the drop) is, the smaller the contact angle is.

20.6.2   Measurement of Contact Angle

A simple and widely used test of surface tension/energy is the *sessile liquid drop test* for measuring contact angle, which can be sensitive to the presence or absence of a few monolayers of contaminants on a workpiece. A single drop of liquid—usually distilled or de-ionized water—is placed on a horizontal workpiece and allowed to spread. The contact angle of the advancing liquid on the surface, θ_a, known as the *advancing contact angle*, is measured. Sometimes, for complex surfaces such as fabrics or textured films and papers, the *receding contact angle*, θ_r, is also measured, by withdrawing liquid from the sessile drop and noting the contact angle when the wetted contact area or 'footprint' begins to decrease in size.

Probably the most commercially important surface characteristic is the wettability, discussed in section 20.6.1. The wettability is related to the advancing contact angle through the spreading coefficient defined by equation (20.40). Qualitatively, the sessile liquid drop test provides a very clear measure of wettability: if the surface is non-wettable, the liquid forms hemispherical beads with a high contact angle; if wettable, the drop spreads over a large area in a thin layer with a low contact angle. Except in special applications (such as printing inks), distilled/de-ionized water is used as the test liquid, and it is applied with a pipette or eyedropper.

In quantitative measurements of the contact angle, the advancing contact angle is first measured, then liquid is drawn from the drop by a fine pipette, and the receding contact angle is measured. The difference between the advancing and receding contact angles is the *contact-angle hysteresis*. The magnitude of this hysteresis may be related to such surface characteristics as roughness, surface contamination, or the anisotropic characteristics (such as weave, fiber orientation, or polymer 'polarization') of some fabrics and polymers.

The *contact angle* can be measured quantitatively by projecting on a screen the enlarged image of a droplet adhering to a horizontal workpiece, where it can be measured directly with a protractor. A higher surface energy results in lower contact angles. Water, which has the relatively high surface tension of 0.073 N/m (73 dynes/cm) in air, is usually used because it provides easily measured contact angles over a wide range of workpiece materials with different surface energies. Alternatively, the surface energy can be measured by daubing the surface of a workpiece with a series of liquids having progressively higher surface energies, until they wet the surface of the workpiece.

20.6.3   Measurement of Thin-Film Adhesion

An important characteristic of thin films, whether for microelectronic applications, paints on plastics, or brightwork on consumer items, is how well the film adheres to a surface. Standard tests have been developed to test thin-film adhesion, of which we describe in the following a few of the simplest and most widely used.

20.6.3.1   *Wettability and Adhesion*

In the majority of industrial plasma surface treatment applications, simple exposure to a plasma will remove contaminants and result in a significant improvement in both surface energy and adhesion. As a general rule, increases in surface energy and contact angle are positively correlated with improvement in adhesion of thin films and bondability of solids (see Tira 1987, Hansen *et al* 1989, Strobel *et al* 1994).

20.6.3.2   *Practical Adhesion of Thin Films*

The characteristic of *practical adhesion* of thin films does not always correlate with wettability. Surface contamination or the presence of a wettable layer poorly bonded to the substrate can lead to poor adhesion. The practical adhesion is measured with a variety of simple tests, as follows.

The *Scotch*® *tape test* is often used as an indicator of how well a thin film adheres to a surface. This test consists of two standard configurations, in which Scotch® brand tape is applied to and firmly rubbed on the surface, then removed by pulling the tape off at 90° from the surface (the 90° test), or at 180°, back upon itself (the 180° test), as illustrated in figure 20.22. Thin films that do not peel off under one or both of these tests are sufficiently adherent for a variety of microelectronic and consumer product (brightwork) applications.

Sometimes the adhesion of a thin film to a surface must be of a much stronger and more robust level than that required to pass the Scotch® tape test. A more robust thin-film adhesion test, used, for example, to test the adhesion of electroplated layers to steel, is the *bending test*. In this test, a flat workpiece is bent on progressively smaller radii of curvature until the bend is of small enough radius to break loose the film, or to demonstrate its adhesion on the smallest radius of curvature of interest. This procedure is illustrated in figure 20.23.

A second robust test of thin-film adhesion is the *bandsaw test*. In this test, a flat workpiece is cut with a bandsaw, with the thin film facing away from the direction in which the saw blade is approaching the workpiece, as illustrated in figure 20.24. If the saw teeth do not lift away the thin film, it is adherent enough for nearly all applications.

20.6.4   Measurement of Surface Bonding

Surface bonding or *bulk adhesion* occurs when two solid materials are joined to each other by an adhesive. Prior exposure of surfaces to be bonded to the active species of a plasma may be an effective method to improve the resulting bond. Such active species as atomic oxygen and oxygen-containing species that react with the surfaces to be bonded are most effective for the improvement of bulk adhesion (Egitto and Matienzo 1994). Good bulk adhesion is associated with good wettability and low contact angles of the adhesive (and water) with

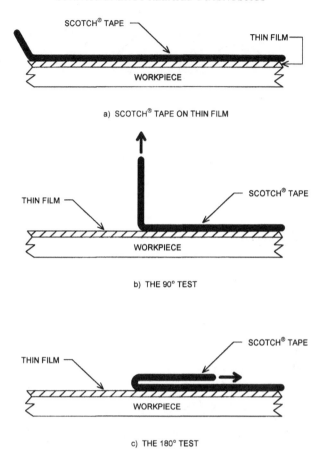

Figure 20.22. The Scotch® tape test for the adhesion of a thin film or coating: (*a*) placement of Scotch® tape on a surface; (*b*) the 90° Scotch® tape test of adhesion; (*c*) the 180° Scotch® tape test of adhesion.

the surface. Thus, sessile liquid drop measurements of wettability and contact angle are often good screening tests for bulk adhesion. It is true, however, that practical bulk adhesion or *bondability* may not be correlated with wettability in the presence of contaminants or a weakly bonded surface layer.

The performance of a bulk adhesive is measured by standard testing protocols, specialized for the material, the specific application, and the industrial sector in which it is to be applied. These protocols are beyond the scope of this text. Bondability or bulk adhesion is tested by bonding two flat slabs of material with the adhesive under test, then pulling them in the plane of the bonded layer until they separate, as illustrated in figure 20.25. Results can range from very little adhesion, to separation of the adhesive from the material at the contact

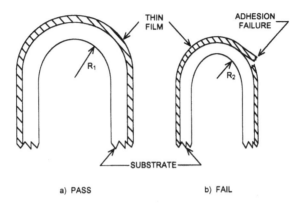

Figure 20.23. The robust sample deformation test for the adhesion of a thin film or coating on a surface: (*a*) thin surface coating adheres; (*b*) surface coating breaks free and test fails.

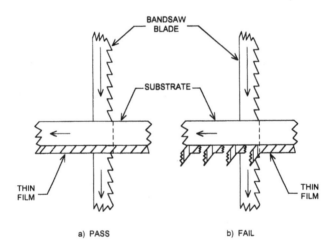

Figure 20.24. The robust bandsaw test for the adhesion of a surface coating or thin film: (*a*) coating passes, does not break away from substrate; (*b*) coating fails, shearing by bandsaw blades breaks loose coating.

plane (*adhesive failure*), to fracture of the adhesive between the bonded surfaces (*cohesive failure*), to bonding so effective that the material fractures instead of the adhesive or the bond (*material failure*) (see, for example, Hansen *et al* 1989, Kaplan and Rose 1991).

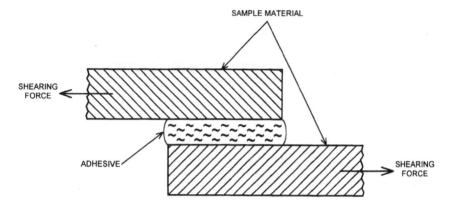

Figure 20.25. Bulk adhesion test in which a test adhesive is placed between two pieces of sample material to which a horizontal shearing force is applied.

20.7 MEASUREMENT OF ELECTRICAL PROPERTIES

Plasma processing, and particularly the deposition of thin films, may have the objective of increasing or decreasing the electrical resistance or other electrical properties of a surface. In this section, we discuss some definitions related to the electrical resistivity and charge density of plasma-treated surface layers.

20.7.1 Measurement of Surface Charge

For such applications as face masks and air filters, it is useful to permanently embed electrostatic charge in or on the surface of a material (usually a polymer), to form a *quasi-electret*. A *true electret* has charges of both signs embedded in it, with opposite charges on opposing surfaces; *quasi-electrets* are fibers or fabrics with charges of a single sign permanently embedded in their bulk or on their surface. Such quasi-electret fabrics are effective in removing dust, pollen, and other particulate contaminants from air flowing through them. These fabrics are formed by spraying electrostatic charge on the surface of the fibers of which they are composed while the latter are still molten (see Deeds 1992, Tsai and Wadsworth 1995).

A procedure for measuring the surface charge density of a film or fabric quasi-electret is the *sheet levitation test*, applied with the apparatus shown in figure 20.26. A positively charged sample is illustrated, with a charge density of σ_s C/m^2, and an areal mass density of ρ kg/m^2. This sample is placed on the lower of two horizontal parallel plates separated by a distance d, and the plates connected to a DC power supply. The voltage V_0 on the positive lower plate is

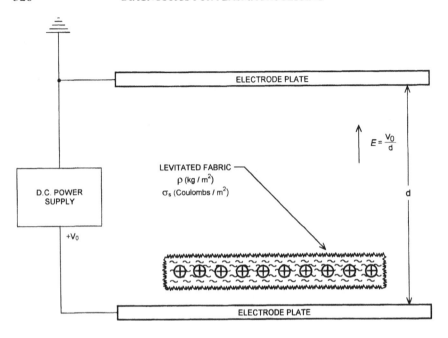

Figure 20.26. Measurement of the surface charge density of a quasi-electret by electrostatic levitation.

increased until the electric field between the plates:

$$E = \frac{V_0}{d} \qquad (20.41)$$

is just strong enough to levitate the charged sample. At that point, the gravitational and electrostatic forces per unit area will be equal:

$$F' = \rho g = \sigma_s E = \sigma_s \frac{V_0}{d} \qquad \text{Pa.} \qquad (20.42)$$

Solving equation (20.42) for the surface charge density,

$$\sigma_s = \frac{\rho g d}{V_0} \qquad \text{C/m}^2. \qquad (20.43)$$

All the variables on the right-hand side of equation (20.43) are easily measured, yielding the areal charge density of the quasi-electret.

20.7.2 Measurement of Electrical Resistivity

For many applications, including carpeting, clothing, and printed sheets, the tendency of fabrics, films, or papers to hold a surface charge is undesirable. The

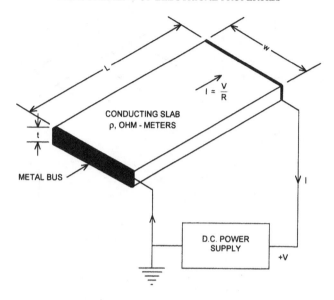

Figure 20.27. Bulk resistivity measurement of a sample slab by passing a known current at a known voltage through a sample of material of known dimensions.

surface of the material may therefore be plasma treated to make it electrically conducting, so the charge will drain away. The effectiveness of such treatment is judged by a measurement of the electrical resistivity, as follows.

Following the discussion in section 4.1 of Volume 1, the *bulk resistivity* of a rectangular slab t meters thick, w meters wide and L meters long can be measured with the arrangement illustrated in figure 20.27. A measurement of the total current drawn between the two ends of the sample by a voltage, V, yields the resistance, R Ω. The bulk resistivity is ρ Ω-m (not to be confused with the mass density), given by

$$\rho = \frac{RA}{L} = \frac{Rwt}{L} \qquad \Omega \text{ m.} \qquad (20.44)$$

This procedure is useful if the sample has well defined dimensions, and if the electrical conductivity is uniform throughout the bulk of the material. In many practical situations, however, the sample thickness is indeterminate (as in fabrics or carpeting), and/or the electrical conduction may be confined to a thin, plasma-treated surface layer of unknown thickness. In such situations, it is useful to measure the surface resistivity with the apparatus illustrated in figure 20.28. Here, one measures the resistance R between metal contacts on the ends of the sample, and uses it as a basis to calculate the *surface resistivity, ρ'*,

$$\rho' = \frac{\rho}{t} \qquad \Omega \qquad (20.45)$$

where ρ is the bulk resistivity. The surface resistivity can be difficult to use

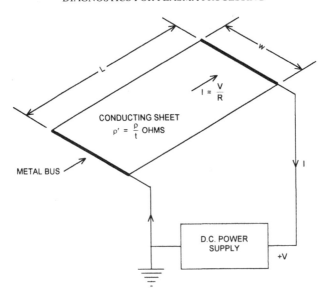

Figure 20.28. The measurement of surface resistivity in a sample of indeterminate thickness.

properly because it has units of ohms, and yet is not a resistance. Substituting equation (20.45) into equation (20.44), the surface resistivity is

$$\rho' = \frac{Rw}{L} \quad \Omega. \tag{20.46}$$

Specific standard protocols are available for measuring these resistivities, depending on the nature of the sample material and the industrial sector of application. These are beyond the scope of this discussion.

20.8   *IN SITU* PROCESS MONITORING

The Latin term '*in situ*' means 'in place', and therefore *in situ* process monitoring refers to diagnostic procedures conducted on the processed layer while the workpiece is 'in place' in the plasma reactor. Such procedures are in contrast with, for example, the surface diagnostic methods that require a workpiece to be removed from a plasma reactor for analysis. *In situ* process monitoring diagnostics are usually conducted while processing is still underway, although some procedures may be conducted after the process is complete, and the plasma turned off. An example of the latter would involve the removal of an upper electrode of a parallel-plate glow discharge reactor for better access of diagnostic instruments to the surface of the workpiece.

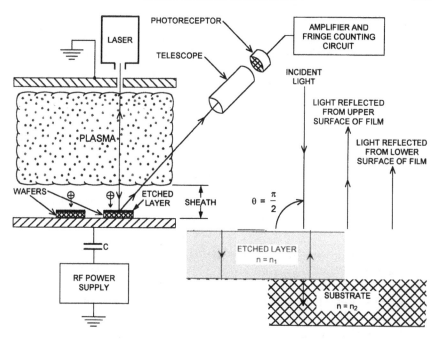

Figure 20.29. An experimental arrangement for *in situ* measurement of the thickness of an etched layer.

20.8.1 Laser Interferometer

A class of *in situ* process monitoring diagnostics is based on interference patterns of visible light produced when the wavelength is comparable to, but smaller than, the thickness of layers deposited or etched in microelectronic circuit fabrication. A characteristic apparatus for *in situ* process monitoring by laser interferometry is shown in figure 20.29. With this apparatus, the thickness of a layer can be monitored continuously during an etching (or deposition) process. Alternatively, laser interferometry can be used at intervals or upon completion of processing by turning off the plasma, and removing the upper electrode to achieve clear access of the laser radiation to the workpiece.

When laser radiation strikes a transparent layer at normal incidence, as shown at the lower right of figure 20.29, the layer will have an index of refraction n_1 and a thickness d. This layer will lie on a substrate with an index of refraction n_2. The laser interferometric technique will work only in those situations for which the layer is transparent to the incident laser radiation, and for which the substrate at least partly reflects the incident laser light. The incident laser light, and the light reflected from the upper and lower surface of the etched layer, will interfere either constructively or destructively, depending on the layer thickness d.

The illustration at the lower right of figure 20.29 shows the incident laser light normal to the etched layer at $\theta = \pi/2$. When the incident laser light arrives with an arbitrary angle θ, the *interference thickness period*, δ, is given by

$$\delta = \frac{\lambda_0}{2\sqrt{n_1^2 - \cos^2\theta_0}} \quad \text{m} \tag{20.47}$$

where n_1 is the index of refraction of the layer, and λ_0 is the free space wavelength of the radiation. Constructive interference maxima will be observed when the thickness of the etched layer, d, is given by

$$d = k\delta \qquad k = 1, 2, 3, \ldots \tag{20.48a}$$

and destructive interference minima will be observed when the layer thickness is given by

$$d = (k + \tfrac{1}{2})\delta \qquad k = 0, 1, 2, 3, \ldots. \tag{20.48b}$$

By monitoring the reflected light with a telescope and photodetector, one can follow a deposition or etching process in time, through successive interference fringes. When the last half-wavelength of thickness is removed in an etching process, the reflected light will reach a constant level associated with the substrate, thus yielding endpoint detection for the etch. Because of the relatively small size of most microelectronic etched structures, a relatively large surface area of etched material is often provided as a 'dummy' for process monitoring. This diagnostic is useful on layers of SiO_2, but is not useful on metallic layers or layers opaque to the laser radiation.

20.8.2   Ellipsometry

For surfaces opaque to the laser radiation of figure 20.29, *ellipsometry* is used. A schematic diagram of an ellipsometry apparatus is shown in figure 20.30. This is a more sophisticated method that relies on measuring the intensity *and* polarization of both incident and reflected light from a surface. The surface need not be transparent to the probing radiation. As shown in figure 20.30, a light source produces radiation at a single wavelength that passes through a polarizing filter, thus establishing its degree of polarization. This polarized radiation approaches the surface of the etched layer at an angle θ_1, is reflected at an angle θ_2, and is monitored where the total amplitude and degree of polarization of the reflected radiation can be measured.

These data can be used to measure the thickness of the etched layer, down to depths much less than the thickness period given by equation (20.48). In ellipsometry, the angle of incidence, θ_1, is usually equal to the angle of reflection, θ_2. The thickness can be determined down to a level of a few monolayers, far less than the free space wavelength of the visible radiation normally used for ellipsometry. In addition, one can measure the index of refraction of both the layer and the substrate. Software programs are available for analysis of ellipsometric data, and further information may be found in Smith (1995).

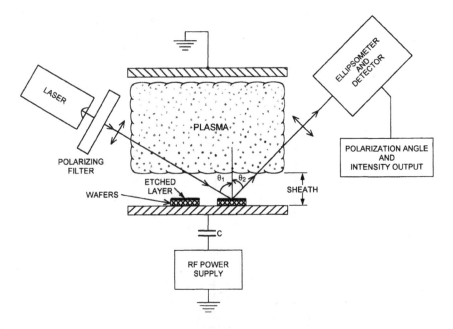

Figure 20.30. A schematic drawing of an experimental arrangement for ellipsometry of the etched layer in an RF plasma reactor.

20.8.3 Quartz Crystal Microbalance

A fourth *in situ* process monitoring diagnostic is the use of resonant crystal loading (the *quartz crystal microbalance*) to measure the deposition or etching rate of materials. This diagnostic method is not usually applied to workpieces during a deposition or etching production run. This method is sometimes applied to a specially prepared test workpiece, the surface of which is covered with a layer of the same material of interest on the production workpieces.

An experimental setup for measuring the etching rate of a material is shown in figure 20.31. In such an apparatus, an ion source provides ions of a known energy that impinge, usually normally, on the surface of the layer to be etched. The etching gas is supplied to a nozzle that distributes the gas over the surface of the layer. A thin layer of the material to be etched, comparable to or thicker than the layer on the workpieces themselves, is applied to the quartz crystal. The addition of the deposited film to the quartz crystal will change its resonant frequency as a result of the additional mass loading.

The crystal is driven at its resonant frequency, and this frequency is monitored as a function of time during the etching process. The resonant frequency of a given crystal will increase as the mass is reduced by etching. The etching rate can be calibrated by measuring the resonant frequency of the crystal before and after depositing a layer of known thickness. The precision of

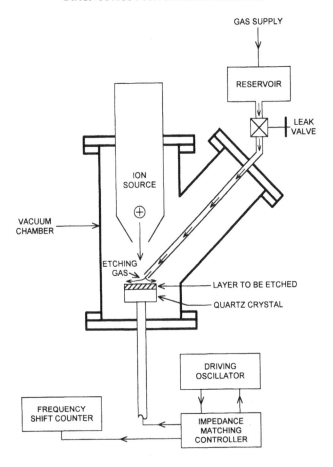

Figure 20.31. Schematic drawing of a quartz crystal microbalance to measure the etching rate of a thin layer mounted on a resonant quartz crystal.

this resonant crystal loading method may be less than 10^{-10} m, less than one monolayer, thus making it possible to monitor the deposition or etching of very thin layers.

20.9 ENDPOINT DETECTION

Probably the most important milestone in an etching process is the *endpoint*, when the etching trench reaches the substrate, and thus completely transfers the pattern on the mask to the etched layer. A number of things happen at the endpoint that allow endpoint detection with the instruments described earlier. The etching process should stop at the endpoint, and the evolution of chemical etching reaction products should cease, or be greatly reduced. If the etching reactor is heavily loaded so that most of the input etching gas is consumed during the etching process, the concentration of etching gas should rebound to levels appropriate to little or no consumption of the etching gas.

Changes in the concentration of the etching reaction products can be detected with optical spectroscopy, as they become excited by electron neutral impact in the plasma, and by mass spectrometry. The mass spectrometry of etching reaction products is a particularly sensitive indicator of the endpoint, since the partial pressure of reaction product gases is directly proportional to the number of etching reactions per unit time that occur on the workpiece.

Other indices of the endpoint include the plasma impedance for RF coupling, and the self-generated DC bias on the workpieces. In addition, when a constant flow of etching gas is maintained, changes in the total pressure sometimes reflect the endpoint of the etching process, usually as the result of changes in the partial pressure of the etching reaction products.

REFERENCES

Barren B J, Green M J, Suez E I and Corn R M 1991 Polarization modulation Fourier transform infrared reflectance measurements of thin films and monolayers at metal surfaces utilizing real-time sampling electronics *Anal. Chem.* **63** 55–60

Bohm C and Peril J 1993 Retarding-field analyzer for measurements of ion energy distributions and secondary electron emission coefficients in low-pressure radio frequency discharges *Rev. Sci. Instrum.* **64** 31–44

Cali F A, Herbert P A F and Kelly W M 1995 Automated Langmuir probe characterization of methane/hydrogen low-pressure radio frequency discharges in a production reactor *J. Vac. Sci. Technol.* A **13** 2920–3

Chatterton P A, Rees J A, Wu W L and Al-Assadi K 1991 A self-compensating Langmuir probe for use in RFD (13.56 MHz) plasma systems *Vacuum* **42** 489–93

Deeds W E 1992 Charging apparatus for meltblown webs *US Patent* 5,122,048

Egitto F D and Matienzo L J 1994 Plasma modification of polymer surfaces for adhesion improvement *IBM J. Res. Dev.* **38** 423–39

Frazier D A, Kolluri O S, Wagner G and Fahner A 1991 Successful TPO painting—cold gas plasma advances painting application *Himont/Plasma Science Technical Note* 7/91

Green M J, Barren B J and Corn R M 1991 Real-time sampling electronics for double modulation experiments with Fourier transform infrared spectrometers *Rev. Sci. Instrum.* **62** 1426–30

Hansen G P, Rushing R A, Warren R W, Kaplan S L and Kolluri O S 1989 Achieving optimum bond strength with plasma treatment *Proc. SME Adhesives '89 (Atlanta, GA)*

Hope D A D, Monnington G J, Gill S S, Borsing N, Smith J A. and Rees J A 1993 Ion energy distributions in $SiCl_4$ and Ar/O_2 dry etching discharges *Vacuum* **44** 245–8

Hutchinson I H 1987 *Principles of Plasma Diagnostics* (New York: Cambridge University Press) ISBN 0-521-32622-2

Iwata H and Asakawa K 1991 *Direct Observation of GaAs Surface Cleaning Process Under Hydrogen Radical Beam Irradiation INSPEC No B91076819 (AIP Conf. Proc.*

227) (Clearwater, FL: Advanced Processing and Characterization Technologies) pp 122–5

Kaplan S L and Rose P W 1991 Plasma surface treatment of plastics to enhance adhesion *Int. J. Adhesion Adhesives* **11** 109–13

Lochte-Holtgreven W 1995 *Plasma Diagnostics* (New York: American Institute of Physics) ISBN 1-56396-388-4

Manos D M and Flamm D L (ed) 1989 *Plasma Etching* (New York: Academic) ISBN 0-12-469370-9

Matieno L J, Emmi F and Johnson R W 1989 Electron spectroscopy *Principles of Electronic Packaging* ed D P Seraphim, R Lasky and C-Y Li (New York: McGraw-Hill) ch 24

Mutsukura N, Fukasawa Y, Machi Y and Kubota T 1994 Diagnostics and control of radio-frequency glow discharge *J. Vac. Sci. Technol.* A **12** 3126–30

Olthoff J K, Van Brunt R J, Radovanov S B, Rees J A and Surowiec R 1994 Kinetic-energy distributions of ions sampled from argon plasma in a parallel-plate, radio frequency reference cell *J. Appl. Phys.* **75** 115–25

Roth J R 1995 *Industrial Plasma Engineering Vol I—Principles* (Bristol: Institute of Physics Publishing) ISBN 0-7503-0318-2

Roth J R and Clark M 1969 Analysis of integrated charged particle energy spectra from gridded electrostatic analyzers *Plasma Phys.* **11** 131–43

Smith D L 1995 *Thin-Film Deposition: Principles and Practice* (New York: McGraw-Hill) ISBN 0-07-058502-4

Strobel M, Lyons C S and Mittal K L (ed) 1994 *Plasma Surface Modification of Polymers: Relevance to Adhesion* (Zeist: VSP) ISBN 90-6764-164-2

Swift J D and Schwar M J R 1970 *Electrical Probes for Plasma Diagnostics* (New York: Elsevier) ISBN 444-19694-3

Tira J S 1987 Adhesive and surface preparation evaluation for stainless steel used in electrical assemblies *SAMPE J.* **18** 18–22

Tsai P P-Y and Wadsworth L C 1995 Electro-static charging of meltblown webs for high-efficiency air filters *Adv. Filtration Separation Technol.* **9** 473–91

Winters H F 1980 *Elementary Processes at Solid Surfaces Immersed in Low Pressure Plasmas (Topics in Current Chemistry 94, Plasma Chemistry III)* ed S Veprek and M Venugopalan (New York: Springer) pp 69–125 ISBN 0-387-10166-7

Zheng J, Brinkmann R P and McVittie J P 1995 The effect of the presheath on the ion angular distribution at the wafer surface *J. Vac. Sci. Technol.* A **13** 859–64

21

Plasma Treatment of Surfaces

The industrial applications covered from this chapter to chapter 25 use plasma active species that are of higher concentration, different in kind, more reactive, and/or more energetic than those produced in chemical reactors. These active species make it possible to produce effects on the surface of materials that can be accomplished in no other way, or which are economically prohibitive by other methods. Such industrially useful active species are produced by corona or glow discharges, which have power densities that range from below 10^{-4} to a few watts per cubic centimeter.

In this chapter, we consider the plasma treatment of surfaces in which plasma active species interact only very superficially with the material, i.e. with the adsorbed surface monolayers or with the first few monolayers of the surface material itself. Such interactions can significantly affect the surface energy, wettability, printability, adhesion, and other commercially important properties of surfaces.

21.1 OBJECTIVES OF PLASMA SURFACE TREATMENT

A variety of industrial purposes are served by exposing the surface of materials to plasma active species. Here we briefly describe these industrial applications, and later in this chapter consider specific examples.

21.1.1 Plasma Surface Treatment

Plasma surface treatment is the use of active species produced by a plasma to modify the surface characteristics of solid materials. Plasma surface treatment may add or remove adsorbed monolayers; it may involve chemical reactions with the surface; it may add or remove surface charge; and it may change the physical or chemical state of the superficial monolayers of a material.

335

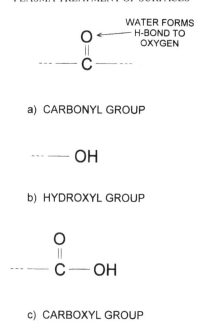

Figure 21.1. Three chemical species capable of enhancing the wettability of a hydrocarbon surface to which they attach: (a) the carbonyl group; (b) the hydroxyl group; and (c) the carboxyl group.

Plasma surface treatment does not usually employ energetic plasmas with power densities above a few watts per cubic centimeter, the maximum characteristic of glow discharges. Plasma surface treatment does not damage or alter the bulk material; it does not inject ions or atoms below the surface (this would be ion implantation, discussed in chapter 22); it does not remove bulk material (this would be *sputtering*, discussed in chapter 23, or *etching*, discussed in chapter 25); and plasma surface treatment also does not add more than a few monolayers to the surface (this would be *deposition*, discussed in chapter 24).

Plasma surface treatment may be classified by the nature of the material treated, such as thin films, fabrics, solids, natural fibers, and polymer fibers. Plasma surface treatment may also be classified by the surface property altered. These surface property effects may include cleaning; decontamination and sterilization; increasing the surface energy; improving cohesive or adhesive properties; and altering the electrical characteristics or surface finish of a material. Finally, plasma surface treatment can have the objective of chemically grafting such functional groups as those shown in figure 21.1 on the surface to achieve improved or more durable surface characteristics. These functional groups and their effects will be discussed in greater depth in section 24.2.5.

Plasma surface treatment can be effected by either active or passive plasma exposure. Treatment by *active exposure* occurs when a workpiece serves as an electrode to draw real currents of ions or electrons from a plasma. Treatment by *passive exposure* occurs when plasma active species bombard the surface of a workpiece without it drawing real currents. Passive exposure is essential for electrically insulating workpieces.

Abnormal glow discharge treatment is a form of active plasma treatment in which an electrically conducting workpiece is made the cathode of a DC abnormal glow discharge, thus ensuring complete coverage of the surface by the plasma. The DC cathode current actively treats the surface by ion bombardment. The ion energies may range from the cathode fall voltage at pressures less than 13 Pa (100 mTorr), to less than 1 eV at pressures above 1.3 kPa (10 Torr).

21.1.2 Plasma Cleaning of Workpieces

The usual objective of *plasma cleaning* is the removal of adherent layers of adsorbed contaminants. These layers may consist of many monolayers of atoms of the ambient gas; monolayers of an ambient gas to which the surface was previously exposed; a single or a few monolayers of oxides or other chemical reaction products of the surface material; hydrocarbons, such as vacuum pump or machining oils; and micro-organisms. Such contaminants can have a profound effect—usually detrimental—on such properties as surface energy, wettability, and adhesion. The removal or neutralization of these surface layers of contaminants is required by many industrial processes.

Surfaces can be cleaned by exposure to the plasma active species. Specific applications of plasma surface cleaning include reduction of the base pressure of vacuum systems by plasma cleaning of interior surfaces; improving the wettability, wickability, and printability of fibers, films, and fabrics; the improvement of adhesive bonding to surfaces; the dewaxing of wool or other animal fibers; the sterilization or disinfection of surfaces; and improving the coverage and adhesion of painted and electroplated surfaces.

21.1.3 Altering Surface Energy

Many commercial applications of plasma surface treatment are intended to increase the surface energy of materials. Increasing the surface energy can be accomplished by plasma cleaning as discussed earlier; by altering the chemical nature of the surface with plasma active species; by embedding or removing charge; or by otherwise using active species to change the physical characteristics of the surface at the atomic scale of size.

21.1.3.1 Phenomenology of Surface Energy

The nature of surface tension forces in liquids, and the use of these forces to measure the surface energy of solid materials has been discussed in sections 20.6.1–20.6.2 in the previous chapter. The *free surface energy* is measured in SI units of J/m^2, or in dynes/cm in the CGS system of units. The surface energy is the work done against surface tension forces in creating a unit area of liquid on the surface at constant temperature, and is associated with the ability of water or inks to wet surfaces. When the surface energy of a material is low, from 30 to 40 dynes/cm, it is relatively *unwettable*, and water will bead up on its surface. If the surface energy is above 60 dynes/cm, the surface will be very *wettable*, and a water drop will spread over a large area, with a contact angle below 10°.

21.1.3.2 Surface-Energy-Related Characteristics

When the surface energy of a material is changed—and plasma exposure normally causes it to increase—a wide range of commercially important properties may be altered in the direction of greater utility. These properties include the following:

Wettability

Probably the most important surface energy-related characteristic is *wettability*, which is a requirement for other important functional characteristics of surfaces. Wettability is the ability to adsorb a liquid on a solid surface, or to absorb the liquid in the bulk of fibrous materials such as paper or fabrics. Wettability implies a high surface energy (50–70 dynes/cm) and a low contact angle, as illustrated previously in figure 20.21.

Wickability

The *wickability* is a measure of the bulk absorptivity of porous or fibrous materials, such as papers or fabrics. It is usually defined in terms of the upward speed of the wetted boundary in a vertical sample dipped in the liquid of interest. Alternatively, the wickability can be characterized by the distance a wetted boundary travels upwards in a vertical sample in a fixed time.

Printability

Printability is the ability to take a printed pattern or ink on the surface of a paper, film, or fabric. It is a surface-energy-dependent property related to wettability. The printability of papers and plastic films for food wrappings has recently re-emerged as an issue in the printing industry. In the past, commercial printers have used hydrocarbon-based inks that adhere well with surface energies of only 30–40 dynes/cm. However, the hydrocarbon ink solvents produce volatile organic compounds (VOCs) which pollute the atmosphere. For environmental reasons, the industry has been made to shift to water-based inks that require surface energies in the range of 40–60 dynes/cm. The surface energies required by water-based inks are higher than the natural surface energy of many papers and packaging materials, so plasma surface treatment may be required.

Dyeability

Dyeability is the ability of a porous or fibrous material such as a fabric to be dyed in bulk. It requires both a high degree of surface adsorptivity and bulk absorptivity. Dyeing affects the surfaces of all fibers throughout a fabric, and may involve chemical reactions with the bulk material of the fibers as well. Dyeing is most effectively done if the fabric to be dyed is wettable and wickable to the dyeing solution.

21.1.3.3   Washability

Washability is the ability to shed dirt and stains and withstand laundering. It is related to the strength and cohesive properties of a fabric as well as its surface adhesion-related characteristics. To demonstrate washability, a material may be required to withstand 50 washing cycles without significant degradation of its strength, color, etc.

21.1.3.4   Functional Applications of Increased Surface Energy

A wide range of industrial processes can benefit from the increase in surface energy brought about by plasma surface treatment. These processes include improving adhesion and bonding of adhesives; improving the adherence of paint; improving the printability of surfaces; fixing ink to reduce smudging; improving bonding for potting/encapsulation of electrical/electronic components; cleaning microelectronic components for ball grid array (BGA) and wire bonding; improving the adherence and resistance to pitting corrosion of electroplated layers; improving the bond and reducing flux requirements in soldering and brazing; and improving the strength and cohesion of composite materials.

21.1.4   Altering Cohesive Properties

Cohesion is the property of a material that enables it to cling together and oppose forces tending to tear it apart. Plasma treatment can effect changes in cohesion among fibers by cross-linking of parallel polymer chains; by increasing surface-to-surface contact cohesion; and by increasing three-dimensional cross-linkages among fibers. These changes on the microscopic scale often lead to changes in *strength* and *washability*, the ability of fabrics to resist disruptive stresses; and to improving the *hand* of fabrics by altering their surface characteristics. The *hand* of a fabric is its tactile feel to the human hand, a property which distinguishes silk from wool, if manually manipulated in an enclosed box, without other sensory input.

21.1.5 Altering Adhesive Properties

Adhesion is the interaction of two surfaces, close to each other or in contact, which causes them to stick together. Adhesion results from a combination of factors, which may include mechanical, electrostatic, chemical, permeation, diffusive, surface roughness, or micro-profile contributions. Adhesion as well as surface energy is normally increased by exposing a surface to the active species of a plasma. For adhesion improvement, atomic oxygen appears to be the most important plasma active species (Egitto and Matienzo 1994).

A common problem in industry is getting paints to adhere to plastic surfaces, such as automotive bumpers and body panels. The conventional treatment uses a spray or bath of *adhesion promoters*, which contain VOCs that may present environmental and occupational safety hazards. It has been shown that in some cases, exposure to plasma active species will produce paint adhesion equal or superior to that of conventional adhesion promoters.

Another area of application of plasma processing exists because the industrial and medical uses of many materials are limited by the extent to which they can be bonded to themselves or other solid materials by bonding agents. This bonding can be greatly improved over that of conventional treatments by exposure to plasma active species. Finally, *composite materials* achieve their most desirable properties when the fibers are well bonded to the matrix material. This adhesive bond can be improved by exposing the fibers to plasma active species, a process that can make the fibers wettable and wickable by the matrix material.

21.1.6 Altering Electrical Characteristics

Many polymeric fabrics are capable of strongly retaining surface static charge. Their utility for carpeting, upholstery, and clothing is greatly enhanced if the static charge drains off quickly. This can be accomplished through an increase in surface electrical conductivity induced by, for example, bonding copper compounds to plasma-treated fibers to produce conductive fabrics.

The anti-static treatment of thin films may be necessary because the utility of paper and plastic films is enhanced if adjacent layers do not cling together electrostatically during processing or use, a phenomenon known as *blocking*. As a second example, static electrical discharges that result from the frictional motion of film through a camera can expose and ruin the film. This static charge build-up can be prevented by treating the film during manufacture with plasma active species, most frequently an atmospheric corona discharge. Procedures for measuring the surface conductivity of materials are discussed in section 20.7.2 in the previous chapter.

It is necessary to temporarily or permanently impart or embed electrostatic charges on a surface in many industrial products and processes. In photocopying, charges are deposited on the surface of paper by contact with a photoelectrically charged drum, and discharged by an atmospheric corona. In a further example, the production of breathing masks and air filters uses atmospheric corona to supply charges that become permanently embedded below the surface of individual fibers of a fabric, to form a quasi-electret. Such charge embedding is possible during the formation of non-woven fabrics, when the individual fibers are still in a molten state (Tsai and Wadsworth 1995). The amount of charge embedded in a fabric is measured by the methods described previously in section 20.7.1.

21.1.7 Altering Surface Finish

Prolonged exposure on a time scale of minutes to a sufficiently energetic glow discharge plasma can cause surface damage, removal of material, and etching or erosion to depths of microns. This can alter the surface adhesion of liquids and adhesives, as well as removing adsorbed monolayers that affect the surface finish.

In addition to microscopic physical damage, plasma exposure can remove adsorbed monolayers, induce chemical reactions among constituents on the surface or between active species and the surface, and deposit thin films ranging in thickness from a few monolayers to many microns. These changes can and do affect the smoothness, tribological/wear, scratch resistance, and optical characteristics of surfaces. Changes in optical characteristics can include color, transparency, and visible and/or infrared reflectivity and transparency. The outcomes of thin-film deposition by plasma exposure will be discussed further in chapter 24.

21.1.8 Altering Bulk Properties

In fabrics, papers, and other porous materials, bulk properties such as tensile and compressive strength, elasticity, wickability, density, and 'hand' of the material can be changed by plasma exposure. Most of these changes occur as the result of changes in the surface energy and/or cohesive properties of fibers on a microscopic scale. Examples of the ability of plasma treatment to affect the bulk properties of fabrics and composite materials will be discussed in subsequent sections of this chapter.

21.2 PASSIVE PLASMA CLEANING

Plasma exposure is capable of removing contaminants from the surfaces of solids or thin films to effect *plasma cleaning*. The contamination to be removed may take the form of adherent monolayers of hydrocarbons, a thin layer of chemical reaction products of the surface material (e.g. oxides), radioactive contamination in processing equipment, micro-organisms, or a layer of contaminants (dirt). Plasma cleaning does not include removing thick layers (more than a few microns) of surface dirt, hundreds of nanometers of surface material (this would be sputtering or etching), or changing the chemical state of the bulk material below the surface.

Plasma-cleaning reactors can be divided into two broad classes: those that rely on *passive exposure*, and those that rely on *active exposure*. In passive exposure, the workpiece does not draw real currents from the plasma. In active exposure, an electrically conducting workpiece serves as an electrode of the discharge and draws real currents from the plasma. We first consider surface cleaning by passive plasma exposure.

21.2.1 Passive Plasma Exposure Regimes

The *independent* (control) *variables* of glow discharges used for surface cleaning, whether by active or passive exposure are, in very rough order of decreasing importance:

- gas type,
- gas pressure (for intermediate pressure discharges),
- power level,
- gas flow rate,
- gas flow geometry,
- operating voltage,
- operating frequency (for RF discharges),
- discharge geometry,
- electrode characteristics,
- magnetic induction.

It is important to control and be able to measure these independent variables, particularly in exploratory research or process development applications. Once optimum conditions are established, in-line processing reactors usually require adjustment of only one or a few of these variables.

These independent variables are usually intended to affect one or more of the following *dependent variables* that are important in industrial plasma cleaning applications:

- treatment time (flux of active species),
- uniformity of treatment,
- chemical/physical nature of treatment.

21.2.2 Passive Plasma Cleaning Mechanisms

On the atomic scale, the phenomenology of surface behavior is influenced— and sometimes dominated—by multiple stacked *monolayers* adsorbed on the surface, as illustrated schematically in figure 21.2. These layers usually consist of atoms of the ambient gas, although the first few more tightly bound monolayers nearest the surface may consist of an exotic substance. Such substances might be hydrocarbon vacuum pump or machining oils, or a working gas used earlier in the history of the workpiece.

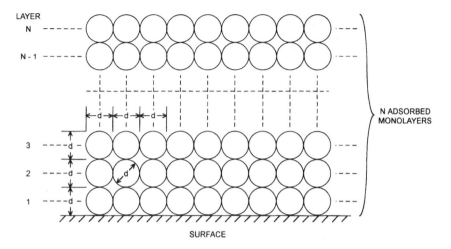

Figure 21.2. Schematic drawing of N adsorbed monolayers on a surface.

The number of adsorbed monolayers, N, can be (and usually is in vacuum systems) as high as several hundred. The lowest few monolayers near the surface are bound with energies which may approach the work function of the surface material ($4 \leq \phi_w \leq 5$ eV for most metals). Such tightly bound monolayers are very difficult to remove by chemical attack or by heating the surface (1 eV $\approx$ 11 600 K, above the melting point of all elemental solids). The outermost monolayers, however, are very loosely bound to the surface. They require only modest heating of the surface, or a modest increase of the incident particle flux, to be removed.

The simplified model for N stacked monolayers on a surface shown in figure 21.2 consists of identical spheres of diameter d, stacked in the most space-consuming configuration, with each occupying a cubical volume with a length d on each side. These monolayers play an important role in the pump-down of vacuum systems, as illustrated in the example in figure 21.3 from Wilson and Brewer (1973). At early times and high pressures, the gas is removed volumetrically, with a 3/2 slope for pressure versus time. Near 1.33×10^{-4} Pa (10^{-6} Torr), the principal gas load becomes surface outgassing, or loss of the lightly bound outer monolayers, which yields the shallower slope on the pump-down curve. In addition to this loss of adsorbed gas, diffusion to the surface of absorbed gases may also contribute to surface outgassing. The monolayers adsorbed on the inner walls of a vacuum system represent a large reservoir of gas atoms. These atoms can *outgas* and raise the pressure of the entire vacuum system when only a small part of the surface is disturbed by heating or plasma bombardment.

The *minimum plasma–wall interaction time* is determined by how long it takes a neutral atom, knocked off the wall by faster moving active species from

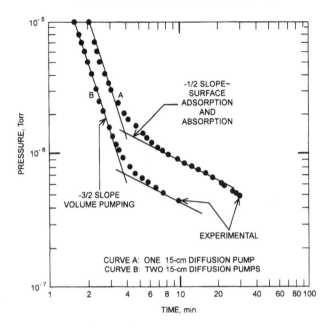

Figure 21.3. An example of two vacuum system pump-down curves: (*a*) one 15 cm diffusion pump; (*b*) two 15 cm diffusion pumps. These curves show changing slopes from volume pumping at short durations, to surface adsorption and absorption outgassing after long times (Wilson and Brewer 1973).

the plasma, to return to the plasma and interact with it. Consider the plasma shown in figure 21.4 of radius r_p, much less than the wall radius r_w, which is created at time $t = 0$ by a spark or laser. If the energetic active species take a negligible time to reach and interact with the wall, the time available before desorbed atoms from the wall reach the plasma axis is

$$\tau_1 = \frac{r_w}{v_0} \qquad \text{s} \qquad (21.1)$$

where v_0 is the most probable velocity of the desorbed atoms, at the wall temperature, T_w. The velocity v_0 is given by

$$v_0 = \sqrt{\frac{2kT_w}{m}} \qquad \text{m/s.} \qquad (21.2)$$

Substituting equation (21.2) into equation (21.1) yields

$$\tau_1 = r_w \sqrt{\frac{m}{2kT_w}} \qquad \text{s} \qquad (21.3)$$

for the minimum plasma–wall interaction time. This time is plotted in figure 21.5

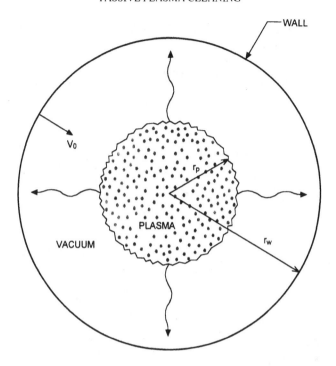

Figure 21.4. A cylindrical plasma of radius r_p interacting with a cylindrical vacuum vessel of radius r_w. Disturbed atoms leave the wall with a thermal velocity v_0 characteristic of the wall temperature T_w.

for the walls at room ($T_w = 293$ K), and liquid nitrogen ($T_w = 77$ K) temperature, and for hydrogen, deuterium, and nitrogen as the dominant working gas.

When surfaces are passively exposed to a plasma, they receive a flux of neutral species (derived in section 2.4.2 of Volume 1) given by

$$\Gamma = \frac{1}{4}n_0\bar{v}_0 = \frac{n_0}{4}\sqrt{\frac{8kT_0}{\pi m}} \qquad \text{particles/m}^2\text{-s} \qquad (21.4)$$

where n_0 is the number density of the neutral species, and T_0 its kinetic temperature, in Kelvin. If charged active species reach a passively exposed surface, their flux is *ambipolar*, as discussed in section 4.7.8 of Volume 1, and this ambipolar flux is driven by the electron number density gradient in its vicinity, given by equation (4.102),

$$\Gamma_i = \Gamma_e = -D_a\nabla n_e = -\frac{D_i\mu_e + D_e\mu_i}{\mu_e + \mu_i}\nabla n_e. \qquad (21.5)$$

The ion and electron mobility and diffusion coefficients in equation (21.5) are those appropriate to the sheath between the plasma and the wall.

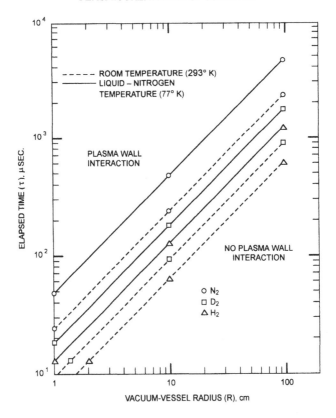

Figure 21.5. Interaction onset time, or the lower bound plasma wall interaction time, as a function of vacuum vessel radius R for hydrogen, deuterium, and nitrogen gas, with walls at room and liquid nitrogen temperatures.

21.2.3   Examples of Passive Plasma Cleaning

In this section we consider three published examples in which passive plasma exposure was used to increase the surface energy or to clean the surface of a wall or workpiece. RF glow discharges were used, and real currents to the surface appeared to play little, if any, role in the cleaning process.

21.2.3.1   Treatment for Adhesive Bonding

J S Tira (1987) exposed type 321 stainless steel samples to intermediate-pressure argon and dry-air plasmas. The objective of these exposures was a paired comparison of the lap shear strength of adhesive bonds on the cleaned surfaces, with the strength of similar adhesive bonds on samples cleaned by conventional means. The results are summarized in table 21.1. The first two rows show the

Table 21.1. Effect of plasma surface treatment on the lap shear strength[a] for adhesive bonding of 321 stainless steel (Tira 1987).

Surface treatment	FM-123-5 film strength (MPa)	Standard deviation (MPa)	Mode of failure
1 (a) Wet blast with 500 mesh Al_2O_3 (b) Vapor degrease in TCE	33.8	1.3	100% cohesive
2 (a) Vapor degrease in TCE (b) Treat with Pasa-jell 101 (c) Rinse in cold water	24.3	2.7	90–100% cohesive
3 Plasma treatment, argon gas, 75 W power, 30 min	35.1	1.24	100% cohesive
4 Plasma treatment, dry air gas, 75 W power, 30 min	36.7	1.32	100% cohesive

[a] Mean tensile shear strength; average of five specimens for each surface treatment.

results of conventional cleaning methods; the equivalent or better results in the last two rows of table 21.1 resulted from a 30 min exposure to argon and dry air plasmas, respectively.

It is clear from table 21.1 that plasma cleaning reduces, modifies, or removes whatever oxides, contaminants, or hydrocarbons may have been on the surface and interfered with adhesive bonding. Plasma cleaning is at least as effective as conventional physical and chemical surface treatments in producing a satisfactory adhesive bond. These results were obtained at a cost in environmental and occupational hazards very low by comparison with conventional cleaning processes, which may produce VOCs, unwanted by-products, and raise occupational safety issues (see Kaplan and Hansen 1991).

21.2.3.2 Hydrocarbon Removal at Intermediate Pressure

The removal of thin layers (up to 4 μm) of hydrocarbon oils applied to several metals by passive exposure to an intermediate-pressure pure oxygen plasma has been studied by Korzec et al (1994). Some characteristics of their experiment are

Table 21.2. Cleaning surface layers consisting of N62 vacuum pump oil by low-pressure RF oxygen plasma (Korzec *et al* 1994).

Characteristic	Low value	Nominal value	High value
RF power level (W)	100	300	800
RF frequency (MHz)		13.56	
Gas pressure (mTorr)	15	150	400
Cleaning time (s)			
hollow cathode	300	400	700
flat cathode	120	120	300
Layer thickness (μm)	0.4	4	4
Removal rate (μm/min)	1.2	1.6	2.0
Dirty contact angle ($^\circ$)	—	64	—
Clean contact angle ($^\circ$)	0	6	10
Surface temperature ($^\circ$C)	50	50	150

listed in table 21.2. The metals used included stainless steel, brass, and aluminum, all of which produced similar results.

Three hydrocarbon vacuum pump oils (T12, N62, and SJ27) and poly-a-olefine (C_nH_{2n}) were applied as surface films, with a thickness that ranged from 0.4 to 4 μm. The progress of the cleaning process was spectroscopically monitored by measuring the emission intensity of the 483.5 nm carbon monoxide line, a reaction product of the hydrocarbon oils and the active species of oxygen (O, O_3). This technique allowed Korzec *et al* to follow the surface cleaning process in time. In all cases reported, the cleaning process appeared to be complete after approximately 300 s (5 min).

The time dependence of cleaning stainless steel by a pure oxygen plasma (Korzec *et al* 1994) is shown in figure 21.6. A thin layer of three vacuum pump oils and poly-a-olefine, each with an initial thickness of 4 μm, were applied to the surface. The surface was exposed at an RF power of 300 W, and an oxygen pressure of 10 Pa (76 mTorr). The stainless steel surface in figure 21.6 was in the 'hollow cathode' configuration of Korzec *et al*. The oily surface was located 2 cm above and faced the powered RF electrode, and was at the same potential as the powered electrode. The process of adsorbed gas removal represented in figure 21.6 has three principal regimes. First an initial peak, as volatile components of the oils are driven off; then, a short plateau of relatively constant removal rate; and finally, a decay as areal coverage decreases and the molecular weight of the remaining hydrocarbons increases.

The effect of varying the thickness of N62 oil (this time monitoring the $\lambda = 308.9$ nm line (OH)) is shown in figure 21.7. Note that, even at a thickness of 4 μm, the cleaning process on the stainless steel surface is essentially complete

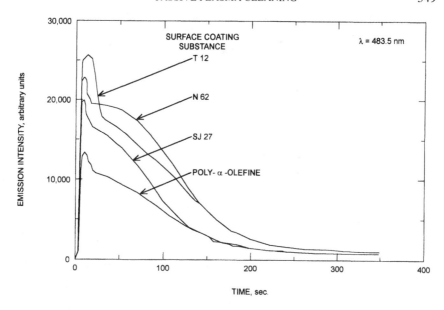

Figure 21.6. Emission intensity of the 483.5 nm line (carbon monoxide) in a hollow-cathode low-pressure RF cleaning discharge as a function of process time for four different lubricants with a starting film thickness of 4 mm, RF power of 300 W, and vacuum pressure of 10 Pa (76 mTorr) (Korzec *et al* 1994).

within 5 min. It is also the case that the cleaning process is nonlinear, because the 4 μm layer requires less than ten times as long to reach a given removal rate as the 0.4 μm layer. The effect of various degrees of coverage of a 4 μm thickness of N62 oil is shown in figure 21.8. The more complete the coverage, the more reaction product of cleaning (OH) is observed, and the longer the clean-up time. This proportionality is intuitively expected, with the clean-up process taking longer (but less than $\approx$5 min) at 100% coverage than for partial coverage.

The contact angle and surface energy of the metal surface changes with time, following the reaction product evolution curves. An example of this change in contact angle is shown in figure 21.9 for two temperatures of the metal sample during exposure in the related 'flat cathode' configuration of Korzec *et al* (1994). Prior to RF oxygen cleaning, the film of N62 oil caused the surface to be non-wettable, with sessile water-drop advancing contact angles above 60°. After about 2 min of exposure, the contact angle was below 10°, indicating a very wettable surface that had been cleaned of its hydrocarbon layers.

The operating conditions used by Korzec *et al* (1994) for their intermediate-pressure RF surface cleaning in an oxygen plasma were summarized in table 21.2. The removal rate was typically in the range of 1–2 μm (10 000 to 20 000 Å) of hydrocarbon oil film per minute, yielding total clean-up times of a few minutes, a

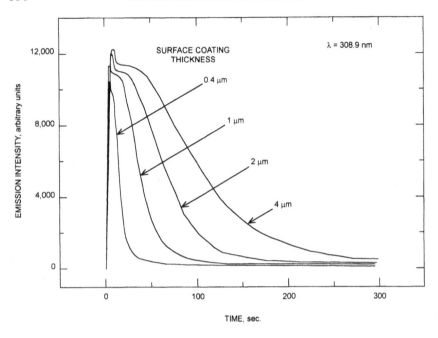

Figure 21.7. Emission intensity of the 308.9 nm line (OH) in the hollow-cathode low-pressure RF cleaning discharge as a function of process time for different thicknesses of the lubricant N62; applied RF power of 300 W and pressure in the process chamber of 10 Pa (76 mTorr) (Korzec *et al* 1994).

duration compatible with many industrial processing requirements. The complete removal of micron-thick hydrocarbon layers by chemical means or the application of detergents is very difficult, or not possible. In addition, only a few monolayers of hydrocarbon oils are sufficient to decrease the surface energy to values between 30 and 40 dynes/cm and render a surface non-wettable. The necessity of removing these last few tightly bound monolayers may be the reason that Korzec *et al* find that the cleaning process requires both high fluxes and high concentrations of active species at the surface. They also find that the maximum emission intensity of hydrocarbon reaction products during the cleaning cycle (CO, OH) correlates with the shortest cleaning time.

21.2.3.3 Surface Cleaning at 1 atm

The expense of vacuum systems and batch processing inhibits the industrial use of glow discharge plasma cleaning at intermediate pressures. The OAUGDP discussed in sections 12.5.2, 15.4, and 17.1.3.3 has been applied to the cleaning of samples of automotive steel and aluminum foil in as-received condition. The surfaces of these samples were contaminated with machining oils (Ben Gadri *et al*

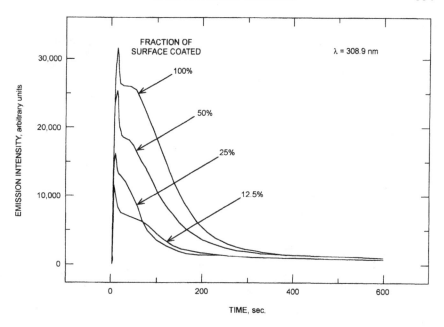

Figure 21.8. Emission intensity of the 308.9 nm line (OH) in the hollow-cathode low-pressure RF cleaning discharge as a function of process time for different lubricated areas covered by 4 mm thick N62 oil film; applied RF power of 300 W and pressure in the process chamber of 10 Pa (76 mTorr) (Korzec *et al* 1994).

2000, Carr 1997). These samples were exposed to an air OAUGDP discharge with the characteristics listed in table 15.1, in the configuration shown in figure 21.10. A photograph of a sample being cleaned was shown in figure 15.7.

The results of passive exposure of automotive steel to an OAUGDP in air are very similar to those discussed in section 21.2.3.2 by Korzec *et al* (1994) for an intermediate-pressure RF glow discharge in oxygen. In the OAUGDP exposure, the as-received samples of automotive steel were unwettable, with a surface contact angle 90° or greater, as determined by the sessile water drop test. After no more than 3 min of exposure to the OAUGDP, the surface energy increased to values approaching 70 dynes/cm. The contact angle of the sample was less than 10°, the wetted 'footprint' area was much larger, and the surface became much more wettable to water, inks, and paints. This wettability displayed an *aging effect*, in which the contact angle decreased toward the original unwettable state over periods of days to weeks, due to re-contamination of the surface.

When surfaces are not contaminated by layers of hydrocarbons microns in thickness, the exposure time to a plasma required to increase the surface energy is greatly reduced. An example of the surface energy of *linear low-density polyethylene* (LLDPE) exposed to an air OAUGDP for various durations is shown

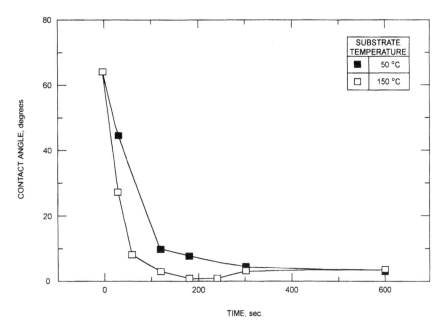

Figure 21.9. Advancing contact angle in degrees measured with water on the metal surface cleaned in the flat-cathode low-pressure RF cleaning discharge. Data are shown as a function of the cleaning time for two temperatures of the cathode during the process; the RF power was 300 W, no magnetic field, and the operating pressure was 10 Pa (76 mTorr) (Korzec *et al* 1994).

in figure 21.11(*a*) (Carr 1997, Ben Gadri *et al* 2000, Roth *et al* 2000a). The untreated surface energy of 31 dynes/cm is increased to values of 70 dynes/cm, as high as can be measured, after only one second of exposure. The surface energy displays an *aging effect* as the LLDPE surface becomes recontaminated over periods of days to weeks, as shown in figure 21.11(*b*). This ability of plasma exposure to rapidly increase the surface energy of materials that were not deliberately covered with micron-thick layers of contaminants was observed in a variety of metal, plastic, and polymer fabrics and films (Carr 1997, Ben Gadri *et al* 2000, Roth *et al* 2000a).

21.3 ACTIVE PLASMA CLEANING

Active plasma cleaning is characterized by the collection of real currents on the workpiece, which also serves as an electrode of a DC electrical circuit. There is an intuitive tendency to assume that active plasma cleaning and other plasma-processing effects are faster and more effective than passive exposure to a plasma

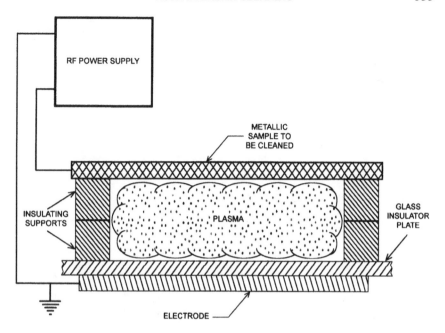

Figure 21.10. A configuration used to clean metal surfaces in the UT Plasma Science Laboratory with the surface to be cleaned as a bare electrode of the OAUGDP.

because a surface is receiving a flux of energetic charged particles in the form of a real current. In many situations, this assumption is not justified.

21.3.1 Active Plasma Exposure Configurations

Active plasma exposure can be accomplished by placing a workpiece on an electrode (usually the cathode of an abnormal DC discharge), where it will receive a real current of ions and/or electrons. This is in contrast to the flux of neutral active species or zero net current, *ambipolar* flux of charged species resulting from passive exposure. Active exposure also can occur on the surface of powered RF electrodes that are DC biased to draw real currents. In some applications, the intensity of the ion and/or electron flux is enhanced by making the surface to be exposed part of a *hollow-cathode* configuration, discussed in section 5.4 of Volume 1. In some applications, the surface is bombarded by a beam of ions generated by a source external to the plasma.

21.3.2 Active Plasma Cleaning Mechanisms

Once passive exposure of the wall to the plasma exceeds the duration τ_1, given by equation (21.3), active plasma–wall interactions become dominant. Active

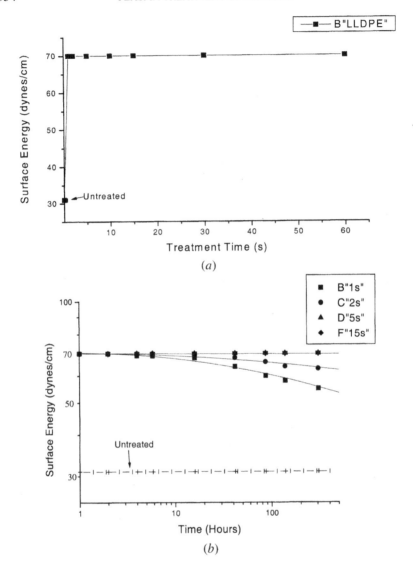

Figure 21.11. Effects on linear low-density polyethylene (LLDPE) of exposure to an OAUGDP operating in air at 5 kHz and 6 kV$_{rms}$: (*a*) surface energy of the LLDPE as a function of time of exposure to the OAUGDP; (*b*) aging effect showing decrease of surface energy as a function of time after exposure (Carr 1997).

interactions proceed until a steady state is reached for the flux of energetic ions or electrons, and the resulting efflux, or outgassing, from the adsorbed monolayers on the wall. The time required to establish this steady state is the *plasma–wall equilibration time*, τ_u.

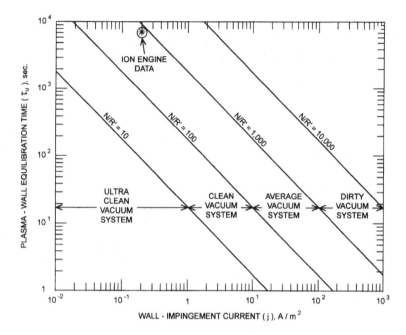

Figure 21.12. The maximum plasma–surface interaction time due to active plasma cleaning as a function of the wall impingement current in A/m² of energetic ions, for four different values of the 'cleanness ratio' N/R'.

A rough estimate of the number of adsorbed neutrals removed per unit area from a wall by energetic particle bombardment may be written as

$$D_1 = \frac{J\tau_u R'}{e} \qquad \text{atoms/m}^2 \qquad (21.6)$$

where J is the current density of energetic ions or electrons impinging on the wall in A/m². The parameter τ_u is the duration of bombardment, and R' is the *outgassing coefficient* for adsorbed gas, not to be confused with equation (14.4) for the reflection coefficient of individual incident particles. The coefficient R' measures the removal of adsorbed monolayers by energetic active species from the plasma.

Assume that monolayers on a surface consist of atoms stacked in a cubical array, with a unit cell of dimension d, as shown in figure 21.2. If there are N monolayers available to be removed, the number of neutral atoms available to be removed from the wall per unit area is approximately

$$D_2 = \frac{N}{d^2} \qquad \text{atoms/m}^2. \qquad (21.7)$$

The duration required to establish a steady-state balance between desorption and redeposition is found by equating equation (21.6) to equation (21.7), and solving

Table 21.3. The state of a vacuum system as a function of the cleanliness parameter N/R'.

Range of N/R'	State of vacuum system
$N/R' \leq 10$	Ultraclean walls
$10 \leq N/R' \leq 100$	Clean walls
$100 \leq N/R' \leq 1000$	Average walls
$N/R' > 1000$	Dirty walls

for the *plasma–wall equilibration* (or clean-up) *time*,

$$\tau_u = \frac{eN}{d^2 J R'} \qquad \text{s.} \qquad (21.8)$$

For the average atom deposited in a monolayer on the surface, $d \approx 3 \times 10^{-10}$ m, which, when substituted into equation (21.8), yields

$$\tau_u = 1.78 N/J R' \qquad \text{s.} \qquad (21.9)$$

Equation (21.9) is plotted in figure 21.12 for vacuum system conditions characterized by table 21.3 in terms of the *cleanliness parameter N/R'*. Figure 21.12 implies that unless vacuum systems used in industrial plasma-processing applications are kept very clean, typical current densities to the surface of 0.10–1.0 mA/cm^2 for active exposure could result in plasma–wall interaction times of hours. This duration is much longer than the etching and deposition run times typically used in the microelectronic industry.

21.3.3 Examples of Active Plasma Cleaning

In this section, we will consider four examples of *active plasma cleaning*, in which the surface to be cleaned collects real DC currents flowing from a plasma or an ion source.

21.3.3.1 Ion-Beam–Wall Interaction

The first example of plasma–wall interaction/cleaning is taken from the aerospace ion propulsion literature (Richley and Cybulski 1962). In this experiment, a space-charge-limited ion beam from a *Kaufman ion source* (see section 6.5 of Volume 1) interacted with adsorbed contaminants on a vacuum tank wall. Figure 21.13, taken from Richley and Cybulski (1962), shows the experimental arrangement, with an ion engine located in the bell jar on the right, and an energetic ion beam impinging on a liquid-nitrogen-cooled target to the left. The pressure near the target was monitored by ion gauges, and is plotted as a function of time in figure 21.14.

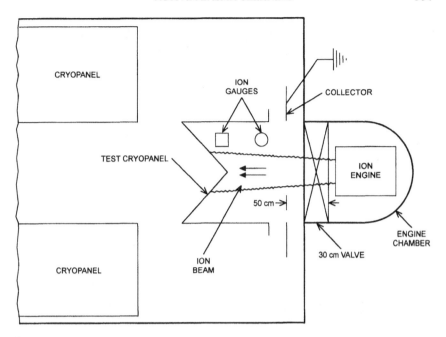

Figure 21.13. Test configuration in which energetic mercury ions from an electrostatic ion engine were incident on a liquid-nitrogen-cooled cryopanel in a large vacuum system (Richley and Cybulski 1992).

The graph in figure 21.14 shows that, after pumping down the vacuum system for three hours, a small rise in pressure occurs as the propellant (mercury vapor) is turned on. When the ion beam is turned on, the stripping of adsorbed monolayers by ion bombardment produces an increase of a factor of 25 in the neutral gas pressure. Equilibration to a steady-state balance between the stripping of monolayers from the surface by the ion beam, and re-adsorption of neutrals on the walls of the vacuum tank, takes approximately three hours. This equilibration time is the upper limit on *plasma–wall interaction time* given by equation (21.9).

These long durations are consistent with the surface outgassing times indicated in figure 21.12. The data of figure 21.14 were taken for an ion current density of $J = 0.2$ A/m^2 with an observed (from figure 21.14) plasma–wall interaction time of $\tau_u \approx 800$ s. This point is plotted on figure 21.12, and is consistent with the prediction of equation (21.9) and table 21.3 for a vacuum system with average-to-dirty walls.

21.3.3.2   Removal of Oxygen from GaAs

Iwata and Asakawa (1991) have shown that active bombardment by an atomic beam of hydrogen atoms (H) will remove a superficial oxide layer from a gallium

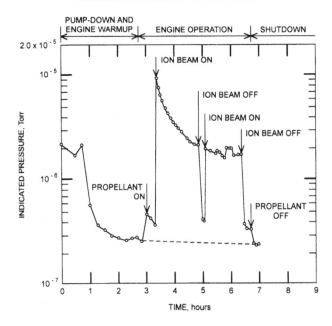

Figure 21.14. Background pressure as a function of time in the vacuum tank of figure 21.13, in which an electrostatic ion engine using mercury propellant at 2500 eV per ion was under test. The beam current was 0.125 A, and the beam target area was 0.65 m^2 (Richley and Cybulski 1992).

arsenide (GaAs) surface. This process is important for certain microelectronic circuit fabrication operations. In this work, they found that the level of H$_2$O partial pressure in the vacuum system was a good index of the removal rate of the oxide layer. Working gas pressures near the sample were below 1.3×10^{-2} Pa (10^{-4} Torr). While bombardment with molecular hydrogen (H$_2$) had little effect on the oxide layer, hydrogen *atom* (H) bombardment removed the oxide layer in 5 to 10 min. The amount of oxide (oxygen) on the surface was measured by Auger electron spectrometry (AES), and was found to decrease exponentially with H-atom exposure time.

21.3.3.3 Cleaning by DC/60 Hz Abnormal Glow Discharge

The aluminum surfaces that form part of the structure of an intense lithium ion-beam diode have been cleaned of carbon and oxygen impurities by a DC/60 Hz *abnormal glow discharge*. This configuration was described in section 16.1.1.1 (Struckman and Kusse 1993). The operating conditions of this discharge are listed in table 21.4. The impurity removal rate and clean-up time are comparable to other active plasma cleaning methods.

Table 21.4. Abnormal glow discharge cleaning of electrodes for an intense ion-beam source (Struckman and Kusse 1993).

Gas	Argon
Pressure	100 mTorr
Species	Al, bombardment by Ar^+
Operating voltage	300 V
Current density	5.0 A/m^2
Cleaning time	300 s
Removal rate	1–10 monolayers/min
Species removed	Carbon, oxygen impurities

Table 21.5. Characteristics of DC-glow-discharge-cleaned accelerator beam tubes (Hseuh et al 1985).

Cathode current density	20 μA/cm^2 = 0.20 A/m^2
Operating voltage	350 V
Gas mixture	Argon, Ar + 10% O_2
Tube material	Stainless steel 304 LN
Area of tubing	1.5 m^2
Dosage of ions	2×10^{18}/cm^2
Duration of treatment	4–6 hr
Surface temperature	225 °C @ 20 μA/cm^2
Pressure	35 mTorr
Total removal	550 monolayers of CO, equivalent
Total carbon	80 monolayers @ 2×10^{18} ions/cm^2
Removal rate	1.5–2.3 monolayers/min

21.3.3.4   DC Glow Discharge Cleaning of Accelerator Beam Tubes

Hseuh et al (1985) used a glow discharge plasma to clean the interior of 304LN stainless steel accelerator beam tubes. These tubes were 8.8 cm inside diameter and 5.5 m long, as shown in figure 21.15. A mixture of argon with 10% O_2 was used, under the conditions listed in table 21.5. The glow discharge was maintained by a 350 V DC potential, with the 1.5 m^2 interior surface of the tube serving as cathode. Time-dependent, mass spectrometric data on the evolution of reaction products indicate that up to 550 monolayers equivalent of CO were removed over a duration of 4–6 hr. This implies a removal rate of 1.5–2.3 monolayers/min, values in agreement with other quantitative reports of plasma cleaning by active plasma exposure.

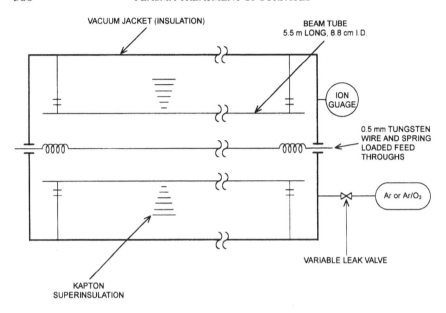

Figure 21.15. Schematic drawing of a DC glow discharge cleaning configuration for 5.5 m long accelerator beam tubes using active ion bombardment of the beam tube inner surface (Hseuh *et al* 1985).

21.3.4   Comparison of Active and Passive Plasma Cleaning

Both *active* and *passive plasma cleaning* appear capable of getting metallic surfaces so clean that their surface energy increases to values approaching 70 dynes/cm, and they become wettable, as measured by the *sessile water drop test*. However, the rate of removal of oil films with an oxygen-containing plasma is up to a thousand times faster with passive intermediate-pressure or atmospheric-pressure exposure than with active exposure. Reported removal rates are μm/min for passive exposure, and monolayers per minute for active exposure and/or ion bombardment.

21.4   PLASMA STERILIZATION

In addition to removing adsorbed monolayers and layers of surface contaminants up to several microns thick, plasma exposure also can kill micro-organisms on a surface. The size of intact micro-organisms ranges from less than 0.1 to more than 10 μm, comparable to the thickness of the oil films removed by plasma cleaning reported by Korzec *et al* (1994), and discussed before in section 21.2.3.2. When plasma sterilization is possible, it offers many advantages over conventional sterilization methods now used in hospitals and the healthcare industry. These advantages can include sterilization at room temperature in times as short as 5 min, using energy-conservative methods that pose no significant occupational or environmental hazards.

21.4.1 Objectives of Plasma Sterilization

The objective of plasma sterilization is to kill and/or remove all micro-organisms which may cause infection of humans or animals, or which may cause spoilage of foods or other goods. *Micro-organisms* are living creatures too small to be visible to the unaided human eye, and characteristically have dimensions that range from 50 nm to several tens of microns. Target micro-organisms for sterilization include *bacteria*, a class of microscopic plants, some of which cause disease in humans; *fungi*, plant-like micro-organisms which lack chlorophyll; *spores*, a dormant, seed-like phase of many micro-organisms classed as plants that are very difficult to kill; and *viruses*, submicroscopic micro-organisms that are much smaller than bacteria or individual cells, and prey on the latter two by invading and replicating within them. All these micro-organisms are composed of proteins, and support their existence by metabolizing food and (usually) oxygen from their environment.

There are various degrees of removal of micro-organisms from a surface. These range from the crudest to the most thorough, as follows:

(1) *cleaning*, the removal of dirt and contamination that foster the growth of micro-organisms;
(2) *antisepsis*, the killing or removal of a significant fraction of micro-organisms that cause infection or decay, and/or the removal of food and nutrients which support them;
(3) *disinfection*, the killing or removal of all but a few individual micro-organisms that cause infection; and
(4) *sterilization*, the complete removal or killing of all micro-organisms.

21.4.2 Test Micro-organisms

In the fields of medicine, biomedicine, healthcare, and the food-processing industry, a relatively small group of micro-organisms has been selected as *test species*. These test species are used to determine the ability of a sterilization method to kill a much broader range of micro-organisms that can cause food spoilage, food poisoning, infection, and death.

An important characteristic of any sterilizing mechanism is its ability to kill bacterial and fungal spores, since they are the most resistant to lethal agents. Bacteria are classified as either Gram positive or Gram negative, depending on the outcome of the *Gram test*, a relatively simple starting point for the classification of bacteria. The micro-organisms alphabetically listed are all relatively harmless, and can be tested with P1 or P2 containment, the lowest levels of biohazard precaution.

21.4.2.1   Aspergillus fumigatus

This fungus is used as a test organism for vegetative mycelia and fungal spores. Fungi are a source of concern in both hospital settings and in the food industry.

21.4.2.2   Aspergillus niger

A second test fungus used as a test organism, similar to A. *fumigatus*.

21.4.2.3   Bacillus stearothermophilus

A spore-forming, Gram positive bacterium that is unusually resistant to heat and other killing agents. It is often used as a test of autoclaving or other sterilizing systems, the killing effects of which are based on high temperatures.

21.4.2.4   Bacillus subtilis

The *Bacillus subtilis* endospore or forespore provides a model to study the molecular mechanism for resistance to environmental stress. These dormant cells are well protected against such environmental threats as desiccation, heat, oxidizing agents, and ionizing and non-ionizing radiation.

21.4.2.5   Deinococcus radiodurans

The *Deinoccocceae* are a Gram negative bacterium, they do not form spores, and they show an unusual capacity to survive high doses of gamma radiation. Indeed, they were originally discovered in the core cooling water of nuclear reactors. This radiation resistance is based on its ability to quickly repair DNA damage caused by ionizing radiation, a characteristic that may be related to its ability to withstand complete dehydration for long periods in its natural arid environment. It is a useful test organism for any sterilization method based on ionizing radiation or DNA breakup.

21.4.2.6   Escherichia coli

E. *coli* is a Gram positive bacterium, and one of the most widely studied and best known of all micro-organisms. It is useful as a test organism for sterilization because of this large database. In particular, its genome is well characterized, and it therefore makes an excellent subject to study possible DNA damage by a sterilization mechanism.

21.4.2.7   Phi X 174

The bacteriophage Phi X 174 is a bacterial virus (it attacks only bacteria), and is used as a safer surrogate for pathogenic or potentially pathogenic human or animal viruses. It is used in many standard tests because its size is comparable to several pathogenic viruses, and it has a spherical shape like many viruses. Phi X 174 is environmentally stable, it does not infect humans, and it is easy to detect and measure in the laboratory.

21.4.2.8 Pseudomonas putida

Two species of *P. putida* are used to assess environmental stresses. *P. putida* Idaho is solvent resistant, as the result of fatty acids added to the outer membrane of its cell envelope. A second species, *P. putida* MW 1200, does not have this protection, and can be used as a paired comparison.

21.4.2.9 Salmonella typhimurium

This Gram negative bacterium is used in the Ames test to detect mutagenic and carcinogenic agents. A histidine auxotroph of *S. typhimurium* that lacks the capacity to make its own histidine is exposed to a killing agent or environmental stress, and placed on a medium that lacks histidine. The only cells that survive are those that have mutated back into a form that can manufacture their own histidine.

21.4.2.10 Staphylococcus aureus

This Gram positive bacterium, like *E. coli*, has been very widely studied and its response to a variety of environmental stresses has been documented in a large database.

21.4.3 Killing Mechanisms

A wide range of killing mechanisms has proven effective against micro-organisms that are human or food pathogens. These methods include environmental stress, both by heat and by poisoning; physical disruption of the structure of the micro-organism; and chemical attack of the cell proteins by oxidation or other reactions.

21.4.3.1 Environmental Stresses

The food-processing and biomedical industries have employed a variety of environmental stresses to kill micro-organisms. These stresses include at least four mechanisms.

(1) *Heat*, in which micro-organisms are raised to a temperature (usually above 126 °C) at which their proteins and other molecular structures are denatured or destroyed, and they die.
(2) *Poisoning* or *disruption of metabolism*, in which a chemical destroys the protein structure of micro-organisms or disrupts their metabolism to the point that they die.

(3) *Suffocation*, in which oxygen or another energy-releasing metabolite is withheld from micro-organisms until they die.
(4) Finally, *starvation*, in which a key food or nutrient required for growth or reproduction is withheld until the organism dies.

Micro-organisms will not survive a sufficiently long exposure to the first two killing mechanisms listed above. When exposed to the last two mechanisms some micro-organisms, including certain bacteria, go into a dormant state as *spores* and survive in that form until their surroundings become more favorable.

21.4.3.2   Physical/Chemical Disruption

The advent of *plasma sterilization* has made available additional killing mechanisms in the form of active species of intermediate pressure and atmospheric plasmas (Ratner *et al* 1990, Young 1997, Rutala *et al* 1998, Kelly-Wintenberg *et al* 1998, 1999, Montie *et al* 2000). One of these mechanisms is disruption of the lipid cell wall of micro-organisms, which allows the cell contents to leak into the surroundings, killing the micro-organism. An example of this mechanism is illustrated in figure 21.16, a paired comparison of scanning electron microscopy (SEM) images of treated and untreated *E. coli*. Untreated *E. coli* are shown on the left, and the disrupted cells resulting from 30 s of exposure to an OAUGDP on the right. The leakage of cell contents was confirmed by spectroscopic analysis of the surrounding fluid (Montie *et al* 2000).

Another plasma-related mechanism for killing micro-organisms is poisoning by oxidizing active species (Rutala *et al* 1998, Vassel *et al* 1998, Herman *et al* 1999, Montie *et al* 2000). These oxidizing species may include oxygen in the form of atomic oxygen and ozone, and nitrogen oxides that form in air plasmas. Atomic oxygen is a particularly useful killing agent because it has one of the smallest atomic radii in the periodic table of the elements. It therefore can readily diffuse into gelatinous media, and permeate through sterilization bags, wrappings, and into the interstices of hospital and biomedical equipment. In addition to its small size, atomic oxygen has a chemical rate constant for oxidation at room temperature that is about one million times that of ordinary molecular oxygen, O_2.

Other killing agents in plasma sterilization include molecular disintegration by active species, in which proteins or other molecules essential for life are broken apart or irreversibly modified by highly chemically reactive active species. Finally, if atomic oxygen or ozone is present, the basic molecular structure of micro-organisms can be destroyed by oxidation or incineration. In these processes, the elemental constituents of the micro-organism's proteins are made to form carbon dioxide, water vapor, and other simple products of oxidation. This process is particularly potent in the presence of atomic oxygen, with its high chemical rate constant.

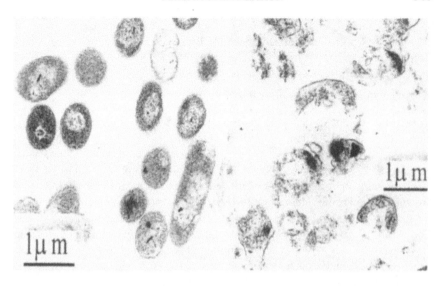

Figure 21.16. Scanning electron microscopy (SEM) images of *E. coli* (*a*) before and (*b*) after 30 s of exposure to an OAUGDP operating with air as the working gas in a parallel-plate reactor for the conditions 10 kV$_{rms}$ and 7.1 kHz (Montie *et al* 2000). ©2001 IEEE.

21.4.4   Sterilization Methods

Conventional methods of sterilization used in hospitals and other major biomedical and healthcare settings are slow and costly, and there is a strong safety and economic motivation to replace them (see Young 1997, Rutala *et al* 1998).   The $50 000+ inventory of surgical instruments for an open-heart operation requires at least an overnight delay to be sterilized for re-use with conventional non-plasma-related methods.   If sterilization could be ensured in an hour or less, such a set of instruments could be used two or three times a day. In addition to the issue of turn-around time, the cumbersome and dangerous technologies used by the sterilization departments of large hospitals represent a major commitment of capital investment and staff time.

21.4.4.1   Conventional Sterilization Methods

The two principal conventional methods of sterilization are *autoclaving* (heat), and *ethylene oxide* (EtO, a toxic poison to micro-organisms).   Ethylene oxide sterilizers require 12–19 hr to do their work; leakage of ethylene oxide gas is hazardous to employees because it is carcinogenic, mutagenic, and teratogenic; it is expensive to comply with environmental and governmental regulatory requirements; large isolated installations are required, usually remote from operating rooms, because of plumbing, drainage, safety, and ventilation requirements; and the cost of the EtO gas is relatively high.   Autoclaving in hospital settings takes about as long as EtO treatment, and shares some of the other drawbacks enumerated above for EtO, including expensive regulatory requirements.

21.4.4.2 Plasma Sterilization Methods

At least two companies have put intermediate-pressure plasma sterilizers on the market (Young 1997, Rutala *et al* 1998). One of these sterilizers uses hydrogen peroxide as the working gas, and the other uses peracetic acid. These units are single, free-standing equipment racks that will accept a few tens of liters of instruments or material to be sterilized. Sterilization in these intermediate-pressure plasma units characteristically takes 75 min, they can be made safe for patients and users, and the units are easily installed near the point of use. The plasma sterilizers employ safe and relatively inexpensive working gases, and they have relatively few regulatory requirements compared to EtO sterilization. However, these units each cost more than $100 000, and are too large and expensive for medical and dental clinic applications.

Some of the size and cost constraints on vacuum plasma sterilizers may be removed by using atmospheric plasmas for sterilization applications (Kelly-Wintenberg *et al* 1998, 1999, 2000, Roth *et al* 1996, 2000b, Roth 1999, Hermann *et al* 1999, Montie *et al* 2000). Sterilization using the OAUGDP may offer further advantages over the well documented vacuum plasma sterilization approaches. The OAUGDP approach has demonstrated the ability to kill several decades of a wide variety of micro-organisms in a few tens of seconds (Kelly-Wintenberg *et al* 1999); to kill micro-organisms in direct contact with the plasma (Montie *et al* 2000); to kill micro-organisms with active species convected to a location remote from the plasma (Roth *et al* 2000b); and to kill micro-organisms embedded in air filters (Kelly-Wintenberg *et al* 2000).

21.4.5 Sterilization Criteria and Measurement

If micro-organisms are subject to such environmental stresses as high temperature or a toxic chemical, they do not all die at once or after a single, fixed duration. At a constant level of environmental stress, the number of micro-organisms remaining at a time t after imposing the stress decreases exponentially with time. Such curves are straight lines on a semilogarithmic plot of number versus time, and are known as *killing curves* or *survival curves*. The *survival time* is the *Napier time* required for the number of micro-organisms to be reduced by a factor of $1/e = 0.368$. An environmental stress such as plasma exposure will cause an initial population of micro-organisms to die off exponentially with time, but each species of micro-organism will have its own characteristic survival time.

In industrial and medical sterilization applications, a widely used measure of killing effect is the *decimal reduction time*, or *D value*. The *D* value is the time it takes to reduce the population of micro-organisms by a factor of 10 or by

90%. Another widely used measure of killing effect is the *kill ratio*, the ratio of the initial to the final population of a micro-organism. In cases of *sterilization*, with no viable micro-organisms left, the kill 'ratio' is taken to be equal to the initial population of micro-organisms. *Commercial sterilization* is not considered to have been achieved unless a kill ratio of at least one million to one has been demonstrated.

To illustrate the survival curves and sterilization criteria just discussed, we take examples from the killing of *E. coli* by an OAUGDP operating with air as the working gas at 1 atm of pressure. The apparatus used for sterilization by direct exposure to the plasma is illustrated in figure 21.17. The plasma is generated in the parallel-plate configuration discussed in section 15.4.1 and illustrated schematically in 21.17(*a*), with the seeded microbiological test specimen exposed directly to the plasma and its active species on the lower electrode. A photograph of the plasma in operation is shown in figure 21.17(*b*). The results of a test on an initial population of 6×10^6 micro-organisms is shown in figure 21.18, which shows the survival curves for *E. coli* (Kelly-Wintenberg *et al* 1999). After only 30 s, sterilization was complete (all organisms were killed).

In some sterilization tests, the test specimens were in sealed medical sterilization bags. These bags consist of a material that is permeable to ethylene oxide, the conventional sterilizing agent. It was found that when exposed to an OAUGDP in air, sterilization of samples in these bags occurred about as rapidly as when the sample was directly exposed to the plasma. These results indicated that the active species diffused through the bag material at least as fast as ethylene oxide. The active species responsible is not likely to be ozone, which would recombine to form O_2 before diffusing through the bag. This observation is one of several lines of evidence that the dominant active species for sterilization in the OAUGDP is atomic oxygen.

In many situations, it is inconvenient or not possible to put the workpiece to be sterilized in the narrow gap between parallel-plate electrodes for direct exposure. For both vacuum and atmospheric plasmas, workpieces have been sterilized in *remote exposure* configurations, in which the active species responsible for sterilization are convected from the plasma to a remote workpiece. An example of an OAUGDP remote exposure reactor is illustrated in figure 21.19. In this device, a serpentine air flow entrains active species from plasma generated on a series of flat OAUGDP panels (see section 15.4.1), and convects these species to the workpiece. This airflow can be single passage, or recirculated through a closed pneumatic loop (Roth *et al* 2000b).

Survival curves for *E. coli* exposed in the Remote Exposure Reactor (RER) are shown in figure 21.20 for samples exposed at the remote location indicated in figure 21.19, and for both single-pass and recirculating air flow. A survival curve for *E. coli* under direct exposure in the parallel-plate configuration of figure 21.17 is shown for comparison. Figure 21.20 demonstrates that several decades of killing can be accomplished in a few tens of seconds by either direct or remote exposure to the active species of the OAUGDP. The data of figure 21.20 also

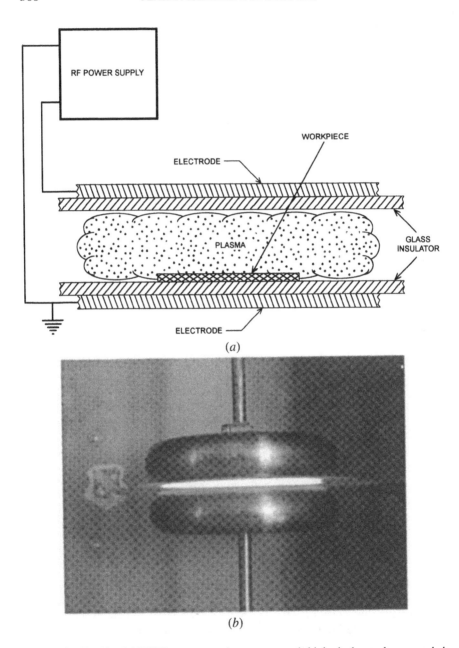

Figure 21.17. The OAUGDP reactor used to expose such biological samples as seeded filter paper, fabrics, Agar medium, and these media sealed in commercial sterilization bags (Brickman *et al* 1996): (*a*) schematic diagram showing position of sample; (*b*) photograph of parallel-plate reactor in operation.

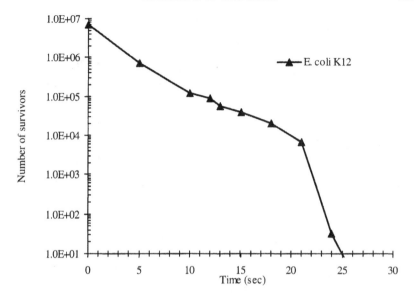

Figure 21.18. Survival curve for an initial loading of 6×10^6 *E. coli* K12 cells on a polypropylene sample, exposed for the times shown on the abscissa to an air plasma in an OAUGDP parallel-plate reactor for the conditions 10 kV$_{rms}$ and 7 kHz.

indicate that at a given exposure time, killing is about ten times more potent in the recirculated gas flow than with single-pass air flow. This observation demonstrates that the lifetime of active species responsible for killing is long enough to make several transits around the flow loop, a distance of about 1.5 m in the device of figure 21.19.

21.5   TREATMENT OF THIN FILMS

Passive exposure to plasma active species is extensively employed to treat papers and thin films used for food wrappings and containers. Such treatment increases their surface energy and improves (or even makes possible) their printability for decoration, advertising matter, and consumer product information. The market for thin-film plasma processing is expected to grow as plasma-related methods are further developed to meet environmental requirements for water-based inks and to serve new industrial and commercial markets.

21.5.1   Hydrocarbon and Polymeric Materials

The large and competitive market for plastic/polymeric thin films and solid materials has led to the development of a large variety of commercially available materials, the common, trade, and chemical names of which can cause confusion.

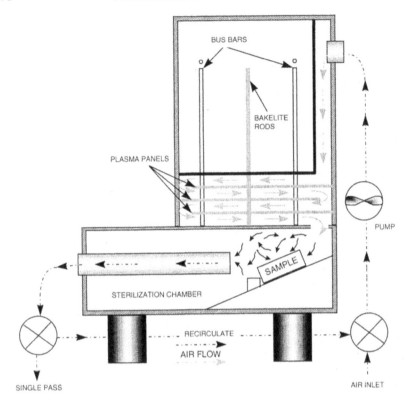

Figure 21.19. Schematic drawing of an OAUGDP remote exposure reactor with three plasma-covered flat panels, and a serpentine flow of air to entrain active species. After active species are delivered to the sample in the remote chamber, the gas flow can be exhausted, or recirculated for build-up of active species (Montie *et al* 2000, Roth *et al* 2000b).

To better identify the plastic/polymeric materials discussed in this section and later in this volume, selected information on the nomenclature of these materials is listed in table 21.6. The plastic/polymeric materials included are divided into *hydrocarbons, hydrocarbon polymers* (*polymers* are long, repetitive molecular chains), and *fluorocarbon polymers*.

21.5.2 Thin-Film Plasma Reactors

A characteristic plasma treatment of thin films is the direct exposure of plastic film intended for food packaging. Such exposure can render the surface printable for decoration, consumer product information, and advertising. Accomplishing this with vacuum plasma surface treatment involves batch processing with reactor systems or tools such as those discussed in section 17.1.1, and illustrated in

Table 21.6. Nomenclature for common hydrocarbons and polymers.

Chemical name	Common acronym	Common trade name(s)	Corporate origin of trade name
Hydrocarbons			
Cellulose acetate			
Polycarbonate	PC	Lexan	General Electric Co.
Polydimethysiloxane			
Poly(ethylene terephthalate)	PET	Mylar	E I duPont de Nemours and Co.
Polyhexamethylamine		Nylon	E I duPont de Nemours and Co.
Poly(methy methacrylate)		Plexiglas, Lucite	Rohm and Haas Co., E I duPont de Nemours and Co.
Poly(oxymethylene)		Delrin, Celcon	Celanese Plastic Co.
Polyurethane			
Poly(vinyl chloride)	PVC		
Hydrocarbon polymers			
Polybutadiene			
Polybutene-1			
Polyethylene	PE		
Polyisobutylene			
Polypropylene	PP		
Polystyrene			
Polyvinylcyclohexane			
Fluorocarbon polymers			
Polyperfluoropropylene		FEP Teflon	E I duPont de Nemours and Co.
Polytetrafluoroethylene		Teflon	E I duPont de Nemours and Co.
Polytrifluorochloroethylene		Kel-F	Minnesota Mining and Manufacturing Co.
Poly(vinyl fluoride)		Tedlar	E I duPont de Nemours and Co.
Poly(vinylidene fluoride)		Kynar	Pennwalt Corp.
Tetrafluoroethylene		TFE Teflon	E I duPont de Nemours and Co.

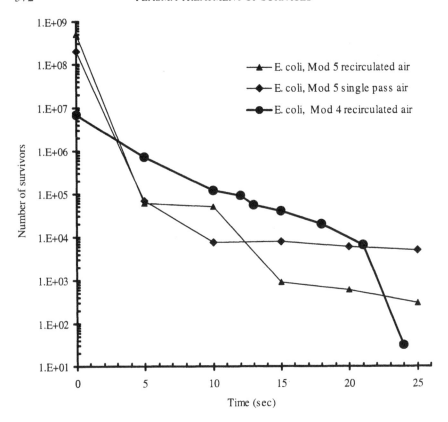

Figure 21.20. Survival curves for *E. coli* cells exposed in an OAUGDP remote exposure (Mod 5) reactor with single-pass and recirculating gas flow, under the operating conditions of 10 kV$_{rms}$ and 7.1 kHz. Both are contrasted with the parallel-plate (Mod 4) direct exposure data of *E. coli* cells from figure 21.18 under the same operating conditions (Montie *et al* 2000).

figures 17.1 and 17.25. In this arrangement, a feed roll of thin film is installed in a vacuum chamber, and unwound onto a second, take-up roll after passing through or near an intermediate-pressure glow discharge plasma generated by any of the mechanisms discussed in section 16.1.

A common plasma source for the treatment of electrically insulating thin films is the DC or RF planar magnetron configuration of figure 16.9, in which neither real nor displacement currents flow through the film under treatment. The batch processing required by this form of treatment is economic only when the end result can be accomplished in no other way, or when the roll replacement and vacuum pumping time is much less than the time required to expose the film to the plasma.

In many industrial applications, such as the processing of textiles in which the webs move with velocities up to 10 m/s, batch processing is neither economic nor practical. A few workers (see Rakowski 1989) have developed differential pumping configurations, such as that illustrated in figure 14.1. These configurations allow the continuous feed of a web, thin film, or fiber into and out of a vacuum system for intermediate-pressure plasma processing through several stages of differential pumping. However, such continuous feed systems have not been widely adopted.

For the continuous processing of thin films, atmospheric coronas or an OAUGDP reactor may be used in the configuration illustrated in figure 17.2. An enclosure is generally required to control the composition of the working gas (if other than air), to protect workers from x-ray or ultraviolet radiation, and to protect them from ozone if the working gas is air or an oxygen-containing mixture. In plasma reactors operating at 1 atm, thin films may be easily processed by feeding the film through a gas-tight seal similar to that illustrated in figure 17.2. If ozone or NO_x production is an issue, a slight negative pressure should be used; if purity of the working gas is an issue, a small positive pressure is appropriate.

21.5.3 Objectives of Thin-Film Plasma Treatment

Probably the largest single application of plasma surface treatment of thin films is to plastic food bags and food wrappings, a market with a value-added cash flow of several hundred million dollars a year in the United States. Such treatment allows the printing of background color, opaque coatings, decorations, consumer product information, and advertising matter on surfaces that are unwettable in their as-manufactured state. For other uses, such films can be processed to make them sufficiently wettable to be written or printed upon, or to take paints, coatings, or adhesives.

Other major applications of thin-film plasma treatment include anti-static processing, and the deposition or neutralization of surface charge. Anti-static treatment of thin films is discussed by Ouellette *et al* (1980), and is important in the manufacture of photographic film. In addition, plasma exposure also can be used to improve the adhesive properties of thin films, to make them bond better to adhesives, paints, or deposited coatings. Plasma exposure can also alter the surface roughness and *tribological* (wear-related) characteristics of thin films.

21.5.4 Mechanism of Effect

In the plasma treatment of thin films and other solid surfaces to increase their surface energy, two kinds of outcome have been observed with respect to the duration of effect. The first and most frequent outcome is greatly increased surface energy and wettability with decreased contact angle immediately after exposure (within the first few minutes or hours). In metals and non-polar solids particularly, this high surface energy is not durable and fades with time, almost

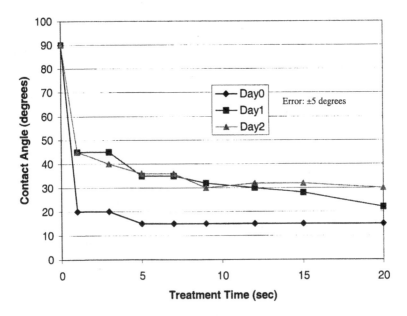

Figure 21.21. Contact angle of polyethylene terephthalate (PET) exposed for various durations to an OAUGDP, as a function of the time in days after exposure (Roth *et al* 2000a).

reaching the initial surface energy after a few days or weeks. This behavior is known as the *aging effect*.

An example of the aging effect has been shown previously in figure 21.11, and is illustrated for polyethylene terephthalate (PET) film in figure 21.21. The data of figure 21.21 (Roth *et al* 2000a) show the contact angle, in degrees, of samples of PET thin film exposed for selected durations to an OAUGDP with air as the working gas. The surface energy is plotted as a function of exposure duration and time in days after exposure.

These data illustrate the very high initial contact angles for the untreated material, the very wettable low contact angles that can be achieved by plasma exposure, and the tendency of longer-exposed samples to take longer to reach a given contact angle. These data also illustrate the tendency of the contact angle of this material to remain at values below 40° (that is, to remain wettable) for periods up to several days after exposure. The trends in the data of figure 21.21 are characteristic of a wide range (but not all) of polymeric, plastic, amorphous, and metallic surfaces.

The second kind of outcome, which can be achieved by exposing many plastic films and some fabrics to an atmospheric plasma, is long-term retention of wettability over periods of many days. An example of this behavior is illustrated in figure 21.22, from Roth *et al* (2000a). These surface energy measurements

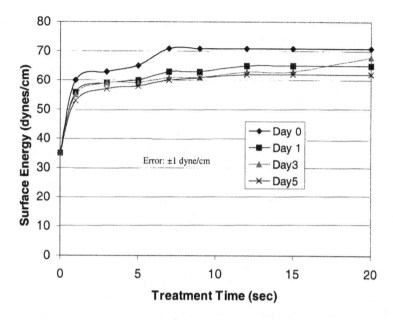

Figure 21.22. Surface energy of 34 g/m^2 (PP) meltblown fabric exposed to an OAUGDP for various durations as a function of the time in days after exposure (Roth *et al* 2000a).

were taken on samples of 34 g/m^2 PP meltblown fabric exposed to an OAUGDP for selected durations, and plotted as a function of time after exposure. The surface energy rises to values above 70 dynes/cm after a few seconds of exposure, and decays after a few days to a value of 62 dynes/cm. Such a durable effect at this high value is sufficient, for example, for in-line printing with practically any water-based ink.

Research into this issue of permanence of effect is recent and ongoing (Wadsworth *et al* 1994, Roth, 1999, 2000a, b). Available evidence indicates that non-wettability of metals and many other materials is due to only a few monolayers of oils or other contaminants on the surface. According to the *plasma cleaning model*, these monolayers are tightly bound to the surface, and cannot be removed with detergents or chemical cleaners. Plasma active species, however, are energetic enough to remove these monolayers, increase the surface energy, and render the surface wettable. The high surface energy and wettability is lost gradually as the cleaned surface is re-contaminated by the surrounding air or by diffusion from within the material, consistent with the data in figure 21.21. This re-contamination takes a period of hours to weeks, a duration that may be sufficient for many in-line industrial applications that take place shortly after plasma exposure, such as printing, coating, or adhesive bonding.

Many potential applications of induced wettability to consumer products such as diapers and clothing require a long shelf life (at least six months) and a durable effect, similar to that illustrated by the data in figure 21.22. There exist two models for the physical processes that may be responsible for this durable effect. The first model is the *polar group attachment model*, in which a durable effect is achieved with plasma surface treatment if an electrophilic atom (e.g. atomic oxygen) or molecular fragment (e.g. OH, CO) is chemically bonded to the carbon chain of hydrocarbon or polymeric materials, in the manner illustrated in figure 21.1. Such chemical additions to the surface molecules provide permanent sites for the attachment of water molecules, and hence permanent wettability.

The second model for polymer or hydrocarbon materials is the *polar group rotation model*, in which polar groups imbedded in the material are rotated to the surface by electric fields or other plasma-related processes. These polar groups increase the surface energy and provide attachment sites for water molecules. If, according to this model, the polar groups rotate or diffuse back into the material over time, the effect is temporary; if the polar groups remain on the surface, the effect is durable.

Polar groups are not found in or on metallic surfaces, so the effect in such cases is likely to result from plasma cleaning alone. In plastic and hydrocarbon-based polymeric materials, polar groups may play a role. Data such as those plotted in figure 21.22 may represent an initial effect due to plasma cleaning, followed by a durable effect due to the attachment or rotation of polar groups to the surface of the material.

21.6 TREATMENT OF POLYMERIC OR ORGANIC SOLIDS

Plasma active species can remove contaminants and adsorbed monolayers from solid materials, and modify their surface structure and composition to increase the surface energy and improve wettability, printability, adhesion, bonding, and other characteristics of commercial importance. The plasma treatment of polymeric/organic solid, three-dimensional workpieces has been under active research since approximately 1960 (Hudis 1974). Such industrial uses have been restricted to relatively high-value items by the requirement for batch processing in vacuum systems. More recent monographs on plasma treatment of polymers include those edited by Yasuda (1984, 1988, 1990), d'Agostino (1990), and Strobel *et al* (1994). A survey article on adhesion improvement of solids by plasma exposure has been written by Egitto and Matienzo (1994).

21.6.1 Plasma Reactor Systems

Recent industrial practice has been to plasma treat polymeric solids using batch processing in intermediate-pressure ($p < 133$ Pa $= 1$ Torr) *glow discharge* plasmas. Such plasmas may be generated by RF at 13.45 MHz, or the DC abnormal glow or microwave plasmas discussed in section 17.1.2. Such reactors usually expose solid components in plasma-filled vacuum chambers with a characteristic volume that may range from 50 liters for small, high-value items, to several cubic meters.

It is normally desirable to minimize the total process time in intermediate-pressure reactors, so the vacuum and gas handling subsystems are capable of pumping down to the operating pressure in a few minutes (Mlynko *et al* 1988). The atmospheric-pressure plasma reactors discussed in section 17.1.3, and particularly the more recently developed OAUGDP have not been extensively applied to the industrial treatment of polymeric solids. However, the absence of a vacuum system and potential elimination of batch processing recommend them for such applications.

21.6.2 Surface Cleaning of Polymeric Solids

It is necessary to clean dust, contamination, or unwanted films from surfaces in many high-technology manufacturing processes. An example is the plasma cleaning of microelectronic circuit boards (Anonymous 1986). After a mechanical step such as drilling through circuit boards, and after etching or stripping, it may be necessary to clean the surface prior to the next step in manufacturing. This can be done with the *hollow-cathode* RF glow discharge configuration shown in figure 21.23, in which an array of parallel circuit boards serves as the powered electrode of an RF discharge. The 'hollow cathode' results in a plasma density increased up to ten times by multiple reflections of electrons between the parallel boards. This increased density causes increased fluxes and higher concentrations of active species, and faster cleaning of the board surface. The plasma operates in the 0.1 Pa (mTorr) pressure range.

21.6.3 Altering Surface Energy

Plasma treatment can affect the surface energy and contact angle of solid polymeric materials, usually in a direction that improves the wettability and printability of their surfaces. Improvements in these characteristics facilitate the bonding of adhesives, paints, electroplated layers, and thin films to a surface.

An example of the use of plasma treatment to improve the adhesion of paints to *thermoplastic olefins* (TPOs) has been published by Frazier *et al* (1991). TPOs are used in the automotive industry to form auto body side panels, bumper covers, grills, etc. Use of TPOs conventionally requires the use of *adhesion promoters* that contain *volatile organic compounds* (VOCs) and that present potential occupational hazards to workers using them. The VOCs and the adhesion promoters can be eliminated, with at least equally good bonding results, by using intermediate-pressure plasma surface treatment.

Frazier *et al* (1991) reported on flat samples of TPO resin that were directly exposed to an intermediate-pressure plasma, and then covered with a standard multilayer coating of paint. The resulting painted samples were compared with

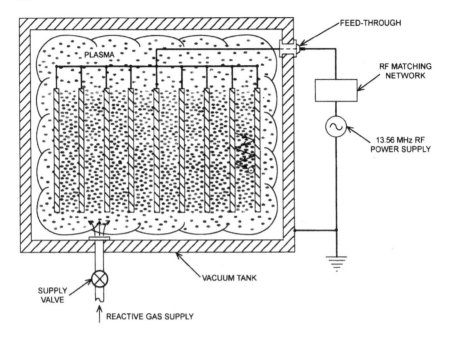

Figure 21.23. A multiple-parallel-plate, 'hollow-cathode' arrangement for cleaning circuit boards by passive exposure to a low-pressure RF generated plasma (Anonymous 1986).

Table 21.7. Adhesion of paints to TPOs with and without plasma surface treatment (Frazier *et al* 1991).

Sample test	Untreated	Adhesion promoter	Plasma treated
Initial adhesion	Fail	Pass	Pass
96 hr humidity	N/A	Pass	Pass
Heat age (7 days @ 70 °C)	N/A	Pass	Pass
Chip resistance (28.9 °C)	N/A	Pass	Pass
Gasoline soak	N/A	≤ 3 min	>96 hr[a]
Gasoline immersion	N/A	<3 cycles	>50 cycles

[a] Test was concluded at 96 hr with no failures.

conventionally painted samples, the surfaces of which were untreated, or treated with a conventional adhesion promoter containing VOCs. The results are shown in table 21.7.

All untreated samples failed an initial Scotch® tape adhesion test, while all plasma and adhesion promoter-treated samples passed it. In further testing, the plasma-treated samples performed as well as or better than the samples treated with the conventional adhesion promoter. The plasma-treated samples had a clearly superior resistance to gasoline exposure. A case has been made by Frazier *et al* (1991) that application of a single plasma reactor production unit to TPO automotive bumpers can save thousands of dollars per year in direct production costs. Plasma exposure can also save the additional cost of dealing with the regulatory requirements of VOCs from conventional adhesion promoters.

21.6.4   Improving the Bonding of Solids

Plasma surface treatment also has been used to improve the adhesive bonding of plastics and other solid materials to each other. In Kaplan and Rose (1991), the authors compared the effects of oxygen and ammonia plasmas on samples of common engineering plastics bonded by two adhesive methods. The engineering plastics were supplied by their manufacturers as injection-molded 'dog-bone' style tensile test samples. The control samples used the conventional surface preparation procedure recommended by the manufacturer of the adhesive. Each sample was cut midway, plasma treated, and bonded to provide a 12.7 mm × 12.7 mm overlap. At least three samples were tested for each condition, with the results summarized in table 21.8. Plasma treatment improves adhesion in all cases. The failure mode changes from separation of the adhesive–plastic interfacial bond (*adhesive failure*), to rupture of the adhesive (*cohesive failure*), and, most impressive of all, to fracture of the plastic sample itself (*material failure*).

Another adhesive bonding application is *composite materials*, in which fibers are embedded in a matrix (Kolluri *et al* 1988). Plasma treatment of the fibers can greatly improve the properties of the composite materials of which they are part, as shown by the data of Kaplan and Rose (1991) in figure 21.24. The shear strength and flexural modulus are shown for unidirectional composites in which the fibers were untreated, corona treated, or plasma treated. Again, plasma treatment greatly improved the characteristics of the product.

A final example of the plasma treatment of solid surfaces, from Hansen *et al* (1989), is shown on table 21.9. These data show the improvement in bonding made possible by treating Tefzel$^{\circledR}$ with an RF oxygen plasma. The plasma was operated at pressures up to 67 Pa (500 mTorr), and with an input of 300–550 W of 13.5 MHz RF power. Tefzel$^{\circledR}$ is a copolymer of tetrafluoroethylene and ethylene, has good resistance to most solvents, and a 'non-stick' surface as the result of its low surface energy and coefficient of friction. It is very difficult to bond Tefzel$^{\circledR}$ to other polymers, to metals, or to itself; also, one cannot paint or print on its surface without treatment. A caustic sodium etch process, which is potentially occupationally hazardous, has been developed to bond this material.

The lap shear bond strength of Tefzel$^{\circledR}$ samples, prepared and tested in the same manner as those in table 21.8, is shown in table 21.9 for untreated, sodium-

Table 21.8. Lap shear strength, in MPa, of the adhesive bonding of untreated and plasma-treated solid surfaces (Kaplan and Rose 1991).

Material	Manufacturer	Type of material	Preferred plasma gas	Control (MPa)	Plasma (MPa)	Improvement factor	Failure mode
Valox™ 310	General Electric	Polyester themoplastic	O_2	3.6	11.3	3.1×	From adhesive to material
Noryl™ 731	General Electric	Polyphenylene ether	NH_4	4.3	12.4	2.9×	From adhesive to material
Durel™	Hoechst Celanese	Polyarylate	NH_4	1.7	14.9	8.6×	From adhesive to cohesive
Vectra™ A625	Hoechst Celanese	Liquid crystal polymer	O_2	6.5	8.6	1.3×	From adhesive to material
Delrin™ 503	Du Pont	Acetal homopolymer	O_2	1.1	4.5	3.9×	From adhesive to cohesive
Ultem™ 1000	General Electric	Polyetherimide	NH_4	1.3	14.4	11.3×	From adhesive to cohesive
Lexan™ 121	General Electric	Polycarbonate	O_2	11.8	15.5	1.3×	From adhesive to cohesive

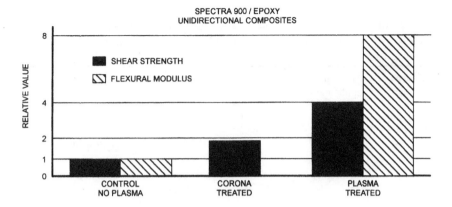

Figure 21.24. The effects of plasma treatment on the shear strength and flexural modulus in a unidirectional composite material without plasma treatment, treated by a corona discharge and plasma treated (Kaplan and Rose 1991).

Table 21.9. Adhesion of Tefzel® after various plasma surface treatments (Hansen *et al* 1989).

Sample test	MPa	SD
Untreated	0.07	0.01
Sodium etch	0.78	0.13
Plasma–O_2/SF_6	2.31	0.11
Plasma–NH_3	1.40	0.10

etched, and plasma-treated samples. The bonding strength of the plasma-treated samples is up to three times that of the conventional, sodium-etched samples.

21.6.5 Biomedical Applications

The normal surface characteristics of polymeric and plastic materials are not suitable for many potential applications to medical/biological products. Since 1980 however, it has been found that plasma surface treatment can facilitate the biomedical uses of common polymer and plastic materials (Ratner *et al* 1990). Biomedically desirable changes to the surface of such materials include *activation*, in which the surface chemistry is changed by adding or forming molecular fragments on the surface of the material, which affect its wettability or biological compatibility; plasma-induced *grafting*, in which plasma active species produce free radicals on the surface and provide a basis for biologically compatible surface coatings; and *deposition*, in which a few monolayers of

Table 21.10. Three examples of biomedical products for which plasma surface treatment has been used successfully to create biocompatible surfaces (Kaplan 1994).

Medical product area	Medical application	Biological process	Bulk material	Desired surface characteristic(s)
Disposable laboratory hardware	Cell culture dishes	Cell growth and attachment	Polystyrene	• Oxidized surface chemistry • Hydrophilic
Medical devices	Catheters	Blood flow/transport	Polyethylene Teflon^{TM}	• Bind heparin • Non-thrombogenic
Diagnostic products	Immunoassay tubes	Protein/antibody binding	Polypropylene	• Absorption • Hydrophobic • Hydrophilic • Covalent bonding

biologically compatible molecules are created or deposited on the surface by active species. Many of these desirable changes can be induced by exposure to an intermediate-pressure oxygen plasma, or to an air plasma at atmospheric pressure.

Table 21.10 (Kaplan 1994) lists three classes of biomedical products for which plasma surface treatment has been used successfully to create biocompatible surfaces. In cell cultures, it is essential for most cells to have intimate contact with the surface in order for the culture to proliferate *in vitro* (in non-living surroundings). The chemical properties of the surface and its degree of wettability determine the viability of the culture. Both of these surface properties can be greatly improved by plasma treatment.

In catheters, blood-compatible surfaces can minimize clot formation by blocking fibrinogen binding at the blood–catheter interface, as illustrated in figure 21.25 (O'Connell *et al* 1981). Plasma processes have been shown to be effective at binding heparin to the inner surface of catheters, thus minimizing blood clotting (Williams *et al* 1986, O'Connell *et al* 1981). Immunoassay tubes, usually molded from polypropylene (PP), require that protein or antibody be bound in an active form on the tube surface. The wettability induced by plasma surface treatment may make possible such applications.

In addition to biocompatibility, plastic and polymeric materials used for biomedical applications must be adhesively bonded to themselves or to other materials. Frequently the untreated surface energy of such materials is so low that it results in poor adhesion and bonding. The effects of plasma surface treatment on adhesive bonding of three materials used for catheters and other biomedical applications are listed in table 21.11 (Kaplan and Rose 1991). In this table, the

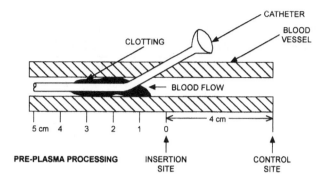

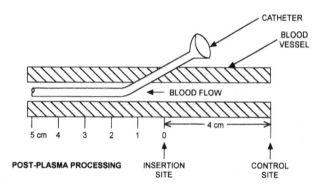

Figure 21.25. Using plasma surface treatment to block the fibrinogen binding at a blood/catheter interface (Kaplan *et al* 1994, O'Connell *et al* 1981).

improvement in adhesive bonding of high-density polyethylene (HDPE), nylon-6, and polypropylene (PP) are listed before and after two kinds of plasma exposure. The improvement in bonding strength of these materials, after exposure to either an intermediate-pressure helium or an oxygen plasma, is a factor of five or ten. The plasma-treated bonds were so strong in many cases that the test sample failed, not the adhesive/polymer interface.

21.7 TREATMENT OF FABRICS AND FIBERS

In the US, each citizen uses about 27 kg of fiber per year for such applications as garments, towels, bedding, carpet, upholstery, and rope. Worldwide, the per capita annual consumption of fiber is about 2 kg, and the annual world consumption of textiles is about 30 million tonnes. Dyeing these textiles, normally an energy-intensive process, requires about 700 000 tonnes of dye per

Table 21.11. Improved bond strength following plasma surface treatment for three polymers used in biomedical applications (Kaplan and Rose 1991).

		Bond strength (MPa)		
Polymer	Treatment	Average	Low	High
HDPE	None	2.2	1.8	2.5
	He plasma	21.6	21.0	22.3
	O_2	17.4	17.0	18.0
Nylon 6	None	5.5	4.4	7.0
	He plasma	19.0	16.5	23.9
	O_2	24.1	21.6	26.1
Polypropylene	None	2.5	—	—
	He plasma	17.9	—	—
	O_2	21.2	—	—

year, and the dyes sometimes contain heavy metals and/or are environmentally toxic.

The US produces approximately 80% of its own textiles. It has been necessary for US textile manufacturers to spend approximately $1.3 billion in the 1980s, and at least $2 billion from 1990 to the year 2000, for environmental controls. The cost of these environmental controls has led to a re-evaluation of traditional, pollution-producing manufacturing processes within the textile industry, and increased interest in cost-effective and less polluting new methods, including various forms of plasma treatment.

21.7.1 Industrial Fabrics and Fibers

Fibers used in fabrics for garments and other domestic purposes can be classified by their origin as either *natural* or *artificial*, as summarized in table 21.12. The *natural fibers* can be subdivided into those of animal origin, which contain protein, and those of plant origin, which contain cellulose. Examples of *protein-based natural fibers* include wool, cashmere, mohair, camel, alpaca, silk, and leather; examples of *cellulose-based natural fibers* include cotton, linen, hemp, jute, and ramie.

Artificial fibers can also be divided into two groups: *regenerated fibers*, which are based on cellulose and manufactured from chemically treated wood pulp; and *synthetic fibers*, which are manufactured from petroleum as a feedstock. Examples of regenerated fibers include rayon, lyocell, acetate, and triacetate. Examples of synthetic fibers include polyester, nylon, olefin, acrylic, modacrylic, spandex, and aramid.

Fabrics, whether made of natural or artificial fibers, can be classified by the

Table 21.12. Classification of fibers used by the textile industry by their origin.

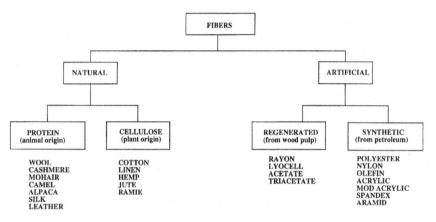

Table 21.13. Classification of fabrics produced by the textile industry.

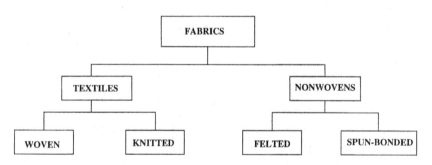

method of assembling the fibers into a fabric, as either *textiles* or *non-wovens*, as shown in table 21.13. *Textiles* can be *woven*, in which two orthogonal layers of parallel fibers are assembled in such a way that the fibers of one set are interlaced alternately above and below the fibers of the second, orthogonal set. Textiles can also be *knitted*, in which a (usually) single fiber is knotted to itself to form a seamless fabric. *Non-wovens* can be *felted*, in which a fabric is made of matted, pressed fibers, usually fur or wool. Non-wovens also can be *spun-bonded*, in which artificial fibers are pressed together in an overlapping, non-parallel array while still molten so that bonding at the fiber contact points produces a cohesive fabric.

21.7.2 Plasma Reactor Systems

In the textile industry, fabric webs are typically 2 m or more in width, and come off looms or spin-bonding equipment at speeds up to 18 m/s. The batch processing of such webs in vacuum systems by intermediate-pressure plasma surface treatment is prohibitively cumbersome and expensive. Direct, continuous feed of a web into and out of a vacuum system is also difficult, although this has been reported by Rakowski (1989) for the intermediate-pressure plasma treatment of wool toe, using the apparatus illustrated in figure 17.3.

More suitable for high-speed in-line industrial applications is the batch processing arrangement illustrated in figure 17.1, and discussed by Griesser (1989), in which the web is passed from a supply to a take-up reel in a vacuum system. Batch processing of fabric webs by intermediate-pressure glow discharge plasma also can be accomplished with DC or RF planar or co-planar magnetron plasmas of the types illustrated in figures 17.24 or 17.25 respectively, or in electrodeless microwave plasmas of the kind described by Neusch and Kieser (1984), and illustrated in figures 17.26 and 17.27. The intermediate-pressure plasma surface treatment of fabric webs has seen only limited industrial use because of the required batch processing and/or expensive vacuum systems.

More recently, fabrics have been exposed to atmospheric plasmas under conditions that do not require vacuum systems or batch processing, and for which the web can be treated continuously as part of an in-line manufacturing process (Roth *et al* 1995a, b). The atmospheric plasmas available for the surface treatment of webs include corona discharges, filamentary or dielectric barrier discharges (DBDs) generated between parallel plates, and OAUGDPs. These atmospheric plasma sources have been discussed in chapter 15.

Atmospheric corona discharges have been shown to be useful for embedding charges in non-woven fabrics intended as particulate filters by Tsai *et al* (1995), and they can have a temporary effect on such surface characteristics as wettability and printability. Filamentary and DBDs offer higher average power densities and active species concentrations in the exposure volume than corona discharges, but are non-uniform, and are capable of puncturing or damaging a continuous film or web under treatment. The OAUGDP offers plasma and active species uniformity, with high average concentrations of the latter.

Two modalities for the plasma treatment of webs can be distinguished. One modality is *surface treatment*, in which wettability or printability only of the exposed surface is desired, for example to print a pattern on a fabric. A second modality is *bulk treatment*, in which the surface of all fibers comprising a fabric is to be affected, as would be the case in improving the dyeability, cohesion, or strength of a fabric. Surface treatment can be done by passive exposure to a plasma; bulk treatment may require that the active species be forced through a web by suction or blowing.

21.7.3 Artificial Fabric Webs

Some objectives of plasma treating the *surface* of artificial fabric webs are listed next in approximate decreasing order of commercial interest. These objectives include improving wettability and wickability of fabrics; improving printability (as in tee-shirt patterns); increasing adhesive strength (as in composite or matrix materials); and increasing superficial charge (as in dust filters). Related objectives for plasma-treating the *bulk* of artificial fabric webs are, again in approximate decreasing order of commercial interest, increasing wettability, increasing wickability, improving dyeability, and imbedding electrical charge.

Non-woven polymer fabric webs were surface and bulk treated by Tsai *et al* (1997) in the OAUGDP reactor illustrated in figure 17.11. This reactor operated with air and other gases, and has the capability of passively exposing only the surface of the fabric or exposing the bulk of the fabric by blowing active species through its thickness.

During exploratory investigations with this reactor, at least eight factors were identified that affect the wettability of non-woven polymeric fabrics undergoing plasma treatment at atmospheric pressure. These factors include the rms voltage applied to the RF electrodes; the RF frequency; the gap distance between the electrodes; the gas flow pattern over the fabric surface; the type of working gas; the temperature of the working gas; the humidity of the working gas; the type of fabric; and the treatment time. Three of these factors are plasma reactor related; four are working gas related; and two are fabric or fabric-processing related.

It has been found by Tsai *et al* (1994) that oxygen-containing gases were the most effective of those investigated. It was also found that the attachment of oxygen or oxygen-containing molecular fragments to surface hydrocarbon molecules appears to be required for long-term or permanent wettability. In later work by Roth *et al* it was found that exposures of tens of seconds may be required to reach a plateau of wettability. This behavior was observed also in the passive cleaning of metallic surfaces, discussed previously in section 21.2.

One of the most significant issues in the plasma surface treatment of fabrics (as well as the surfaces of other materials) is *aging* or the durability of effect. For nearly all applications, permanent wettability is most desirable, but is often elusive. Unless special measures are used (such as using oxygen-containing gases), most plasma-treated surfaces exhibit a wettability that decreases with time after exposure.

Data from Tsai *et al* (1994) on the surface energy of a 71 g/m^2 meltblown polypropylene (PP) non-woven fabric, as a function of time after exposure are shown in figure 21.26. A more wettable surface is in the direction of increasing surface energy; an untreated surface is unwettable, and has a surface energy below 40 dynes/cm (40×10^{-3} N/m). The operating conditions that produced these data are given in table 21.14. Most conditions of exposure show a monotone decrease in wettability over a period of a few days after exposure.

The degree of wettability shown on figure 21.26 may be sufficient for many

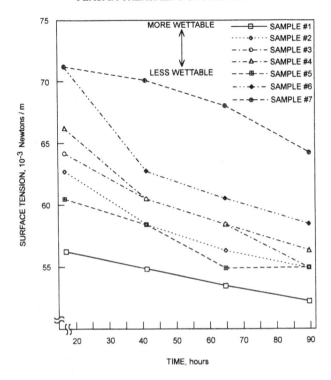

Figure 21.26. Decay of the surface energy with time for a 71 g/m² meltblown polypropylene web exposed to the OAUGDP. The seven samples shown were exposed under conditions shown in table 21.14. Untreated web material has a critical surface tension below 40×10^{-3} N/m.

industrial applications, such as printing, dyeing, or adhesive bonding to the surface, where these operations are performed in-line within a few seconds or minutes after plasma exposure. For consumer items such as clothing, diapers, and other hygiene products, durable wettability and wickability with a shelf life of at least six months is required. On table 21.15 are shown some results of treating 71 g/m² non-woven PP webs with CO_2, or $CO_2 + O_2$, by passive exposure to an OAUGDP (Tsai *et al* 1994). In many cases, the wettability (critical surface tension) stabilized at values well above the untreated values of ≈ 40 dynes/cm, or 40×10^{-3} N/m. The surfaces of these stabilized samples remained wettable and re-wettable after more than one year.

Similar results were obtained by Neusch and Kieser (1984) for the time dependence of adhesion of Scotch® tape to the plasma-treated surface of polyethylene and polyamide. These authors used a variant of the Scotch® tape test in which the 180° peeling time of a fixed length of tape under a fixed load is used as a measure of adhesion. In figure 21.27(*a*) is shown the peeling time

Table 21.14. Operating conditions of the UTK MOD-III OAUGDP reactor for the treatment of 71 g/m^2 meltblown webs (Tsai *et al* 1994).

| | | | | Critical surface tension (10^{-3} N/m[a]) | | | |
| | Treatment time (min:s) | Working gas | Gas temp. (°C) | 3-7-94 17:30 | 3-8-94 17:30 | 3-9-94 17:30 | 3-10-94 19:00 |
Sample							
1	3:00	CO_2, H_2^b	21.7	56.3	54.9	53.5	52.2
2	1:00	CO_2	36.1	62.8	58.4	56.3	54.9
3	2:00	CO_2	36.7	64.2	60.5	58.4	54.9
4	3:00	CO_2	35	66.2	60.5	58.4	56.3
5	1:00	CO_2	20.6	60.5	58.4	54.9	54.9
6	2:00	CO_2	20	71.2	62.8	60.5	58.4
7	3:00	CO_2	21.1	71.2	70.1	68.0	64.2

[a] Fabric was treated at 00:15 am on 3-7-1994 at 1 atm pressure, applied voltage = 8.0 kV$_{rms}$, frequency = 2 kHz, gap between electrodes = 4.3 mm, pyrex thickness = 3.2 mm, and E_{rms} = 16 kV/cm.
[b] Volume ratio CO_2:H_2 = 93:7.

as measured by this test for polyamide, as a function of aging time, after low-pressure treatment in a 6 Pa (45 mTorr) $CHClF_2$ microwave plasma. This shows a very slow monotonic decrease in surface adhesion, probably asymptotic to the untreated peeling time of 5 s.

In figure 21.27(*b*) are further data, also from Neusch and Kieser (1984), on peeling time as a function of aging time for polyethylene treated in a microwave plasma. The working gases of this plasma consisted of a mixture of 4 Pa (30 mTorr) of $CHClF_2$, and 2.6 Pa (20 mTorr) of O_2. The adhesion of the plasma-treated surface remained stable for more than a day, then dropped off monotonically toward its untreated peeling time of one second. In both sets of data in figure 21.27, the exposure time to the plasma of the web tested, t_t, was only 4 s.

The mechanisms responsible for durable wettability observed in the polypropylene (PP) fabrics exposed to an OAUGDP reported in table 21.15, in CO_2, or a mixture of CO_2 and O_2, were investigated by applying selected surface diagnostic techniques. Figure 21.28 shows scanning electron microscope (SEM) images of the polypropylene fiber surfaces. Figure 21.28(*a*) is an untreated PP fiber, and figure 21.28(*b*) is a treated PP fiber, showing a roughened, etched surface on the individual fibers after plasma exposure. Figure 21.29 shows atomic force microscopy (AFM) profile scans of the surface of a fiber before and after plasma treatment. These AFM scans also indicate a greatly roughened post-treatment state, consistent with the SEM photomicrographs on figure 21.28.

Probably most significant to the issue of permanent wettability are the ESCA

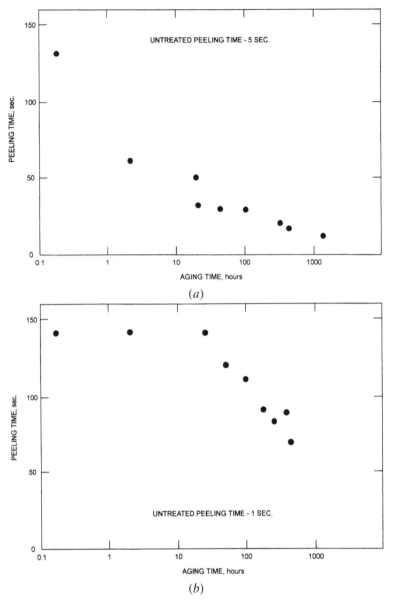

Figure 21.27. An adhesion test devised by Neusch and Kieser (1984) plotting the peeling time of Scotch[®] tape under a fixed load as a function of time after plasma exposure (*a*). The peeling time of untreated polymeric material was 5 s; the force applied to the Scotch[®] tape was 3.6 N, the treatment time in the plasma was 4 s, the power level was 2 kW, frequency of the microwave power 2.45 GHz, and the plasma was a 6 Pa $CHClF_2$ plasma. (*b*) The peeling time versus aging time for polyethylene exposed under the same conditions as (*a*), except that the plasma consisted of 4 Pa of $CHClF_2$ plus 2.6 Pa of O_2.

Table 21.15. Operating conditions of the UTK MOD-III OAUGDP reactor in carbon dioxide gas for treatment of 71 g/m^2 meltblown polypropylene webs (Tsai *et al* 1994)[a].

Surface	Treatment time (min:s)	Voltage (kV_{rms})	Frequency (kHz)	Working gas	Gas temp. (°C)	Critical surface tension $(10^{-3}$ N/m) 3-8-94 17:30	3-9-94 17:30	3-10-94 19:00
1	1:00	9.32	1.2	CO_2, O_2^b	21.1	70.1	60.5	58.4
2	2:00	9.4	1.2	CO_2, O_2^b	21.1	71.2	66.2	64.2
3	2:00	9.4	1.2	CO_2, O_2^c	21.1	71.2	71.2	68.0
4	3:00	9.4	1.2	CO_2, O_2^c	22.2	71.2	71.2	68.0
5	4:00	9.4	1.2	CO_2, O_2^c	23.9	71.2	71.2	68.0
6	1:00	9.6	1.2	CO_2, O_2^c	54.4	71.2	60.5	58.4
7	2:00	9.4	1.2	CO_2, O_2^c	54.4	71.2	60.5	58.4
8	3:00	9.35	1.2	CO_2, O_2^c	55	71.2	60.5	58.4
9	1:00	9.52	1.2	CO_2	53.9	58.4	56.3	54.9
10	2:00	9.52	1.2	CO_2	53.3	58.4	58.4	54.9
11	3:00	9.52	1.2	CO_2	54.4	58.4	58.4	56.3
12	4:00	8.92	1.5	CO_2	48.9	60.5	58.4	58.4
13	1:00	8.82	1.5	CO_2	20.6	58.4	56.3	54.9
14	4:00	8.8	1.5	CO_2	20.6	60.5	58.4	58.4
15	3:00	8.8	1.5	CO_2	21.1	60.5	58.4	56.3
16	2:00	8.92	1.5	CO_2	20	58.4	56.3	56.3

[a] Fabric was treated at 3:15 am on 3-8-94 at 1 atm pressure: gap between electrodes = 2.7 mm, Pyrex thickness = 3.2 mm.
[b] Volume ratio $CO_2:O_2$ = 85:15.
[c] Volume ratio $CO_2:O_2$ = 88:11.4.

scans of the polypropylene web shown in figure 21.30. Untreated PP, which normally contains no oxygen, shows only a contaminant trace at the characteristic oxygen value of ~530 eV. The PP web treated by an OAUGDP shows a very significant bound oxygen peak. Such a peak also appears in the ESCA scans of other durably wettable, OAUGDP-treated polymeric materials.

21.7.4 Natural Fabric Webs

Plasma surface treatment can have significant effects on natural fabric webs, of both animal (protein) and plant (cellulose) origin. To date, most of these studies

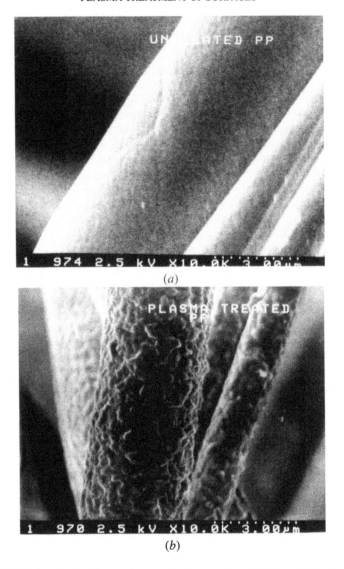

Figure 21.28. A scanning electron image of polypropylene (PP) fibers (*a*) untreated and (*b*) fibers exposed for several minutes to an OAUGD CO_2 plasma (Tsai *et al* 1997).

involved exposure to intermediate-pressure glow discharge or atmospheric corona plasmas. Early work in this area is discussed by Pavlath (1974). Plasma surface treatment of natural materials has thus far seen few commercial applications, but this technique is attracting growing interest as the potential of the atmospheric corona, dielectric barrier, and uniform glow discharges are exploited.

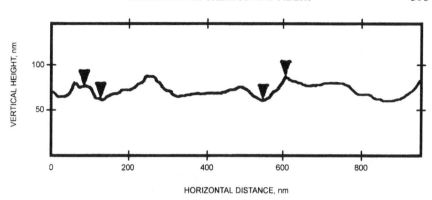

MARKER POSITIONS	1st SET	2nd SET
HORIZONTAL DISTANCE [nm]	42.86	57.15
VERTICAL DISTANCE [nm]	14.04	26.14

(a)

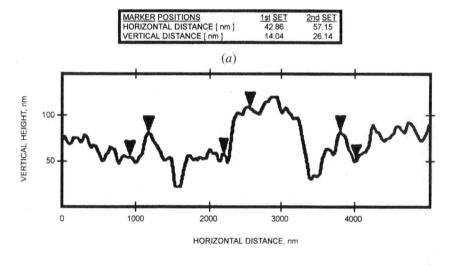

MARKER POSITIONS	1st SET	2nd SET	3rd SET
HORIZONTAL DISTANCE [nm]	250.00	350.00	225.00
VERTICAL DISTANCE [nm]	27.57	50.90	31.40

(b)

Figure 21.29. Atomic force microscope (AFM) surface roughness measurement of polypropylene fibers: (a) an untreated polypropylene fiber; (b) a polypropylene fiber treated with a carbon dioxide OAUGDP. Vertical units on the profiles are in nanometers (Tsai *et al* 1997).

In general, artificial fabrics possess greater stability, wear resistance, recyclability, and strength than natural fabrics, while the latter, as a class, offer

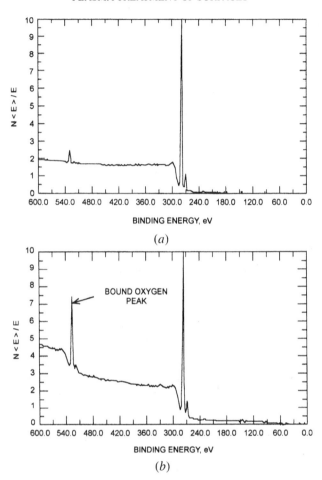

Figure 21.30. ESCA scans of meltblown 450 MFR polypropylene (PP) web: (*a*) untreated fabric, with very small oxygen peak due to contaminants; (*b*) an OAUGDP-treated PP fabric, showing a significant bound oxygen peak (Tsai *et al* 1997).

better 'hand' and greater wettability, wickability, dyeability, printability, and wearing comfort than artificial fabrics. The objective of plasma surface or bulk treatment of natural fabrics is to enhance the latter characteristics of natural fabrics, while, if possible, making them more competitive with the stability, wear resistance, and strength of artificial fabrics.

Natural wool and furs are coated with a layer of animal oil that reduces their surface energy, is hydrophobic, and compromises their wickability, wettability, printability, and dyeability. In addition, wool consists of fibers (hairs) that are coated with overlapping scales, which affect the cohesion and strength of fabrics made from them.

Plasma treatment normally does not affect the bulk of the individual fibers; only the outermost few tens of monolayers (≈ 10 nm) should be affected. Surface treatment of a natural fabric web might improve the printability of the material, as in the work of Rakowski with wool discussed in section 14.1.3. Surface treatment can also improve the adhesive characteristics of the surface to itself, and to bonding agents. Plasma treatment of the surface of all the fibers in a natural fabric (bulk treatment) can be expected to improve its dyeability, wickability, cohesion, and strength.

The survey article by Pavlath (1974) documented exposure of natural fabrics to an atmospheric corona that resulted in large increases (factors of 7 to 25) in the wet-bonding strength of cellulose strips treated by a corona discharge. In addition, the bond strength of cellulose to polymer sheets increased by up to 15 times after atmospheric corona exposure. Pavlath also reported that exposure to an intermediate-pressure glow discharge increased the breaking strength of cotton yarn up to a factor of 1.7. Intermediate-pressure plasma treatment of wool resulted in a 20% increase in fiber tensile strength and improved its wettability, dyeability, abrasion resistance, and resulted in shrink-proofing as well. As was true of polymeric solids, thin films, and artificial fabrics, some of these plasma-induced effects display an aging effect, and fade out after a few tens of minutes or hours. The cohesive force of cotton or wool fabric behaves in this way, for example.

The mechanisms responsible for the effects of plasma treatment on natural fabrics appear to be generally similar to those documented for artificial fabrics in section 21.7.3. Among natural fabrics, permanent effects appear to be those that result from reacting oxygen atoms or other oxygen-containing molecular fragments with the proteins or cellulose molecules of the fibers. Permanent effects also result if the surface of the fiber is permanently roughened or pitted, or if a film of animal oils is removed from the fibers. Temporary effects, which last from days to weeks, appear to be associated with the removal of surface monolayers of hydrophobic contaminants, which re-establish themselves after plasma treatment.

The practical advantages of plasma exposure of natural textiles have been documented by Rakowski (1989), who compared two processes used to achieve the printability of wool toe (partially processed woolen cloth). The economic and environmental implications of these processes were discussed above in section 14.1.3. Rakowski's results also illustrate the potential advantages of conducting plasma surface treatment at 1 atm, as opposed to intermediate pressures that require a vacuum system. If no vacuum pumping were required, wool toe could be treated at an energy cost limited to the power required to maintain the plasma, about 5–15% of the vacuum pumping power.

Atmospheric DBDs, also known as filamentary discharges, have been used to increase the strength and improve other properties of wool for commercial production. The successful treatment process involves control of the temperature of the wool, and exposure to an atmospheric filamentary discharge as the wool executes a serpentine path between large parallel, insulated electrode plates.

REFERENCES

Anonymous 1986 Board-cleaning technique using hollow cathode plasma discharge *IBM Tech. Disclosure Bull.* **29** 1848–50

Ben Gadri R, Roth J R, Montie T C, Kelly-Wintenberg K, Tsai P P-Y, Helfritch D J, Feldman P, Sherman D M, Karakaya F and Chen Z 2000 Sterilization and plasma processing of room temperature surfaces with a one atmosphere uniform glow discharge plasma (OAUGDP) *Surf. Coatings Technol.* **131** 528–42

Brickman C, Yu Y, Roth J R, Tsai P P-Y, Wadsworth L C, Montie T C and Kelly-Wintenberg K 1996 Room temperature sterilization of materials with a one atmosphere uniform glow discharge plasma *Proc. 1996 Ann. Meeting* (Research Triangle Park, NC: Society for Industrial Microbiology)

Carr A K 1997 Increase in the surface energy of metal and polymeric surfaces using the one atmosphere uniform glow discharge plasma *MS in EE Thesis* University of Tennessee

d'Agostino R (ed) 1990 *Plasma Deposition, Treatment, and Etching of Polymers* (New York: Academic) ISBN 0-12-200430-2

Egitto F D and Matienzo L J 1994 Plasma modification of polymer surfaces for adhesion improvement *IBM J. Res. Dev.* **38** 423–39

Frazier D A, Kolluri O S, Wagner G and Fahner A 1991 Successful TPO painting—cold gas plasma advances painting application *Himont/Plasma Science Technical Note* 7/91

Griesser H J 1989 Small scale reactor for plasma processing of moving substrate web *Vacuum* **39** 485–8

Hansen G P, Rushing R A, Warren R W, Kaplan S L and Kolluri O S 1989 Achieving optimum bond strength with plasma treatment *Proc. SME Adhesives '89 (Atlanta, GA)*

Hermann H W, Henins I, Park J and Selwyn G S 1999 Decontamination of chemical and biological warfare (CBW) agents using an atmospheric pressure plasma jet (APPJ) *Phys. Plasmas* **6** 2284–9

Hseuh H C, Chou T S and Christianson C A 1985 Glow discharge cleaning of stainless steel accelerator beam tubes *J. Vac. Sci. Technol.* A **3** 518–22

Hudis M 1974 Plasma treatment of solid materials *Techniques and Applications of Plasma Chemistry* ed J R Hollahan and A T Bell (New York: Wiley) pp 113–47 ISBN 0-471-40628-7

Iwata H and Asakawa K 1991 *Direct Observation of GaAs Surface Cleaning Process Under Hydrogen Radical Beam Irradiation INSPEC No B91076819 (AIP Conf. Proc. 227)* (Clearwater, FL: Advanced Processing and Characterization Technologies) pp 122–5

Kaplan S L 1994 Private communication

Kaplan S L and Hansen W P 1991 Plasma—the environmentally safe treatment method to prepare plastics and composites for adhesive bonding and painting *Proc. 1st Int. SAMPE Environmental Symp. (San Diego, CA)*

Kaplan S L and Rose P W 1991 Plasma surface treatment of plastics to enhance adhesion *Int. J. Adhesion Adhesives* **11** 109–13

Kelly-Wintenberg K, Hodge A, Montie T C, Deleanu L, Sherman D M, Roth J R, Tsai P P-Y and Wadsworth L C 1999 Use of a one atmosphere uniform glow discharge plasma (OAUGDP) to kill a broad spectrum of microorganisms *J. Vac. Sci. Technol.* A **17** 1539–44

Kelly-Wintenberg K, Montie T C, Brickman C, Roth J R, Carr A K, Sorge K, Wadsworth L C and Tsai P P-Y 1998 Room temperature sterilization of surfaces and fabrics with a one atmosphere uniform glow discharge plasma *J. Indust. Microbiol. Biotechnol.* **20** 69–74

Kelly-Wintenberg K, Sherman D M, Tsai P P-Y, Ben Gadri R, Karakaya F, Chen Z, Roth J R and Montie T C 2000 Air filter sterilization using a one atmosphere uniform glow discharge plasma (the Volfilter) *IEEE Trans. Plasma Sci.* **28** 64–71

Kolluri O S, Kaplan S L and Rose P W 1988 Gas plasma and the treatment of advanced fibers *Proc. SPE Advanced Polymer Composites '88 Tech. Conf. (Los Angeles, CA)*

Korzec D, Rapp J, Theirich D and Engemann J 1994 Cleaning of metal parts in oxygen radio frequency plasma: process study *J. Vac. Sci. Technol.* A **12** 369–78

Mlynko W E, Cain S R, Egitto F D and Emmi F 1988 Plasma processing *Principles of Electronic Packaging* ed D P Seraphim, R Lasky and C-Y Li (New York: McGraw-Hill) ch 14

Montie T C, Kelly-Wintenberg K and Roth J R 2000 An overview of research using a one atmosphere uniform glow discharge plasma (OAUGDP) for sterilization of surfaces and materials *IEEE Trans. Plasma Sci.* **28** 41–50

Neusch M and Kieser J 1984 Surface activation of polymers in a microwave plasma *Vacuum* **34** 959–61

O'Connell J P, Dunn T S, Rumaks A and Williams J L 1981 Detection of thrombus formation on intravascular catheters using 125I-fibrogen *Thrombosis Res.* **21** 111–20

Ouellette R P, Barbier M M and Cheremisinoff P N (ed) 1980 *Electrotechnology Volume 5 Low-Temperature Plasma Technology Applications* (Ann Arbor, MI: Ann Arbor Science) ISBN 0-250-40375-7

Pavlath A E 1974 Plasma treatment of natural materials *Techniques and Applications of Plasma Chemistry* ed J R Hollahan and A T Bell (New York: Wiley) pp 149–75 ISBN 0-471-40628-7

Rakowski W 1989 Plasma modification of wool under industrial conditions *Melliand Textilberichte* **70** 780–5

Ratner B D, Chilkoti A and Lopez G P 1990 Plasma deposition and treatment for biomaterial applications *Plasma Deposition, Treatment, and Etching of Polymers* ed R d'Agostino (Boston, MA: Academic) pp 463–516 ISBN 0-12-200430-2

Richley E A and Cybulski R J 1962 Experimental effects from cesium and mercury ion beams *National Aeronautics and Space Administration Report* NASA TND-1217

Roth J R 1995 *Industrial Plasma Engineering: Vol I—Principles* (Bristol: Institute of Physics Publishing) ISBN 0-7503-0318-2

——1999 Method and apparatus for cleaning surfaces with a glow discharge plasma at one atmosphere of pressure *US Patent* 5,938,854

Roth J R, Chen Z, Sherman D M, Karakaya F, Tsai P P-Y, Kelly-Wintenberg K and Montie T C 2000a Increasing the surface energy and sterilization of non-woven fabrics by exposure to a one atmosphere uniform glow discharge plasma (OAUGDP) *Proc. Int. Nonwovens Technical Conf. INTC-2000 (Dallas, TX)*

Roth J R, Ku Y, Tsai P P-Y, Wadsworth L C, Sun Q, Montie T C and Kelly-Wintenberg K 1996 A study of the sterilization of nonwoven webs using one atmosphere glow discharge plasma *Proc. 1996 TAPPI Conf. (NC)* pp 225–30 ISBN 0-89852-658-2

Roth J R, Sherman D M, Ben Gadri R, Karakaya F, Chen Z, Montie T C, Kelly-Wintenberg K and Tsai P P-Y 2000b A remote exposure reactor (RER) for plasma processing and sterilization by plasma active species at one atmosphere *IEEE Trans. Plasma Sci.* **28** 56–63

Roth J R, Tsai P P-Y, Liu C and Wadsworth L C 1995a Method and apparatus for glow discharge plasma treatment of polymer materials at atmospheric pressure *US Patent* 5,456,972

Roth J R, Tsai P P-Y, Wadsworth L C, Liu C and Spence P D 1995b Method and apparatus for glow discharge plasma treatment of polymer materials at atmospheric pressure *US Patent* 5,403,453

Rutala W A, Gergen M F and Weber D J 1998 Comparative evaluation of the sporicidal activity of new low-temperature sterilization technologies: ethylene oxide, 2 plasma sterilization systems, and liquid peracetic acid *Am. J. Infect. Control* **26** 393–8

Strobel M, Lyons C S and Mittal K L (ed) 1994 *Plasma Surface Modification of Polymers: Relevance to Adhesion* (Zeist: VSP) ISBN 90-6764-164-2

Struckman C K and Kusse B R 1993 High-purity intense lithium-ion-beam sources using glow-discharge cleaning techniques *J. Appl. Phys.* **74** 3658–68

Tira J S 1987 Adhesive and surface preparation evaluation for stainless steel used in electrical assemblies *SAMPE J.* **18** 18–22

Tsai P P-Y and Wadsworth L C 1995 Electro-static charging of meltblown webs for high-efficiency air filters *Adv. Filtration Separation Technol.* **9** 473–91

Tsai P P-Y, Wadsworth L C and Roth J R 1997 Surface modification of fabrics using a one-atmosphere glow discharge plasma to improve fabric wettability *Textile Res. J.* **67** 359–69

Tsai P P-Y, Wadsworth L C, Spence P D and Roth J R 1994 Surface modifications of nonwoven webs using one atmosphere glow discharge plasma to improve web wettability and other textile properties *Proc. 4th Ann. TANDEC Conf. (Knoxville, TN)*

Vassal S, Favennec L, Ballet J-J and Brasseur P 1998 Hydrogen peroxide gas plasma sterilization is effective against *Cryptosporidium parvum* oocysts *Am. J. Infect. Control* **26** 136–8

Wadsworth L C, Roth J R, Tsai P P-Y and Spence P D 1994 Fiber surface modifications using one atmosphere glow discharge plasma *Proc. 20th ACS National Meeting (San Diego, CA)*

Williams J L, Dunn T S, O'Connell J P and Montgomery D 1986 Heparinization of plasma treated surfaces *US Patent* 4,613,517

Wilson R G and Brewer G R 1973 *Ion Beams with Applications to Ion Implantation* (New York: Wiley) ISBN 0-471-95000-9

Yasuda H K (ed) 1984 *Plasma Polymerization and Plasma Treatment (J. Appl. Polymer Sci.: Appl. Polymer Symp. 38)* (New York: Wiley)

——1988 *Plasma Polymerization and Plasma Treatment of Polymers (J. Appl. Polymer Sci.: Appl. Polymer Symp. 42)* (New York: Wiley)

——1990 *Plasma Polymerization and Plasma Interactions with Polymeric Materials (J. Appl. Polymer Sci.: Appl. Polymer Symp. 46)* (New York: Wiley)

Young J H 1997 New sterilization technologies *Sterilization Technology for the Health Care Facility* ed M Reichert and J H Young (Gaithersburg, MD: Aspen) ch 26, pp 228–35 ISBN 0-8342-0838-5

22

Surface Modification by Implantation and Diffusion

Energetic ion interactions with surfaces can be categorized into four types:

(1) ion interactions with a surface layer that produce *sputtering* (considered in chapters 14 and 23);

(2) ion-induced *chemical reactions* between the solid surface and a background gas (*etching*, considered in chapter 25);

(3) *ion implantation* below the surface to produce a thin layer of alloyed or 'doped' material; and

(4) thermal diffusion into a material of plasma ions that reach its surface.

In this chapter, we will consider the last two of these interactions, and examine the technology and results of industrial applications of ion implantation and thermal diffusion.

22.1 ION IMPLANTATION TECHNOLOGY

Ion implantation consists of directing ions onto surfaces with enough energy that they penetrate the atomic structure of the material and come to rest many atomic layers below the surface. The ion energies used for implantation are in the range 10–300 keV and result in penetration depths of less than a micron. Perhaps surprisingly, such thin layers can have significant effects of commercial importance on the corrosion and wear resistance, and on the electrical, optical, and tribological properties of surfaces.

22.1.1 Historical Development

Ion-beam implantation technology was developed in parallel with the ion sources used for physics research and space propulsion in the 1960s. During this period, ion-beam implantation became widely used for '*doping*' semiconductors (the mixing or alloying of a thin surface layer), and many additional applications became apparent after it was developed for this purpose. These applications included the treatment of high-value items such as aerospace bearings and medical implants to improve their hardness and wear characteristics (Buchanan *et al* 1988). In the late 1980s, Conrad (1988) introduced *plasma ion implantation*, a less expensive and complex method of implanting ions into surfaces that is receiving growing industrial acceptance.

22.1.2 Areas of Application

The implantation of ions into a thin surface layer has been found to have important effects on the surface-related properties of materials. These effects have made possible several areas of application.

(1) Ion implantation of suitable ionic species can improve the hardness of metals, greatly increasing the life of impact tools and tool bits.
(2) It can improve the hardness of ceramics and their resistance to wear and abrasion.
(3) Ion implantation can improve the tribological (wear) properties of metals. This has been particularly important in aerospace applications to bearings, and medical prostheses such as replacement hip and knee joints.
(4) Ion implantation can increase the corrosion resistance and pitting potential of metals (Williams *et al* 1991).
(5) Ion-beam implantation can be used to dope semiconductors to produce particular gates and transistors needed in microelectronic circuits.
(6) Ion implantation can alloy surface layers of metals or insulators to change their electrical and optical characteristics.
(7) Ion implantation can be used for controlled defect production in crystalline materials, and controlled radiation damage.

22.1.3 Mechanisms of Action

Much needs to be learned about the mechanisms by which ion implantation improves such characteristics as the corrosion resistance or the hardness and wear characteristics of metals, but some broad understanding of these phenomena is already available. Carbon and nitrogen, when implanted, induce hardening and improved wear resistance in many common engineering metals. Implantation many microns below the surface would normally be required to ensure a sufficient depth to accommodate the effects of wear, but such depths would require ions with an energy in excess of 1 MeV. Ions of 40 or 50 keV, a more practicable energy, will implant to a depth of only a few tenths of a micron.

Surprisingly, such shallow implantation depths have been found sufficient to improve the wear characteristics of metals over long periods, because ions appear to migrate into the metal ahead of the wear surface, as illustrated in figure 22.1.

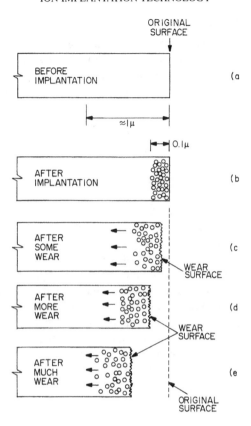

Figure 22.1. A sample material (*a*) before ion implantation, (*b*) after ion implantation, (*c*) after some wear, (*d*) after more wear, and (*e*) after much wear. Note the manner in which the implanted ions migrate ahead of the wear surface.

In this model, the original workpiece (*a*) is implanted to a depth of approximately 0.1 μm (*b*). After some wear (*c*), the implanted ions are found to have moved into the bulk of the metal ahead of the wear surface. This continues (*d*), and finally, after much wear (*e*), the material that occupied the original implantation depth is worn away. However, sufficient implanted ions have migrated ahead of the wear surface and are present below the newly exposed surface to maintain the improved hardness and wear resistance of the metal. The mechanism by which implanted ions move ahead of the wear surface is not well understood. Possible mechanisms include local heating of the surface during the wear process, which allows ions to thermally diffuse into the metal ahead of the wear surface; or, alternatively, migration of the ions ahead of the wear surface along defects or crystallographic boundaries.

The mechanism by which implanted nitrogen ions increase the pitting potential and corrosion resistance of metals is also poorly understood. Smith (1995) has found that the plasma ion implantation of 10–20 keV nitrogen ions into 304L stainless steel at doses ranging from 10^{14}–10^{15} ions/cm^2 increases its salt-water *pitting potential* by a factor of two or three, a significant improvement in its corrosion resistance. Higher energies and higher doses do not work as well; presumably the implanted layer must lie on or very near the surface of the metal, where the pitting corrosion process takes place.

22.1.4 Doping of Semiconductors

The most widespread application of ion implantation is the *doping* of semiconductors in the microelectronic industry. *Ion-beam implantation* is presently the method of choice for this purpose. This technique has been developed over a period of more than 40 years to the point where a wide choice of dopant ions is available. It also offers the ability to program the depth profile of dopants, thereby modifying and controlling the electrical characteristics of the implanted layers. This profiling produces electronic circuit elements capable of a variety of logical functions in their imbedded circuits. The types of implanted profiles possible by ion implantation include controlled superficial doping, the production of buried layers of dopants, and insulating layers. In addition to microelectronic circuitry, ion-implanted semiconductors also are used as sensors in nuclear radiation detectors.

22.1.5 Other Industrial Applications

Many other industrial applications of ion implantation have been developed, some of very recent origin. Ion implantation can be used to improve the characteristics of *superconducting thin films*. It is possible to chemically modify surface layers to change their adhesive bonding characteristics, wettability, or the electrophoretic properties of polymeric materials. Phosphors can be implanted on the surface of video displays and lighting devices, as can anti-reflective coatings on optical elements. Finally, ion implantation is becoming more widely used to alter the hardness and wear resistance of metals. In recent years some very promising results have been achieved in which ion implanted tools and tool bits have shown improvements in their useful lifetime by factors of two to more than 50. It is also the case that ion implantation is capable of greatly improving the corrosion resistance of metals.

Unfortunately, the ion-beam implantation process is expensive, and is thus far confined to small, high-value items like microelectronic circuits, and to vital applications where corrosion and wear resistance are of the utmost importance, such as medical implants and components for space vehicles.

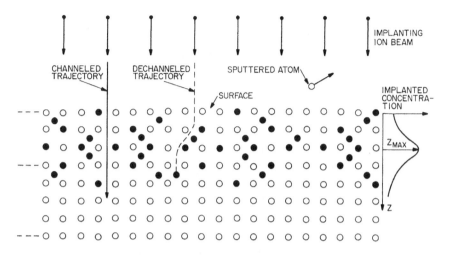

Figure 22.2. Ion implantation of the surface of a crystalline material. Implanted ions are shown as full circles.

22.2 ION IMPLANTATION DOSE AND DEPTH PROFILES

This section covers selected theoretical aspects of ion implantation that are important to industrial applications of this technology. More complete coverage of the theory and practice of ion implantation and related material may be found in monographs listed at the end of the chapter. Dearnaley *et al* (1973), Wilson and Brewer (1973), Carter and Grant (1976), and H F Winters (1980) are excellent resources on this subject.

22.2.1 Ion Implantation Phenomenology

A characteristic implantation process was illustrated in figure 14.21, and is reproduced in figure 22.2. A beam of energetic ions is directed vertically downward on a crystalline lattice. A few ions travel long distances through tubes or channels in the orderly structure of the lattice, in *channeled trajectories*. Most ions scatter immediately and perform a random walk into the interior as *dechanneled trajectories*.

There is some interaction of the energetic ions with adsorbed surface monolayers of the working gas (not shown in figure 22.2). This interaction is an example of *active plasma cleaning*, discussed in section 21.3.2. This process desorbs the outermost monolayers of the adsorbed gas (*outgassing*), and reaches a steady state in a time comparable to the *plasma–wall equilibration time* given by equation (21.8). In addition to becoming implanted, incident ions can sputter the workpiece surface, driving off ε atoms per incident ion. One normally wishes to avoid sputtering more material than is implanted. This can happen for $\varepsilon > 1.0$,

Table 22.1. Atomic and ionic radii for selected species.

Element	Atomic radius ($\text{Å}, \times 10^{-10}$ m)	Ionic radius ($\text{Å}, \times 10^{-10}$ m)
Si	1.20	0.65
N	0.72	0.20
O	0.60	0.16
C	0.80	0.24
Al	1.41	0.72
He	1.30	0.80
Ne	1.63	1.15
Ar	1.88	1.52
Kr	2.00	1.75
Fe, Co, Ni	1.25	0.95
Mo	1.75	—

particularly for oblique incidence of the implanting ion beam.

Channeled trajectories occur along unobstructed directions in a crystalline lattice. The channel radii depend on the ion energy through the elastic scattering cross sections, and are comparable to the atomic radii of the lattice atoms. Some selected atomic and ionic radii are listed on table 22.1. These radii are drawn to scale in figure 22.3, along with a $\langle 111 \rangle$ planar representation of silicon, with its effective *Thomas–Fermi scattering radius*. This classical physics picture indicates why some ions are easier to implant than others. The scale of figure 22.3 is in ångstroms, where $1 \text{ Å} = 10^{-10} \text{ m} = 0.1 \text{ nm}$.

In the ion implantation process, one usually wishes to avoid channeling in order to implant ions a known, controlled distance below the surface. Methods to avoid channeling by the incident ions (*dechanneling*) are illustrated schematically in figure 22.4. These methods include:

(1) dechanneling by aiming the ion beam at a large angle to all lattice planes;
(2) dechanneling by stacking fault or (poly)crystalline boundaries in the material;
(3) dechanneling by interstitial, dopant, or dislocated atoms in the lattice, the result of radiation damage, or impurities; and
(4) dechanneling by an amorphous or surface oxide layer.

As is true of sputtering, energetic ions undergoing implantation can interact with a lattice in two ways: (1) at low energies, the ions interact with the lattice as a collective whole, the *baseball-on-mattress* model; or (2) at higher energies, ions can interact by sequential binary collisions with individual atoms of the lattice, the *billiard ball* model. The second, binary collisional model, is the most productive theoretically, and the most accurate when compared with experiments. As was

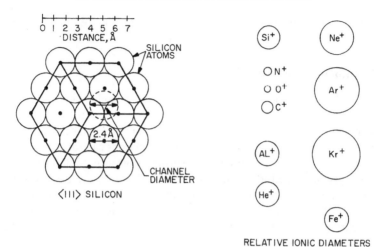

Figure 22.3. The ⟨111⟩ plane of silicon drawn to scale, with the approximate channel diameter indicated by the dotted lines. The relative diameters of several characteristic ions are shown on the right of the figure.

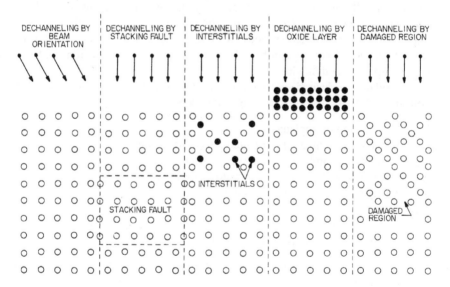

Figure 22.4. Various methods to prevent channeling of an implanting ion beam in a crystalline lattice. The adsorbed surface monolayers are not shown.

discussed in section 14.6.3, interactions of energetic ions with a lattice can be characterized by two physics regimes: (1) interactions described by classical mechanics; and (2) interactions requiring a quantum-mechanical description.

Most industrial applications of ion implantation are in the classical physics regime.

22.2.2 Implantation Dose

The required ion *dose* (or *fluence*), implantation depth, and ion energy depend on the application of the implanted material. Ion doses range from $10^{14}/cm^2$ (for corrosion inhibition applications), to nearly $10^{18}/cm^2$ (for enhanced hardness and wear resistance applications). The required ion energies can range from 10 keV for the thin surface layers required for corrosion resistance, to 300 keV or more for the deeply implanted ions sometimes required to improve the hardness and wear characteristics of tool bits. The depth of implanted ions can be as little as 50 nm (0.05 μm) for corrosion resistance, where implanted species must be near the surface to be effective, to as much as 500 nm (0.5 μm), for the thicker implanted layers required by hardness and wear applications.

Ordinary foundry hardening processes are based on melting an entire charge of metal, and mixing alloying materials or dopants throughout its bulk. This process produces a mole fraction consistent with thermodynamic equilibrium at the melting point, only about 1% for carbon or nitrogen atoms in molten metals. This is normally much too low to induce significant improvements in hardness and wear characteristics of metals. Ion implantation, however, makes possible non-equilibrium mole fractions above 40% for carbide or nitrogen–metal layers. These high mole fractions can be achieved to depths up to a micron, with ion energies up to several hundred keV.

To relate the desired *mole fraction X*, to the required *dose* or *fluence* in atoms/m^2, D, we define the following parameters:

X Mole fraction of implanted species
M_1 Atomic mass number of implanted ions (kg/kg-atom)
M_2 Atomic mass number of metal (kg/kg-atom)
A_0 Avogadro's number (6.02 $\times$ 10^{26} atoms/kg-atom)
ρ Density of implanted material (kg/m^3)
N Number density of implanted ions (ions/m^3)
ϕ_i Ion flux in ions/m^2-s
D Dose or fluence in ions/m^2 (or ions/cm^2).

The number density of implanted ions with a mole fraction X in the implanted metal is given by

$$N = \frac{X A_0 \rho}{X M_i + (1 - X) M_2} \quad \text{ions/m}^3. \tag{22.1}$$

If the thickness of metal implanted is l, the *total dose* or *fluence* is given by

$$D = \phi_i t_0 = Nl = \frac{X A_0 \rho l}{X M_1 + (1 - X) M_2} \quad \text{atoms/m}^2 \tag{22.2}$$

where t_0 is the total exposure time.

The ion *flux*, ϕ_i, can be written in terms of the ion current density J_i, or the total current I falling on an area A,

$$\phi_i = \frac{J_i}{e} = \frac{I}{eA} \qquad \text{ions/m}^2\text{-s.} \qquad (22.3)$$

In the published literature, the *dose* or *fluence* is usually expressed in units of atoms or ions/cm^2.

22.2.3 Dechanneled Ion Implantation

Dechanneled ion implantation results when an energetic ion with an initial energy E_1' (eV) undergoes a random walk through the atoms of a solid to reach an average depth R_p, the *range* of the ion. The stochastic nature of the random walk results in a spread of depths about the range. This spread is designated σ, and is the *standard deviation* or, as it is sometimes referred to in the ion implantation literature, the *straggle*.

Because of the stochastic nature of the collisions as an ion moves into a workpiece, a *Gaussian distribution* is a good approximation to the depth profile of ions implanted with a given energy E_1'. The probability density function, $P(x)$, of finding an ion at the depth x from the surface is given by

$$P(x) = \frac{dN}{dx} = \frac{1}{\sigma\sqrt{2\pi}} \exp\left[-\frac{1}{2}\left(\frac{x - R_p}{\sigma}\right)^2\right] \qquad (22.4)$$

where R_p is the range, σ is the standard deviation or 'straggle', and the distance from the maximum concentration at $x = R_p$ is

$$\Delta x = x - R_p. \qquad (22.5)$$

The normalizing constant, $(\sigma\sqrt{2\pi})^{-1}$ is chosen such that

$$\int_{-\infty}^{\infty} P(x)\,dx = 1.0. \qquad (22.6)$$

The profile can be normalized to the standard deviation,

$$X \equiv \frac{\Delta x}{\sigma} = \frac{x - R_p}{\sigma} \qquad (22.7)$$

in terms of which equation (22.4) may be written as

$$P(X) = \frac{dN}{dX} = \frac{1}{\sqrt{2\pi}} \exp\left[-\frac{X^2}{2}\right]. \qquad (22.8)$$

Equation (22.8) is plotted in figure 22.5 in semi-log and linear representations, with a maximum $P(0) = 1/\sqrt{2\pi} = 0.399$ at $X = 0$.

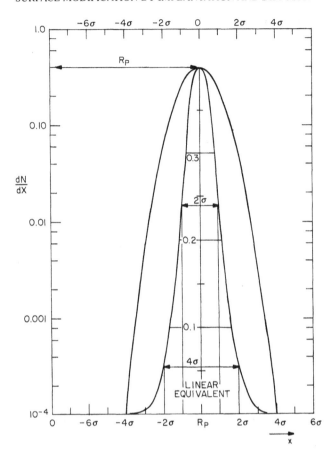

Figure 22.5. The implantation density distribution function of dechanneled ions in a solid material as a function of depth below the surface. The maximum of the implanted density distribution is located at the range R_p. The distribution of ions with distance x below the surface is shown on semi-logarithmic and linear scales. The parameter σ is the standard deviation or 'straggle' of the ion population.

If the workpiece is bombarded for a time t_0 with a flux ϕ_i ions/m²-s, the *fluence* or *dose* is

$$D_i = \phi_i t_0 \qquad \text{ions/m}^2. \tag{22.9}$$

In terms of this dose, the *depth profile* of implanted ions is, using equation (22.4),

$$n_i(x) = \frac{D_i}{\sigma\sqrt{2\pi}} \exp\left[-\frac{1}{2}\left(\frac{x - R_p}{\sigma} \right) \right] \qquad \text{ions/m}^2. \tag{22.10}$$

The relation between the dose and the maximum number density of implanted

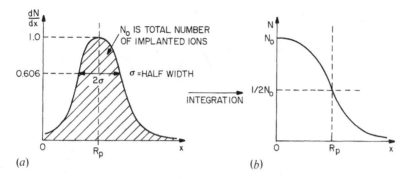

Figure 22.6. A schematic diagram of (*a*) the density distribution function of implanted ions as a function of depth; and (*b*) the integral profile, showing the fraction of ions implanted below the depth *x*.

ions is, for $x = R_p$,

$$n_{i\,max} = \frac{D_i}{\sigma\sqrt{2\pi}} = \frac{0.399\,D_i}{\sigma} \qquad \text{ions/m}^3. \qquad (22.11)$$

Sputtering of the surface material must be accounted for because it may decrease the range R_p, or even remove the implanted material as a competing process.

Equation (22.10) differs fundamentally from the results of *low-energy plasma thermal diffusion treatment*, or *thermal doping*, since the latter is achieved by heating the workpiece and allowing the dopant atoms to thermally diffuse into the metal. This thermal diffusion process always results in a maximum concentration at the surface of the workpiece, with a monotonically decreasing number density with increasing depth.

Data on ion implantation are generally presented in one of two ways, both illustrated in figure 22.6: a *differential profile*, like figure 22.5 and shown in 22.6(*a*); or an *integral profile*, like figure 22.6(*b*), in which the fraction of implanted ions deeper than the distance *x* is plotted as a function of *x*. For a Gaussian profile, this fraction is $\frac{1}{2}$ at the range R_p.

An example of the depth profile of boron ions implanted on silicon for six energies is shown in figure 22.7, due to T E Seidel, taken from Wilson and Brewer (1973). These profiles have been normalized to the maximum density in each case and a Gaussian distribution fitted to the experimental data. The ranges and standard deviations for several ion–workpiece combinations important in microelectronic circuit fabrication are presented in Wilson and Brewer (1973). Figure 22.8, taken from this reference, illustrates how a desired doping profile is achieved by several distinct ion implantation steps, each with a different dose and energy. The accompanying table (the *implantation schedule*) indicates the ion energy, the dose, and the other implantation parameters required to produce the desired depth profile with only four implantation runs.

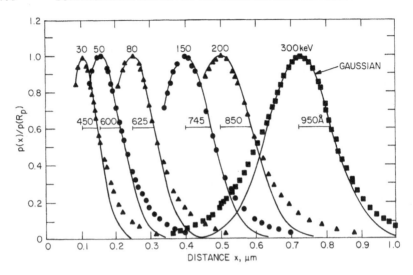

Figure 22.7. The normalized depth profiles of B^+ ions, implanted with various energies into silicon, and fitted with Gaussian curves. The standard deviation, in ångstroms, is indicated on each depth profile (Wilson and Brewer 1973).

22.2.4   Channeled Ion Implantation

The Gaussian implantation profiles resulting from random walk slowing down of ions may not occur in crystalline materials where channeled trajectories are possible. Some orientations of the crystalline lattice may present clear pathways, or *channels*, along which ions travel for long distances without scattering. This situation is illustrated in figure 22.9. The depth profile illustrates a peak near the surface that results from a population of random-walked ions like those in figure 22.7. Further into the workpiece are *dechanneled ions* that were scattered by small angles and came to rest at moderate depths. Finally, there exists a second peak in the concentration that results from well-channeled ions that penetrated deep into the workpiece before coming to rest.

When the ion velocity vector is misaligned by more than a *critical angle* θ_c with respect to a channeling direction, the scattering process will revert to random-walk binary scattering. This critical angle is given by

$$\theta_c \approx \frac{a}{d} \left(\frac{2eZ_1Z_2}{4\pi\varepsilon_0 E'_i d} \right)^{1/4} \tag{22.12}$$

where d is the lattice spacing, in meters, parallel to the beam direction, a is the effective radius of the electron cloud given by equation (14.43), and E'_i the energy in electronvolts of the incident ion.

Figure 22.10, taken from Carter and Grant (1976), shows a characteristic example of the depth profile of 40 keV ^{32}P ions implanted into silicon at three

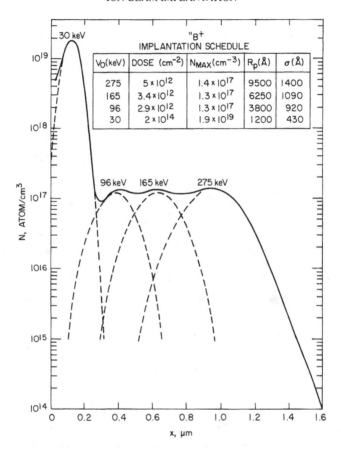

Figure 22.8. A doping profile synthesized by calculation from four separate implantations at the energies and doses indicated on the implantation schedule. The example shown is $^{11}B^+$ ions implanted into silicon (Wilson and Brewer 1973).

beam angles. The effect of channeling is to carry the ^{32}P ions much further into the silicon, as is illustrated by the 'aligned' data in the top curve of figure 22.10. With only a $2°$ misalignment from the channeling direction, the depth profile in the interior of the workpiece decreases significantly. With an $8°$ misalignment, the depth profile is much more shallow, approximately Gaussian in form, and characteristic of dechanneled scattering.

22.3 ION-BEAM IMPLANTATION

Ion-beam implantation is accomplished by accelerating ions in a unidirectional beam to energies that range from approximately 20 keV to as much as 600 keV,

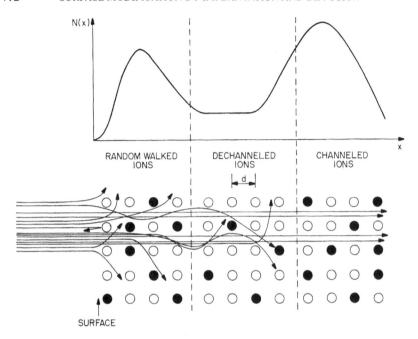

Figure 22.9. A schematic diagram of ion implantation under conditions for which populations of both channeled and dechanneled ions are apparent.

which are then implanted in a workpiece in a high-vacuum system. Some aspects of the physics of the interaction of these ions with the crystalline lattice of the workpiece material have been covered in section 14.6; here we discuss the technology of producing and manipulating a beam of ions for implantation.

22.3.1   Ion-Beam Implantation Systems

Ion-beam implantation is accomplished by the two types of system shown in figure 22.11, depending upon the ion energy required. The first, lower-energy space-charge-limited, ion implantation systems have been discussed in chapter 6 and are the most widely used (see also Brown 1989). The second, high-energy implantation system utilizes cyclotrons, Van De Graaff generators, or other particle accelerators. These sources are less widely used because of their expense, and the less compelling need to implant to the greater depths made possible by these sources.

The space-charge-limited ion sources described in chapter 6 include those used in the microelectronic industry for wafer doping. These sources are not always operated at the full space-charge-limited current density, either because of a desire to prolong the implantation process over a longer duration so it can be more easily controlled, or to limit power deposition and attendant heating

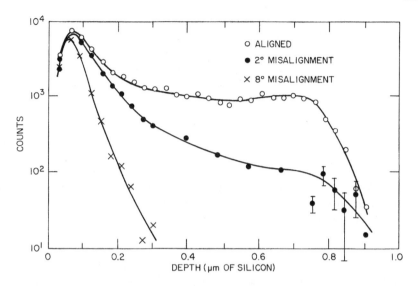

Figure 22.10. An experimental illustration of channeling and dechanneling by deliberate beam misalignment for 40 keV ^{32}P ions implanted into silicon. The curves represent three different alignment angles with respect to the ⟨110⟩ axis of the silicon crystal (Carter and Grant 1976).

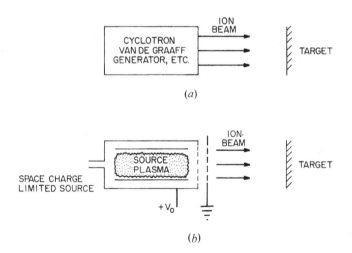

Figure 22.11. Schematic diagrams of ion-beam implantation systems: (a) high-energy ($E' > 300$ kV) implantation system; and (b) intermediate-energy ($10 \leq E' \leq 300$ keV) implantation system.

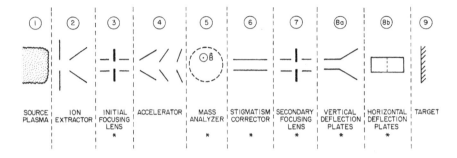

Figure 22.12. A characteristic ion-beam handling system for high-energy ion-beam implantation. The components with the asterisk may be omitted or arranged in a different sequence depending on the application.

of the workpiece surface. Surface heating is normally avoided, because it may cause thermal diffusion of implanted atoms deep into the material, away from the surface layers where implantation is desired.

Figure 22.12 shows the components of a characteristic ion-beam implantation system. Many of these individual components have been discussed in chapter 6, and will only be reviewed here. Proceeding from left to right, these components may include:

(1) The *source plasma* in which the ions required are created from the appropriate neutral gas. The power and mass utilization efficiencies of these sources are usually low compared with a Kaufman source, but source power efficiency is seldom an issue in ion implantation apparatus.

(2) The *ion extractor* is characteristically a Pierce geometry designed to form a unidirectional, paraxial beam with an energy of a few kilovolts.

(3) An *initial focusing lens* is sometimes used to limit beam divergence, and keep the beam to a small diameter as it passes through the remainder of the beam handling system.

(4) A *final accelerating stage* raises the ions to the desired energy, which may range from 20 to 300 keV in microelectronic applications.

(5) A *mass analyzer* is used when the ion source produces more than one species or if isotopic separation of a particular atomic mass is desired.

(6) An *astigmatism corrector* is used to correct for astigmatism in the ion optics of the beam. *Astigmatism* is the tendency of beam ions to come to a focus along two perpendicular lines each at right angles to the direction of the beam, rather than to focus at a single point.

(7) A *secondary focusing lens* is used to maintain the paraxial nature of the beam, control the beam divergence, or focus the ion beam on the workpiece.

(8) *Deflection plates* are used to deflect or raster the beam electrostatically over the surface of the workpiece. By installing two sets of plates at right angles,

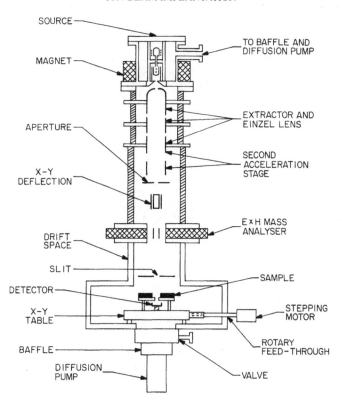

SOURCE ——

MAGNET ——

APERTURE ——

X–Y
DEFLECTION ——

DRIFT
SPACE ——

SLIT ——

DETECTOR ——

X–Y
TABLE ——

BAFFLE ——

DIFFUSION
PUMP ——

TO BAFFLE AND
DIFFUSION PUMP

EXTRACTOR AND
EINZEL LENS

SECOND
ACCELERATION
STAGE

E x H MASS
ANALYSER

SAMPLE

STEPPING
MOTOR

ROTARY
FEED–THROUGH

VALVE

Figure 22.13. A low-energy ion-beam implantation system (Wilson and Brewer 1973).

it is possible to sweep out, in a controlled manner, the entire surface area of
the workpiece, or move the beam from one fixed workpiece to another for
treatment.

(9) *Workpiece.*   The object to be implanted, usually a silicon wafer in
microelectronic applications.

A schematic diagram of a low-energy ion implantation system is shown in
figure 22.13, taken from Wilson and Brewer (1973). This system consists of
an ion source at the top, with an extractor electrode and multiple accelerating
electrodes, a pair of electrostatic deflection plates, an E/B mass analyzer, and
finally a workpiece on a movable platform in the vacuum chamber.

In production runs, it is essential to monitor and control the beam current
and energy in order to obtain the desired depth profile and dose. The ion energy is
controlled by adjusting the total voltage through which the ion is accelerated. The
dose is controlled by adjusting the time of exposure and the beam current density.
It is important to monitor the total current in the ion beam to ensure that the dose
is reproducible and under control.

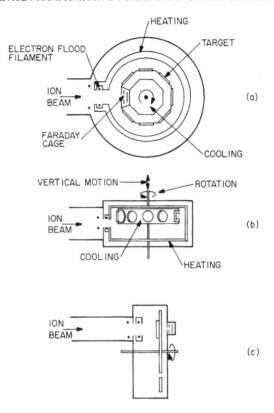

Figure 22.14. Three configurations of chambers for multiple workpieces used in production ion-beam implantation systems.

The beam current may be monitored by including a *retarding potential energy analyzer*, or a *Faraday collector*, among an array of workpieces that are successively moved into the beam by a rotating holder, as shown in figure 22.14 (Wilson and Brewer 1973). This figure also illustrates several methods used to simultaneously or serially expose multiple workpieces in order to reduce the time required to batch process them in a vacuum system.

A schematic diagram of a commercial high-energy ($E_i = 300$ keV) ion implantation system is shown in figure 22.15, taken from Wilson and Brewer (1973). This system has three separate beam lines, each for a different application of the ion beam. This arrangement allows more diverse uses and less down time for the facility as a whole. The unit shown is capable of producing a wide range of ion species over a voltage range 30–300 kV, and at beam currents ranging from 0.01–100 μA. Mass separation is accomplished by a 60 cm Bainbridge E/B mass filter, and the beam line in use is selected by electrostatic deflection of the source beam.

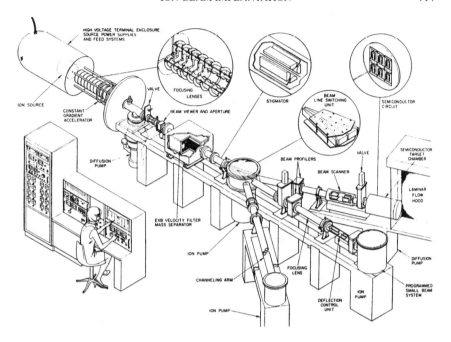

Figure 22.15. A commercial 300 kV ion-beam implantation system with three beam lines for separate applications of the ion beam (Wilson and Brewer 1973).

Ensuring uniformity of dose is an issue in ion-beam implantation systems, as is the necessity of keeping the surface approximately at room temperature to avoid uncontrolled loss of implanted atoms by thermal diffusion into the bulk of the workpiece. To avoid these problems, the workpiece may be rotated through the beam using 'epicyclic' motional averaging, in which the workpiece repeatedly passes through the ion beam, usually with each passage occurring at a different angle with respect to the direction of the workpiece motion through the beam.

22.3.2 Constraints on Ion-Beam Implantation

In ion implantation applications, the sources operate at or below the space-charge-limited current given by the *Child law*, discussed in section 3.6 of Volume 1. The flux of singly charged ions of mass M_1 from a source such as that shown in figure 22.16 is given by

$$\phi_i \leq \frac{J_e}{e} = \frac{4\varepsilon_0}{9e}\sqrt{\frac{2e}{M_1}}\frac{V_0^{3/2}}{d^2}. \tag{22.13}$$

In equation (22.13), V_0 is the accelerating voltage and also the ion implantation energy, and d is the width of the gap between the accelerating electrodes. The

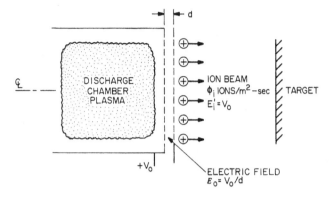

Figure 22.16. Production of a space-charge-limited ion beam using an electron bombardment (Kaufman) ion source.

average electric field between the accelerating electrodes is given by

$$E_0 = \frac{V_0}{d} \quad \text{V/m.} \tag{22.14}$$

In the development of space-charge-limited electrostatic ion engines for space propulsion, it has been found that, in the presence of a plasma, it is difficult to maintain electric fields higher than

$$E_0 < 10 \,\text{kV/cm} = 1 \,\text{MV/m} \tag{22.15}$$

without excessive electrical arcing during the startup process (*conditioning*). This relationship can be used to estimate an upper bound on ion current densities at various energies $V_0 = E_i'$. Substituting equation (22.14) into equation (22.13), one obtains

$$\phi_i < \frac{4\varepsilon_0}{9e} \sqrt{\frac{2e}{M_1 V_0}} E_0^2 \quad \text{ions/m}^2\text{-s.} \tag{22.16}$$

An estimate of the maximum practicable ion flux in these space-charge-limited sources is found by substituting $E_0 = 1$ MV/m into equation (22.16), to yield

$$\phi_i \approx \frac{3.4 \times 10^{23}}{\sqrt{A E_i'}} \quad \text{ions/m}^2\text{-s} \tag{22.17}$$

in which A is the atomic mass number of the ion, and $E_i' = V_0$ is its energy in eV. This estimate is shown in figure 22.17 for several ions of interest in the field of microelectronics, with the flux given in ions/cm²-s.

At doses from 10^{12} to 10^{15} ions/cm², values appropriate to many microelectronic implantation applications, an implication of figure 22.17 is that

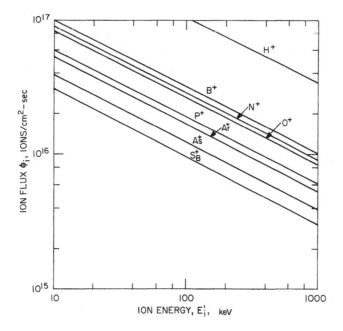

Figure 22.17. The ion flux, as a function of ion energy in eV, for several ions used in commercial microelectronic implantation applications.

a space-charge-limited beam can supply such doses in much less than 1 s of exposure. Such durations are so short that they are difficult to control. For this reason, the ion implantation of semiconductors may be accomplished with beam currents below the space-charge-limited value or by passing multiple wafers through the beam only briefly in a rotating fixture. However, when ion implantation is used to improve the hardness or wear resistance of metals, doses on the order of 10^{16} to more than 10^{17} ions/cm² are required, making the use of space-charge-limited beams desirable for these applications.

The power flux on a surface from ions of energy $E_i' = V_0$ is given by multiplying the flux of figure 22.17 by the energy of the individual ions,

$$P_i = e V_0 \phi_i = \frac{4\varepsilon_0}{9} \sqrt{\frac{2e}{M}} \frac{V_0^{5/2}}{d^2} \qquad \text{W/m}^2. \tag{22.18}$$

If the source is limited to the inter-electrode electric field E_0 given by equation (22.16), the maximum power flux is

$$P_i \approx \frac{4\varepsilon_0}{9} \sqrt{\frac{2V_0 e}{M}} E_0^2 \qquad \text{W/m}^2. \tag{22.19}$$

The power fluxes may range from several hundred to several thousand W/cm² for space-charge-limited currents, as shown in figure 22.18. Such heating rates,

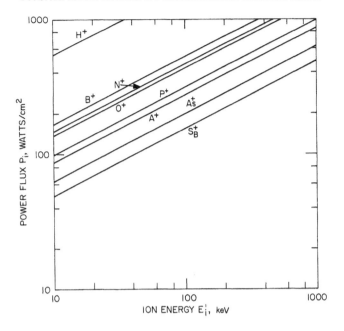

Figure 22.18. The space-charge-limited power flux, in W/cm², of beams of selected ions, as a function of ion energy.

if continued for many seconds, can raise a surface to temperatures that result in thermal diffusion of the implanted ions deep into the material. For this reason, ion implantation usually requires active cooling of workpieces, rotation of the workpiece through the beam by a rotating fixture or ion fluxes well below space-charge-limited values.

22.3.3   Discussion of Ion-Beam Implantation

Ion-beam implantation offers several advantages as a means of depositing ions in the surface layers of materials. The ion energy is easily adjusted above levels of approximately 10 keV, to obtain a programmable implantation depth or range. At a given ion energy, the dose is easily controlled by adjusting the exposure time, t_0. These times are moderate for most applications, on the order of minutes. On table 22.2 is a list of over 60 species that are commercially available for ion implantation.

Unfortunately, while ion-beam implantation is well developed and economic for the high-value products of the microelectronic industry, the process has disadvantages that have limited its application elsewhere. Ion-beam implantation is expensive and capital intensive; the process requires normal incidence to the workpiece; implantation of three-dimensional objects requires masking and

Table 22.2. Species available for commercial ion-beam implantation.

H	O_2	Rb
D	S	Sr
He	Cl	SiF_3
Li	Ar	Y
Be	K	Zr
B	BCl	Mo
C	Ca	Ru
^{13}C	SiF	Ag
N	Sc	Cd
^{15}N	Ti	In
O	BF_2	Sn
^{18}O	V	Sb
H_2O	Cr	Te
F	Mn	I
Ne	Fe	Xe
Na	Co	Cs
Mg	Ni	Ba
Al	Cu	Ce
N_2	Zn	Er
BeF	BF_3	Ta
CO	Ga	W
Si	Ge	Pt
^{29}Si	As	Au
^{30}Si	Se	Hg
BF	Br	Pb
P	Kr	Bi

movable fixturing; and since the process is done at intermediate pressures, vacuum or batch processing is required.

22.4 PLASMA ION IMPLANTATION

Plasma ion implantation is a recently introduced process that can improve the hardness, wear resistance, and corrosion resistance of metals by implanting energetic ions up to a few tenths of a micron below the surface. This technology has been developed to overcome at least two disadvantages of ion-beam implantation, the high cost of ion-beam implantation equipment, and the necessity for masking and fixturing of workpieces.

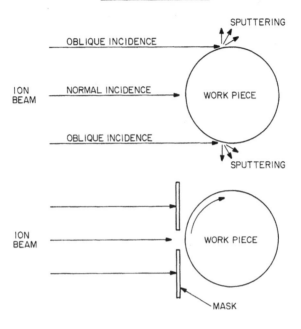

Figure 22.19. The role of sputtering in ion-beam implantation: (*a*) sputtering by ions impinging on the sides of the workpiece at oblique angles of incidence; and (*b*) masking to avoid oblique sputtering of the workpiece.

22.4.1 Process Characteristics

To understand the necessity of masking and fixturing of workpieces, consider figure 22.19(*a*), in which an ion beam is directed toward a cylindrical workpiece on the right. Ions that approach the cylinder at normal incidence will be implanted in the manner discussed previously. Ions that approach at an oblique angle will cause greatly enhanced sputtering of the cylindrical workpiece, sometimes at such high rates that the implanted surface is removed faster than it can be implanted. Masks are used to limit the impact angle of the ion beam to near-normal incidence on the workpiece, as shown in figure 22.19(*b*). To ensure uniform coverage, the cylinder in figure 22.19(*b*) is rotated past the slit in the mask, to achieve uniform, normally incident implantation of its surface. If the workpiece is approximately spherical, as are such medical prostheses as hip joints, a complicated fixturing arrangement is necessary that is designed to rotate the workpiece past a hole in the mask.

 Plasma ion implantation, sometimes called '*plasma immersion ion implantation*' (PIII), or '*plasma source ion implantation*' (PSII®, registered trademark of ASTeX, Incorporated) consists of inserting the workpiece to be

PLASMA ION IMPLANTATION

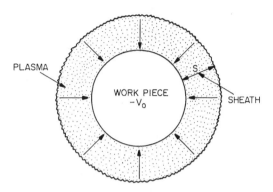

Figure 22.20. Plasma ion implantation of a workpiece by negative biasing and the attraction of ions from the plasma.

implanted into a plasma, as illustrated in figure 22.20. The workpiece is pulsed to a high negative voltage, V_0, such that it acts like a Langmuir probe in ion saturation, and draws ions from the plasma. These ions hit the surface of the workpiece with an energy $E'_i = V_0$ and, if the sheath is sufficiently thin, uniformly over its surface.

If the sheath thickness, S, is small compared to the radius of curvature of the workpiece, uniform implantation of the surface should occur without masking or fixturing. The implantation voltage, V_0, can range from 10 to 200 kV, far higher than the ion energies reaching the surface in *low-energy plasma thermal diffusion treatment*, discussed in section 22.5 of this chapter. In plasma ion implantation, the high voltage is usually switched off after tens of microseconds. This brief pulsing avoids quenching the plasma, or overheating the surface of the workpiece to the point that the implanted atoms thermally diffuse into its interior. Like ion-beam implantation, plasma ion implantation can be used, under conditions that will be discussed later, to improve the hardness and wear (tribological) characteristics of metals.

22.4.2 Plasma Ion Implantation Systems

Plasma ion implantation is accomplished in the apparatus shown schematically in figure 22.21. Characteristically, a large-volume (at least several hundred liters), uniform plasma is generated by microwave power, or a DC or RF glow discharge. Sometimes the plasma electrons are magnetized, or surface magnetic confinement is provided by permanent magnets in a cusp array (see section 3.4 of Volume 1) to improve the plasma source efficiency or increase its electron number density. The electron number density of a plasma intended for implantation may range from 10^{14} to 10^{17} electrons/m^3. Descriptions of plasma ion implantation apparatus

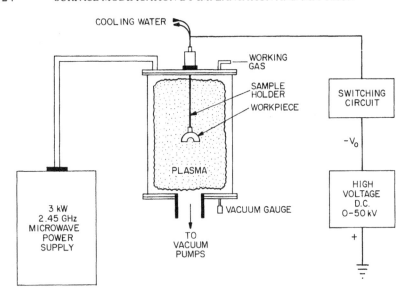

Figure 22.21. A plasma ion implantation apparatus in which the workpiece to be implanted is immersed in a large uniform plasma.

may be found in Conrad (1988), Bernius and Chutjian (1989), Keebler (1990), Spence *et al* (1991), and Anders *et al* (1994).

An electrically conducting workpiece is inserted in this plasma, and a pulsed, high negative DC voltage is applied by a fast switching circuit. DC voltages as low as 10 kV have been used for imparting corrosion resistance; voltages as high as 200 kV have been used for implantation and wear applications where deep implantation is desired. The power supply and the switching circuit must be capable of providing peak currents from one to several hundred amperes during an implantation pulse, depending on the plasma density and the area of the workpiece.

The switching circuit must be capable of limiting the duration of the current pulse, as well as adjusting the pulse repetition frequency in order to control the total dose, avoid quenching the plasma, and to avoid overheating the surface. Pulse rise times are characteristically a microsecond, their durations are 10–30 μs, and their repetition rates are usually below 500 Hz, with 200 Hz being typical. There is evidence (Matossian and Goebel 1992) that pulse repetition rates of 1 kHz or higher lead to surface heating and thermal diffusion of the ions into the implanted material.

The UT microwave plasma facility (MPF), an example of a plasma ion implantation system, is illustrated in figure 22.22. The MPF is a large-volume, high-density, steady-state, non-resonant microwave plasma reactor intended for plasma ion implantation and other industrial plasma-processing studies (Spence *et*

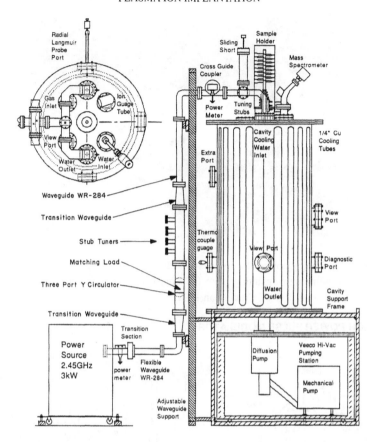

Figure 22.22. Elevation view of the vacuum tank and microwave hardware of the UT microwave plasma facility (MPF), used for plasma ion implantation.

al 1991, Keebler 1990). The MPF consists of a cylindrical stainless steel vacuum tank with a vertical axis, approximately 50 cm in diameter and 1 m high. The outer surface of its cylindrical sidewall is fitted with permanent magnets that produce a multipolar cusp of no more than 30 mT at the inner surface of the vacuum tank. The presence of the multipolar magnetic field improves the confinement and electron number density of the plasma by a factor of 1.7.

The plasma is maintained in the 206 liter volume by 2.45 GHz non-resonant microwave power at levels between 1 and 2 kW. An elevation view of the vacuum tank is illustrated in figure 22.22, along with the microwave components needed to sustain the plasma. The working gas in this facility is commercial dry nitrogen at background pressures ranging from 2.6–13 Pa (20–100 mTorr). During plasma ion implantation, a high negative DC voltage is applied to the workpiece in pulses from 5–50 μs in length, with a repetition rate of 100–2000 Hz (Keebler *et al*

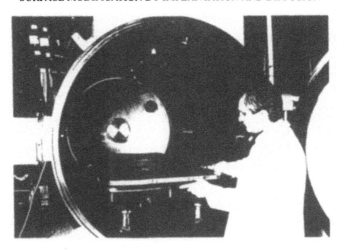

Figure 22.23. Photograph of the Hughes Research Laboratory plasma ion implantation facility. The aluminum mounting plate for workpieces is 1 m^2, and can be pulsed to voltages of more than 100 kV (Matossian 1994).

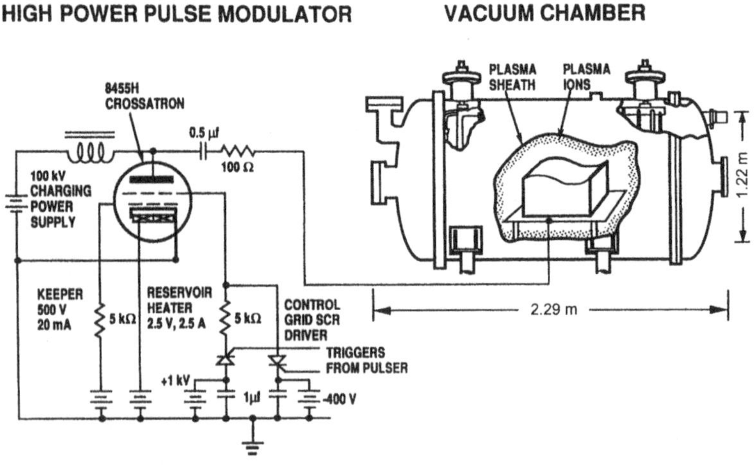

Figure 22.24. Schematic diagram of the Hughes Research Laboratory plasma ion implantation facility (Matossian 1994).

1989, Keebler 1990). Ions are delivered to the workpiece during each pulse, and the process repeated until the required dose accumulates on the workpiece.

A second example is the Hughes Research Center's plasma ion implantation facility, a photograph of which is shown in figure 22.23 (Matossian 1994). This large vacuum tank, approximately 1.22 m in diameter and 2.29 m long, is used to plasma ion implant workpieces with dimensions up to 1 m for studies of their hardness and wear characteristics. The components of this facility are shown on the schematic diagram of figure 22.24. Peak currents up to 100 A can be supplied at voltages up to 200 kV, in pulses with rise times on the order of a microsecond, durations of tens of microseconds, and at repetition rates up to 1 kHz.

22.4.3 The Implantation Sheath

It was pointed out in section 9.4 of Volume 1 that DC plasma sheath theory is not well understood, and attempts at quantitative understanding of sheath characteristics (especially predicting their thickness) may be unsuccessful. This difficulty is worsened by the conditions encountered in plasma ion implantation sheaths, including their rapid time dependence, and an applied voltage thousands of times the plasma potential.

22.4.3.1 The Role of Sheath Thickness

An important issue in plasma ion implantation is how well the surface of complex objects is covered by the plasma. Complex objects with compound curvature, such as gear teeth and turbine blades, will have a surface undulation or roughness with a characteristic radius of curvature R. If these objects are placed in a plasma which, during implantation, produces a sheath of thickness S, one of the two situations shown in figure 22.25 will result.

In order to accelerate ions across a sheath to an energy V_0 and implant normal to the surface without shadowing, it is necessary that the ratio of the sheath thickness to the smallest characteristic radius of curvature R of the workpiece be much less than one, as indicated in figure 22.25(b). If the opposite occurs, as shown in figure 22.25(a), with the sheath thickness greater than the characteristic undulation of the surface, shadowing of surfaces, uneven doses, and oblique incidence may result. Oblique incidence is particularly undesirable, since it can cause local sputtering at a rate faster than that at which the ions can be implanted.

22.4.3.2 Sheath Phenomenology

An important characteristic of the sheath surrounding a workpiece undergoing plasma ion implantation is that it functions as a bipolar diode that contains oppositely flowing charges of opposite sign. When ions are accelerated across the sheath and impact on the workpiece, they produce electrons by secondary emission, with the secondary emission coefficient γ. At energies above 10 kV of interest in plasma ion implantation, ions are capable of knocking off one or more electrons per incident ion from aluminum or stainless steel. This yields a significant enhancement of the total current flowing to the workpiece (Shamim et al 1991). This bipolar flow is illustrated in figure 22.26. Its effect is to lower the net density of space charge that acts to limit the current, and hence to increase substantially the space-charge-limited current density.

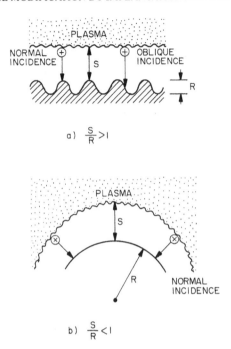

Figure 22.25. Two extreme limits of the ratio of sheath thickness to the dimensions of surface structure on the workpiece: (*a*) sheath thicker than the characteristic scale size of surface irregularities of the workpiece; and (*b*) sheath thickness small compared to the radius of curvature of the workpiece.

Beyond the issue of bipolar flow is the fact that the ions may begin their transit across the sheath with a significant initial velocity, rather than the zero initial velocity assumed in conventional Child law diode theory. The bipolar diode with finite initial velocity has been examined by Howes (1964, 1966a, b), who showed that the effects of bipolar flow and finite initial velocity both increase the current density that flows across a diode (sheath), relative to that possible if only unipolar flow with zero initial velocity is considered.

22.4.4   Ion Implantation Current

The ambipolar space charge in an implantation sheath limits the current density to a maximum value in a manner not different in principle from the simple unipolar *Child law* mechanism. In this section, we examine two principal models for the current density flowing to the workpiece.

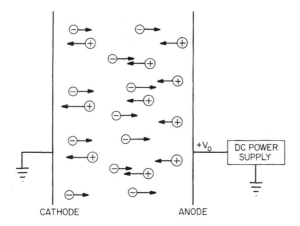

Figure 22.26. The bipolar diode, in which counter-streaming currents of opposite charge carry the current, and reduce the net charge density between the electrodes below that of a unipolar diode.

22.4.4.1 Models for Current Flow

A major factor in determining the current density in plasma ion implantation applications is the existence of *bipolar flow*. In this bipolar flow, the flux of ions on a workpiece is bounded by two models, which span the likely range of ion fluxes to be expected. The lower limit is the *ion saturation model*, in which the workpiece is regarded as a Langmuir probe deep in ion saturation. The ion flux is the random ion current from the plasma,

$$\Gamma_{ir} = \tfrac{1}{4} n_i \bar{v}_i \qquad \text{ions/m}^2\text{-s}. \tag{22.20}$$

The ion thermal velocity is given by equation (2.9), and when substituted into equation (22.20) yields

$$\Gamma_{ir} = \frac{1}{4} n_i \sqrt{\frac{8 e T_i'}{\pi M}} \qquad \text{ions/m}^2\text{-s}. \tag{22.21}$$

The second, upper limit is the *Bohm sheath model* for the workpiece current density, which predicts a much higher flux of ions to the surface. In this model, ions reach the workpiece with the *Bohm velocity* given by equation (9.146). Using equation (9.148) for the Bohm ion flux to a surface, we have

$$\Gamma_{ib} = n_i v_b \sqrt{\frac{e T_i'}{M}} \qquad \text{ions/m}^2\text{-s}. \tag{22.22}$$

Taking n_i as the number density of ions at the sheath edge in both cases, the ratio of the fluxes predicted by these two models is

$$\frac{\Gamma_{ib}}{\Gamma_{ir}} = 4\sqrt{\frac{\pi T'_e}{8T'_i}} = 2.51\sqrt{\frac{T'_e}{T'_i}}. \tag{22.23}$$

In a typical plasma used for plasma ion implantation, the electron kinetic temperature is $T'_e = 10$ eV; the ion kinetic temperature in the plasma is poorly known, but is much less than the electron kinetic temperature; it might be $T'_i \approx 0.10$ eV. For these conditions, the range of surface fluxes predicted by these two models is

$$1 \le \Gamma_{ib}/\Gamma_{ir} < 25. \tag{22.24}$$

Thus, the ion flux is equal to the random ion flux at the surface of the workpiece if the ion saturation model is appropriate, and to approximately 25 times this value if ions reach the sheath with the Bohm velocity.

These considerations make it clear that the presence of a Bohm-like sheath will greatly increase the implanted ion current. Experimental measurements of the ion fluxes to a surface during implantation, however, are even higher than those predicted by the Bohm model in equation (22.22). This excess ion flux can be understood as the result of the additional contribution of bipolar space-charge-limited flow with non-zero initial ion velocities, as discussed by Howes (1964, 1966a, b).

22.4.4.2 Observed Current Characteristics

The UT microwave plasma facility (MPF) shown in figure 22.22 produced some characteristic voltage and current waveforms for nitrogen ion implantation into a stainless steel workpiece (Kamath and Roth 1993, Kamath 1994). Figure 22.27 shows simultaneous profiles of voltage and current, with each averaged over ten pulses, for an implantation pulse of approximately 20 μs duration. The current and voltage rise times are controlled by the workpiece capacitance. The current termination is abrupt, due to the vacuum tube switching element; the voltage decay is determined by the leakage of charge from the workpiece into the plasma. On figure 22.28 a similar set of profiles from Matossian (1994) for an implantation voltage of $V_0 = 95$ kV, and a peak current of approximately 200 A is shown.

In figure 22.29 two simple-sweep current waveforms for implantation voltages of 10 kV, taken from the MPF during ion implantation runs intended to impart corrosion resistance are shown (Kamath and Roth 1993). These profiles exhibit characteristic features normally seen in current pulses associated with plasma ion implantation, and are consistent with the computational models of Calder *et al* (1993), Malik *et al* (1993), and Amin *et al* (1993). The current first rises on a time scale of a microsecond or less, determined by the electrical circuit characteristics, and reaches an initial value I_1. The current then drops monotonically to an asymptotic value I_0 at time $t = \infty$. Note in figure 22.29(b) that the asymptotic current remains constant beyond $t = 40$ μs, the approximate

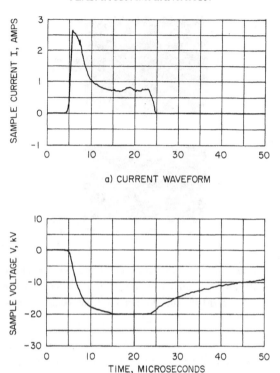

a) CURRENT WAVEFORM

b) VOLTAGE WAVEFORM

Figure 22.27. Simultaneous traces of voltage and current, with each averaged over ten pulses, during an implantation pulse on the UT Microwave Plasma Facility. The current reaches an asymptotic value of 0.75 A before being cut off at $t = 23$ ms. The slow fall of voltage after the end of the pulse was due to charge leaking into the plasma.

time an ion-acoustic disturbance would require to travel from the workpiece to the boundary of the MPF plasma.

In figure 22.30 single-sweep waveforms of the ion current as a function of time for eight pulse lengths, with 10 kV of applied potential during the pulse are shown. During normal plasma ion implantation, pulse lengths of 20 μs and repetition rates below 200 Hz were used in order to avoid heating the surface of the water-cooled workpieces. In all the traces in figure 22.30, the abrupt termination of the current pulse is due to the cut-off of the current by the switching circuit. The long pulse durations shown in figure 22.30 are incompatible with a model in which the sheath boundary moves out into the plasma without stabilizing at a fixed thickness. If the sheath boundary propagated into the plasma, one would expect a progressively larger volume of plasma to be engulfed, the collected current to increase, and then drop to zero as the plasma is quenched.

The long duration of constant asymptotic current shown in figure 22.30

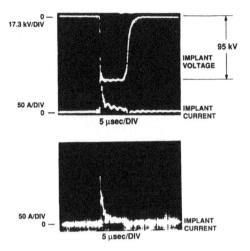

Figure 22.28. Current and voltage waveforms in the Hughes Research Laboratory's plasma ion implantation facility, for N_2^+, implantation voltage $V_1 = 95$ kV; pulse repetition rate, 200 Hz; duty factor 0.2%, and $10^{17}/cm^2$ dose in 4 hr (Matossian 1994).

is consistent with a model in which the sheath boundary has moved to a final thickness a factor of two to four larger than the initial thickness, and stabilized at that value. The *slowest* velocity at which the sheath boundary might propagate outward is the *ion-acoustic velocity*, approximately 5.2 km/s in this plasma. A sheath boundary moving with this speed would reach the outer radius of the MPF plasma, at $R = 25$ cm, in 48 μs. After sheath stabilization, the workpiece acts as the cathode of a DC abnormal glow discharge. The sheath is its cathode region, with a cathode fall distance determined by the minimum of the *Paschen curve*, as discussed in sections 8.6.4 and 9.2.1.

22.4.5   Ion Implantation Exposure Time

Plasma ion implantation normally occurs as the result of a repetitive series of pulses, each similar to those illustrated in the previous section. Without pulsing, the ion current in the asymptotic, constant part of the waveform may cause overheating of the surface of the workpiece and thermal diffusion of the ions into its interior. Such deeply diffused ions may not be useful for microelectronic, hardness, wear, or corrosion resistance applications. Surface heating is reduced by pulsing the implantation current, and adjusting the repetition frequency to a value that does not overheat the surface. A repetition frequency of 200 Hz is adequate for most applications, while 1 kHz may lead to thermal diffusion of the implanted ions. The pulsing of the implantation voltage is terminated when the required dose accumulates on the workpiece (see, for example, Conrad (1988) and Shao *et al* (1995)).

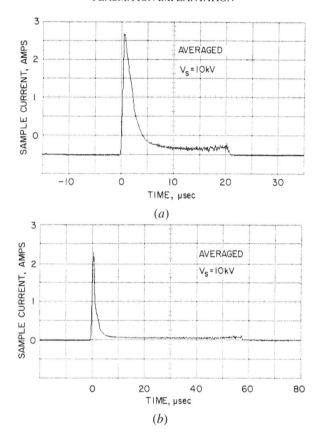

Figure 22.29. Simple-sweep plasma ion implantation current waveforms from the UT MPF in a nitrogen plasma, with an implantation voltage $V_0 = 10$ kV: (*a*) standard $t = 20$ ms pulse length; and (*b*) same conditions, but with a pulse length $t = 57$ ms.

To determine the exposure time t_0 for such a series of repetitive pulses, it is useful to define the following parameters:

A_s the exposed workpiece area subject to ion implantation, in m^2 or cm^2;

D the required dose or fluence on the workpiece, in ions/m^2 or ions/cm^2;

I_i the instantaneous true current of ions flowing to the workpiece, in amperes; and

I_m the instantaneous metered current flowing to the workpiece, in amperes.

Ion bombardment of the workpiece produces secondary electron emission, as a result of which the metered and true ion currents are related by

$$I_m = I_i(1 + \gamma_i) \quad \text{A.} \tag{22.25}$$

The exposure time may be calculated with the following additional parameters:

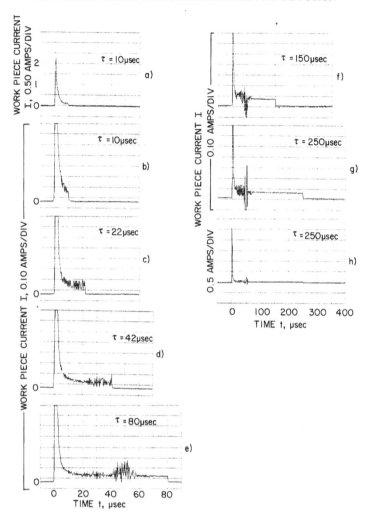

Figure 22.30. Experimentally measured single-sweep profiles of workpiece ion current as a function of time for eight different pulse durations, taken with an implantation voltage of 10 kV. The microwave power to the plasma was 1100 W. Background pressure was 2.6 Pa (20 mTorr) of nitrogen gas, electron number density $n_e = 2.9 \times 10^{16}/m^3$, plasma potential $V_p = +117$ V, electron kinetic temperature $T_e \approx 15$ eV.

N the total number of ions required to expose a workpiece;

Q the total charge required to expose a workpiece, in coulombs, assuming that all ions are singly charged;

t_0 the required exposure time of the workpiece, in seconds;

T the duration of an individual implantation pulse, in seconds;

Ω the repetition rate of pulses, in Hz;

I_0 the average true ion current during the pulse, in amperes; and

I_m the *time-averaged* metered current flowing to the workpiece, in amperes.

The total charge in coulombs delivered to the workpiece can be written as

$$Q = \int_0^{t_0} I_t(t)\, dt = \frac{\bar{I}_m t_0}{1 + \gamma_i} = T I_0 \Omega t_0 \qquad (22.26)$$

and the total number of ions delivered to the workpiece is

$$N = DA_s = \frac{Q}{e} = \frac{\bar{I}_m t_0}{e(1 + \gamma_i)} \qquad \text{ions.} \qquad (22.27)$$

Solving the two right-hand terms of equations (22.26) and (22.27) for the required exposure time yields

$$t_0 = \frac{eDA_s(1 + \gamma_i)}{\bar{I}_m} = \frac{Q}{T i_0 \Omega} \qquad \text{s.} \qquad (22.28)$$

22.4.6   Outcomes of Plasma Ion Implantation

This section contains experimental data reported in the literature in three areas of concern relating to the performance of plasma ion implantation:

(1) *dose uniformity* over the surface implanted;
(2) *dose profile* below the surface; and
(3) dose *implantation mechanism*, whether pure plasma ion implantation or a hybrid in which thermal diffusion into the interior also occurs.

22.4.6.1   Dose Uniformity

The issue of *dose uniformity* was addressed by Matossian (1992) with the results shown in figure 22.31. In this experiment, a square aluminum table approximately 91 cm in each dimension was nitrogen-ion implanted, and the implanted dose measured across two orthogonal, centered traverses. These indicated a uniform dose to within approximately 5 cm of the edge, where fringing electric field effects come into play.

A second approach to the issue of dose uniformity was reported by the Astex corporation (Anonymous 1989), using the four target spheres shown in figure 22.32. To assess shadowing or interference effects, four target spheres made of Ti–6Al–4V alloy were placed at the vertices of a square with side d, as shown in figure 22.32(a). After plasma ion implantation with nitrogen, the implanted dose of nitrogen ions was measured over the upper hemisphere of one of the spheres, at the locations shown on figure 22.32(b). The doses at these nine locations were measured, and are presented in bar-graph form in figure 22.33. With the possible exception of the top ('north pole') of the sphere, the dose was uniform within experimental error, demonstrating no interference effects from the surrounding spheres.

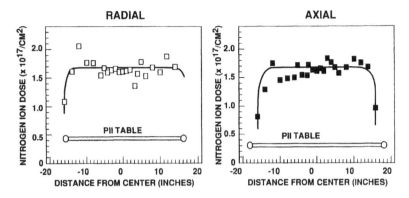

Figure 22.31. Dose uniformity of implanted nitrogen on a 91 cm × 91 cm aluminum table implanted to $V_0 = 95$ kV in the Hughes plasma ion implantation facility, along two orthogonal dimensions of the table (Matossian 1994).

22.4.6.2   Dose Profile

A second issue is the dose implantation profile, and whether it has the range, standard deviation, and profile with depth expected of high-energy implanted ions. On figure 22.34 is a characteristic profile reported by Conrad *et al* (1988) for the plasma ion implantation of N_2^+ implanted into Ti–6Al–4V alloy. This was done with a 50 kV implantation voltage, and a total dose of 3×10^{17} cm^{-2}. The maximum of this distribution is at 50–100 nm below the surface.

A further comparison of plasma ion implantation of N^+ at an implantation voltage of 50 kV into a Ti–6Al–4V alloy has also been published by Conrad *et al* (1988). These data were compared with the implantation profile predicted by a standard computer modeling code, with the results shown in figure 22.35. There is relatively good agreement, indicating that the plasma ion implanted profiles do not differ from those generated by other implantation processes, such as ion-beam implantation.

22.4.6.3   Dose Implantation Mechanism

Another issue associated with plasma ion implantation is whether the ion flux is so high that the implanted surface overheats. The increased surface temperature may allow ions to diffuse into the interior of the metal rather than remaining near the surface. Some relevant data for 50, 90, and 100 keV N_2^+ ions implanted into 304 stainless steel by Matossian and Goebel (1992) are shown in figure 22.36. The 90 and 100 keV profiles had similar doses (7×10^{16} and 1×10^{17}/cm^2, respectively), indicating that ion-beam implantation and plasma ion implantation produced equivalent results, for a 200 Hz pulse repetition rate. This figure also contains two 50 keV implantation profiles at pulse repetition frequencies of 200

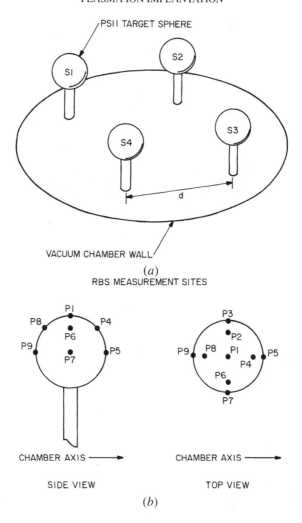

Figure 22.32. Bell jar arrangement to assess the uniformity of the nitrogen plasma ion implantation dose on a spherical Ti–6Al–4V workpiece: (*a*) array of four spheres to test interference and shadowing effects; and (*b*) nine test locations on one sphere at which the nitrogen dose was measured, using Rutherford backscattering spectroscopy (RBS) (Anonymous—Astex Corp., 1989).

and 1000 Hz. The faster repetition rate evidently heated the stainless steel surface to the point that the nitrogen thermally diffused much deeper into the workpiece than the 200 Hz data.

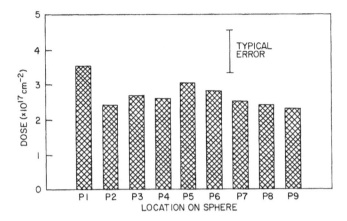

Figure 22.33. The implanted nitrogen dose, measured with Rutherford backscattering spectroscopy (RBS), at the nine locations shown in figure 22.32(*b*), along with a typical error bar.

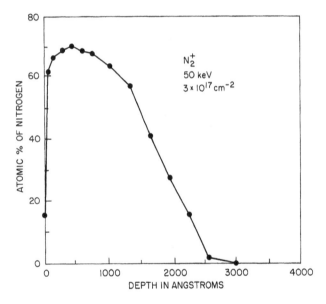

Figure 22.34. The profile of nitrogen concentration determined by Auger electron spectroscopy (AES), in nitrogen plasma ion implanted Ti–6Al–4V alloy with an implantation voltage of $V_0 = 50$ kV, and a total dose $D = 3 \times 10^{17}/\text{cm}^3$ (Conrad *et al* 1988).

22.4.7   Applications of Plasma Ion Implantation

The commercial applications of ion-beam and plasma ion implantation outside the microelectronics industry are at an early stage in their development. Thus far

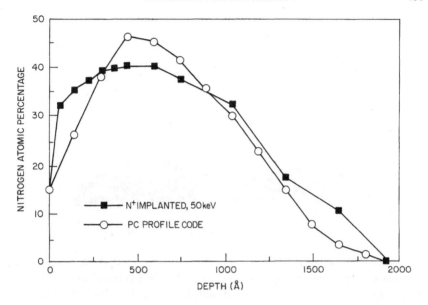

Figure 22.35. Auger electron spectroscopic depth profile of nitrogen concentration resulting from 50 kV implantation into Ti–6Al–4V alloy, compared with the prediction of the PC profile code (Conrad *et al* 1988).

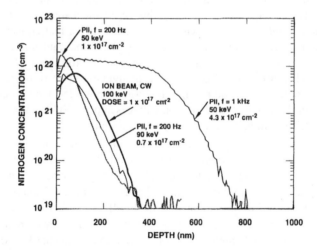

Figure 22.36. Profiles of nitrogen plasma ion implanted into 304 stainless steel at 50 and 90 kV, at repetition rates of 200 and 1000 Hz, compared with an ion-beam-implanted nitrogen profile at a comparable ion energy and dose (Matossian 1994).

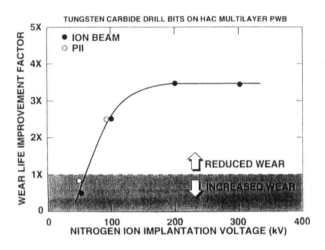

Figure 22.37. Dependence of the wear life of tungsten carbide drill bits as a function of nitrogen ion implantation voltage, for ion-beam implantation and plasma ion implantation. Wear life improvement factor is relative to unimplanted samples (Matossian 1994).

these applications are limited to relatively high-value items, or to items like tool bits, the lifetimes of which are increased by a large factor.

In figure 22.37 some data from Matossian (1994) on the improvement in the wear life of tungsten carbide drill bits used on a material of interest to the automotive industry are shown. These data indicate that equivalent results are obtained for ion-beam and plasma ion implantation, and that implantation voltages above 150 kV are required. The poor performance of plasma ion implantation below 60 kV in figure 22.37 is thought to be due to competition from sputtering, where the maximum of the sputtering curve is located for this material. These data indicate that at implantation voltages above 150 kV, improvements of a factor of 3.5 in drill bit lifetimes should be possible.

Similar results have been reported for improvement of the lifetime of ion-beam-implanted tools for fabricating aluminum. These results are shown in table 22.3. The table lists the particular tool, the implanted material, the aluminum alloy worked, and the factor by which the lifetime of the implanted tool was increased, compared to an untreated tool. In all cases, the lifetime of the tool was increased by at least a factor of two, and in some cases, by more than a factor of 10. Only the relatively high cost of ion-beam implantation has kept this process from being more widely used.

The volumetric material loss of pierce punches (used for making holes in thick metal plate) reported by Conrad et al (1988) is presented in figure 22.38. Data are shown for untreated and nitrogen plasma ion-implanted workpieces, along with the wear loss of a workpiece with a conventional TiN coating. The lifetime of nitrogen plasma ion-implanted pierce punches can be increased by as

Table 22.3. Improvement in lifetime of ion-beam-implanted tools for fabricating aluminum (Anonymous 1993).

Tools	Materials	Application	Results
Chases	HSS	380 & 384 alloy	5 × life
Taps	M2	384 alloy	2.5 × life
Thread hob	M2	355 alloy	3 × life
Forming rolls	L6	5052 alloy	15 × life
Pierce punches	M2	6101 alloy	20 × life
Pilot dies	A2	5454 alloy	Pickup reduced
Reamers	Carbide	384 alloy	2 × life

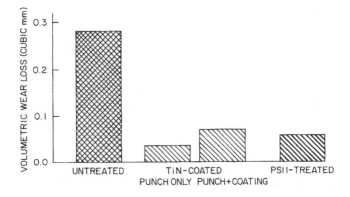

Figure 22.38. Volumetric wear of M2 pierce punches after 20 000 operations for untreated, conventional, and plasma-ion-implanted samples.

much as a factor of 100 in some instances. In addition to industrial tools, plasma ion implantation has been used successfully to harden femoral knee components used as medical prostheses (Chen *et al* 1991).

Finally, in addition to these applications, plasma ion implantation can improve the corrosion resistance and increase the *pitting potential* of aluminum and stainless steel (Keebler *et al* 1989, Gupta 1991, Smith *et al* 1994, Smith 1995). An illustration of this is the *potentiodynamic polarization curves* shown in figure 22.39 (Smith *et al* 1994). Curve B is an unimplanted workpiece of 304L stainless steel; curve A is a workpiece of nitrogen plasma ion-implanted 304L stainless steel with a dose of $1 \times 10^{15}/cm^2$. The *pitting potential* is that voltage at which the sudden increase in workpiece current occurs as pitting initiates: this is about 750 mV for the unimplanted workpiece, and 1450 mV for the implanted workpiece—a significant increase.

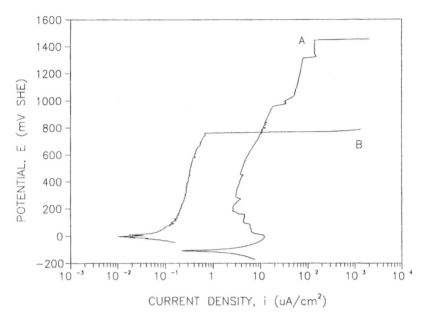

Figure 22.39. Potentiodynamic polarization behavior of (*a*) nitrogen plasma ion implanted 1×10^{15} ions/cm^2 304L stainless steel, and (*b*) control 304L stainless steel (Smith *et al* 1993).

Figure 22.40 shows the measured pitting potential as a function of implanted dose, compared to an unimplanted workpiece at the horizontal dotted line. It was found by Smith (1995), and Smith *et al* (1994) that the best corrosion resistance occurs at doses of 10^{14}–10^{15}/cm^2 and at implantation voltages of 10–20 kV. This contrasts with the doses of 10^{17}–10^{18}/cm^2 and the 50–200 kV needed for deeper implantation in hardness and wear applications. Since corrosion occurs at and very near the surface of materials, low ion energies and low doses near the surface are appropriate.

22.4.8   Characteristics of Plasma Ion Implantation

Relative to ion-beam implantation, plasma ion implantation has several advantages:

(1) a relatively small capital investment;
(2) small, complex objects can be implanted without fixturing or masking;
(3) dose control is relatively easy through changing the pulse width, T, and the repetition rate, Ω; and
(4) since ion fluxes are relatively high in plasma ion implantation, moderate exposure times, tens of minutes, will normally be required.

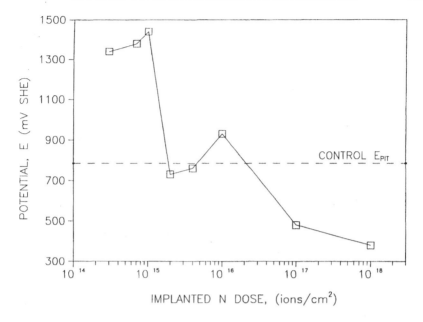

Figure 22.40. Measured pitting potentials in 1 wt% NaCl as a function of implanted dose (Smith *et al* 1993).

Disadvantages of plasma ion implantation may include the following:

(1) batch processing, imposed by the requirement that the entire process occur in a vacuum;
(2) a more restricted range of ion species than ion-beam implantation, because of the difficulty of creating a plasma consisting of some ionic species of interest; and
(3) a more restricted energy range, since it is difficult to do plasma ion implantation above 50 kV.

22.5   LOW-ENERGY PLASMA THERMAL DIFFUSION TREATMENT

Low-energy plasma thermal diffusion treatment is most frequently accomplished by inserting a metallic workpiece into a low-energy, intermediate-pressure DC abnormal glow discharge plasma, and using the workpiece as a cathode of the discharge. This process is widely used in foundries and the metalworking industry to harden the surfaces of workpieces used as gear teeth, bearings, cutting tools, and in other applications where hardness and wear resistance are desired.

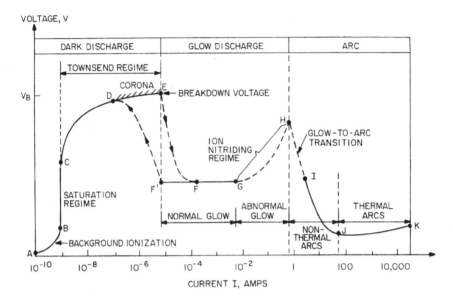

Figure 22.41. The current–voltage curve of the classical DC low-pressure electrical discharge tube, with the low-energy plasma thermal diffusion treatment region indicated in the abnormal glow regime between points G and H.

22.5.1   Process Characteristics

In low-energy plasma thermal diffusion treatment, the workpiece (cathode) surface is heated, either by ion bombardment or by placing the entire workpiece in an external furnace. Ions that bombard the heated surface thermally diffuse into the interior of the workpiece over periods of time from tens of minutes to more than ten hours, and reach depths which may be 1 or 2 mm below the surface. The addition of such diffusive atoms can greatly improve the hardness and wear characteristics of ferrous workpieces. Because the process is diffusive, the atom density profile is always a maximum at the workpiece surface and decreases monotonically with depth. The mole fraction of diffusive atoms can greatly exceed thermodynamic equilibrium values, since this process is not in thermal equilibrium in any of its stages. The process shares this characteristic with ion implantation.

Low-energy plasma thermal diffusion treatment has been given many names in the literature, usually dependent upon the particular species involved. Examples include *plasma nitriding* with nitrogen atoms; *plasma boronizing*, when boron is the species of interest; *plasma carbonitriding*, when the atoms are supplied by a carbon-containing nitriding gas; or *plasma carborizing*, when carbon atoms are involved. The process also has been referred to as *plasma implantation* or even *plasma ion implantation*. These terms are not used here to avoid confusion with the high-energy plasma ion implantation process discussed previously in section 22.4 of this chapter.

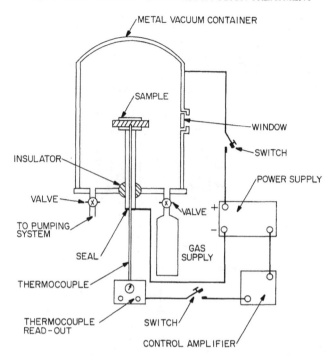

Figure 22.42. A typical system for low-energy plasma thermal diffusion treatment of metals (Hollahan and Bell 1974).

22.5.2 Mechanism of Treatment

Low-energy plasma thermal diffusion treatment is accomplished by making the metal to be treated the cathode of an *abnormal glow discharge*. The region of the classical low-pressure DC electrical discharge *I–V* characteristic in which low-energy plasma thermal diffusion treatment is accomplished is shown in figure 22.41. In the *abnormal glow* region from G to H on the current–voltage curve, the plasma completely covers the cathode, exposing the entire cathode surface to a flux of ions.

Low-energy plasma thermal diffusion treatment uses the apparatus illustrated in figure 22.42, taken from Hollahan and Bell (1974). In this example, the workpiece is mounted in the center of an evacuated metal bell jar and is electrically connected to the cathode of a DC power supply, usually at a potential from 500 V to a few kV. Ions arrive at the surface of the cathode, which acts like a negatively biased Langmuir probe in ion saturation in contact with the plasma, with a flux determined by the random ion flux. This flux depends on the ion density at the sheath edge and the ion thermal velocity, given by

$$\phi_{ir} = \tfrac{1}{4} n_i \bar{v}_i \qquad \text{ions/m}^2\text{-s}. \qquad (22.20)$$

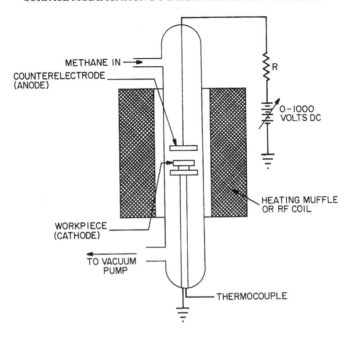

Figure 22.43. A schematic diagram of a plasma carbonizing apparatus with an external heater to maintain the workpiece temperature during treatment.

Substituting for the ion thermal velocity in equation (22.20), one obtains the random ion flux as a function of the ion number density, the ion kinetic temperature, and the ion mass M,

$$\phi_{ir} = \frac{1}{4} n_i \sqrt{\frac{8eT_i'}{\pi M}} \qquad \text{ions/m}^2\text{-s.} \qquad (22.29)$$

In the DC abnormal glow discharges used for this process, the ion kinetic temperature is typically at or slightly above room temperature. Higher fluxes may result if the ion flux reaching the sheath boundary is determined by the Bohm velocity.

Once the ions reach the surface they do not accumulate, but thermally diffuse into the metal. In order that this thermal diffusion occur at a sufficiently rapid rate to be of commercial interest, the entire workpiece must be heated to temperatures of many hundreds of degrees centigrade. In the low-energy plasma thermal diffusion treatment process, the diffusive atoms may reach depths of a millimeter or more, significantly modifying the surface layer.

Sometimes the power flux of bombarding ions is sufficient to raise the workpiece to temperatures adequate for thermal diffusion into the surface.

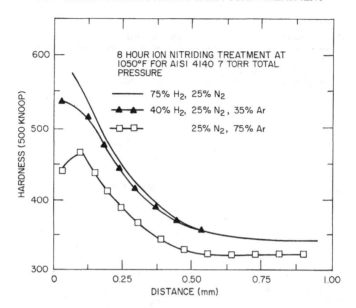

Figure 22.44. Effects of gas composition on depth profile of hardness produced by ion nitriding (Hollahan and Bell 1974).

Figure 22.42 (after Hollahan and Bell 1974) is an example of an apparatus that does not require an external furnace. The temperature of the workpiece is monitored with a thermocouple, and controlled by switching the discharge voltage on and off, as indicated on the schematic diagram in figure 22.42. Usually, however, the workpiece must be heated by such external means as inserting the entire discharge in a furnace, as illustrated in figure 22.43 (Grube and Gay 1978).

22.5.3 Applications

The low-energy plasma thermal diffusion treatment has the potential advantage that, if it is operated in the abnormal glow discharge regime, the entire surface of complex workpieces with compound curvature can be covered with plasma, and treated simultaneously. This process has been applied to the molds for plastic automotive grilles. The most widely used version of this process is *plasma nitriding*, in which the operating gas in the abnormal glow discharge is either nitrogen or ammonia. When boronizing of the surface layer is desired, the working gas is usually BCl_3.

The possibilities for tailoring the depth profile of diffused atoms are limited in the low-energy plasma thermal diffusion treatment process, since thermal diffusion necessarily results in a monotonic profile with a maximum at the surface, and decreases with depth. The depth at which a given atomic density of atoms is found may be adjusted by changing the temperature of the workpiece and the

Table 22.4. Examples of low-energy plasma thermal diffusion treatment (Buecken 1978).

Process	Material	Plasma parameters
Nitriding	Steel	NH_3, N_2, $N_2 + H_2$ 0.1–1.3 Pa (8×10^{-4}–10^{-2} Torr) 673–873 K max. 40 hr
	Cr–steel–TiC	85% N_2 + 15% H_2 1 kPa (7.5 Torr) max. 1023 K 6 hr
	Ti alloy	1073 K 20 hr
Borizing	Fe	95% BCl_3 + 5% H_2 650 Pa (4.8 Torr) 1073 K
Carbonitriding	Steel	C-containing nitriding gas 983–1143 K 5 hr
Carburizing	Steel Nb, W	1123–1273 K

duration during which the thermal diffusion process takes place.

Figure 22.44 (Hollahan and Bell 1974) illustrates a depth profile of hardness resulting from eight hours of plasma nitriding treatment. A mixture of hydrogen and nitrogen gas at 567 °C was used in an abnormal glow discharge at a pressure of 930 Pa (7 Torr). Figure 22.44 exhibits the characteristic monotone decrease of hardness (proportional to nitrogen atom concentration) with depth. The profile depth was approximately a quarter of a millimeter, a much deeper penetration of the nitrogen atoms than occurs in the ion-beam or plasma ion implantation processes discussed previously.

Some characteristic operating conditions for various forms of low-energy plasma thermal diffusion treatment are listed on table 22.4, from Buecken (1978). Note the relatively long times required, a major disadvantage in industrial processing applications. The DC abnormal glow discharges used for these applications usually operate at the upper limit of the DC discharge regime, at pressures of several hundred Pascal (several Torr).

REFERENCES

Amin A, Aossey D, Nguyen B T, Kim H-S, Cooney J L and Lonngren K E 1993 Sheath evolution in a negative ion plasma *Phys. Fluids* B **5** 3813–18

Anders A, Anders S, Brown I G, Dickinson M R and MacGill R A 1994 Metal plasma immersion ion implantation and deposition using vacuum arc plasma sources *J. Vac. Sci. Technol.* B **12** 815–20

Anonymous 1988 *The Ion Implanter: a Newsletter for Ion Surface Technology Customers* (Clawson, MI: Ion Surface Technology)

——1989 *Plasma Source Ion Implantation: PSII* (Woburn, MA: Applied Science and Technology)

Bernius M T and Chutjian A 1989 High-voltage, full-floating 10 MHz square-wave generator with phase control *Rev. Sci. Instrum.* **60** 779–82

Brown I G (ed) 1989 *The Physics and Technology of Ion Sources* (New York: Wiley) ISBN 0-471-85708-4

Buchanan R A, Lee I-S and Williams J M 1988 Iridium ion implantation of surgical titanium alloy: corrosion inhibition and charge injection effects *Proc. ASM Conf. on Ion Implantation and Plasma-Assisted Processes for Industrial Applications (Atlanta, GA)* ed R F Hochman, H Solnick-Legg and K O Legg (American Society for Materials) pp 53–6

Buecken B B 1978 Erzeugung verschleissfester schichten in einer stromstarken glimmentladung *Technik* **33** 395–9

Calder A C, Hulbert G W and Laframboise J G 1993 Sheath dynamics of electrodes stepped to large negative potentials *Phys. Fluids* B **5** 674–90

Carter G and Grant W A 1976 *Ion Implantation of Semiconductors* (New York: Wiley) ISBN 0-470-15125-0

Chen A, Scheuer J T, Ritter C, Alexander R B and Conrad J R 1991 Comparison between conventional and plasma source ion-implanted femoral knee components *J. Appl. Phys.* **70** 6757–60

Conrad J R 1988 Method and apparatus for plasma source ion implantation *US Patent* 4,764,394

Conrad J R, Dodd R A, Worzala F J, Qiu X and Post R S 1988 Plasma source ion implantation—a new, cost-effective, non-line-of-sight technique for ion implantation *Proc. ASM Conf. on Ion Implantation and Plasma-Assisted Processes for Industrial Applications (Atlanta, GA)* ed R F Hochman, H Solnick-Legg and K O Legg (American Society for Materials) pp 185–91

Dearnaley G, Freeman J H, Nelson R S and Stephen J 1973 *Ion Implantation* (New York: Elsevier) ISBN 0-444-10488-7

Grube W L and Gay J G 1978 High rate carburizing in a glow discharge methane plasma *Metall. Trans.* A **9** 1421–9

Gupta A 1991 Corrosion inhibition through nitrogen implantation into aluminum by ion beam and plasma source ion implantation *MS Thesis* University of Tennessee at Knoxville, TN

Hollahan J R and Bell A T (ed) 1974 *Techniques and Applications of Plasma Chemistry* (New York: Wiley) ISBN 0-471-40628-7

Howes W L 1964 Effect of initial velocity on one-dimensional, bipolar, space-charge currents *J. Appl. Phys.* **36** 2039–45 (see also *NASA Technical Note* TN D-2425)

——1966a One-dimensional space-charge theory: I. Generalization *J. Appl. Phys.* **37** 437–8

——1966b One-dimensional space-charge theory: II. Relativistic Child law *J. Appl. Phys.* **37** 438–9

Kamath S G 1994 Factors affecting the plasma ion implantation of metallic samples *MS*

Thesis University of Tennessee at Knoxville, TN

Kamath S G and Roth J R 1993 Observation of sheath characteristics on a sample undergoing plasma ion implantation *1993 IEEE Int. Conf. on Plasma Science (Vancouver, BC)* Conference Record, IEEE Catalog No 93CH3334-0, p 101

Keebler P F 1990 A large volume microwave plasma facility for plasma ion implantation studies *MS Thesis* Department of Electrical and Computer Engineering, University of Tennessee, Knoxville, TN

Keebler P F, Roth J R, Buchanan R A and Lee I-S 1989 Corrosion-related characteristics of plasma ion implanted samples exposed in a steady-state Penning discharge *APS Bull.* **34** 2020–1

Malik S M, Fetherston R P, Sridharan K and Conrad J R 1993 Sheath dynamics and dose analysis for planar targets in plasma source ion implantation *Plasma Sources Sci. Technol.* **2** 81–5

Matossian J N 1992 Private communication

——1994 Plasma ion implantation technology at Hughes Research Laboratories *J. Vac. Sci. Technol.* **12** 850–3

Matossian J N and Goebel D M 1992 High voltage (100 kV), large scale (3 m^3) plasma ion implantation (PII) facility *34th Ann. Meeting APS Division of Plasma Physics (Seattle, WA)* post-deadline paper

Powell R A (ed) 1984 *Dry Etching for Microelectronics* (New York: Elsevier) ISBN 0-444-86905-0

Shamim M M, Scheuer J T, Fetherston R P and Conrad J R 1991 Measurement of electron emission due to energetic ion bombardment in plasma source ion implantation *J. Appl. Phys.* **70** 4756–9

Shao J, Round M, Qin S and Chan C 1995 Dose–time relation in BF_2 plasma immersion ion implantation *J. Vac. Sci. Technol.* A **13** 332–4

Smith P P 1995 Ion implantation applications for pitting corrosion inhibition—(tungsten beam implantation of aluminum and nitrogen plasma ion implantation of 304L stainless steel) *PhD Dissertation* University of Tennessee, Knoxville, TN

Smith P P, Buchanan R A, Roth J R and Kamath S G 1994 Enhanced pitting corrosion resistance of 304L stainless steel by plasma ion implantation *J. Vac. Sci. Technol.* B **12** 940–4

Spence P D, Keebler P F, Freeland M S and Roth J R 1991 A large-volume, uniform unmagnetized microwave plasma facility (MPF) for industrial plasma-processing applications *Proc. Workshop on Industrial Plasma Applications and Engineering Problems, 10th Int. Symp. on Plasma Chemistry (Bochum)*

Williams J M, Gonzales A, Quintana J, Lee I-S, Buchanan R A, Burns F C, Culbertson R J, Levy M and Treglio J R 1991 Ion implantation for corrosion inhibition of aluminum alloys in saline media *Nucl. Instrum. Methods* B **59/60** 845–50

Wilson R G and Brewer G R 1973 *Ion Beams, with Applications to Ion Implantation* (New York: Wiley) ISBN 0-471-95000-9

Winters H F 1980 *Elementary Processes at Solid Surfaces Immersed in Low Pressure Plasmas (Plasma Chemistry III)* ed S Veprek and M Venugopalan (New York: Springer) ISBN 0-387-10166-7

23

Thin-Film Deposition by Evaporative Condensation and Sputtering

This chapter is concerned with the deposition by evaporative condensation and sputtering of thin films, the thickness of which may range from approximately ten nanometers to several tens of microns. These plasma-related deposition technologies are essential to several major industries, as well as a wide range of niche applications.

23.1 APPLICATIONS OF THIN FILMS

Plasma-related sputtering and evaporative condensation technologies are used to deposit thin films for microelectronic circuit fabrication, optical coatings, recording media, ornamental brightwork, protective coatings, and many other applications. For the sake of completeness, some evaporative condensation technologies which do not require plasmas, but which use plasma-related methods, are also included.

23.1.1 Early History of Thin-Film Deposition

Some processes used to deposit thin films, including gilding and wet chemical deposition methods, originated before the Industrial Revolution. The deposition of *electroplated* thin films using electrotechnology originated with Michael Faraday (1791–1867) in the early 19th century, and thermal evaporation methods that require high-vacuum systems were introduced commercially in the early 20th century.

Plasma-sputtered thin films have been used industrially for over 100 years. An early application of plasma-sputtered thin films is illustrated in figure 23.1, taken from Thomas A Edison's 1902 patent (Waits 1997). In this application, a

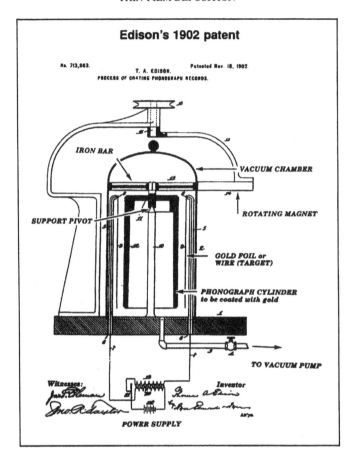

Figure 23.1. Edison's 1902 patent on a plasma-assisted gold sputtering apparatus (Waits 1997).

thin coating of gold is sputtered on a wax master copy of a cylindrical Edison phonograph record. The process illustrated in figure 23.1 shows an electrical discharge produced under vacuum by excitation with a high-voltage induction coil. The sputter cathodes in figure 23.1 consist of two gold wires or foil strips, which, in the presence of the plasma, sputter gold on the outer surface of the cylindrical wax master cylinder. The cylinder is suspended on a pivot and rotated past the gold wires by magnets external to the glass bell jar, to achieve uniformity of coating by *motional averaging*. This metalized master cylinder was later further electroplated to form a metal master mold from which large numbers of secondary cylindrical phonograph records were made.

This method was used from approximately 1901 to 1921, and allowed the reproduction of phonographic grooves less than 25 μm deep at a density of approximately four grooves per millimeter. In the period from 1913 to 1921, 25 cm diameter flat Edison diamond disc phonograph records were made using the same method. This method of making master molds was later abandoned, in a foreshadowing of current microelectronic wafer scale-up issues, because the process could not produce the 30 cm and larger discs introduced commercially in 1927. This process was a forerunner of the compact discs (CDs) which have sputtered aluminum coatings. Since 1950, a variety of additional plasma-related deposition methods has been developed.

23.1.2 Product Applications of Thin Films

Thin films may be deposited by the purely *physical evaporation* or *sputtering* processes discussed in this chapter, or by the *plasma chemical vapor deposition* (PCVD) processes discussed in chapter 24. Irrespective of how they are deposited, thin films serve a variety of functions that include the following:

(1) electrically conductive or insulating layers for microelectronic chip fabrication;
(2) reflective and anti-reflective coatings for architectural and automotive glass;
(3) optical and magnetic recording media;
(4) decorative and ornamental brightwork;
(5) protection against abrasion and corrosion;
(6) oxygen barrier coatings for food packaging;
(7) optical components and coatings; and
(8) biomedical products.

We will discuss in detail here a few of the most commercially important of these thin-film applications.

The industrial sputtering of thin films is done under vacuum at pressures low enough that sputtered atoms are transported directly from the target to the workpiece with few or no collisions. The long mean free paths required imply pressures below 6.7 Pa (50 mTorr). Plasma-based sputtering methods are rarely operated at pressures lower than 0.13 Pa (1.0 mTorr) because it is difficult at such pressures to generate a plasma dense enough to induce sputter deposition at rates of commercial interest. Other technologies that use space-charge-limited ion-beam–target sputtering are carried out at pressures as low as $\sim 10^{-3}$ to $\sim 10^{-4}$ Pa (10^{-5} to 10^{-6} Torr).

23.1.2.1 Microelectronics

The microelectronics industry could not exist in its present form without the plasma-assisted deposition of conducting and insulating thin films. In terms of total value-added cash flow of the product, the plasma-assisted deposition of thin films for microelectronic circuit fabrication is the most important application of this technology (Anonymous 1991). These films are characteristically from a few

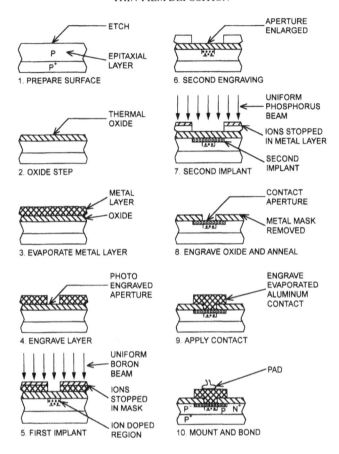

Figure 23.2. Steps in the fabrication of an n–p–n device, including implantation of an n^+–p^-–p^+ diode (modified from Wilson and Brewer 1973).

tenths to several microns thick, and may cover the surface of wafers up to 30 cm in diameter.

Microelectronic circuits are fabricated in a series of steps that build up multiple layers, an example of which (modified from Wilson and Brewer 1973) is illustrated in figure 23.2. This figure shows steps in the fabrication of an n–p–n device, including the ion implantation of an n–p–p diode. An example of the structure of microelectronic circuit elements in a wafer base layer is shown in figure 23.3, which has been modified from Shaw (1993). This figure illustrates the manner in which thin-film and ion-beam implantation technologies are used to produce resistors, transistors, diodes, and capacitors.

The characteristic microelectronic circuit illustrated in figure 23.4 (Pramanik 1995) consists of several layers of insulators and two layers of electrical

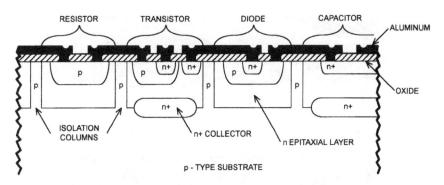

Figure 23.3. Structure of basic microelectronic circuit elements (modified from Shaw 1993).

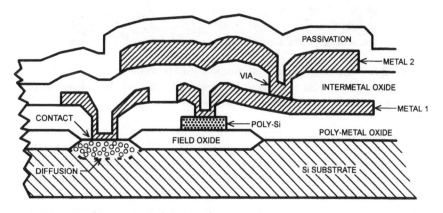

Figure 23.4. Schematic cross section of a microelectronic circuit showing two metal layers connected by vias (from Pramanik 1995).

conductors. The latter must be connected by electrically conducting *vias* (vertical penetrations through a layer) at strategic points through adjacent insulating layers. The lowermost layer is connected by additional vias to semiconducting elements mounted in or on the silicon base layer. The insulating layers have been penetrated by plasma etching as a first step in forming vias that connect conductors above and below them. A scanning electron image of the cross section of a five-level system of interconnects in a microelectronic circuit is shown in figure 23.5, taken from Ohba (1995). The white bar at the lower center is 1 μm long. This illustration shows five metallic layers, M1 through M5, four of which are interconnected by the tungsten vias W1, W2, and W3.

The characteristic size of the smallest electronically active element in circuits such as those illustrated in figures 23.2 through 23.5 is called the *design rule*, about which more will be said in chapter 25. Until the advent of the Pentium[R]

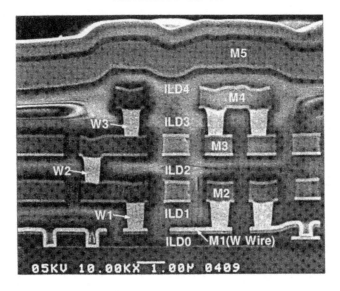

Figure 23.5. Scanning electron microscope (SEM) sectional image of five-level interconnects using blanket tungsten plugs. Features labeled W1, W2, etc are tungsten vias between metal layers M1, M2, etc (from Ohba 1995).

chip, the design rules for microelectronics were about 0.5 microns or larger. Competitive pressures within the microelectronic industry have pushed these design rules to values in the range from 0.2–0.3 μm at the time of writing. These smaller design rules permit faster, more compact chips to be fabricated, each of which contains many more active circuit elements than preceding generations of microelectronic chips. The issues raised by progressively decreasing design rules affect etching processes more profoundly than deposition, so their discussion will be reserved for chapter 25.

Sputter deposition is used to produce a variety of thin films needed in microelectronic circuit fabrication. Both non-reactive argon sputtering and reactive nitrogen sputtering have been employed in depositing a variety of metals that include titanium, titanium nitride, and aluminum–copper alloys. Increased collisional scattering of sputtered atoms at pressures above 1.3 Pa (10 mTorr) slows the deposition rate and leads to glancing angles of deposition as well as a porous, less dense film. These sputter deposition processes therefore take place at operating pressures less than 0.7 Pa (5 mTorr) in order to maximize the film density at an acceptable deposition rate.

A widely used barrier material for interconnects has been titanium nitride, which can be deposited either by sputter deposition or by chemical vapor deposition (CVD). These barriers prevent mobile ions within the microelectronic chip from traveling to sensitive junctions, and prevent reactive gases from reaching exposed oxides. CVD and plasma chemical vapor deposition (PCVD) are competitive technologies with plasma or ion-beam sputtering, and the particular approach chosen depends on a variety of technical and institutional factors.

23.1.2.2 Architectural and Automotive Glass

Many important applications of thin-film deposition exist outside the microelectronic industry. A major application of plasma-related sputtering technology is the coating of glass and plastics to improve their appearance or reflective characteristics. One of the largest applications in terms of annual cash flow is the production of coated architectural and automotive glass, using plasma-assisted sputtering. Aesthetic and energy conservation considerations have led architects of major buildings, as well as automotive designers, to specify sputter-coated window glass. Occasionally, coated architectural glass is chosen primarily for its decorative value in tall office buildings. However, many plasma sputter-coated thin films have the ability either to reflect or absorb infrared radiation, thus leading to effective control of solar energy and improved efficiency of heating and air conditioning systems.

Plasma-deposited thin films with the appropriate characteristics can reflect incident infrared radiation, thus reducing heat loads in warm climates, and saving air-conditioning expense. Other plasma-deposited thin films are either transparent to or absorb infrared radiation, a characteristic which can be used to trap solar energy and reduce demands on the heating system of buildings in cold climates. Some architectural glass is even more sophisticated, with multiple coatings between parallel panes that in cold climates absorb solar radiation on the outdoor-facing surface, and reflect infrared radiation from the indoor-facing surface, thus effectively trapping solar energy. In warm climates, such multiple coatings can reflect infrared radiation from the outdoor-facing surface, and transmit it through the indoor-facing surface.

The coatings that provide such reflectivity have a thickness that is a significant fraction of the wavelengths absorbed or reflected, in the range 0.5–1.5 μm. The coatings of architectural glass are rarely thicker than this, since they must be thin enough to take advantage of daylight, and reduce the power demands for interior lighting. Coatings that are reflective in the visible part of the spectrum may also be used for decorative or privacy purposes.

An example of double window glazing for architectural applications is shown in figure 23.6(a). In a characteristic application, two panes of glass are separated by a dead space, usually filled with an inert gas such as argon. The inner and/or outer surfaces of the two panes of glass are sputter coated to enhance or reduce the thermal reflectivity of the glass, depending on whether the window is intended for a warm or cold climate.

An industrial flat glass coater manufactured by the BOC Group, Inc. is shown in figure 23.7. This unit is capable of producing several square kilometers of coated architectural window glass per year. Such glass coaters are

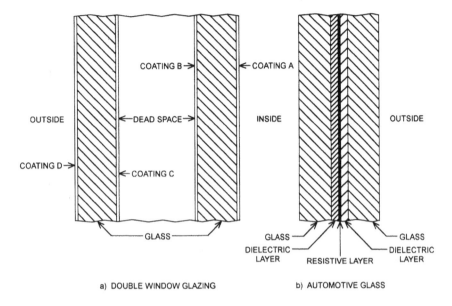

Figure 23.6. Sputtered coatings applied to double window glazing and automotive glass.

among the largest rectangular vacuum systems built for commercial applications. Figure 23.7 illustrates the array of box beams welded to the flat surfaces of the vacuum vessel to withstand atmospheric pressure. The rollers on which the glass moves after being coated are visible at the end of the vacuum chamber, to the lower left.

23.1.2.3 Resistive Coatings for Electrical Heating

A further application of thin-film technology is the deposition of opaque or transparent electrically conductive coatings for resistive heaters. Transparent electrically conductive layers can be included in a sandwich of specialty automotive windshield glass to facilitate defrosting. Such automotive windshields typically have one electrode at the bottom of the windshield and a second running along the top. An electrical current flows through the transparent, electrically conducting thin film, which warms up by ohmic heating to provide immediate defrosting of the windshield.

This application is illustrated in figure 23.6(*b*), in which the interior layer of a sandwich is electrically resistive and optically transparent. This layer heats up when a current flows from the top to the bottom of the windshield, as illustrated in figure 23.8. The surface of automotive glass also can be coated with sputtered thin films designed to reduce the influx of solar energy to the car interior. Such coatings reduce the air-conditioning load on the engine, and passively protect the

Figure 23.7. Photograph of a magnetron sputtering vacuum coater for architectural window glass (courtesy of the BOC Group, Inc., of Fairfield, CA).

interior from high temperatures that result from solar heating of an unoccupied vehicle.

23.1.2.4 Protective and Tribological Applications

Another application of thin-film sputtering is improvement of the *tribological*, or wear characteristics of surfaces. The sputter deposition of a harder or more wear-resistant surface layer protects softer substrates, as in the wax phonograph masters discussed in section 23.1.1. Such coatings can protect easily damaged materials like plastic rods or automotive headlight covers against scratching, sintering, pitting resulting from sliding or rotary motion, or gravel impact.

Further applications of sputtered thin films include the prevention of corrosion on ferrous or other metallic substrates that are used indoors or in other protected settings. The sputter coating of brightwork for ornamental and decorative items is particularly useful when the coatings not only improve the corrosion resistance of the substrate, but also improve the marketability of the item by imparting a more lustrous and pleasing appearance.

Figure 23.8. Photograph of resistively heated automotive window glass under test in a cold chamber (courtesy of the BOC Group, Inc., of Fairfield, CA).

23.1.2.5   Recording Media

Recording media, including compact discs, floppy discs, and magnetic tape, are major markets for a variety of evaporative, ion-beam sputtering, plasma-assisted ion-beam sputtering, and PCVD technologies. The large numbers of individual discs and the long lengths of magnetic tape involved make the economics of the deposition technologies particularly sensitive to the batch processing required by vacuum deposition methods.

23.1.2.6   Niche Applications

Small mirrors and reflective surfaces can be produced for scientific instruments as well as industrial and consumer applications by sputter coating a surface with a thin layer of highly reflective metal. Plasma-assisted thin-film deposition is used to coat plastic toys, automotive brightwork, and other consumer products with highly reflective coatings to enhance their marketability and appearance. Sputter deposition must be conducted at pressures of 1.3 Pa (10 mTorr) or less, which requires relatively expensive vacuum systems and batch processing of the materials being coated. Many potential uses of plasma-based sputter deposition for the decorative and protective coating of consumer goods are economically infeasible, because the requirements for a vacuum system and batch processing may limit the economics of thin-film deposition to relatively high-value items.

23.2 THIN-FILM CHARACTERISTICS

Thin films deposited for ornamentation, protection, oxygen barrier coatings, microelectronic circuit fabrication, or for modifying the optical properties of a surface have characteristic physical properties that establish their utility. Many of these properties are quantifiable, and some can be found in standard references such as the *Handbook of Chemistry and Physics* (Weast 1988). These quantifiable properties include the thickness, composition, density, electrical resistivity, magnetic properties, internal stress, and emissivity of the film. Sputtered thin films have other difficult to quantify characteristics of commercial importance, including their morphology, color, luster, adhesion, and corrosion resistance. Those properties most important to such large industrial markets as microelectronics, recording media, and glass coating include the film density, electrical resistivity, magnetic properties, and emissivity.

23.2.1 Film Thickness

There is ambiguity in the industrial processing literature about the distinction between a 'thin' film and a 'thick' film. In this text, *thin films* are those less than 10 μm thick, and *thick films* are those between 10 μm and 1 mm in thickness. Thin films, as defined here, may be deposited by the plasma-related processes with which this chapter is concerned. Characteristic applications of microelectronic thin films usually range from 0.2 to 10 μm thick, with most such films having a thickness between 0.5 and 3 μm. Thin plastic films made for industrial and consumer use (an example of which is Saran$^{\circledR}$ wrap) typically are between 10 and 100 μm thick. 'Thick' films of refractory materials which can be deposited by thermal plasma flame spraying or related methods will be discussed in Volume 3.

23.2.2 Film Morphology

The *morphology* of a thin film refers to its nature, roughness, and configuration, and includes as factors the *topography*, *coverage*, and *crystallography* of the film. Thin films are required to have uniformity, constant thickness, and freedom from pinholes and defects for most applications, and these characteristics are related to the film morphology.

23.2.2.1 Topography

The *topography* of a thin film is its three-dimensional surface configuration, analogous to the features of a relief map. In thin-film deposition, size scales from 1 mm down to 1 nm may be of interest in various applications. The different types of microscopy used to examine and record the topography are discussed in section 20.4.

23.2.2.2   Coverage

The *coverage* of a surface by a thin film refers to how it conforms to the substrate on which it is deposited. Three major forms of coverage are illustrated in figure 23.9. Figure 23.9(a) illustrates *conformal coverage*, in which the deposited material may approach the surface either *unidirectionally* (as from a beam or directional bombardment of ions), shown at the upper right; or *isotropically* (as in deposition by a gas), shown at the upper left. In conformal coverage, the deposited material flows after deposition, like paint. During this process, the deposited material is pulled by surface tension forces to form a uniform coating before it hardens. Conformal coverage is rarely observed in the plasma-assisted deposition of thin films.

A second form, *unidirectional coverage*, is illustrated in figure 23.9(b). In this form of coverage, the deposited material approaches the substrate unidirectionally, usually normally, but does not flow after contact with the surface. For normal incidence on a surface, a deposited layer adheres only on horizontal surfaces, without coating the vertical sidewalls. These unidirectional adherent coatings are encountered, for example, in evaporative condensation or sputtered thin-film deposition.

A third form of coverage is *isotropic coverage*, illustrated by figure 23.9(c), in which deposited species approach the surface isotropically and stick where they hit. Isotropic coverage may be encountered in plasma-assisted ion-beam sputtering, in some forms of plasma/cathode sputtering, and in the PCVD process discussed in chapter 24. For a discussion of the angular distribution of sputtered atoms in a magnetron sputtering tool, see Eisenmenger-Sittner *et al* (1995).

The isotropic coverage illustrated in figure 23.9(c) occurs, for example, when monomers produced in a glow discharge plasma arrive at a surface isotropically, and polymerize to form an adherent film on the surface. Such films build up their maximum thickness on the extensive horizontal areas of the substrate. However, when a step or trench exists in the substrate, the thickness of material on the sidewall will be proportional to the arrival angle of the species responsible for forming the layer.

At the top of a long vertical trench like that illustrated in figure 23.9(c), the film has a thickness, d, in the horizontal section where the deposited species arrive isotropically over a solid angle 2π. The layer thickness will be $d/2$ at the top sidewall of the trench where the precursor molecules arrive over the smaller solid angle π. The thickness of the deposited film at the bottom of the trench is limited by the arrival angles of species that enter the top of the trench and are able to reach the bottom.

In figure 23.9(c), the arrival angle at the bottom of a long, horizontal trench,

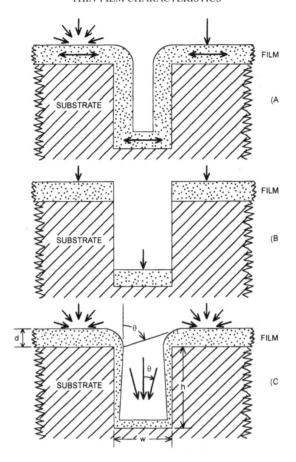

Figure 23.9. Modes of thin-film deposition: (a) a conformal adherent coating; (b) a unidirectional adherent coating; and (c) an isotropic, adherent coating.

with width w and height h, is given by

$$\tan\theta = \frac{w}{h}. \tag{23.1}$$

At the bottom of the trench, the thickness of the deposited layer is t_b, and at the top, the thickness is $t_t = d/2$, where the cylindrical (not solid) arrival angle in a plane normal to the trench is $\theta = \pi/2$. On the upper horizontal surface of the substrate the arrival angle is $\theta = \pi$, and the thickness of the deposited layer is $t_s = d$.

From the geometry of figure 23.9(c), the ratio of the thickness at the bottom

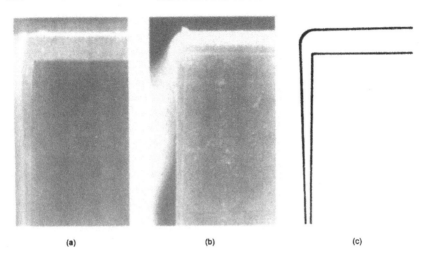

(a)   (b)   (c)

Figure 23.10. Photomicrographs and computer model of step coverage of a plasma-deposited thin film: (*a*) scanning electron microscope (SEM) cross section of SiO$_x$ deposited at 200 °C; (*b*) SEM cross section of SiN$_x$H deposited at 330 °C; (*c*) step coverage calculated for deposition rates proportional to arrival angle (magnification > 5000), from Adams (1983). ©1983 by PennWell Publishing Company.

of the trench to that on the upper horizontal surface is given by

$$\frac{t_b}{t_s} = \frac{\tan^{-1}(w/h)}{\pi} \qquad (23.2)$$

from which it follows that the thickness of the deposited layer at the bottom of the trench is given by

$$t_b = d \frac{\tan^{-1}(w/h)}{\pi}. \qquad (23.3)$$

In the example of isotropic adherent deposition shown in figure 23.9(*c*), the coating of the sidewall ranges from a thickness $d/2$ at the top of the trench, to a thickness given approximately by equation (23.3) at the bottom.

Isotropic adherent films can result from the deposition of polymeric or oxidized thin films for microelectronic applications. An example of the formation of an adherent film is shown in the scanning electron micrographs of the coverage of a step in figure 23.10 (Adams 1983). Image (*a*) is SiO$_x$ deposited at 200 °C, (*b*) is a SiN$_x$:H layer deposited at 330 °C, and image (*c*) is the step coverage calculated for deposition rates proportional to the arrival angle, i.e. isotropic adherent coverage.

Isotropic adherent coverage can lead to undesirable non-uniformities, and even voids in the deposited layer, by mechanisms illustrated in figure 23.11, also taken from Adams (1983). These illustrations are computer simulations of an

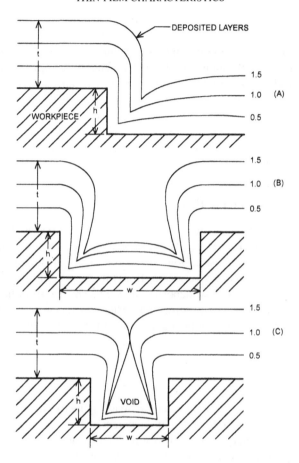

Figure 23.11. Calculated step coverage for isotopic, adherent deposition with the layer thickness *t* normalized to the step height *h*, from Adams (1983): (*a*) an isolated step; (*b*) a wide trench; and (*c*) a narrow trench, illustrating closure of the surface and void formation when *t* equals the trench width, *w*.

isotropic adherent layer deposited on: (*a*) a step, (*b*) a wide trench, and (*c*) a narrow trench. As the upper opening of the trench fills in, the arrival angle and the thickness deposited at the bottom of the trench becomes progressively smaller. Finally, as illustrated in figure 23.11(*c*), closure of the trench occurs when the thickness of the deposited layer is equal to the trench width. In the computer simulation of figure 23.11, the thickness of the deposited layer was normalized to the step height or to the depth of the trench. Voids created by the isotropic adherent trough filling illustrated in figure 23.11 are undesirable in any application, and are particularly so in the fabrication of microelectronic circuits, where heat transfer and reliability problems result.

23.2.2.3   Crystallography

An important element of the morphology of thin films is their *crystallography*, or degree of order on the atomic scale of size. Thin films may be *amorphous* or completely lacking in ordered, crystalline structure. If *polycrystalline*, the film is organized as many small, crystalline grains packed together as a solid that presents crystalline facets at the surface. If the film is a *single crystal*, it is highly organized into a single-crystalline solid without grain boundaries. If material continues to be deposited on a surface to form a *thick film* or *ingot* in single-crystalline form, the process is referred to as *epitaxy*.

23.2.3   Thin-Film Composition

Thin films resulting from ion-beam sputtering or plasma-assisted deposition may consist of a variety of materials. These can include *crystalline solids*; *metal films* in either a *crystalline* or *amorphous* state; thin layers of polymers; *insulating oxides*, either deposited as active species from the plasma or formed by oxidation of the substrate; *nitrides*, also deposited as active species from the plasma, or formed by chemical reaction of nitrogen with the substrate material; and *dopants*, or *alloys*, created by adjusting the composition of the active species responsible for deposition.

23.2.4   Film Density

Nearly all applications of thin films require that they be free of *voids*, and possess a density as close as possible to that of the parent, solid material of the sputtering target. A factor that may result in the formation of voids in sputtered material is operation at pressures above a few Pascal (tens of milliTorr). At this pressure, collisions of the sputtered atoms occur between the sputtering target and the workpiece. Such collisions result in the arrival of sputtered atoms at oblique angles on the workpiece surface, illustrated in figure 23.12(*a*), and the formation of voids over trenchlike topography, as illustrated in figure 23.11(*c*).

   Voids in a thin film can degrade its surface morphology, reflectivity, and such bulk transport properties as diffusivity and thermal conductivity. Voids may result in poor adhesion of a thin film to its substrate or to the layer above it and/or create small-scale stress concentrations that cause undesirable changes in its physical properties with time. Voids in a deposited layer may be minimized by several methods. One method is to operate at pressures low enough that sputtered atoms are collisionless between the target and the workpiece. Another method is to ensure normal incidence of the sputtered atoms by using collimation, as in figure 23.12(*b*). Another approach is to maintain the film and substrate at elevated temperatures during deposition (*annealing*). Once formed, voids can be eliminated by annealing or melting the thin film. The use of aluminum reflow sputtering to deal with voids associated with steps and trenchlike features in via

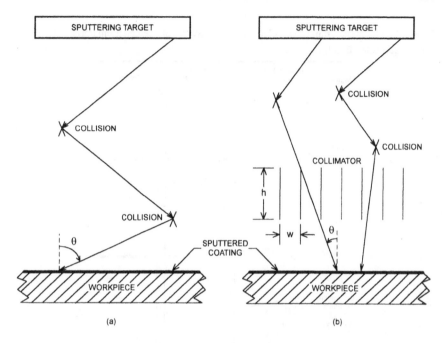

Figure 23.12. Collimation of sputtered atoms in a vacuum sputter coater operating at pressures such that the atom mean free path is less than the distance from the target to the workpiece: (*a*) collisional deposition; (*b*) collisional deposition with collimation.

and interconnect metallization on microelectronic chips has been discussed by Kikuta (1995).

23.2.5 Electrical Resistivity

An important characteristic of thin films, particularly in microelectronic applications, is their *electrical resistivity*, ρ. The material of thin films can be divided into insulators, semiconductors, and metals, as indicated in table 23.1, where these materials are plotted on a vertical scale of bulk electrical resistivity, measured in Ω-m. The metals are those materials with electrical resistivity below 10^{-2} Ω-m; semiconductors are materials with an electrical resistivity between 10^{-2} and 10^8 Ω-m; and insulators are those materials with an electrical resistivity above 10^8 Ω-m. Materials with the highest known electrical resistivities, like Teflon[R], are in the range of 10^{15} Ω-m.

The electrical resistivity of thin films is important to such applications as the heated automotive windshields illustrated in figure 23.8, and in microelectronics. The effective electrical sheet resistivity of silver deposited on an amorphous zinc stannate layer for glass coating applications is shown as a function of silver thickness in figure 23.13 (Arbab 1997). Note that the resistivity decreases faster

Table 23.1. The range of electrical resistivity of selected materials used in plasma-deposited thin films.

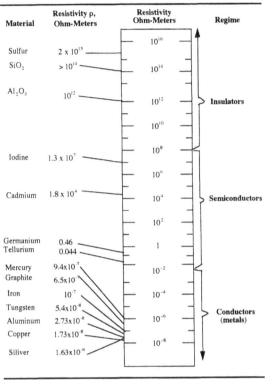

Material	Resistivity ρ, Ohm-Meters	Resistivity Ohm-Meters	Regime
		10^{16}	
Sulfur	2×10^{15}		
SiO₂	$> 10^{14}$	10^{14}	
Al₂O₃	10^{12}	10^{12}	Insulators
		10^{10}	
		10^8	
Iodine	1.3×10^7	10^6	
Cadmium	1.8×10^4	10^4	Semiconductors
		10^2	
Germanium	0.46	1	
Tellurium	0.044		
Mercury	9.4×10^{-7}	10^{-2}	
Graphite	6.5×10^{-7}		
Iron	10^{-7}	10^{-4}	
Tungsten	5.4×10^{-8}		
Aluminum	2.73×10^{-8}	10^{-6}	Conductors (metals)
Copper	1.73×10^{-8}		
Siliver	1.63×10^{-8}	10^{-8}	

than linearly at small thickness, and does not approach the bulk resistivity of silver, 1.63×10^{-8} Ω-m, until the film thickness is much greater than 35 nm.

23.2.6 Stresses

An important characteristic of thin films is their level of *tensile* or *compressive stress*. Mechanical stress, which is often the result of unrelieved stresses between the film and its substrate, can have a number of undesirable effects on thin films. These effects include the *brittle failure* of materials; the *delamination* or *peeling* of a thin film as the result of separation at its interface with its substrate; and the peeling of thin films when their thickness exceeds a critical value. A serious stress-induced problem is the formation of voids in deposited metal *vias* between metallic layers of a microelectronic circuit.

Compressive or tensile mechanical stresses in thin films, and between thin films and their adjacent layers, originate from a variety of sources. One source in

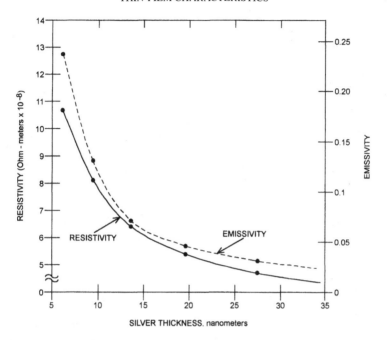

Figure 23.13. Emissivity and resistivity of a thin silver film as functions of the film thickness in nanometers (Arbab 1997).

microelectronic circuits is a difference in thermal expansion coefficients between a thick silicon substrate and the material of a thin film. Thin-film materials with a thermal expansion coefficient smaller than that of a thick silicon substrate will be forced into tension on heating, and into compression on cooling. Thin-film materials with thermal expansion coefficients greater than that of silicon will show the opposite effect.

The deposition of thin films under non-equilibrium conditions may also cause mechanical stress. When the conditions of deposition are such that a new layer of film is deposited before the underlying atoms have had time to diffuse into their equilibrium positions, an *intrinsic stress* will be produced in the film. This stress may be either compressive or tensile depending on the details of the process and materials. Changes in the physical or chemical composition of the thin-film material also may result in changes in stress. Silicon-based glasses, for example, can absorb relatively large amounts of water from ambient air, and develop high compressive stresses. Solid state chemical reactions or phase changes in a thin film normally involve volume changes that also cause stresses. Gross changes of the material of the thin layer, such as plastic flow in metals, can relieve stresses to some extent. Most effects of stress on thin films are detrimental, and are of primary relevance to microelectronic applications. Stresses, however, can produce

Table 23.2. Properties of selected materials used in vacuum sputter-deposited low emissivity coatings (Arbab 1997).

Function	Material
Protective overcoat	Titanium oxide, silicon nitride, tin oxide
Anti-reflection	Indium oxide, indium tin oxide, zinc oxide, tin oxide, zinc stannate, bismuth oxide, silicon nitride
Primer	Copper, titanium, nickel–chromium alloys, zinc, indium
Low emissivity	Silver, gold

delamination of films used for decorative or protective purposes, as in coated glass, and automotive bumpers or fender panels.

23.2.7 Emissivity

The *emissivity* is an important characteristic of automotive and architectural glass, which has been well documented (see, for example, Arbab 1997). When an electromagnetic wave propagates through a material, reflects from its substrate, and is incident on its surface from beneath, the wave may experience constructive or destructive interference. Matching boundary conditions at the surface yields a relation between the emissivity and the electrical resistivity of the film.

Low-emissivity window glass has a complex, multilayered structure that serves multiple functions. These functions include achieving a low emissivity (the *noble metal layer*); preventing damage to the metal layer by active species in the sputtering plasma (the *primer*); providing a satisfactory coloration by optical interference effects in the visible; preventing reflection (the *dielectric* or *anti-reflection* layers); and protecting the surface from marring and scratching (the *protective overcoat*). The adhesion of these layers to each other and the substrate is effected by choosing compatible materials from table 23.2, and by exposing the surfaces to active species of the plasma.

Figure 23.14 illustrates the sequence of layers used in some commercial *noble metal* (copper, silver, or gold) coatings of architectural glass. Early commercial products featured a single noble metal layer as in figure 23.14(a); more recently, dual-layer coatings with improved performance have been developed and are illustrated in figure 23.14(b). The materials widely used in vacuum sputter-deposited low-emissivity coatings are listed in table 23.2 (Arbab 1997). An example of an experimental low-emissivity single-layer metallic coating is illustrated in the cross-sectional transmission electron microscope (TEM) image in figure 23.15 (Arbab 1997). Note the 30 nm scaling bar in the lower right-hand corner.

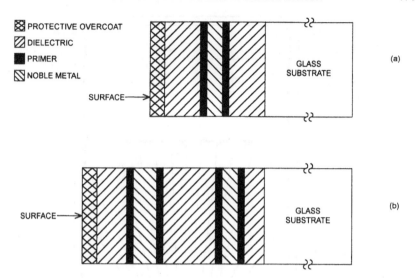

Figure 23.14. Thin coatings applied to window glass for emissivity control: (*a*) single noble metal coating; and (*b*) dual noble metal coating.

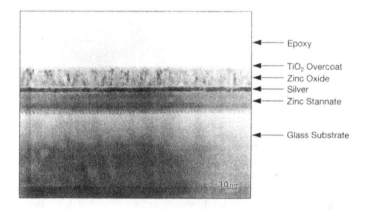

Figure 23.15. A cross-sectional transmission electron microscope (TEM) image of an experimental low-emissivity single-layer metallic coating (Arbab 1997). Note the 30 nm scaling bar in the lower right-hand corner.

23.3   DEPOSITION BY EVAPORATIVE CONDENSATION

Over the past 200 years, thin-film deposition methods have been developed which do not require the assistance of a plasma. Many of these methods, however, use vacuum systems, ion sources, and other plasma-related technologies in their operation. An industrial plasma engineer may be called upon to use these methods

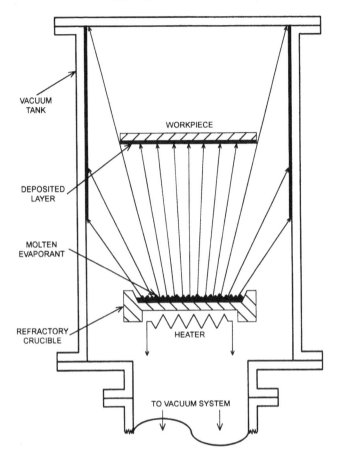

VACUUM
TANK

WORKPIECE

DEPOSITED
LAYER

MOLTEN
EVAPORANT

REFRACTORY
CRUCIBLE

HEATER

TO VACUUM SYSTEM

Figure 23.16. Schematic drawing of a vacuum system for the evaporative deposition of thin films, containing a molten evaporant and a heated crucible.

for special applications. For completeness therefore, selected deposition methods that use many of the same technologies as those used in plasma-assisted thin-film deposition are included in this and the next section of this chapter.

23.3.1 Evaporative Deposition Systems

After the introduction of large-volume vacuum systems suitable for industrial use in the early 20th century, high-temperature *evaporative deposition* systems such as that illustrated in figure 23.16 were developed. These systems rely on putting the workpiece in a large vacuum tank below 1.3 Pa (10 mTorr), at which pressures the mean free paths of individual atoms are comparable to or greater than the dimensions of the vacuum chamber. As illustrated in figure 23.16, the *evaporant* (normally a pure metallic element) is heated in a refractory crucible to a temperature at or above its melting point. At such temperatures, individual atoms evaporate at a sufficiently rapid rate to coat the interior of the vacuum chamber, including the workpiece, with a thin film of the evaporated material. The temperature required for evaporative deposition is normally above the melting point but below the boiling point of the material.

23.3.2 Physical Data for Evaporative Deposition

Table 23.3 lists selected elements, taken from Weast (1988) which may be used for high-temperature evaporative deposition, with their chemical symbol, atomic number, atomic weight, and melting and boiling points in Kelvin. The vapor pressure of the molten evaporant illustrated in figure 23.16 is governed by the *Clausius–Clapeyron equation*,

$$\frac{\mathrm{d}p}{\mathrm{d}T} = \frac{pH}{RT^2} \tag{23.4}$$

where p is the pressure in Pascal, T is the temperature in Kelvin, R is the gas constant, and H is the change in latent enthalpy of vaporization per mole. Equation (23.4) can be rearranged into the form

$$\frac{\mathrm{d}p}{p} = \frac{H\,\mathrm{d}T}{RT^2}. \tag{23.5}$$

Both sides of equation (23.5) can be integrated to yield a relationship between the vapor pressure and the temperature of the heated material,

$$\ln\left|\frac{p}{p_0}\right| = -\frac{H}{RT}\bigg|_{T_1}^{T_2} - \frac{H}{R}\left(\frac{1}{T_2} - \frac{1}{T_1}\right) \tag{23.6}$$

or

$$p = C\exp(-H/RT) \tag{23.7}$$

where C is a constant for a given material and reference pressure.

23.3.3 High-Temperature Evaporative Deposition

The materials used for evaporative deposition are placed in an electrically resistance heated refractory crucible, as illustrated in figure 23.16. The temperature of these materials is raised to the point at which evaporated atoms form a uniform coating of the desired thickness on the workpiece during an economically acceptable length of time. Unless direct sublimation of the evaporant takes place, the evaporant is molten, and the workpiece must be located above it. The evaporated atoms travel in straight lines, so that only workpieces with a relatively flat or weakly undulating surface can be uniformly coated with a layer of constant thickness.

Table 23.3. Melting and boiling points of selected elements (Weast 1988).

Element	Symbol	Atomic Number	Atomic Weight	Melting Point (K)	Boiling Point (K)
Aluminum	Al	13	26.9815	933.52	2740
Beryllium	Be	4	9.012 18	1551	3243
Bismuth	Bi	83	208.980	544.5	1833
Boron	B	5	10.81	2352	2823
Carbon	C	6	12.011	3925	—
Chromium	Cr	24	51.966	2130	2945
Cobalt	Co	27	58.9332	1768	3143
Copper	Cu	29	63.546	1351	2840
Gold	Au	79	196.967	1337.58	3081
Iodine	I	53	126.905	386.7	457.5
Iridium	Ir	77	192.22	2683	4403
Iron	Fe	26	55.847	1808	3023
Lead	Pb	82	207.2	600.65	2013
Lithium	Li	3	6.941	453.69	1615
Magnesium	Mg	12	24.305	922	1363
Manganese	Mn	25	54.938	1517	2235
Mercury	Hg	80	200.59	234.28	629.73
Molybdenum	Mo	42	95.94	2890	4885
Nickel	Ni	28	58.69	1726	3005
Niobium	Nb	41	92.9064	2741	5015
Platinum	Pt	78	195.08	2045	4100
Potassium	K	19	39.0983	336.4	1033
Silicon	Si	14	28.0855	1683	2628
Silver	Ag	47	107.868	1235	2485
Sodium	Na	11	22.9898	370.96	1156
Sulfur	S	16	32.06	385.95	717.82
Tantalum	Ta	73	180.9479	3269	5698
Tin	Sn	50	118.71	505.12	2543
Titanium	Ti	22	47.88	1933	3560
Tungsten	W	74	183.85	3683	5933
Vanadium	V	23	50.9415	2163	3653
Zinc	Zn	30	65.39	692.73	1180
Zirconium	Zr	40	91.224	2125	4650

When coating uniformity is an issue, the workpiece can be moved on the bottom of a rotating turntable or horizontally past the evaporative crucible to obtain a more nearly uniform coating by motional averaging. The requirement for batch processing in a high-vacuum system tends to limit this method of thin-film deposition to relatively high-value items such as mirrors, metalized wrapping papers, and optical components.

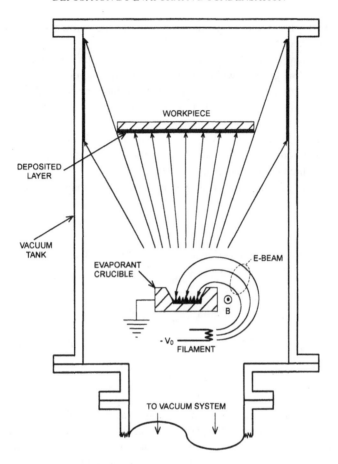

Figure 23.17. Schematic diagram of an *evaporative condensation* coating system containing a molten evaporant heated by an intense electron beam.

23.3.4   Electron-Beam Evaporative Deposition

The *electron (e)-beam evaporative deposition* system illustrated in figure 23.17 has been developed more recently.   It operates at pressures below 1.3 Pa (10 mTorr), where the mean free paths of evaporated atoms are comparable to or larger than the distance between the evaporant crucible and the workpiece. In this device, the material to be evaporated is heated by an energetic electron beam emitted by a filament below the crucible illustrated in figure 23.17. A magnetic induction of about 10 mT near the crucible causes the electron beam from the filament to follow a curved 270° trajectory. This trajectory has a radius of curvature that increases as the electrons gain energy, until they finally arrive at the molten metal in the evaporant crucible, which serves as an anode. Impurities emitted by the hot filament do not contaminate the molten evaporant or the workpiece in this geometry.

The filament is maintained at potentials of several kilovolts to several tens of kilovolts below ground, while the evaporant crucible and the molten metal it contains are at ground potential. The magnetic induction responsible for bending the electron beam can be provided by permanent magnets, but many electron-beam evaporative deposition systems use electromagnets. Electromagnets allow the magnetic induction to be varied, and this makes it possible to sweep the electron beam across the surface of the molten evaporant, thus improving the uniformity of the heating and evaporation process.

The radius of curvature of an electron beam in a magnetic induction B is given by

$$R = \frac{mv}{eB} \quad \text{m} \tag{23.8}$$

where the velocity of the electrons, v, in the magnetic field is given in terms of the electron energy, E', in eV as

$$v = \sqrt{\frac{2eE'}{m}} \quad \text{m/s.} \tag{23.9}$$

Substituting equation (23.9) into equation (23.8) gives the instantaneous radius of curvature for electrons

$$R = \frac{1}{B}\sqrt{\frac{2mE'}{e}} \quad \text{m.} \tag{23.10}$$

For a typical e-beam evaporative deposition system, the required magnetic induction is about 10 mT. The power dissipated in such e-beam deposition systems is characteristically on the order of a few kilowatts to a few tens of kilowatts. Electron-beam evaporative deposition systems similar to those in figure 23.17 are widely available commercially, and are used for coating relatively high-value items with uniform reflective and/or electrically conducting thin films.

23.3.5   Thermal Plasma Spraying

Other physical methods for depositing coatings that operate at 1 atm will be discussed more fully in Volume 3. These methods use high-energy density DC arcs or inductive RF plasma torches, such as the arcjet plasma spraying tool illustrated in figure 23.18. In this tool, an axisymmetric arcjet of the kind discussed in section 10.7.3 of Volume 1 (Roth 1995) is operated with helium or argon as the working gas. A fine power of the material to be deposited is fed into the plasma, where it melts. The molten droplets travel downstream, and impinge on a workpiece that moves past the exhaust plume at a rate calculated to produce a deposited layer of the desired thickness. The thickness of this layer builds up as molten feed material re-condenses on the surface of the workpiece, to produce a thick film ranging from 0.1–2.0 mm thick.

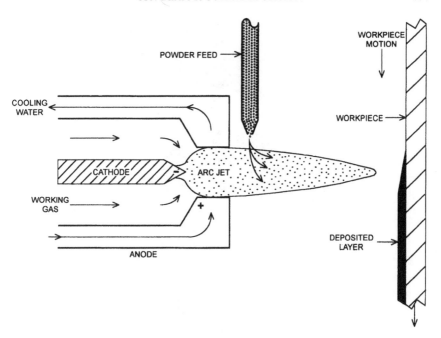

Figure 23.18. Schematic diagram of deposition by plasma spraying. Powder is fed into the arcjet plasma, melted, and deposited in a layer of the desired thickness on a workpiece moving past the arcjet.

A related thermal plasma spraying method that produces thick films is illustrated in figure 23.19. This approach uses an inductive plasma torch, discussed in section 11.4 of Volume 1, in which powdered feed material is injected at the top, where it is melted as it flows through the plasma. The molten feed material coats a workpiece at the bottom, which moves by at a rate determined by the desired thickness of the deposited layer. Plasma coating with the inductive plasma torch is conducted at background gas pressures that range from 1 atm to below a tenth of an atmosphere. More coverage of thermal plasma coating processes will be presented in Volume 3.

23.4 ION-BEAM SPUTTER DEPOSITION

Another physical method to deposit thin films that does not require a plasma is *ion-beam sputter deposition*, illustrated in figure 23.20. In such a system, a space-charge-limited ion source of the kind discussed in chapter 6 generates a beam of ions with energies between 1 and 20 keV. This energy is selected to lie near the sputtering yield maximum of the *sputter plate* (*target*) on which the ions impinge, as discussed previously in section 14.5.

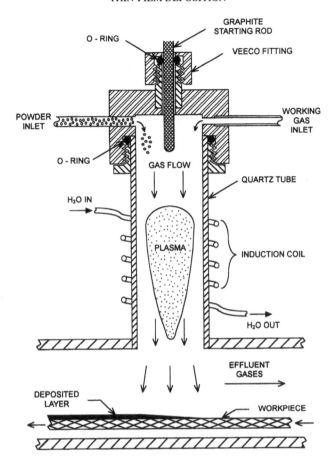

Figure 23.19. Schematic diagram of an RF plasma torch for thin-film deposition, with a workpiece moving past the torch position at the bottom of the illustration. Fine powder is injected into the gas flow at the top, melted in the plasma, and deposited on the workpiece below.

The resulting *sputtering yield* (sputtered atoms per incident ion) may be greater than unity, and can be further increased by causing the ion beam to impinge on the target at a shallow angle. As was pointed out in section 14.5.4, an angle of incidence exists at which the sputtering yield curve reaches a maximum. Ion-beam sputtering deposition is conducted at the same low pressures as other vacuum deposition methods just discussed, typically below 1.3 Pa (10 mTorr). At these pressures, the mean free paths of the beam ions and sputtered atoms are equal to or greater than the distance from the ion source to the target, and from the target to the workpiece.

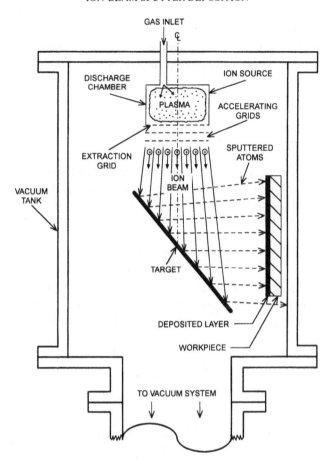

Figure 23.20. Schematic diagram of an ion-beam sputtering system for thin-film deposition. An energetic ion beam impinges on an angled target, sputtered atoms from which are deposited on the workpiece to the right.

23.4.1 Ion-Beam Sputter Deposition Rate

The high-temperature and e-beam evaporative deposition methods have deposition rates that are a strong function of the temperature of the evaporant. The vapor pressure of the evaporant is, in turn, a strong function of absolute temperature given by equation (23.7). The deposition rate due to ion-beam sputter deposition, illustrated in figure 23.20, is limited by the ion flux and hence by the space-charge-limited current density of ions available.

23.4.1.1   Ion Flux

From chapter 3, the Child law space-charge-limited current density of the source illustrated in figure 23.20 is given by equation (3.143),

$$J_c = \frac{4\varepsilon_0}{9}\sqrt{\frac{2e}{M}}\frac{V_0^{3/2}}{d^2} \qquad \text{A/m}^2 \qquad (3.143)$$

where V_0 is the potential difference between the source accelerating grids separated by the distance d.

In chapter 6 it was pointed out that the electric field between the accelerating grids of space-charge-limited ion sources is limited by sparking and electrical breakdown to a maximum of 10 kV/cm. This maximum electric field, in the presence of UV photoemission and/or a nearby plasma, has the approximate value

$$E_{max} = \frac{V_0}{d} \approx 10^6 \text{ V/m}. \qquad (23.11)$$

Substituting equation (23.11) into equation (3.143) yields an expression for the maximum ion flux from a space-charge-limited ion source, the accelerating electric field of which is limited to the value given by equation (23.11),

$$\phi_c = \frac{J_c}{e} = \frac{4\varepsilon_0}{9e}\sqrt{\frac{2e}{M}}\frac{E_{max}^2}{V_0^{1/2}} \qquad \text{ions/m}^2\text{-s}. \qquad (23.12)$$

23.4.1.2   Target Erosion Rate

The flux of atoms sputtered from the target (or the *erosion velocity*) by the ion flux of equation (23.12) may be calculated from the atomic number density of the atoms in the target,

$$n_s = \frac{\rho N_A}{A} \qquad \text{atoms/m}^3. \qquad (23.13)$$

In equation (23.13), ρ is the mass density of the target in kg/m^3, N_A is Avogadro's number, 6.022×10^{26} atoms/kg-mole, and A is the average atomic weight of the target material, in kg/kg-mole.

Equations (23.12) and (23.13) may be combined to yield the *erosion velocity* of the target due to sputtering. This velocity is comparable to, but an upper bound on, the deposition rate if a majority of the sputtered atoms are deposited on the workpiece. This target erosion velocity is given by

$$v_E = \varepsilon\frac{\phi_c}{n_s} = \frac{4\varepsilon\varepsilon_0 A}{9e\rho N_A}\sqrt{\frac{2e}{M}}\frac{E_{max}^2}{V_0^{1/2}} \qquad \text{m/s}. \qquad (23.14)$$

The erosion velocity given by equation (23.14) is a function of the energy-dependent sputtering yield, ε.

23.4.1.3 Deposition Rate

Consider the ion-beam sputtering deposition of aluminum, for which $A = 27$ kg/kg-mole, and $\rho = 2702$ kg/m^3. Assume that the maximum electric field between the accelerating grids of the ion source is $E = 1$ MV/m. These assumptions yield for the erosion velocity of aluminum,

$$v_E = 1.085 \frac{\varepsilon}{V_0^{1/2}} \quad \mu\text{m/s}$$

$$= 65 \frac{\varepsilon}{V_0^{1/2}} \quad \mu\text{m/min.} \tag{23.15}$$

If the acceleration potential for ions is $V_0 = 5000$ V, and the sputtering yield for aluminum is one atom per incident ion ($\varepsilon = 1$), the erosion velocity due to sputtering is 0.92 μm (9200 Å)/min. If few of these sputtered atoms are lost on their way to the workpiece, this would represent a high deposition rate in most industrial applications.

23.4.2 Physical Processes In Sputter Deposition

The sputtering process has been discussed in section 14.5 of this volume. Physical processes relevant to sputter deposition have been covered in depth in standard references, including Vossen and Kern (1978), Mort and Jansen (1986), and Smith (1995). As illustrated in figure 23.21, sputtering occurs when an incident particle of mass M_1 approaches the surface of a cathode or target consisting of particles of mass M_2, and knocks one or more sputtered particles off the surface.

The incident particle may either be *implanted* (which is more likely at energies above 10 keV) or *reflected* from the surface after the interaction. Ion-beam sputtering is usually done at energies higher than 1 keV, and sometimes at energies up to 50 keV. At these energies, the incident ions may eject more than one sputtered atom from the target per incident ion.

23.4.2.1 Target/Cathode Sputtering

Plasma-based *target/cathode sputtering*, the subject of section 23.5, is done at lower ion energies, between approximately 50 and 1000 eV, where fewer than one sputtered atom is produced per incident ion. Examples of sputtering yields for a variety of incident and target materials were presented in section 14.5. Figure 23.22, taken from Wilson and Brewer (1973), is a characteristic example of ion-beam sputtering in which argon ions impinge on copper. Data from several authors are plotted over a wide range of energy. The maximum sputtering yield of this combination peaks at about 20 keV. Figure 23.22 also illustrates a common difficulty with sputtering data in that sputtering yields published by different authors may not agree for the same nominal conditions of sputtering, including ion energy.

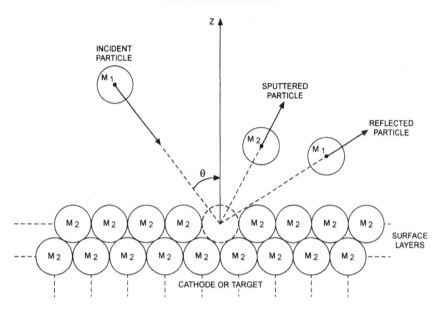

Figure 23.21. The sputtering process when an incident particle of mass M_1 approaches the surface of a cathode or target consisting of particles of mass M_2, and knocks out one or more sputtered particles.

Another important characteristic of the sputtering yield is its strong angular dependence. Figure 14.18, from Molchanov and Tel'kovskii (1961), illustrates the increased sputtering yield that results when 27 keV argon ions impinge on polycrystalline copper at other than normal incidence. In these data, the sputtering yield reaches a maximum at an incidence angle of about 70° from the vertical. To obtain the largest possible sputtering yield, ions should impinge on the sputtering target with an angle near this value.

As discussed in section 14.5.4, the sputtering yield curve can be approximated as a function of incident angle out to angles of about 70° by the following phenomenological relationship:

$$\varepsilon(\theta)\cos\theta \approx \varepsilon(\theta = 0). \qquad (23.16)$$

This angular dependence of the sputtering yield is why, in most industrial *ion-beam* sputtering equipment, the ion beam impinges on the target surface at a shallow angle rather than normal to the surface. In *plasma* sputtering, it is more difficult to arrange for the incident ions to approach the surface at other than normal incidence, and the increased yield from shallow incidence is not normally available.

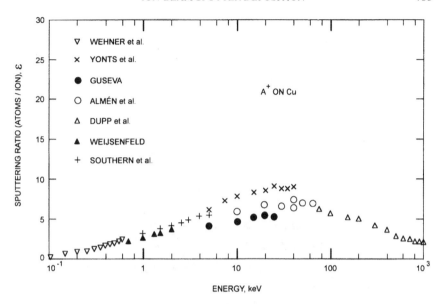

Figure 23.22. A characteristic example of argon ion sputtering of copper. Data from several authors are plotted over a wide range of energy (Wilson and Brewer 1973).

23.4.2.2 Transport To Workpiece

Once sputtered atoms are knocked off the cathode or target, one of two things happens. Either the pressure is so low, as is characteristic of space-charge-limited *ion-beam sputtering* at high vacuum, that a sputtered atom suffers no collisions before it reaches the workpiece. At higher pressures, characteristic of *plasma/cathode sputtering* that results when ions are accelerated across a sheath to a cathode, the gas pressure must be high enough to produce a sufficiently dense DC or RF glow discharge plasma. Such pressures may reach 1.3 Pa (10 mTorr), a level at which a sputtered atom may make more than one collision between the target and the workpiece. This intermediate-pressure plasma sputtering is illustrated in figure 23.12(*a*).

23.4.2.3 Collimated Sputtering

If a sputtered atom collides with the working gas between the cathode and the workpiece, it may arrive with relatively low energies and a wide range of incident angles. These characteristics may produce a sputtered film that is porous or has other undesirable properties, including poor adhesion to the workpiece. A method has been developed to deal with this difficulty in the plasma ion sputtering of titanium or titanium nitride films. In this application, a *honeycomb collimator*

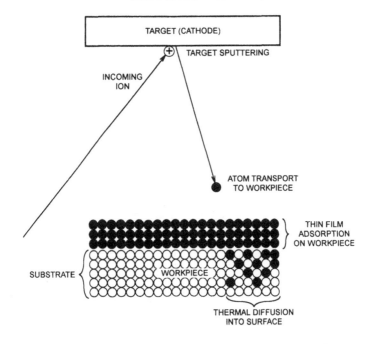

Figure 23.23. Physical processes that may occur after a sputtered atom arrives at a surface.

is inserted between the target and the workpiece, not too far above the latter, as illustrated in figure 23.12(*b*) (see Ryan *et al* 1995).

23.4.2.4 *Adsorption on Surface*

Some physical processes that can occur after a sputtered atom arrives at the surface are illustrated on figure 23.23. The sputtered material may *adsorb* on the workpiece or substrate to form a thin film. After the film has accumulated to a thickness of several monolayers, a physical reaction called *nucleation* may occur. During nucleation, the surface energy of the sputtered material plays an essential role in causing the amorphous stack of sputtered atoms to aggregate into a thin film. This process may proceed further to form a *polycrystalline* or even a *single-crystalline* film.

The newly arrived adsorbed atoms may subsequently *diffuse* into the thin film already present or, in some instances, into the substrate. These atoms may undergo *chemical reactions* that bond the material of the film together or to the substrate. Crystallization or chemical reaction of the sputtered film can be induced by annealing the film at higher-than-ambient temperature or by briefly raising the film above its melting point. The physical processes involved in the formation of thin films are discussed in greater depth by Smith (1995).

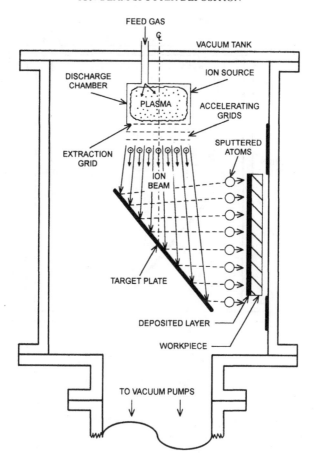

Figure 23.24. An illustration of *inert ion-beam deposition*, in which a Child-law-limited beam of chemically inert ions sputters atoms from a target. These atoms, in a secondary process, impinge on a workpiece and deposit a thin film.

23.4.3 Inert Ion-Beam Deposition

In the process of *inert ion-beam deposition*, a space-charge-limited beam of chemically inert ions (such as argon or helium) at a pressure below 0.13 Pa (1 mTorr) sputters atoms from a target. These atoms, in a secondary process, impinge on a workpiece and deposit a thin film. This technology is illustrated in figure 23.24. The target plate is placed at an angle greater than 45° from the vertical to maximize the sputtering yield, and the workpiece is located to intercept the maximum sputtered atom flux.

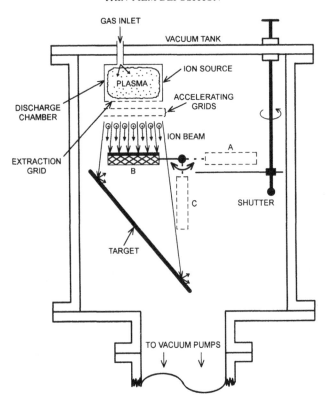

Figure 23.25. A rotating workpiece holder and shutter added to improve film quality in an ion-beam–target sputter deposition system.

23.4.3.1   Conditioning of Target and Workpiece

Additional procedures can be used if ion-beam–target sputter deposition on unconditioned surfaces is not adequate to produce a high-quality layer. A feature added to improve film quality may include a rotating workpiece holder and shutter, illustrated in figure 23.25. After the workpiece is installed in location A, the surface to receive deposition is protected from sputtering by a shutter. In this configuration, the ion beam '*conditions*' the target by removing oxides, adsorbed gases, surface films of vacuum or machining oils, and other contaminants. In the next phase, the shutter is rotated out of the way, and the workpiece is rotated into the ion beam at position B to condition its surface before deposition. The usual result of conditioning the workpiece surface is to greatly increase its surface energy and the adhesion of the deposited film. In the final phase, the conditioned workpiece is rotated to location C that results in the highest deposition rate, where it remains until the film has the desired thickness.

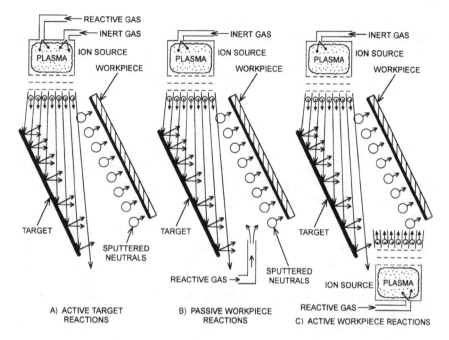

Figure 23.26. Three versions of *ion-beam reactive sputter deposition*: (*a*) deposition of a chemical compound by reactions between the ions and the target material; (*b*) passive target sputtering by an inert ion beam, followed by chemical reactions on the workpiece surface induced by reactive feed gas flowing over its surface; and (*c*) a two-beam configuration that sputters a layer of target atoms with an inert ion beam, and reacts these target atoms with ions reaching the workpiece from a second source.

23.4.3.2   Reactive Sputter Deposition

Sometimes it is desired to create a coating of oxides, nitrides, or other chemical compounds with ion-beam sputter deposition alone, and without the complication of an ambient plasma at the target or workpiece. If the desired heterogeneous chemical reactions take place on the workpiece surface, this can be done by *ion-beam reactive sputter deposition*, three reactor configurations for which were discussed in section 17.3.1.2, and are illustrated in figure 23.26.

23.4.4   Performance Characteristics of Ion-Beam Sputtering

Ion-beam-assisted sputter deposition makes possible independent control of most inputs to the sputtering process. The ion energy, ion current density, angle of incidence on the target, and angle of incidence on the workpiece are all independent variables. The primary ion mass and energy can be adjusted independently while keeping the beam current density constant. In addition, the requirement of low-pressure vacuum operation minimizes the number of collisions of the sputtered atoms before their arrival at the workpiece, thus reducing porosity, and improving the adhesion and quality of the deposited layer. The ion and target characteristics can be changed independently of the workpiece, an advantage in those situations in which several successive layers must be deposited on a workpiece.

Since deposition rates normally must be maximized for economic reasons, the space-charge-limited ion sources discussed in chapter 6 of Volume 1 (Roth 1995) and in section 17.3.1 are used. These sources operate at ion energies from 500 eV to several keV, and are above those readily attained with plasma/cathode sputter deposition. Ion current densities at the target may be on the order of 1.0 mA/cm^2, with power densities on the sputtering target ranging from a few tenths to tens of W/cm^2. This target heat load must be removed by cooling, which may require an additional subsystem. However, this heat load does not appear on the workpiece, which can be maintained near room temperature. The diameter of the widely used *Kaufman ion sources* may exceed several tens of centimeters. The diameter of some other types of ion source used for the production of ions of non-volatile materials (see chapter 6) may be restricted to beams only a few millimeters in diameter.

The deposition rates on the workpiece of these systems are at least several tens of nm/min, otherwise they would not be of commercial interest. Some deposition methods exceed 100 nm (1000 Å)/min, the benchmark for commercial interest in depositional technologies. In microelectronic, but not necessarily in other applications, a uniformity of deposition across the workpiece of 5% or less is desirable. This uniformity may be difficult to achieve with ion-beam-assisted sputtering technology, especially with wafers 30 cm or more in diameter. The uniformity of deposition may be improved by *motional averaging*, as discussed in section 18.4.

23.4.5   Primary Ion-Beam Deposition

An interesting but rarely used ion-beam-assisted deposition technique is *primary ion-beam deposition*, illustrated in figure 23.27. In this process, the 'primary' ion beam is generated in high vacuum by a space-charge-limited source. The beam consists of ions with energies E_c, below the self-sputtering yield of 1.0, of the material that is to be deposited on the surface. *Self-sputtering* is the sputtering of a material by its own atoms, and for solid materials, E_c ranges from 100 eV to several keV. If the ion energy produces a greater sputtering yield than one atom per incident ion, more atoms will be sputtered than arrive in the form of ions, and net deposition will not occur. At energies below E_c, net deposition of metallic and other materials that are solid at room temperature may occur. The ion energy at which the process is operated is determined by competition between the rapidly falling space-charge-limited ion current density at low ion energies, and the increasing self-sputtering yield at energies approaching E_c. In a characteristic

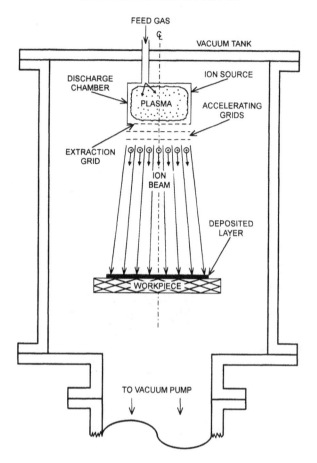

Figure 23.27. Apparatus for *primary ion-beam deposition*, in which a thin film is built up by self-sputtering of the substrate and accumulation of the ion-beam material.

primary ion-beam deposition application, ion energies of about $0.5E_c$ have been found to be optimum.

A reason for continued interest in this technique, in spite of its technical difficulties, is the vision of maskless circuit deposition with the apparatus illustrated in figure 23.28. In this concept, a source of metal or insulator ions produces a beam that is focused to dimensions smaller than the design rule of the circuit being fabricated. The ion beam is then electrostatically deflected (by means discussed in section 5.7.3 of Volume 1) to deposit an electrically conducting circuit or insulating coating on the substrate without masks or etching. An issue with this form of maskless circuit fabrication is the long deposition times that result from the low ion current densities associated with space-charge-limited source operation at the energy E_c. A second issue is the requirement

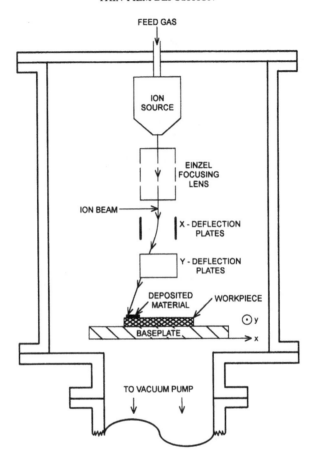

Figure 23.28. Apparatus for maskless circuit deposition using primary ion-beam deposition of an electrostatically deflected and tightly focused ion beam.

for a beam diameter smaller than the current microelectronic design rule. Such a small dimension is very ambitious, since early experiments cited in Harper (1978) achieved an ion-beam diameter no better than 200 μm. However, it may be appropriate to revisit this concept with the application of technology related to modern ion microprobe techniques.

23.5 PLASMA-ASSISTED ION-BEAM SPUTTER DEPOSITION

The broad topic of *plasma-assisted sputter deposition* has been divided into two major groups because of the very different technologies involved. The first technology considered in this section is *plasma-assisted ion-beam sputter*

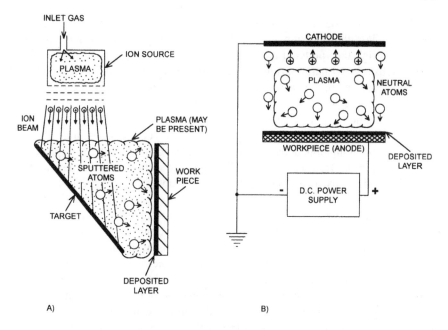

Figure 23.29. Two forms of plasma-assisted sputtering: (a) plasma-assisted ion-beam sputter deposition, in which energetic ions from an external source sputter atoms from a target, which are deposited on a workpiece; and (b) *plasma/cathode sputter deposition*, in which sputtered atoms from a cathode are deposited on the anode/workpiece.

deposition. An apparatus for this process is shown in figure 23.29(a), in which energetic ions from an external source sputter atoms from a target, and the atoms are deposited on the workpiece. An ambient plasma is in contact with either the target or workpiece.

The second major group of sputter deposition technologies is *plasma/cathode sputter deposition*, illustrated in figure 23.29(b) and to be discussed in section 23.6. In this approach to deposition, an energetic ion source and ion-beam target are not present. Instead, the deposited atoms are produced by sputtering a sacrificial cathode with ions accelerated across the sheath of a negative glow plasma. The workpiece may either be mounted on the anode, immersed in the plasma, or located external to the plasma opposite the cathode.

23.5.1   Ion-Beam Plasma Deposition

A variety of heterogeneous chemical reactions may occur during the ion-beam-induced sputtering process, in addition to simple sputtering of an atom from a target and its deposition on a workpiece. The sputter deposition process may require plasma cleaning of the surfaces of the target or workpiece, and may include heterogeneous chemical reactions that involve the plasma active species and/or the working gas. These latter species may react with the target material, the sputtered atoms in transit to the workpiece, or the material of the substrate. Energetic ions or sputtered atoms may additionally induce heterogeneous chemical reactions between adsorbed monolayers of background gas and the target or substrate material. The processes that involve heterogeneous chemical reactions can be characterized as *reactive sputtering*.

Some applications of secondary ion-beam–target sputtering may require surface decontamination, activation, or reactive deposition on the workpiece for which the plasma-free technologies discussed in section 23.4 are not useful. In such applications, the addition of a plasma subsystem may be justified, as illustrated schematically in figure 23.30. In this *ion-beam plasma deposition* configuration, the target and/or the workpiece are surrounded by an ambient glow discharge plasma, characteristically operated at neutral gas pressures of a few tenths of a Pascal (a few milliTorr). The plasma is available to condition and decontaminate the surfaces of the target and workpiece, to increase the surface energy of the workpiece, and to provide highly oxidizing or reducing plasma active species to react chemically with the target and/or the deposited film.

When the working gas pressure exceeds a few Pascal (tens of milliTorr), the *primary beam* ions may have mean free paths great enough to be collisionless before hitting the target, by virtue of their high energy. The *plasma* ions, however (which may have originated by ionization of sputtered target atoms), will suffer multiple collisions before reaching the workpiece, resulting in *ion plating*, to be discussed in section 23.5.5.

23.5.2   Deposition of Sputtered Atoms

For simple sputtering of a target uncomplicated by the presence of a plasma, the energy dependence, angular dependence, yields, sputtering velocity, and other features of the sputtering process were discussed in sections 14.5 and 23.4. Such sputtering is characteristic of ion-beam–target sputtering under vacuum, illustrated in figure 23.20. In such systems, the sputtered atoms are emitted from the target at all angles (but not necessarily isotropically), and as a result, many atoms miss the workpiece and accumulate on the walls of the vacuum vessel. The maximum rate of deposition on the workpiece is no greater than the sputtering rate of the target in beam–target sputtering systems. This maximum deposition rate is given by the *erosion velocity* due to sputtering, equation (23.14) in section 23.4.1.3

23.5.3   Plasma Conditioning of Surfaces

The most elementary form of plasma-assisted sputtering is that in which no chemical reactions occur between the sputtered atoms and the beam ions, the plasma active species, or the workpiece. In *plasma conditioning*, the plasma only

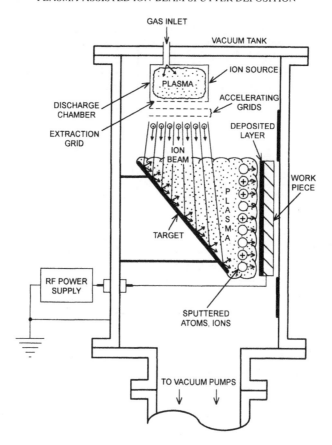

Figure 23.30. An ion-beam plasma deposition configuration, in which the target and/or the workpiece are surrounded by an ambient glow discharge plasma.

affects the target or workpiece surfaces in a way that leads to a more adherent or rapidly deposited thin film. In the *plasma-activated ion-beam-sputtering system* illustrated in figure 23.31, energetic ions arrive at a sputtering target immersed in a background plasma. The presence of a background plasma may improve the sputtering and deposition processes in several ways. These include altering the composition or reducing the thickness of adsorbed monolayers on the surfaces of the target or workpiece, and/or increasing the surface energy of the substrate for deposition.

An additional motivation for using an ambient plasma in thin-film deposition is that the deposition process can be conducted at relatively low temperatures, since the energy flux on the workpiece associated with secondary ion-beam-induced sputtering is relatively low. This is an important factor in microelectronic and other applications where diffusion of the deposited film into the substrate is

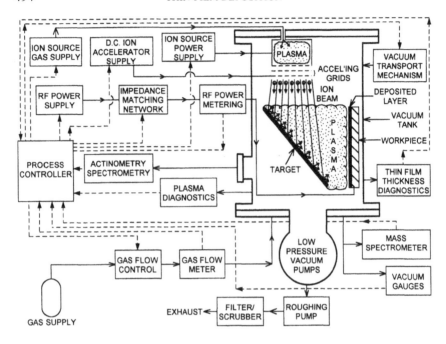

Figure 23.31. A schematic drawing of the subsystems of a plasma-assisted ion-beam sputtering deposition system.

undesirable. It may also be a factor in applications for which thermal stresses have a detrimental effect on adhesion.

23.5.4 Deposition by Heterogeneous Chemical Sputtering

The ambient plasma can affect the energetic ions responsible for sputtering before they arrive at the target, and leave them in an excited state, further ionize them, or disassociate them if in molecular form. These processes can increase the sputtering yield or promote chemical reactions of the incident ions with the target material, the products of which may be sputtered. The ambient plasma also can affect the sputtered species as it travels from the target to the workpiece by fragmenting its molecules, by ionizing individual atoms, or by raising it to excited atomic or molecular states. The usual result of such processes is to make the sputtered material more chemically reactive. This may affect the chemical composition, the density, or the morphology of the deposited layer.

When sputtered material arrives at the surface of the workpiece, it may undergo heterogeneous chemical reactions with plasma active species, with the working gas, with adsorbed monolayers on the surface of the workpiece, or with the material of the workpiece itself. Within the broader category of *heterogeneous chemical sputtering*, the more restricted subject of *reactive sputtering* refers to chemical reactions of the sputtering ions with the substrate of the target or workpiece only. When the chemical reaction products generated at the substrate surface are volatile, the process is sometimes referred to as *chemical sputtering*. However, this term is usually reserved for a form of plasma etching in which energetic ions react directly with the material to be etched.

23.5.5 Deposition by Plasma Ion Plating

A final heterogeneous process that leads to thin-film deposition is *ion plating*. In this process, positive ions that originate as active species of an energetic ambient plasma are deposited on a negatively biased workpiece to form a thin film. This process is analogous to electroplating, with the plasma playing the role of the electrolyte, and the motion of the ions to the cathode that of positive ions in an electroplating bath, hence, *ion plating*.

An ion-beam sputtering apparatus for ion plating is illustrated in figure 23.32, in which atoms of the material to be deposited are produced by sputtering the target, and ionized by the ambient plasma. These ions need not travel directly and without collisions to the workpiece; unlike the other forms of sputter deposition discussed so far, long mean free paths are not required, and ion plating can be (but not always is) operated at relatively high gas pressures. Once the ions are formed in the plasma, they are extracted across its sheath by biasing the workpiece to sufficiently high negative potentials to attract an ion flux that deposits the required thickness in an acceptable time. In this form of ion plating, the target may also be biased to optimize emission of the atoms or ions of interest. A related version of ion plating dispenses entirely with the ion source and target shown in figure 23.32, and relies on a dense glow discharge, arc, or torchlike plasma to produce the required ions from a suitable feed gas.

23.6 PLASMA/CATHODE SPUTTER DEPOSITION

The ion-beam-assisted sputtering technologies for thin-film deposition discussed in the previous section are well established, highly developed, and are found in many research applications and industrial niche markets. However, systems that use ion beams to produce sputtering require an ion source subsystem with its own gas supply, high voltage and discharge chamber power supplies, as indicated in figure 23.31, and operation in the low-pressure regime below about 1 Pa ($\sim$10 mTorr). These requirements can be serious disadvantages in large-scale industrial applications, such as brightwork and architectural and automotive glass production, where output is measured in square kilometers per year.

For these, as well as smaller scale deposition applications, *plasma/cathode sputter deposition* has been developed. This system is illustrated in figure 23.33, and has the plasma-related elements illustrated in figure 23.29(*b*). In this system, the entire ion-beam subsystem is dispensed with, and replaced by ions accelerated

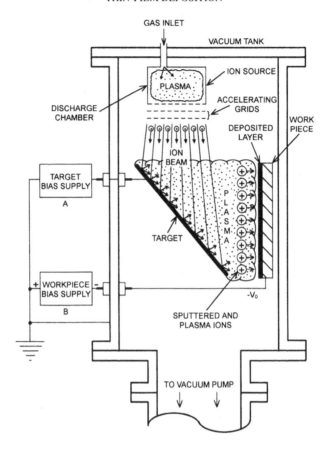

Figure 23.32. An ion-beam-sputtering apparatus for *ion plating*, in which atoms of the material to be deposited are produced by sputtering the target, ionized by the ambient plasma, and are deposited ('plate out') on the workpiece/cathode.

across the cathode sheath, through the cathode fall voltage, to a sacrificial target-cathode where sputtering occurs (Clark 1971, Corbani 1975).

23.6.1 Plasma Sources and Reactors for Plasma/Cathode Sputtering

Except for a few research and minor niche applications, industrial *plasma/cathode sputter deposition* is accomplished with either one of two types of plasma reactor. These reactors are based on the DC *parallel-plate* plasma source discussed in section 16.1.1, or the *magnetron* plasma source discussed in section 16.1.2. The parallel-plate plasma reactors used for plasma/cathode sputtering are discussed in section 17.1.4.2, and the various magnetron plasma reactors are discussed extensively in section 17.4. The application of these sources to industrial

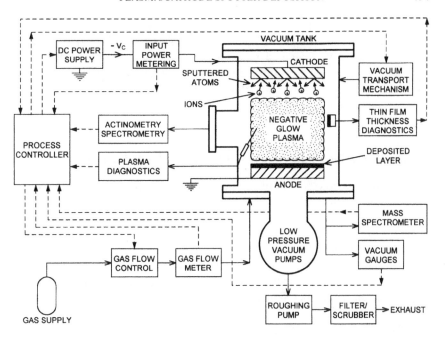

Figure 23.33. A *plasma/cathode sputter deposition system* in which the ion-beam subsystem is replaced by ions accelerated from the glow discharge plasma to a sacrificial target-cathode where sputtering occurs.

sputtering is discussed in Vossen and Kern (1978), Jansen (1986), and Smith (1995).

23.6.2 Plasma/Cathode Sputter Deposition Methods

The industrial utility of plasma/cathode sputtering has given rise to a wide range of operating procedures and configurations. Sputtered atoms from the target/cathode may transit the plasma and become ionized before reaching the workpiece, leading to a version of *ion plating*. These atoms may also be left in excited states that promote a heterogeneous chemical reaction on the surface of the workpiece. The workpiece may be immersed in the glow discharge plasma, located on the anode (*immersed workpiece configuration*), or located at a site remote from the negative glow region of the plasma (*external workpiece* or *remote exposure configuration*). If sputtered atoms become ionized in the glow discharge, the quality and adhesion of the sputtered layer may be improved by *electrode biasing*. Such biasing can lead to *ion plating* or to activation of the workpiece surface by energetic plasma or sputtered ions.

If the sputtered material is ionized or in an excited state, *reactive deposition* may occur on the workpiece surface to form oxides or nitrides from oxygen or nitrogen in the working gas. The sputtered material may also react with the substrate of the workpiece, leading to *activated deposition*. Bombardment of the workpiece by sputtered material in an ionized or excited state, or by active species from the glow discharge plasma, can remove contaminants and/or adsorbed monolayers from the surface of the workpiece. The usual result of activated deposition is a more nearly uniform, better adhering, and higher quality deposited film.

23.6.3 Operating Characteristics of Plasma/Cathode Reactors

The operating characteristics of the plasma/cathode sputter deposition reactors discussed in section 17.4 are similar because in industrial service they use argon as the working gas, and operate at pressures between 0.13 and 1.3 Pa (1 and 10 mTorr). Argon is used because it is the cheapest chemically inert gas available to industry. The pressure range is limited by the requirement of long enough mean free paths for sputtered atoms to travel with minimal collisions from the cathode-target to the workpiece. These reactors use a wide range of metals as cathode-targets to deposit a sputtered coating with the desired characteristics.

The sputtering and deposition rates of various metals relative to copper, when used as cathode-targets in argon plasma/cathode sputter deposition reactors, are roughly indicated in table 23.4. The first column is the relative (to copper) sputtering yield of normally incident 500 eV argon ions on the targets indicated, from Vossen and Cuomo (1978). The second column is the relative deposition rate of a DC planar magnetron, with cathodes of the metals indicated, reported by Waits (1978). Finally, the third column is the deposition reported for a commercial axisymmetric planar magnetron reactor (US Inc. 1989). There is moderately good agreement on the relative sputtering/deposition rates that can be expected from this type of source.

The operating parameters of plasma/cathode sputtering reactors depend on the configuration used and the size of the workpiece to be exposed at one time. In these reactors, the operating voltages can range from a few hundred volts to more than a kilovolt, and the working gas is usually argon. The operating pressure in these reactors ranges between 0.13 and 1.3 Pa (1 and 10 mTorr), and is usually between 0.26 and 0.65 Pa (2 and 5 mTorr).

Magnetic inductions in plasma/cathode magnetron sources must be sufficient to magnetize the electrons; this characteristically requires from 5–20 mT. The average power flux on the cathode target may be from 5–30 W/cm^2. Total power inputs into a single deposition source can range up to several kilowatts. As indicated in figure 17.58, the dimensions of cylindrical magnetrons can range from a few centimeters to several tens of centimeters. Racetrack planar magnetrons used to coat architectural glass can be up to 30 cm wide (including both sides of the racetrack) and 3 m or more long. Deposition rates are characteristically within a factor of five of 1.67 nm/s, or 1000 Å/min. Deposition

Table 23.4. Sputtering and deposition rates of various metals relative to copper, when used as cathodes in argon plasma/cathode sputter deposition reactors (from Vossen and Cuomo 1978, Waits 1978).

Target materials	500 eV Ar$^+$ sputtering yield (Vossen and Cuomuo 1978)	DC planar magnetron deposition rate (Waits 1978)	Commercial S-Gun type deposition (US Inc. 1989)
Copper	1.0	1.0	1.0
Gold	1.02	1.16	—
Silver	1.33	—	2.0
Aluminium	0.45	0.52	0.8
Nickel	0.62	—	0.6
Chromium	0.50	0.64	0.6
Molybdenum	0.34	—	0.55
Titanium	0.22	0.27	0.35

rates tend to be linearly proportional to plasma input power, and to decrease with target–workpiece separation.

23.7 QUALITY ISSUES IN SPUTTER DEPOSITION

Contamination of the deposited layer is reduced by using pure targets or cathodes, by maintaining a non-reactive atmosphere above the workpiece, and by avoiding back-sputtering from surrounding surfaces that may be subject to ion bombardment. In addition, the workpiece can be maintained face-down during deposition, to avoid falling particles of cathode material or the accumulation of dust that might settle under gravity.

Uniform surface coverage by magnetron sputtering reactors can be ensured by motional averaging, and by azimuthal E/B drift of the plasma in axisymmetric and racetrack-shaped configurations. The reactor configurations designed to accomplish this are described in more detail in sections 17.3 and 17.4. Although the principal emphasis is on the uniformity of the *deposited* layer, most sputtering reactors are developed to produce, in addition, a relatively uniform sputtering rate *on their cathodes*. This ensures that the cathode material will be efficiently used, and that the cathodes will not require replacement too frequently. The thin-film density can be maintained at desired levels by operating at sufficiently low pressures that scattering collisions between the target and workpiece are infrequent; or by thermal annealing of the sputtered layer. When film density is a problem with plasma/cathode deposition, a shift to ion-beam–target sputtering may be called for.

Poor adhesion of the deposited layer can be the result of surface contamination or too low a surface energy on the workpiece. Either problem can be dealt with by conditioning the surface of the workpiece with ion-beam bombardment or exposure to the active species of the plasma. Another cause of poor adhesion, as well as subnormal density, may be too high an operating pressure. Finally, the deposited layer can be left with internal stresses that can lead to poor adhesion, flaking, or cracking. One way to minimize these stresses is to anneal the layer after deposition, or to deposit the layer on a heated substrate.

Plasma deposition of electrically conducting thin films is most frequently carried out at pressures below 1.3 Pa (10 mTorr) in order to ensure long mean free paths for the sputtered atoms that deposit the film. Even the production of polymeric or insulating thin films, which do not require long mean free paths, has been conducted in glow discharge plasmas at pressures below 1 Torr. The use of vacuum processing has restricted plasma-assisted deposition of thin films to relatively high-value items that can be handled with a quasi-continuous production process. If the capital costs of vacuum processing could be eliminated by developing plasma-based thin-film deposition processes which function at 1 atm, many additional industrial applications of this technology would be possible.

REFERENCES

Adams A C 1983 Plasma Deposition of Inorganic Films *Solid State Technol.* **26** 135–9
Anonymous 1991 *Plasma Processing of Materials: Scientific Opportunities and Technological Challenges* (Washington, DC: National Academy Press) ISBN 0-309-04597-5
Arbab M 1997 Sputter-deposited low emissivity coatings on glass *MRS Bull.* **22** 27–35
Clark P J 1971 Sputtering apparatus *US Patent* 3,616,450
Corbani J F 1975 Cathode sputtering apparatus *US Patent* 3,878,085
Eisenmenger-Sittner C, Beyerknecht R, Bergauer A, Bauer W and Betz G 1995 Angular distribution of sputtered neutrals in a post magnetron geometry: measurement and Monte Carlo simulation *J. Vac. Sci. Technol.* A **13** 2435–43
Harper J M E 1978 Ion beam deposition *Thin Film Processes* ed J L Vossen and W Kern (New York: Academic) ch II-5, pp 175–206 ISBN 0-12-728250-5
Jansen F 1986 Plasma deposition processes *Plasma Deposited Thin Films* ed J Mort and F Jansen (Boca Raton, FL: Chemical Rubber Company) pp 1–19 ISBN 0-8493-5119-7
Kikuta K 1995 Aluminum reflow sputtering *MRS Bull.* **20** 53–6
Molchanov V A and Tel'kovskii V G 1961 Variation of the cathode sputtering coefficient as a function of the angle of incidence of ions on a target *Sov. Phys.–Dokl.* **6** 137–8
Mort J and Jansen F (ed) 1986 *Plasma Deposited Thin Films* (Boca Raton, FL: Chemical Rubber Company) ISBN 0-8493-5119-7
Ohba T 1995 Chemical-vapor-deposited tungsten for vertical wiring *MRS Bull.* **20** 46–52
Pramanik D 1995 Aluminum-based metallurgy for global interconnects *MRS Bull.* **20** 57–60

Roth J R 1995 *Industrial Plasma Engineering: Vol I—Principles* (Bristol: Institute of Physics Publishing) ISBN 0-7503-0318-2

Ryan J G, Brodsky S B, Katata T, Honda M, Shoda N and Aochi H 1995 Collimated sputtering of titanium and titanium nitride films *MRS Bull.* **20** 42–5

Shaw J M 1993 Overview of polymer for electronic and photonic applications *Polymer for Electronic and Photonic Applications* ed C P Wong (San Diego, CA: Academic) pp 1–65 ISBN 0-12-762540-2

Smith D L 1995 *Thin-Film Deposition: Principles and Practice* (New York: McGraw-Hill) ISBN 0-07-058502-4

US Inc. 1989 *Planar Magnetron Sputter Sources* sales literature on the US Gun II

Vossen J L and Cuomo J J 1978 Glow discharge sputter deposition *Thin Film Processes* ed J L Vossen and W Kern (New York: Academic) ch II-1, pp 11–73 ISBN 0-12-728250-5

Vossen J L and Kern W (ed) 1978 *Thin Film Processes* (New York: Academic) part II, chs 1–5, pp 11–206 ISBN 0-12-728250-5

Waits R K 1978 Planer magnetron sputtering *Thin Film Processes* ed J L Vossen and W Kern (New York: Academic) ch II-4, pp 131–73 ISBN 0-12-728250-5

——1997 Edison's vacuum coating patents *AVS Newsletter* May/June, pp 18–19

Weast R C (ed) 1988 *Handbook of Chemistry and Physics* 69th edn (Boca Raton, FL: Chemical Rubber Company)

Wilson R G and Brewer G R 1973 Ion beam sputtering *Ion Beams—With Applications to Ion Implantation* (New York: Wiley) ch 4, pp 317–52 ISBN 0-471-95000-9

24

Plasma Chemical Vapor Deposition (PCVD)

This chapter covers selected topics on the *plasma chemical vapor deposition* (PCVD) of thin films. *PCVD* takes place when a heterogeneous chemical reaction between a plasma and one or more of the remaining three states of matter results in the deposition of a thin film. Such a reaction may include participation by the working gas, the plasma active species, and the workpiece surface. These heterogeneous *chemical* reactions set PCVD apart from the purely *physical* sputter deposition processes discussed previously in chapter 23.

The characteristics of reactors or 'tools' used for PCVD of thin films have been discussed in chapter 17. Reactors for non-polymeric thin-film deposition are covered in section 17.5.1, and reactors for polymeric thin-film deposition in section 17.5.2. Some specialized reactor features used for PCVD are discussed in chapter 18. The coatings deposited by PCVD can be broadly characterized as *polymeric*, *non-polymeric*, *amorphous*, and *crystalline*. In this chapter, polymeric thin-film deposition is emphasized because of its importance to microelectronics and other major industrial sectors.

24.1 THIN-FILM DEPOSITION BY PCVD

Several excellent books are available on the PCVD of thin films. Many techniques developed for the microelectronic industry have been discussed by Hollahan and Bell (1974), Yasuda (1978), Veprek and Venugopalan (1980), Mort and Jansen (1986), Lieberman and Lichtenberg (1994), and Smith (1995).

24.1.1 Development of PCVD

It has been known since the late 19th century that glow discharge plasmas that contain reactive or polymerizable chemicals would deposit thin films on the inside of vacuum systems. These films were long regarded as a nuisance because

502

Table 24.1. Acronyms used in conventional and plasma-assisted thin-film deposition.

CVD	Chemical Vapor Deposition
APCVD	Atmospheric Pressure CVD
HFCVD	Hot-Filament CVD
PECVD	Plasma-Enhanced CVD
PACVD	Plasma-Assisted CVD
PCVD	Plasma CVD

they were unwanted and difficult to remove. It was not until the late 1960s that the potential value of these films to microelectronic circuit fabrication was appreciated (see, for example, Denaro *et al* 1968). In the period from 1960 to 1975 research and development in this field flourished, and the technical approaches developed during this period established a basis for the thin-film PCVD techniques now used.

These techniques continue to be developed further, and applied to a progressively wider range of uses outside the field of microelectronics. In addition to the comprehensive references previously cited, references on the more restricted but industrially important subject of *polymeric* thin-film deposition are available. Early examples from this large literature include *Plasma Polymerization*, a symposium proceedings edited by Shen and Bell (1979); and *Plasma Polymerization*, a monograph by Yasuda (1985). More recent compilations include *Plasma Deposition, Treatment, and Etching of Polymers*, edited by d'Agostino (1990); and *Plasma Deposition of Polymeric Thin Films*, a symposium proceedings edited by Danilich and Marchant (1994).

24.1.2 Nomenclature of Heterogeneous Deposition Reactions

The literature of thin-film deposition contains numerous acronyms to describe the various forms of thin-film deposition using plasma-assisted chemical vapor deposition. Some of the more widely encountered acronyms are summarized in table 24.1. These processes include *chemical vapor deposition* (CVD), which occurs in the absence of a plasma, and is discussed in the next section. If a purely chemical deposition process occurs at atmospheric pressure without the participation of a plasma, the process is called *atmospheric pressure chemical vapor deposition* (APCVD). This acronym applies to CVD operated at atmospheric pressure, and does *not* refer to PCVD at 1 atm. The confusing acronym (APCVD) will not be used further in this text.

The plasma-processing literature refers to *plasma-enhanced chemical vapor deposition (PECVD)* and *plasma-assisted chemical vapor deposition (PACVD, PCVD)*, which are considered synonymous in this text. In this text also, we will refer to '*plasma-assisted chemical vapor deposition*' as PCVD. Sometimes a

CVD or a PCVD process is augmented by a hot filament, the function of which is to break up or modify molecules that approach or impact its hot surface. These processes are referred to as *hot-filament CVD (HFCVD)* or *hot-filament PCVD (HFPCVD)*, respectively.

24.1.3   Deposition Using Heterogeneous Chemical Reactions

Prior to the development of methods for plasma-assisted thin-film deposition, several purely chemical or electrochemical methods were developed for depositing thin films on surfaces.   These form the basis for current PCVD techniques, and are discussed in the following sections.

24.1.3.1   *Liquid–Solid Chemical Reactions*

Some of the oldest industrial methods of depositing thin films are based on heterogeneous liquid–solid chemical reactions.   These methods include *electroplating*, developed early in the 19th century by Michael Faraday. Electroplating technology has changed little since that time, and is carried out with the apparatus illustrated in figure 24.1. The apparatus consists of a liquid electroplating bath in which ions of the material to be deposited leave the anode, flow through the solution as an ionic current, and form a deposited layer on the cathode. This heterogeneous chemical process is driven by a potential difference of tens to hundreds of volts provided by a DC power supply. The rate of deposition of the electroplated layer is controlled by adjusting the current flowing to the workpiece/cathode, which may range up to 1000 A.

In the past, electroplating has been widely used for brightwork and protective coatings (automobile bumpers), but it is gradually yielding to other technologies (many of them plasma based) capable of directly coating insulating as well as electrically conducting substrates. At least equally important, the electroplating process normally produces significant quantities of toxic or unwanted by-products that pose serious regulatory issues under modern safety and environmental regulations.

24.1.3.2   *Plasma Solid Chemical Vapor Deposition*

Another method sometimes used to produce protective coatings is exothermic heterogeneous chemical reactions that are not driven by an external power supply or the presence of a plasma.   Such CVD reactions can produce metallic or insulating surface coatings. Other conventional methods of depositing thin films by heterogeneous chemical reactions include gas–solid chemical reactions that can be used, for example, to form a protective oxide coating on metals.   An example is the aluminum oxide (Al_2O_3) coating that naturally forms on aluminum exposed to the atmosphere. The surface of materials such as iron, for which an oxide coating is not protective, sometimes can be treated by chemically nitriding or carbonizing the surface, which may form a protective coating.

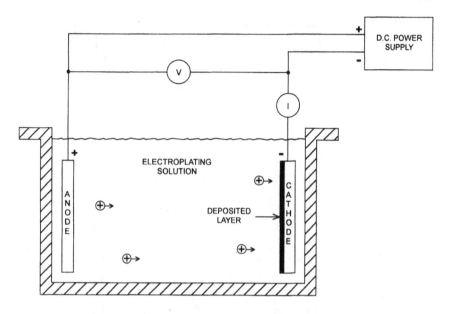

Figure 24.1. Illustration of an electroplating bath used to deposit thin films of the anode material on the cathode/workpiece.

24.1.3.3 PCVD

A variety of PCVD methods have been developed to deposit thin films on surfaces, when methods that involve sputtering are unsatisfactory or not possible. *PCVD* is of three types, each of which finds a variety of uses in microelectronic and other applications. The first is *plasma polymerization*, in which monomers or molecular fragments formed as plasma active species recombine on the surface of the workpiece and form a thin polymeric film. The second type of PCVD is *plasma epitaxy*, in which plasma active species form successive monolayers of the workpiece substrate, and build up a crystalline or polycrystalline solid. The third type of PCVD involves other *heterogeneous chemical reactions*, in which the working gas and/or active species react chemically with the workpiece surface to form a thin film of chemical compounds.

24.1.4 PCVD Processes

A PCVD system is illustrated schematically in figure 24.2. Such systems are usually operated in the intermediate-pressure regime above 1.3 Pa (10 mTorr). In some PCVD processes, including epitaxy, it may be necessary to maintain the workpiece at temperatures of several hundred degrees centigrade. A working gas, or combination of gases, is fed into the vacuum system and forms active species in the plasma. In PCVD, these active species may include monomers, or atoms

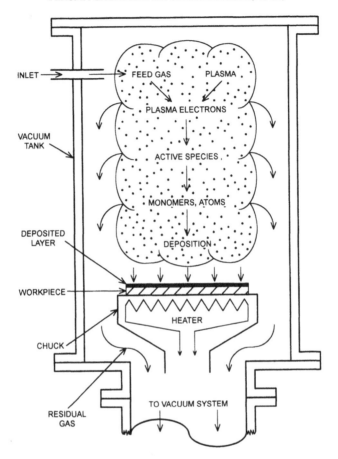

Figure 24.2. A schematic illustration of apparatus for plasma-assisted chemical vapor deposition (PCVD). In PCVD, the feed gas interacts with the plasma to produce active species, including monomers or precursors that are deposited in a thin film on the workpiece.

of elements that are to be deposited by epitaxy. The process of PCVD is usually applied to high-value workpieces that can justify the capital costs of the vacuum system and batch processing.

24.1.4.1 Plasma Polymerization

Plasma polymerization is a widely used deposition method in which monomers supplied to, or produced within a deposition chamber by plasma–chemical reactions, polymerize on contact with the workpiece to build up a polymeric coating. Plasma polymerization can take a variety of forms, depending on the chemical nature of the monomers or molecular fragments deposited on the surface. With the proper monomers and operating conditions, a thin polymeric film can consist of *hydrocarbons*, long polymeric chains consisting of linked carbon atoms with two hydrogen atoms attached to each carbon. It is also possible, with the correct input species, to form *fluorocarbon* polymers in which fluorine replaces the hydrogen atoms of hydrocarbons. Fluorocarbons are much less chemically reactive than hydrocarbons and have a very high electrical resistivity.

Fine *powders* or *oil droplets* can be produced by physical and chemical processes in PCVD reactors that contain gaseous phase monomers or precursors for epitaxy (Barnes *et al* 1992). Powder production by plasma polymerization can be desirable and intentional, as in the production of fine powders for thermal plasma spraying and other thermal plasma-processing applications. These powders characteristically have diameters from 0.1 to tens of microns, and their production will be discussed further in Volume 3. In other areas of application, particularly in the fabrication of microelectronic circuits, powders and oils are unwanted contaminants that can contaminate the surface of a microelectronic wafer and ruin the workpiece.

24.1.4.2   Plasma Epitaxy

A related heterogeneous process is *epitaxy*, in which atoms of a crystalline material are supplied to a reaction chamber or produced by chemical reactions in the immediate vicinity of the workpiece. These atoms or precursors deposit on the surface to produce successive monolayers of a crystalline material. Epitaxy is used, both with and without the assistance of plasma–chemical reactions, to build up single-crystal silicon ingots for microelectronic wafers, polycrystalline diamond films, and crystals of other pure elemental materials.

24.1.4.3   Heterogeneous Chemical Reactions

In many applications, a plasma may be required in a PCVD system like that illustrated in figure 24.2 in order to produce highly reactive or oxidizing species. These species may include ozone, atomic oxygen, or atomic nitrogen, which react chemically with the surface to produce plasma-assisted *oxidation* or *nitriding*. Motivations for producing such thin films on surfaces are protection against further oxidation, improvement of tribological and wear characteristics, as in the hardening of titanium subject to PCVD nitriding, or to produce a decorative surface layer.

24.1.5   Applications of PCVD Thin Films

The deposition of polymeric thin films by low-pressure glow discharge PCVD originated in the 1960s as a technology required for microelectronic circuit fabrication. As often happens in the industrial applications of plasma, this form of PCVD performed functions in microelectronic applications which could be performed in no other way. It soon became apparent, however, that this technology could be applied to many industrial processing applications outside the field of microelectronics.

Some of these applications include macroelectronic devices, optical components, biomedical products, permeable membranes, impermeable barrier coatings, automotive products, composite materials, and the fabrication of three-dimensional micromachines. New industrial uses of this technology continue to be reported on an almost monthly basis, most of them restricted only by the requirement that the process be carried out in a vacuum system by batch processing.

24.1.5.1 Microelectronic Devices

The microelectronic applications of PCVD are essential to that industry and represent an investment in capital equipment and annual cash flow measured in billions of dollars (Anonymous 1991). The uses of this technology within the microelectronic industry include the deposition of insulating, dielectric, and protective coatings and layers on wafers during the fabrication process; the deposition of low-dielectric-constant capacitor dielectrics, as a sandwich material between two metal films in the formation of diodes; and the formation of masks for conventional, x-ray, and electron-beam vacuum lithography with dry resists. The microelectronic applications of PCVD have been reviewed by Morita and Hattori (1990), and a much broader range of applications by Morosoff (1990).

24.1.5.2 Macroelectronic Devices

On a larger scale of size, PCVD methods are used to form *macroelectronic devices* for use in hybrid or ordinary electronic circuits. The number and kind of such devices have continued to expand in recent years. An example is flat and roll capacitors, in which a PCVD film is sandwiched between two metallic (usually aluminum) electrode sheets that may be left flat, or rolled up to form a cylinder. PCVD polymers have the advantage as capacitor dielectrics of a high dielectric strength, up to 100 MV/m, and stability at high temperatures, up to 250 °C or higher. They have the disadvantages of low dielectric constants, from 2.0 to 3.0 in most cases, and a relatively high dielectric loss and heat generation, due to polymeric cross-linking. Additional applications of PCVD films to macroelectronic devices include the fabrication of sensors for pH and moisture (*hygrometers*) which use ion-sensitive field effect transistors (*ISFETs*), fuses, and in the fabrication of liquid crystal displays.

24.1.5.3 Optical Applications

Plasma-chemically deposited thin films, both polymeric and non-polymeric, have found widespread optical applications because of their transparency, robust physical properties, good adherence to the substrate, and resistance to chemical attack. Polymeric thin films are used as transparent protective coatings on digital storage discs, including CDs. Other optical applications include protective coatings against atmospheric moisture, chemical attack, and abrasion for visible and infrared optical components made of friable or chemically reactive materials.

In addition to protective coatings, polymeric PCVD can transform the surface of poly(methylmethacrylate) (PMMA) and other materials used in contact and intraocular lenses from hydrophobic to hydrophilic, thus significantly improving their wettability and biocompatibility (Ratner *et al* 1990). Polymeric PCVD films are also used as wavelength filters on optical components, and in the formation of lenses and fiberoptic waveguides, for reactive index control by varying the properties of the film during deposition.

24.1.5.4 Biomedical Products

Biocompatible polymeric PCVD thin films may be used in a variety of biomedical products in addition to contact and intraocular lenses. *Biocompatibility* is the ability of a material to serve its intended function *in vivo* (in contact with a living organism) without adverse effects and without degradation. Biocompatibility is a characteristic of some polymeric thin films that may include some or all of the following properties:

(1) chemically inert;
(2) having good adhesion to its substrate;
(3) flawless, pinhole-free conformity to substrates;
(4) automatic sterilization during the deposition process;
(5) being mechanically robust; and
(6) providing a barrier to the passage or diffusion of undesirable chemical species.

Some fluorocarbon and other polymers are *blood compatible*: that is, they can be in direct contact with flowing blood without initiating clots, thrombi, or other adverse effects. Blood compatibility is frequently associated with the low surface energies that make PCVD polymers hydrophobic. A potential biomedical application of thin films is in the culture of tissue and micro-organisms, in which the surfaces must be relatively wettable by the fluids and nutrients involved. This can be done by plasma surface treatment (see section 21.1.3) or PCVD of a polymeric material with high surface energy.

24.1.5.5   Permeable and Impermeable Membranes

Polymeric thin films can be *selectively permeable* to specified gases or liquids, or can be *barrier coatings* impermeable to unwanted contaminants (Ratner *et al* 1990). In medicine, selectively *permeable membranes* created by PCVD of appropriate polymers are used for blood oxygenators and dialyzers. Impermeable *barrier coatings* are used in food packaging, to prevent unwanted tastes, odors, water, or oxygen from spoiling food. Preventing the permeation of oxygen is particularly challenging, since its atomic radius is smaller than any other common element in the periodic table. Barrier coatings often serve multiple functions as protective coatings designed to protect underlying films or layers from abrasion or mechanical damage.

24.1.5.6   Emerging Niche Applications

Other developing applications of polymeric PCVD thin films include automotive components, where weight and cost considerations motivate the increasing use of such polymeric materials as high-density polyethylene, Nylon 12, and Nylon 66. These materials are not entirely satisfactory for automotive applications, because they are permeable to the hydrocarbons in gasoline and can exude volatile organic compounds (VOCs). In addition, they are sufficiently good electrical insulators to present *electrostatic discharge (ESD)* safety concerns arising from sparking in the presence of air and gasoline. The PCVD of electrically conducting, organometallic polymeric chemical barrier coatings resulting from the polymerization of tetramethyltin (TMT) or monobutyltrivinyltin (MBTVT) as feedstocks can alleviate both of these concerns.

Another application of polymeric PCVD is the coating of fibers incorporated into *composite materials*. When fabrics or fibers coated with a polymeric film are embedded in an epoxy matrix, the resulting composite material shows a significant increase in tensile and shear strength (Rose *et al* 1989). These effects are thought to arise from the increased adhesion and chemical compatibility of the coated compared to the uncoated fibers embedded in the composite (Morosoff 1990).

Finally, polymeric PCVD coatings have been applied to *microfabrication* processes, an example of which is the production of targets ('microballoons') for laser fusion experiments (Morita and Hattori 1990). In this application, the targets were spheres with a diameter of a few hundred microns, coated uniformly with a polymeric coating about ten microns in thickness.

24.2   PHYSICAL AND CHEMICAL PROCESSES IN PCVD GLOW DISCHARGES

Plasma-assisted CVD is possible because energetic electrons produce molecular fragments and other active species that form thin films superior to those produced

Table 24.2. Energies of primary active species associated with glow discharge plasmas.

Species	Energy (eV)	
	In plasma	On wafer
Electrons	1–30	Not present
Ions	0.025–1	10–100
Metastables	0–30	0–30
Visible/UV photons	1–30	1–30

by purely chemical or electrochemical processes. Such active species are made in significant concentrations by purely chemical processes only with difficulty, if at all. Even when this is possible, CVD processes may produce significant quantities of toxic or unwanted by-products. Plasma-generated active species and their characteristics have been discussed previously in section 14.2, and that discussion will not be repeated here. Some points of particular relevance to the plasma-assisted deposition and etching of polymeric, conducting, semiconducting, and insulating thin films are discussed now.

24.2.1 Formation of Primary Active Species

The *primary active species* include the working gas (or feed gas(es)) and the electrically charged species of the plasma. The primary active species undergo inelastic binary collisions to form *secondary active species* that normally participate in deposition and etching. The charged primary active species receive their energy directly from the energizing electric field; the secondary active species receive their energy indirectly by inelastic collisions with the charged primary active species.

The primary active species responsible for producing molecular fragments and other secondary active species needed for thin-film deposition are listed in table 24.2. This table contains an estimate of the approximate energy range, in electronvolts, of electrons, ions, and other active species in a characteristic glow discharge plasma, and on the surface of a wafer or workpiece exposed to such a plasma. A workpiece at the plasma boundary experiences bombardment by the primary active species, the charged components of which may be accelerated across a sheath, or prevented by the sheath electric field from reaching the workpiece. Most glow discharge plasmas float at a positive potential with respect to their surroundings, and with respect to workpieces at the boundary. As a result, the flux and energy of electrons bombarding the workpiece may be small, whereas ions are accelerated across the sheath by the electric field, and may acquire energies of several or tens of electronvolts.

Table 24.3. Selected diatomic bond strengths for species used in plasma-processing applications.

Bond	Bond strength (eV)
Br–Br	2.0
C–C	6.3
C–Si	4.7
Cl–Cl	2.6
Cl–H	4.5
Cl–O	2.8
F–F	1.6
F–H	5.9
F–O	2.3
F–Si	5.7
F–W	5.7
Fe–O	4.0
H–H	4.5
I–I	1.6
N–N	9.8
N–Si	4.6
O–O	5.2
O–Si	8.3
S–S	4.4
Si–Si	3.4

24.2.2 Formation of Secondary Active Species

Secondary active species are produced in the plasma by binary (and rarely by three-body) inelastic collisions of electrons with the working gas(es), ions, and neutral species in the plasma. The energies involved in chemical reactions and collisional processes in plasmas are sometimes a matter of confusion. Plasma-related processes studied by physicists conventionally have their energies expressed in electronvolts, while chemical bond strengths and reaction potentials are expressed by chemists in kilocalories per mole or kilojoules per mole. In order to establish a common frame of reference, chemical bond strengths and reaction energies in this text have been expressed in electronvolts per molecule, or electronvolts per bond, rather than kilocalories or kilojoules per mole.

In table 24.3 are shown selected diatomic bond strengths, most of which are relevant to microelectronic plasma processing. In the left-hand column are listed the diatomic bonds considered. In the column on the right the energy, in electronvolts, necessary to break that bond is listed. In comparing the energies in the right-hand column with the energies of primary active species listed in table 24.2, it becomes evident why plasmas can produce such highly chemically

Table 24.4.   Selected diatomic and polyatomic bond strengths for species used in plasma-processing applications.

Bond	Bond strength (eV)
Decomposition of methane	
$H–CH_3$	4.5
$H–CH_2$	4.8
$H–CH$	4.4
$H–C$	3.5
Decomposition of silane	
$H–SiH_3$	3.9
$H–SiH_2$	2.8
$H–SiH$	3.6
$H–Si$	≤ 3.1
Decomposition of sulfur hexafluoride	
$F–SF_5$	4.0
$F–SF_4$	2.3
$F–SF_3$	3.6
$F–SF_2$	2.7
$F–SF$	4.0
$F–S$	3.6
Decomposition of water vapor	
$H–OH$	5.2
$H–O$	4.4
Other bonds	
$CF_2=CF_2$	3.3
$O=CO$	5.5
$O–C$	11.2
$Cl–CCl_3$	3.2
$F–C_2F_5$	5.5
$O–N_2$	1.7
$O–NO$	3.2
$O–N$	6.5

reactive species as atomic chlorine, atomic oxygen, and atomic fluorine. The most energetic electrons in the Maxwellian distribution of a glow discharge plasma have energies comparable to or greater than these bond strengths. Significant numbers of highly reactive atomic species therefore can be formed in a typical glow discharge plasma by breaking these bonds.

In table 24.4 additional diatomic and polyatomic bond strengths relevant to microelectronic plasma processing are shown. Selected bonds are listed in the left-hand column of table 24.4, and the bond strength in electronvolts in the right-hand column. These bonds have been organized in terms of the progressive decomposition of common plasma-processing feed gases such as methane (CH_4) silane (SiH_4), sulfur hexafluoride (SF_6), etc.

The decomposition reactions implied by the grouped bond strengths in table 24.4 indicate that the energy required to remove additional hydrogen or fluorine bonds does not increase as the parent carbon, silicon, or sulfur-containing molecule is progressively stripped of its attached atoms. Instead, the energy required to remove successive atoms from these polyatomic molecules may remain approximately the same or even decline as the molecule becomes more highly stripped. This behavior may be intuitively unexpected to someone more familiar with atomic physics, where the energy required to remove successive electrons from high-Z atoms increases with each successive electron.

At the bottom of table 24.4 are shown some selected reactions which indicate that the electrons in a glow discharge plasma possess sufficient energy to produce a variety of highly chemically reactive species, including atomic chlorine, fluorine, and oxygen atoms. Glow discharge plasmas will have the full range of molecular fragments indicated by the progressive breaking and removal of the atomic bonds shown in these decomposition reactions, if the dwell time of the parent molecule in the plasma is sufficiently long for these reactions to take place.

The formation of chemically active *free radicals* (species with an unpaired electron) is also an important process in many glow discharge plasmas used for thin-film deposition. In table 24.5 selected enthalpies of formation of free radicals relevant to plasma processing are shown. The free radical is shown in the middle column, and the formation energy of that free radical, in electronvolts, is shown in the column on the right. As in the previous two tables, positive formation energy means that energy must be added to the parent molecule in order to form the reaction product or to break the bond. Negative formation energy, however, indicates that an exothermic chemical reaction takes place when the species is formed from its parent molecule.

The free radicals in table 24.5 are grouped in progressive stages of either formation or decomposition of the parent molecule shown in the left-hand column. The precursor molecules of silicon tetrafluoride and silicon tetrachloride with negative (exothermic) formation energies are reaction products from the etching of silicon. For the free radicals, the formation energies shown in the right-hand column are comparable to energies of the electrons or etching ions that participate in the formation of these species. These reactions show a trend that, as the parent molecule is decomposed or formed, the respective energies of decomposition or formation tend to decrease as atoms are removed from or attached to the parent molecules. The final atoms that attach to a molecule to complete its structure are more energetically bound to the parent molecule than the first atom or two. This behavior of the bonding energies is the reverse of that which exists among electrons orbiting a high-Z nucleus in atomic physics.

Table 24.5. Selected enthalpies of formation of free radicals for species used in plasma-processing applications.

Parent molecule	Free radical	Formation energy (eV)
Silane	SiH_3	2.0
	SiH_2	2.5
	SiH	3.9
Silicon tetrafluoride	SiF_3	-10.6
	SiF_2	-6.10
	SiF	-0.20
Silicon tetrachloride	$SiCl_3$	-3.3
	$SiCl_2$	-1.7
	SiCl	-2.0
Sulfur hexafluoride	SF_5	-9.4
	SF_4	-7.9
	SF_3	-5.0
	SF_2	-3.1
	SF	$+0.13$
Other radicals	CH	6.2
	$CH{\equiv}C$	5.9
	CN	4.5
	CCl_3	0.82
	OH	0.41

24.2.3 Physical Processes in PCVD Plasmas

There exist phenomena in glow discharge plasmas of primary interest to physicists or those interested in plasma science. These phenomena include *excitation* of an electron in the atomic shell of an atom, resulting from the inelastic collision of an energetic plasma electron; *ionization*, the removal of an electron from an atom as the result of an inelastic collision by an energetic electron or other active species with an atom; *transport*, including diffusion of neutral and charged active species within the plasma and across the sheath between the plasma and the workpiece; and the *accumulation of dust and oils*, which result from the formation of hydrocarbons or quasi-fractal crystalline entities in the glow discharge plasma. The first three of these processes were discussed in section 14.2. The accumulation of dust and oils is a process that can be important in both the deposition and etching of microelectronic circuits. A further discussion of polymerization and dust formation as it applies to plasma etching may be found in chapter 25, sections 25.4.1.5 and 25.4.1.6, respectively.

24.2.4   Chemical Reactions in PCVD Plasmas

In addition to the phenomena already discussed, chemical reactions occur in a deposition reactor that are of primary interest to individuals in the field of chemistry or materials science. These processes include *exothermic* and *endothermic chemical reactions*; *isomerizations*, in which molecules form *isomers*, mirror image molecular structures, when subject to inelastic collisional processes in a glow discharge; *dissociation*, in which a diatomic or small polyatomic molecule is broken into two approximately equal parts by energetic active species; *stripping*, in which a single atom or small molecular fragment is removed from a much larger polyatomic molecule; *eliminations*, in which specific atoms or small molecular fragments are separated from their parent molecules without otherwise affecting the remainder of the parent molecule; *scissioning*, in which long molecules or polymers are broken up into two or more large fragments; *dimerizations*, in which an atom or molecular fragment is removed from a parent molecule, which then combines with a second parent molecule to form a bimolecular compound not originally present; and *polymerization*, in which monomers or other molecular fragments combine in the plasma or on surfaces to form long chainlike polymers. Such polymers, as discussed in section 25.4.1.5, can form oils if the molecular weight of the resulting polymer is relatively low; or quasi-fractal crystalline solids if the molecular weight is high.

In addition to these chemical reactions in deposition plasmas, heterogeneous surface reactions also occur. In these reactions, a solid surface acts like a third body to make possible chemical reactions that cannot occur as binary reactions, because of constraints imposed by energy and momentum conservation laws. Such reactions will be discussed in chapter 25, since they play an important role in etching of thin films.

24.2.5   Chemical Surface Activation

When active species interact with a surface, many heterogeneous reactions are possible. When they react with adsorbed monolayers, the usual effect is to remove the monolayers by *desorption* (outgassing) or the formation of volatile reaction products. Such surface reactions usually result in significant increases in surface energy and wettability, as was discussed in chapter 21. When active species interact with the workpiece material itself (the *substrate*), the surface properties of the workpiece can change. A new compound can be formed or molecular fragments can be attached to a polymeric or hydrocarbon-based substrate material.

As an example of the importance of surface chemical reactions, consider three active species: the *carbonyl group* (C=O), the *hydroxyl group* (OH), and the *carboxyl group* (O + C–OH). These molecular groups are illustrated in figure 24.3. The chemical attachment of these groups to hydrocarbons or polymers is known to affect such surface energy dependent properties as the wettability, wickability, and

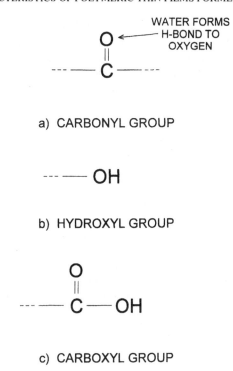

Figure 24.3. Three chemical species capable of enhancing the wettability of a hydrocarbon surface to which they attach: (*a*) the carbonyl group; (*b*) the hydroxyl group; and (*c*) the carboxyl group.

printability of polymeric materials. In figure 24.3(*a*), the double-bonded oxygen (available from the plasma as atomic oxygen), when attached to a hydrocarbon chain, forms an H-bond to water, thus permanently increasing the wettability of the surface of which it is a part. In addition to atomic oxygen bonding to the carbon spine of a long hydrocarbon molecule, as in figure 24.3(*a*), atomic oxygen or the hydroxyl group (OH) can attach themselves to the end of a long hydrocarbon (or other) molecule to form sites for bonding of water or other species, as shown in figures 24.3(*b*) and (*c*). These sites for bonding water make the polymer or hydrocarbon permanently hydrophilic or wettable.

24.3 CHARACTERISTICS OF POLYMERIC THIN FILMS FORMED BY PCVD

The utility of polymeric thin films deposited by PCVD arises from their wide range of physical characteristics. This range allows the surface properties of a substrate coated with a polymeric thin film to be tailored without compromising its bulk properties, and to add value to substrate materials with otherwise unsatisfactory surface properties. Unfortunately, quantitative information on the properties of polymeric thin films is scattered in a very extensive literature, to which we do not attempt to do justice here. The reader is referred for such information to Bell (1980), relevant chapters in Mort and Jansen (1986), and a survey article by Morosoff (1990).

The subject of polymeric PCVD thin films includes a wide variety of morphological structures, including classical *linear polymers*, in which *monomers*, the basic building blocks, are linked in long repeating linear chains. In polymeric solids, these long chains are arranged in parallel array or with some other higher-order (polymer crystalline) structure. PCVD may also produce *quasi-polymeric structures* with linear branches in two and three dimensions that have no consistent repeating pattern. The linear and quasi-polymeric structures may also form complex three-dimensional configurations to produce *cross-linked polymers*. In addition, polymeric coatings or films also may contain *macromolecular structures* formed by monomers and/or molecular fragments with a complex three-dimensional morphology and few or no repeating units in linear arrays. These forms of polymeric PCVD are now discussed.

24.3.1   PCVD Polymerization Processes

The PCVD operations considered in this and the following section are carried out in a *glow discharge* plasma, either DC or RF, hence the occasionally used generic designation of *glow discharge polymerization*. The processes that lead to PCVD are more complex than those associated with CVD, and are illustrated schematically in figure 24.4. In PCVD, the *feed material* may consist either of a *monomer*, which is capable of forming a conventional polymer; or a *polymerizable material*. The latter will not form polymeric structures when subject to conventional CVD, but its plasma-generated molecular fragments will do so.

Referring to figure 24.4, the feed material may undergo direct heterogeneous chemical reactions to form polymers on the surface of the workpiece by *plasma-induced polymerization*, reaction (1). To occur, these reactions require exposure of the workpiece surface to the plasma active species; otherwise they would be an example of conventional CVD. The feed material may also be scissioned or fragmented by the plasma electrons (reaction (2)) to form molecular fragments. These molecular fragments are themselves monomers, or are otherwise capable of forming polymeric structures on the surface of the workpiece by reaction (3). This reaction route, (2) + (3), is referred to as *plasma polymerization*, and the plasma-produced molecular fragments in reaction (3) as *polymerizable species*.

Some of the feed material illustrated in figure 24.4 may also react to form *non-polymerizing reaction products* (reaction (4)). These reaction products may contaminate the workpiece and vacuum system or form unwanted volatile reaction products that are pumped out. The polymerizable species may, instead

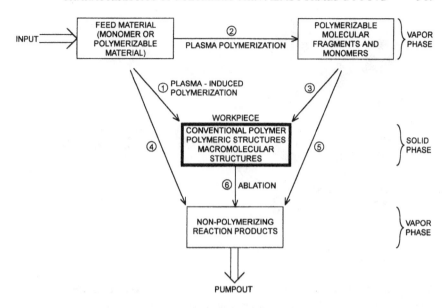

Figure 24.4. A schematic illustration of processes that may take place during polymeric PCVD.

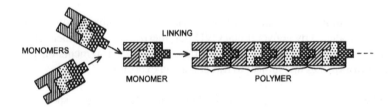

Figure 24.5. A schematic illustration of conventional polymerization, in which monomers link up in a repeating structure to form a long, linear polymer.

of depositing as polymers on the workpiece, undergo further plasma–chemical reactions in reaction (5) to form unwanted volatile reaction products. Finally, the polymeric coating deposited on the workpiece may be simultaneously etched or volatilized by plasma active species, a process known as *ablation* (reaction 6).

   In this section, we are interested only in those polymerization reactions that require the presence of a glow discharge plasma. The polymeric structures formed by glow discharge PCVD may be illustrated by the schematic 'molecular fragments' in figures 24.5–24.8. *Conventional polymerization* is illustrated in figure 24.5, in which *monomers* link up in a repeating structure to form a long, linear *polymer*. Figure 24.6 illustrates *plasma polymerization*, the formation of a polymer chain by the linkage of plasma-scissioned molecular fragments, which

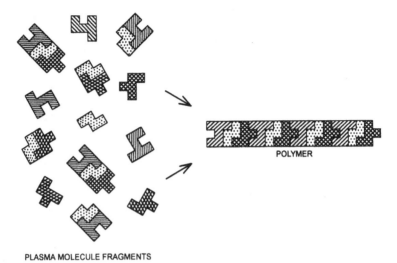

PLASMA MOLECULE FRAGMENTS

Figure 24.6. A schematic illustration of the formation of a polymer chain by the linkage of plasma-scissioned molecular fragments, which may include monomers.

may include monomers. The result is a linear polymeric chain indistinguishable from that produced by conventional CVD polymerization. Such linear chains are preferentially formed from plasma-generated molecular fragments on remotely exposed workpieces not subject to direct plasma exposure.

When a workpiece is exposed directly to the plasma active species, *quasi-polymeric structures* such as those illustrated in figure 24.7 may be formed from molecular fragments generated in the plasma. These structures can include linear polymeric chains, branched polymers, quasi-polymers that form a multibranched, cross-linked two- or three-dimensional structure, and long linear polymers cross-linked in two or three dimensions. Finally, the plasma-scissioned molecular fragments may coalesce into a *macromolecular structure* of very high cross-linkage density of the kind illustrated in figure 24.8.

The actual structure of a quasi-polymer deposited by PCVD from ethylene has been inferred from infrared and spectroscopic data by Tibbit *et al* (1976), and is reproduced in figure 24.9. This structure is characterized by numerous cross-linkages, attached functional groups, and a branching three-dimensional structure that would not be expected in purely CVD-deposited polyethylene. In addition to the structure indicated in figure 24.9, cross-linkages to other sites on this PCVD polyethylene are indicated by the bonds highlighted as feature 'A' to the upper left of this figure.

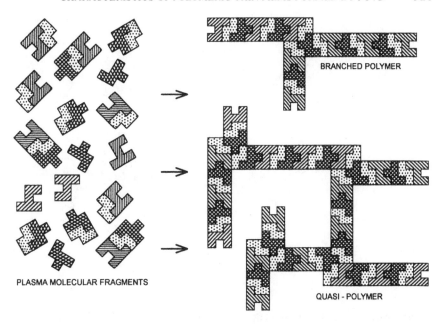

Figure 24.7. A schematic illustration of polymeric structures that may be formed from precursors generated in the plasma. These structures may include linear polymeric chains, branched polymers, quasi-polymers that form a multibranched, cross-linked two- or three-dimensional structure, and long linear polymers cross-linked in two or three dimensions.

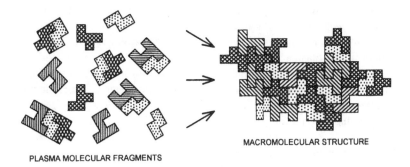

Figure 24.8. A schematic illustration of plasma-scissioned molecular fragments deposited as a macromolecular structure of very high cross-linkage density.

24.3.2 Forms of PCVD Polymers

The objectives of thin-film PCVD include the production of a uniform, adherent, solid film with additional characteristics for particular applications. The

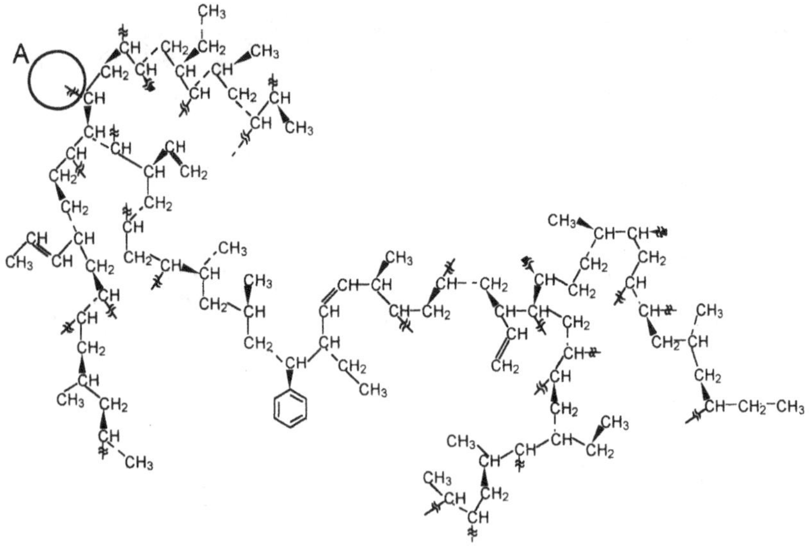

Figure 24.9. The actual structure of a quasi-polymer deposited by PCVD from ethylene reprinted from Tibbit *et al* (1976) by courtesy of Marcel Dekker Inc. This structure contains numerous cross-linkages, attached functional groups, and a branching three-dimensional structure. Cross-linkages to other sites are indicated by the bonds highlighted as feature 'A' in the upper left.

monomers and polymerizable materials used as feed materials, in addition to producing the desired long polymeric structures, are also capable of producing *oils* and *powders*. Oils are liquid polymers too short to assume the solid state at room temperature, and *powders* grow from seed polymers or macromolecular structures in the plasma volume (David *et al* 1990, Barnes *et al* 1992). Both are undesirable and can contaminate a polymeric thin film on a workpiece. The powders are structurally very similar to the thin-film material itself; the material is simply not in the desired form or location. The PCVD-generated oils are polymeric chains, an example of which from Tibbit *et al* (1976) is illustrated in figure 24.10. This structure, in addition to having a higher carbon/hydrogen (C/H) ratio than the polymeric solid illustrated in figure 24.9, has many fewer cross-linkages than the solid material.

The formation of oils, powders, or films during PCVD depends on such operating parameters as the plasma input power (active species/monomer/molecular fragment formation rate); the working pressure (density of feed materials); and the flow rate of the feed materials (dwell time for reaction in the plasma). In figure 24.11 is shown some data reported by Kobayashi *et al* (1974) for the PCVD of polyethylene as a function of gas pressure, ethylene flow rate, and plasma input power. These data indicate that oils were formed when the flow rate was high and the dwell time for reaction low. Figure 24.11 also indicates that powder was

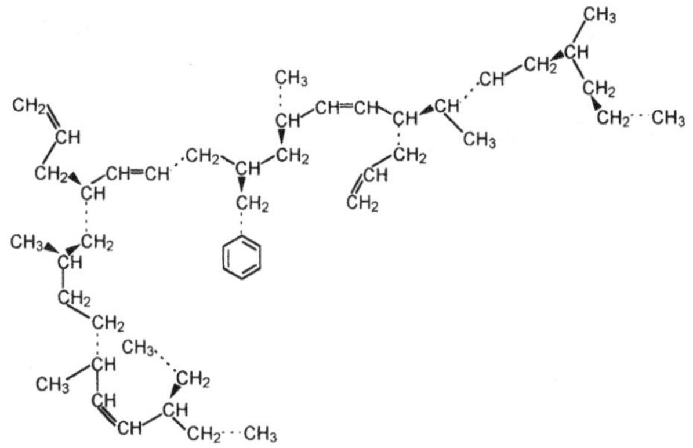

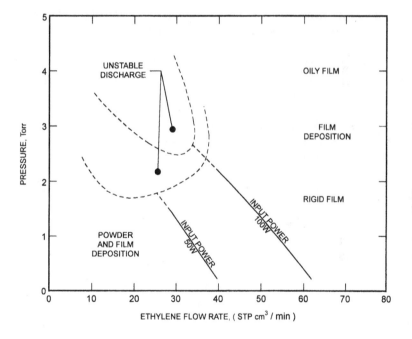

Figure 24.10. An example of a PCVD-generated oil reprinted from Tibbit *et al* (1976) by courtesy of Marcel Dekker Inc. This structure has a higher carbon/hydrogen (C/H) ratio than the polymeric solid illustrated in figure 24.9, and has far fewer cross-linkages.

Figure 24.11. Data reported by Kobayashi *et al* (1974) for the PCVD of polyethylene as a function of gas pressure, ethylene flow rate, and plasma input power.

formed when the flow rates were low and the dwell time of the polymerizable species in the plasma was relatively long. At a given pressure, higher power generates more polymerizable species, and shifts the regime of powder formation to higher flow rates.

It has been shown by I *et al* (1996) that gas phase reactions between O_2 and SiH_4 in a predominantly argon plasma formed micrometer SiO_x dust particles. These particles levitated in the plasma volume when the plasma was energized. When the plasma power is turned off, the dust settles on the lower electrode where workpieces may be mounted. This dust is capable of ruining an otherwise satisfactory thin-film deposition. The production of dust may be avoided by appropriate adjustment of the flow rate, pressure, and plasma power, or by orienting the workpiece to avoid the fallout of dust on its surface.

24.3.3   Surface Properties of Polymeric Films

The surface properties of polymeric and other thin films are determined within the first few tens of nanometers of the surface. These properties include the *surface energy* of the film, discussed in section 21.1.3. The surface energy, in turn, determines the wettability, wickability, printability, adhesion, and bonding of the material. Generally, the higher the surface energy, the better are these latter characteristics for most applications. Some polymeric thin films at the low end of the surface energy scale are *hydrophobic*, with surface energies of 30 dynes/cm and water contact angles greater than 90°. Other materials at the high end of the surface energy scale are hydrophilic, with surface energies approaching 70 dynes/cm and contact angles less than 10°.

The surface *tribological* properties of *hardness* and *wear resistance* are related to the ability to resist abrasion and to remain unchanged by repeated sliding, rolling, and surface-to-surface contact. Unlike most other surface properties, the tribological characteristics of a surface may be influenced by material up to several microns deep. These characteristics include the *friction coefficient*, the resistance of two surfaces to sliding contact, which can be greatly reduced by deposition of appropriate polymeric coatings.

For some applications, the low surface energy of Teflon$^{\circledR}$ or other fluorocarbon thin films is used to create a non-stick surface. Such polymeric coatings may be used in bearings or sliding seals operated in vacuum systems, or under conditions of humidity or chemical vapor attack which might, without such coatings, lead to sintering or scoring of the surfaces in contact. Many of the most widely used polymeric films are selected for their resistance to chemical attack and the related ability to form a protective coating on more chemically reactive substrates.

Highly chemically reactive thin films are seldom of industrial interest, because their thickness is inadequate to resist chemical erosion for very long, or to produce a durable chemical effect. An exception is the highly reactive, long-lived free radicals characteristic of some vacuum-deposited polymeric thin films. These free radicals may be present in measurable amounts more than a year after their formation, although their areal density is subject to an aging effect, and decreases with time after formation. On exposure to the atmosphere, these free radicals react with the components of air (N_2, O_2, and water vapor), as well as hydrocarbon and other contaminants, to form attachment points for water or other molecules. These attached atoms or molecules (including O, C, CO, OH, etc) are durable, and result in permanent wettability and low water contact angles, as was discussed in section 24.2.5. An example of the temporary and durable wettability of polymeric fabrics due to the free radical attachment of oxygen was given in section 21.7.3.

In some applications of polymeric thin films, a high areal density of surface free radicals is undesirable. The areal density of such free radicals may be greatly reduced or eliminated by annealing the surface at temperatures ranging from 200 to 300 °C, although higher temperatures can again increase their density. The areal density of free radicals remaining on the surface of polymeric materials can also be increased or decreased by exposure after deposition to a plasma having a different proportion of carrier gas to monomers or to different polarizable working gases. Such plasma surface treatment can also affect other surface properties, as discussed in chapter 21.

24.3.4   Bulk Properties of Polymeric Films

Thin films deposited either by sputtering or by PCVD (including polymeric thin films) may build up *internal* or *intrinsic stresses* in the bulk of their material. If such stresses are *tensile*, they cause the film to contract and arch concave with respect to its surface; if *compressive* (characteristic of polymeric PCVD thin films), the film tends to expand and arch convex to its surface. These stresses tend to increase with film thickness, and may reach levels at which they cause the film to delaminate from its substrate. This factor is a principal limitation on the thickness of PCVD films, including polymeric thin films.

In polymeric thin films, the stress levels increase with cross-link density, and hence tend to be highest for workpieces directly in contact with the plasma and subject to bombardment by the full range of plasma active species. Because cross-linkage is minimized by PCVD conducted at a site remote from the plasma, polymeric thin films formed by remote deposition tend to have lower intrinsic stress levels and a lesser tendency to delaminate. The intrinsic stress levels in polymeric films can be reduced also by other procedures. These procedures include conducting the deposition process in the presence of inert carrier gases; conducting polymeric deposition in direct contact with the plasma in those cases for which bombardment has an annealing effect; and annealing the film at high temperatures after deposition.

Nearly all applications of polymeric thin films require that they be uniform and adhere to the substrate. The presence of satisfactory *adhesion* is usually determined with the Scotch® tape test, discussed in section 20.6.3. The adhesion of thin films is affected by the internal stresses, discussed above, and by the surface energy of the substrate at the time deposition is initiated. The surface energy is usually great enough to ensure satisfactory adhesion if the substrate is directly exposed to the plasma prior to the initiation of PCVD deposition.

It is relatively difficult to achieve *uniformity* of polymeric PCVD over large workpieces using a static, immobilized workpiece exposed to the relatively small glow discharge plasma sources that are the industry standard. A uniform polymeric coating may be achieved in some cases by a combination of deposition remote from the plasma that provides the active species, and/or motional averaging of the workpiece.

Other important characteristics of polymeric thin films include their *insolubility* in water or other fluids to which the workpiece is exposed, and their *fusibility*, or ability to be bonded or joined to themselves or other materials. These characteristics are particularly important in food packaging, where PCVD thin films are used as an oxygen and moisture barrier. The property of fusibility is essential to the sealing of seams and edges. Most of the polymeric materials selected for use in current applications are insoluble in water and common liquids so they can protect more soluble substrates. Their fusibility may be such that heat treatment or exposure to an appropriate solvent provides bonding to similar or related materials.

24.3.5 Optical Properties of Polymeric Films

Polymeric PCVD films are used as optical coatings because of their transparency, adhesion, and resistance to chemical attack and mechanical abrasion. These characteristics make polymeric PCVD films valuable not only as protective coatings, but also as non-reflective coatings in the visible part of the spectrum for lenses, windows, and other optical components. Polymeric PCVD coatings have indices of refraction of about 1.4. They can produce an anti-reflective effect in the visible portion of the spectrum with a thickness that is an odd multiple of one-quarter the wavelength of interest, with 580 nm representing the middle of the visible spectrum.

Other applications, including optical waveguides, rely on the ability to build up, by adjusting the mixture ratio of feed gases during deposition, a transparent film or cylindrical polymeric fiber with a variable index of refraction across its thickness or diameter. The attenuation of such optical waveguides can reach values below 4 dB/m at visible wavelengths. These values are satisfactory for use in integrated microelectronic circuits, but not for fiber optic cables.

In some applications, polymeric PCVD films serve several functions simultaneously. For example, films resulting from PCVD of tetrafluoroethylene and chlorotrifluoroethylene are infrared transparent and provide a water barrier as well as an anti-reflective coating for lenses and other optical components made of such water-soluble materials as NaCl and CsI. The latter halogen salts are used in such infrared optical components as windows and lenses. PCVD films deposited from vinyltrimethoxysilane not only serve as an anti-reflective coating, but also can protect friable substrates such as polycarbonate from mechanical abrasion (Morosoff 1990).

24.3.6   Electrical Properties of Polymeric Films

The electrical characteristics of polymeric PCVD thin films are of interest because such films can be deposited over large areas with a uniform thickness and good adhesion to their substrate.   Also, very important to high voltage electrical applications, polymeric PCVD thin films are free of such defects as *pinholes* and *voids*. The absence of *pinholes* arises from the nature of the PCVD process, during which electric field lines concentrate at holes or thin spots, resulting in enhanced deposition at such locations.   This feedback mechanism fills in holes or thin areas, while reducing deposition over thicker regions.   The uniformity, adhesion, and absence of pinholes make polymeric PCVD coatings useful for such applications as capacitor dielectrics and electrically insulating layers and coatings.

The electrical properties of polymeric PCVD coatings most germane for microelectronic and industrial applications include their *dielectric breakdown strength*, their *dielectric constant* (or permittivity), their *dielectric loss factor*, their *electrical conductivity*, and, in some applications, their *embedded charge density*.   The dielectric breakdown strength of some PCVD materials can be as high as 1 GV/m (Wrobel and Wertheimer 1990) and are more typically 100 MV/m, values suitable for capacitors and insulating layers in microelectronic applications.   The dielectric constant of most PCVD polymers is relatively low, from 2.0 to 3.0.   Many polymeric films, particularly those with a high degree of cross-linking, have dielectric loss factors about ten times that of conventional linear polymers.   Such high loss factors reduce the utility of PCVD polymers for RF applications or fast switching in microelectronic circuits.

Polymeric PCVD films deposited from organic monomers are good electrical insulators and have a bulk electrical resistivity ranging from 10^{15}–10^{20} Ω-m. Such a high resistivity is desirable for many applications, but may be detrimental in others. For example, the electrical resistivity needed to dissipate static electrical charge on the surface of carpeting, fabrics, or plastic automotive components exposed to gasoline fumes is greater than about 10^{10} Ω-m. The deposition of these relatively conductive coatings requires monomers such as tetramethyltin (TMT) as a feed gas, which deposits organometallic polymers with a bulk electrical conductivity in the range from about 0.01–1 Ω-m. Under the proper operating conditions during deposition, charges can become embedded in the bulk of polymeric PCVD films. The resulting thin film can form a *quasi-electret* of immobilized charge that can leak away only upon annealing the film to a temperature higher than that during deposition.

24.3.7   Barrier Properties of Polymeric Films

Polymeric thin films created by CVD may be either permeable or impermeable to liquids or gases.   Both permeable and impermeable barrier coatings and membranes have important uses in healthcare, biomedicine, and food packaging.

Oxygen and water vapor may permeate through polymeric or other packaging materials lacking a *barrier coating*. Impermeable barrier coatings are essential in food packaging to prevent off-tastes and retard spoilage, and in medical packaging to maintain sterility and prevent degradation of the enclosed products. Off-tastes, spoilage, and product degradation can be caused by the permeation of oxygen, the molecular form of which is among the smallest of all molecules, or water vapor, also a relatively small molecule.

Barrier coatings may be useful in preventing the loss of environmentally polluting volatile organic compounds (VOCs) from containers or tubing that contain gasoline or other volatile hydrocarbons, or from plastic products that contain plasticizers. Polymeric PCVD barrier coatings may, in addition, prevent oxidation, corrosion, or other forms of chemical attack of surfaces, particularly of metals. A polymer film thickness of 1–5 μm can form a water vapor barrier to prevent corrosion and hydroscopic damage of water-sensitive materials; a thickness of less than 50 nm can form an adequate oxygen barrier for food packaging.

Polymeric PCVD films may also form *permeable coatings* for the diffusive separation of mixtures of liquids or gases. Selective permeation of the lighter molecules in a mixture through a permeable thin film occurs with a flux that is a function of the molecular weight, and is inversely proportional to the film thickness. The ability to deposit polymeric PCVD films over large areas uniformly and without pinholes or other defects recommends their application to such selectively permeable coatings. Polymeric PCVD films are used as semi-permeable membranes for gases and liquids in a variety of biomedical applications. These applications include dialyzers, blood oxygenators (artificial lungs, hemoperfusion), and drug delivery devices (nicotine patch, etc) that require a membrane to control the transport rate of body fluids, oxygen, or pharmaceuticals.

24.4 GLOW DISCHARGE POLYMERIZATION

PCVD includes the deposition of thin films with desirable strength or stability characteristics by all forms of plasma, including dark discharges, corona, glow discharges, and arcs. Industrial use of PCVD for the formation of polymeric thin films employs DC, RF, or microwave glow discharges to achieve *glow discharge polymerization*. Glow discharges contain electrons with energies high enough to produce monomers or polymerizable species from more complex molecular feedstocks by scissioning, dissociation, diamerizations, etc. Further coverage of glow discharge polymerization may be found in Denaro *et al* (1968), Poll *et al* (1976), Yasuda (1985), d'Agostino (1990), and Danilich and Marchant (1994).

24.4.1 Choice of Starting Material

The *starting material, reagent,* or *feedstock* for the formation of a polymeric thin film is a monomer when intended for conventional polymeric CVD. In PCVD, the starting material can have a more complex structure, or even be unpolymerizable by conventional CVD, because of the ability of a glow discharge to scission and fragment the input molecular species. The morphological and chemical nature of a polymeric PCVD-deposited coating may, and usually does, differ from that produced by conventional CVD. A PCVD coating may exhibit a more complex three-dimensional structure, cross-linking, an absence of exactly repeating units, and a higher density of free radicals and polar attachment sites. These features are a strong function of the chemical nature of the starting material.

The starting materials most widely used in PCVD can be divided into five classes:

(1) *Hydrocarbons* are compounds formed primarily of hydrogen and carbon (H + C), which may include saturated monomers, multiple carbon chains, double and triple carbon bond structures, and cyclic structures. Hydrocarbons are the most important starting material for the industrial production of PCVD polymers. The PCVD of hydrocarbons has been discussed in several of the general references cited in section 24.1, and by Hollahan and Rosler (1978).

(2) *Hydrocarbons with polar groups.* Polar groups give the resulting PCVD polymers such characteristics as *chemical reactivity, hydrophilicity,* and *durable wettability.* These compounds are used as PCVD feedstocks in applications for which it is desired to combine the characteristics associated with chemically reactive polar attachment sites with the strength, thermal stability, and low cost of hydrocarbons.

(3) *Halocarbons,* of which *fluorocarbons* are the most commercially important example, consist of carbon chains, double- and triple-bonded carbon compounds, and cyclic carbon structures in which hydrogen atoms are replaced by fluorine. In the PCVD of fluorocarbon polymers, incomplete polymeric chains and complex, cross-linked structures can be formed, and these are referred to as *plasma polymerized fluorinated monomer* (PPFM) films. Polymeric fluorocarbon coatings resulting from PCVD adhere well to a wide range of substrates, have a low free energy, and are chemically inert. These characteristics can produce surfaces that are biocompatible, have a low friction coefficient, are relatively unwettable, and protect a substrate against chemical attack. The PCVD of fluorocarbons has been discussed by Kay *et al* (1980) and by d'Agostino *et al* (1990).

(4) *Organosilicon polymers.* The starting materials for *organosilicon polymers* are more complex than those discussed above. The feedstock molecules for hydrocarbon and fluorocarbon PCVD are dominated by two pairs of elements, (H + C) and (F + C), respectively. However, the feedstocks for PCVD of organosilicon compounds are dominated by three (H + C + Si),

four (H + C + Si + O or N), or more elements. Further information on SiO_2-related PCVD can be found in Veprek-Heijman and Boutard (1991). The coatings resulting from PCVD of organosilicon polymers, or silane (SiH_4) with hydrocarbons, are structurally complex. These coatings are used in microelectronics as insulating and passivating layers, and in related applications such as photovoltaic devices, electrophotography, flat panel displays, and particle detectors.

(5) *Organometallic polymers.* The formation of *organometallic polymers* is an even more recent PCVD technology. The starting materials may consist of conventional hydrocarbon monomers, evaporated or sputtered metals, volatile organometallic compounds, or metallic compounds such as trimethyl gallium or AsH_5. Organometallic compounds resulting from PCVD may be electrically conducting, and potentially useful in the microelectronic and related industries.

24.4.2 Precursor Formation

The input working gases used in plasma processing can be characterized as either *polymerizable* or *non-polymerizable*, depending on whether or not they form polymers when exposed to the electrons and other active species of a plasma. Most research and experience with low-pressure glow discharges is with non-polymerizable gases such as oxygen, nitrogen, and noble gases, which behave in characteristic ways that are unlike polymerizable gases such as styrene and benzene. For example, gases that are non-polymerizable, or do not form deposits on the surfaces surrounding a plasma, operate at a constant working gas pressure. Polymerizable gases, however, produce precursors that deposit on the surrounding walls to form polymers. The formation of unwanted polymeric thin films on the interior of the vacuum system acts like a cold trap or vacuum pump to remove molecules from the plasma volume. As a result, the pressure in polymeric and other PCVD reactors can drop significantly when the plasma is turned on.

The removal of precursors from the plasma volume by polymeric or other forms of deposition on surrounding surfaces gives rise to a deposition rate that depends on the surface-to-volume ratio of the PCVD glow discharge plasma, and on the history of the plasma reactor. The total rate of formation of precursors, S_v, in a volume, V, with a number density, n_0, of working gas and an electron number density n_e per cubic meter is the product of the reaction rate R and the volume,

$$S_v = VR = Vn_e n_0 \langle \sigma v \rangle_{se} \qquad \text{precursors/s} \qquad (24.1)$$

where the term $\langle \sigma v \rangle_{se}$ is the *reaction rate coefficient* for electron–neutral impact formation of the precursors from the working gas.

Because the precursors have a sticking probability nearly equal to unity, the deposition rate on the wall area, A, surrounding the plasma is the product of the precursor flux, Γ_p, and this area,

$$S_A = A\Gamma_p = An_p v_p/4 \qquad \text{precursors/s} \qquad (24.2)$$

where n_p is the precursor number density and v_p is the precursor thermal velocity. Setting equation (24.1) equal to equation (24.2) and solving for the precursor number density yields

$$n_p = 4Vn_e n_o \langle \sigma v \rangle_{se} / A v_p \qquad \text{precursors/m}^3. \qquad (24.3)$$

Since the sticking probability is high, the deposition rate will equal the precursor flux, which with the help of equation (24.3) can be written

$$\Gamma_p = n_p v_p / 4 = V n_e n_o \langle \sigma v \rangle_{se} / A \qquad \text{precursors/m}^2\text{-s.} \qquad (24.4)$$

Thus, in polymeric deposition reactors (and in deposition reactors generally), the deposition rate is proportional to the plasma volume-to-surface area ratio, to the electron number density, and to the partial pressure of the working gas (proportional to its density n_o).

24.5 GLOW DISCHARGE REACTORS FOR PCVD

The configuration and operational characteristics of plasma reactors used for non-polymeric PCVD are discussed in section 17.5.1, and those used for polymeric PCVD in section 17.5.2. The glow discharge plasma sources used in these reactors are described in chapter 16, section 16.1. Note, however, that not all plasma sources discussed in section 16.1 are suitable for or are used in industrial polymeric PCVD.

24.5.1 Plasma Sources for Polymeric PCVD

The objective of depositing insulating coatings by polymeric PCVD may be inconsistent with the use of DC glow discharge plasma sources of the kind discussed in sections 16.1.1.1 and 16.1.2.2. Such coatings cover not only the workpieces, but also other surfaces, including electrodes that must draw real currents from the plasma.

24.5.1.1 DC Glow Discharge Reactors

When DC glow discharges are used for polymeric PCVD, the electrodes, but not the workpieces, must be kept free of an insulating coating. The procedures to accomplish this include selectively heating either the workpiece or electrode; suitably arranging the flow of feed gases over the electrode to minimize deposition; or by arranging for remote exposure, rather than *in situ*, deposition. If *in situ* polymeric PCVD is conducted in a DC glow discharge, the workpieces may be mounted on the cathodes to promote cross-linking and other unconventional forms of polymer formation by bombardment of ions and other active species. Such placement of the workpieces may, however, cause contamination of the deposited film by sputtered cathode material.

24.5.1.2   RF Parallel-Plate Deposition Reactors

The RF parallel-plate and related reactors discussed by Raizer *et al* (1995) and in sections 16.1.1–16.1.2, and 16.2.3 can produce glow discharge plasmas with electron number densities greater than that of DC glow discharges by a factor of up to ten. These higher electron number densities lead to higher fluxes of polymerizable and other active species, and shorter processing times during production runs. These capacitive reactors characteristically operate at 13.56 MHz and at pressures between 6.7 and 133 Pa (50 and 1000 mTorr).

24.5.1.3   'Electrodeless' RF Deposition Reactors

The tendency of polymeric PCVD operations to produce coatings that prevent the flow of real currents has led to the widespread use for this application of so-called *'electrodeless' RF sources* of the kinds described in sections 16.1.3 and 16.2.2–16.2.4. In reactors containing these sources, RF plasma generation does not require that a real current be drawn from an electrode in electrical contact with the plasma. The lack of electrodes makes these reactors particularly suitable for both *in situ* and remote exposure polymeric PCVD operations.

24.5.1.4   Microwave Sources for PCVD

Microwave power is relatively inexpensive (i.e. below $1.00/W). For regulatory reasons, the microwave plasma sources described in section 16.2.2 operate at 2.45 GHz in the United States, and sometimes at 915 MHz in other countries. The microwave sources described in section 13.5 of Volume 1 can generate a large volume of uniform plasma suitable for *in situ* polymeric PCVD (Asmussen 1989). The absence of real currents and powered electrodes in microwave plasma sources is consistent with their use for uniform deposition on large or three-dimensional workpieces.

The efficiency with which microwave power is transferred to the plasma is relatively high (above 50%) in extraordinary mode ECR plasmas, but is not as high in the non-resonant and unmagnetized microwave cavity plasmas sometimes used for PCVD. The power levels actually delivered to the plasma in most industrial ECR reactors range from a few hundred watts to a few kilowatts. In deposition plasmas, power levels up to 1 kW are sometimes necessary to maintain the electron number density required to produce a high flux of active species, and thereby to speed up the deposition process.

In ECR microwave plasmas, plasma generation in the resonant region may be non-uniform, and this non-uniformity may be propagated as the plasma expands along diverging magnetic field lines from the resonant surface to the workpiece. Industrial experience indicates that it is more difficult to produce a uniform plasma across a large flat workpiece with ECR microwave reactors than it is with plasmas created in inductive flat-coil or parallel-plate capacitive RF reactors.

A significant disadvantage of ECR microwave reactors in industrial settings is their requirement for a magnetic field permeating a relatively large volume of plasma. Sometimes, as in the distributed ECR reactor illustrated in figure 16.14, the magnetic field can be provided by permanent magnets that do not require electrical power or cooling. In most cases, the magnetic field required for ECR resonance is high enough, and occupies a sufficiently large volume, that electromagnets are required. The use of electromagnets requires a continuous input of electrical power, usually far more than is required to generate the plasma itself. This power to the electromagnets is dissipated as ohmic losses, which must be removed by cooling water or forced convection. The requirement for DC electrical power and cooling for the electromagnets presents reliability and maintenance issues that industrial engineers may prefer not to deal with if other plasma sources can provide the required processing without magnetic fields.

24.5.1.5   PCVD in Atmospheric Pressure Reactors

Until recently, PCVD has been done in vacuum systems with the plasma sources and reactors discussed in the previous four sections and in sections 16.1–16.2 and 17.5.1–17.5.2. The development of atmospheric plasmas, including the OAUGDP has made possible PCVD at 1 atm (Donohoe and Wydeven 1979, Roth 1995, Roth *et al* 1995a, b). The physics of the OAUGDP is described in section 12.5.2 of Volume 1 (Roth 1995).

With suitable feed gases, the OAUGDP reactor can deposit thin films at 1 atm without the requirement for a vacuum system or batch processing. The OAUGDP operates at kilohertz frequencies, below the threshold of regulatory interest (for communications purposes), so the frequency can be adjusted to optimize its operation. This contrasts with other RF and microwave reactors, where regulatory requirements make it extremely difficult to use the frequency as a performance-optimizing parameter.

24.5.2   The Effects of Plasma Pulsing on Deposition

The deposition rate of polymeric films is a function of the pulse width and duty cycle of the RF or microwave power energizing the plasma. Pulsing the power input to microwave generated deposition plasmas is relatively easy (this is how the power input to microwave ovens is controlled) as is pulsing parallel-plate capacitive reactors. It was reported in early work by Vinzant *et al* (1979) on capacitive parallel-plate reactors that the deposition rate is a function of the *duty cycle D*,

$$D = \tau_{on}/(\tau_{on} + \tau_{off}) = 1/(1 + r) \qquad (24.5)$$

where the relative *off time*, r, is

$$r \equiv \tau_{off}/\tau_{on}. \qquad (24.6)$$

These authors find that for polymerization of ethane and ethylene at pressures of approximately 266 Pa (2 Torr), for example, the deposition rate increases monotonically, but more slowly than linearly, with the duty cycle. They also find that if the duty cycle is held constant at 0.5 ($\tau_{on} = \tau_{off}$) while the time on is varied, the deposition rate for τ_{on} less than about 0.2 ms is higher than that for longer, essentially steady-state durations. Understanding in detail the reason *why* a plasma that is energized for a shorter time produces a higher deposition rate will require a thorough understanding of the reaction kinetics.

24.5.3   The Effects of RF Frequency on Deposition

As was discussed in chapters 12, 13, and 16, regulatory requirements limit the frequency of RF power supplies for plasma sources to 13.56 MHz and its harmonics, and to 915 MHz and 2.45 GHz. As a result, the possible effects of varying the RF frequency on the performance of industrial plasma reactors are infrequently investigated. An exception in the field of polymeric PCVD was reported by Morita *et al* (1979). These authors varied the frequency of operation of a parallel-plate RF reactor over the range from 50 Hz to 13.56 MHz. They determined the plasma polymerization rate (or rate of PCVD) of an ethane plasma, and measured the voltage and discharge current of the plasma source as functions of frequency over this range.

The results for the voltage–current characteristic of the RF parallel-plate plasma source used in these investigations is plotted in figure 24.12. The voltage, current, and phase angle between voltage and current (the latter not shown in figure 24.12) are independent of frequency up to approximately 200 kHz. The variation of both voltage and current above 200 kHz is such that the power supplied to the plasma remains approximately constant up to the last datum. Thus, the power input to the plasma was approximately constant over the entire frequency range from 50 Hz to at least 2.0 MHz.

The deposition rate of the polymeric thin film resulting from ethane as the working gas, in units of $\mu g/cm^2$-min, is plotted as a function of frequency in figure 24.13. The deposition rate varies by less than a factor of 1.5 up to 1 MHz, after which it drops to low levels. The deposition rate has two broad peaks, which Morita *et al* (1979) attribute to ion (near 5 kHz) and electron (near 600 kHz) mobility drift trapping between the electrodes during an RF cycle. This suggested trapping mechanism is the same as that analyzed in section 12.5.2 for the OAUGDP.

The deposition rate in the experiment of Morita *et al* (1979) does not vary significantly as a function of frequency until 1 MHz, after which it drops to low levels. This is approximately the *critical sheath frequency* discussed in section 19.5.3, and given by equation (19.37). Below this frequency, the plasma behaves resistively and ions can transit the sheath (although they might be trapped in the plasma volume); above the critical sheath frequency the ions cannot transit the sheath, and the plasma behaves capacitively. These data indicate that

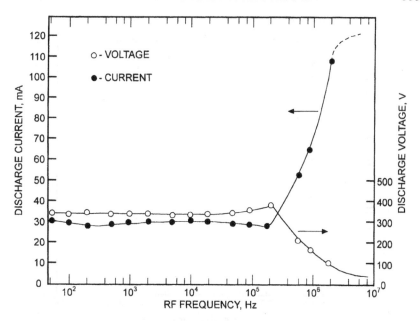

Figure 24.12. Polymeric deposition reactor characteristics reported by Morita *et al* (1979) in which they varied the frequency of a parallel-plate RF reactor over the range from 50 Hz to 13.56 MHz. The plasma source voltage–current characteristic is plotted as a function of frequency. ©1979. Reprinted by permission of John Wiley & Sons, Inc.

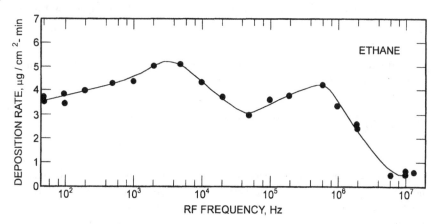

Figure 24.13. The deposition rate of a polymeric thin film reported by Morita *et al* (1979) for the conditions of figure 24.12. The data shown use ethane as the working gas. The deposition rate in units of μg/cm^2-min is plotted as a function of frequency. ©1979. Reprinted by permission of John Wiley & Sons, Inc.

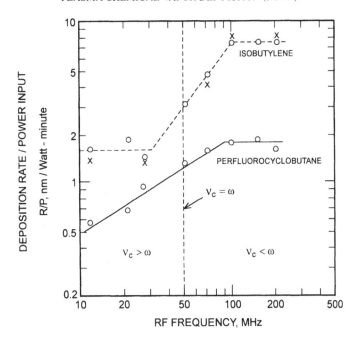

Figure 24.14. The effect of electron collisionality on the polymeric deposition rate from Claude *et al* (1987) is shown for two precursor gases: isobutylene on the upper curve, and perfluorocyclobutane on the lower curve. The data are plotted as functions of the RF frequency on the abscissa, and the ratio of deposition rate/plasma input power on the ordinate.

polymeric deposition reactors, perhaps all deposition reactors, should be operated below the critical sheath frequency of equation (19.37).

24.5.4 The Effects of Electron Collisionality on PCVD

It has been shown by Claude *et al* (1987) that the ratio of the electron collision frequency to the RF driving frequency has a significant and systematic effect on polymeric deposition. In their experiment, they generated an RF plasma, the frequency of which could be varied from 12–400 MHz. The working gas pressure was held at 27 Pa (200 mTorr), with about 80% argon and 20% a monomeric precursor. In their experiments, the electron collision frequency was approximately constant at 50 MHz while the RF driving frequency was varied.

The effect of electron collisionality on the polymeric deposition rate is shown in figure 24.14 for two precursor gases: isobutylene on the upper curve, and perfluorocyclobutane on the lower curve. The data of Claude *et al* (1987) are plotted as functions of the driving frequency on the abscissa, and the ratio of deposition rate/plasma input power on the ordinate. The deposition index on the

Table 24.6. Plasma source characteristics for thin-film deposition.

Source	Section	Neutral gas Pressure p (Torr)	Electron kinetic temperature T_e (eV)	Electron number density (n_e/m^3)
DC parallel plate	16.1.1.1	0.05–1	2–4	10^{15}–10^{16}
RF parallel plate	16.1.1.2	0.05–1	3–6	10^{16}–10^{17}
Magnetron	16.1.2	0.001–0.2	2–4	10^{14}–10^{16}
Electron bombardment	16.2.1	10^{-4}–10^{-2}	2–5	10^{14}–10^{15}
DC arcjet	15.5	10–760	1–2	10^{18}–10^{21}
Inductively coupled	16.2.3	10^{-4}–10^{-1}	1–10	10^{15}–5×10^{18}
Microwave ECR	16.2.2.1	10^{-5}–10^{-2}	5–15	10^{16}–10^{18}
Microwave non-resonant	16.2.2.2	10^{-3}–5	5–15	10^{14}–10^{17}

ordinate is proportional to the thickness of deposited polymer per joule of energy transferred to the plasma.

In figure 24.14, the electrons collide with the argon atoms more than once per RF cycle to the left of the vertical line at 50 MHz; they collide less frequently than once per RF cycle to the right of this line. It is clear from figure 24.14 that electron collisionality with respect to the driving RF frequency has an important effect on polymeric deposition in this case, and perhaps for all deposition reactors. Clearly, polymeric deposition is most rapid and/or most efficiently done when the electrons persist for two or three RF cycles before colliding. The data in figure 24.14 can be understood on the basis that at low frequency, the electrons collide and give up the energy gained from the electric field before they have an opportunity to acquire the full energy corresponding to the maximum RF electric field. These low-energy electrons are presumably less efficient in producing active species. At high frequency, the electrons can acquire the maximum energy associated with the electric field, and their higher energies then produce more active species, including precursors for polymerization.

24.6 SUMMARY OF DEPOSITION REACTOR PLASMA PARAMETERS

Characteristic values of the neutral gas pressure, electron kinetic temperature, and electron number density of plasma sources used for thin-film deposition are listed in table 24.6. These values are approximations only, as the database of well diagnosed industrial plasmas is very small, and the wide range of configurations and operating conditions actually used in industrial applications can lead to large differences in these parameters.

REFERENCES

Anonymous 1991 *Plasma Processing of Materials: Scientific Opportunities and Technological Challenges* (Washington, DC: National Academy Press) ISBN 0-309-04597-5

Asmussen J 1989 Electron cyclotron resonance microwave discharges for etching and thin-film deposition *J. Vac. Sci. Technol.* A **7** 883–93

Barnes M S, Keller J H, Forster J C, O'Nell J A and Coultas D K 1992 Transport of dust particles in glow-discharge plasmas *Phys. Rev. Lett.* **68** 313–16

Bell A T 1980 The mechanism and kinetics of plasma polymerization *Plasma Chemistry* vol III, ed S Veprek and M Venugopalan (New York: Springer) pp 43–68 ISBN 0-387-10166-7

Claude R, Moisan M, Wertheimer M R and Zakrzewski Z 1987 Comparison of microwave and lower frequency discharges for plasma polymerization *Plasma Chem. Plasma Process.* **7** 451–64

d'Agostino R (ed) 1990 *Plasma Deposition, Treatment, and Etching of Polymers* (New York: Academic) ISBN 0-12-200430-2

d'Agostino R, Camarossa F, Fracassi F and Illuzzi F 1990 Plasma polymerization of fluorocarbons *Plasma Deposition, Treatment, and Etching of Polymers* ed R d'Agostino (New York: Academic) ch 2, pp 95–162 ISBN 0-12-200430-2

Danilich M J and Marchant R E 1994 *Plasma Deposition of Polymeric Thin Films (J. Appl. Polymer Sci.: Appl. Polymer Symp. 54)* (New York: Wiley) CCC 0570-4898/94/010001-02

David M, Babu S V and Rasmussen D H 1990 RF plasma synthesis of amorphous AlN powder and films *AIChE J.* **36** 871–6

Denaro A R, Owens P A and Crawshaw A 1968 Glow discharge polymerization—styrene *Eur. Polymer J.* **4** 93–106

Donohoe K G and Wydeven T 1979 Plasma polymerization of ethylene in an atmospheric pressure-pulsed discharge *J. Appl. Polymer Sci.* **23** 2591–601

Hollahan J R and Bell A T 1974 *Techniques and Applications of Plasma Chemistry* (New York: Wiley) ISBN 0-471-40628-7

Hollahan J R and Rosler R S 1978 Plasma deposition of organic thin films *Thin Film Processes* ed J L Vossen and W Kern (New York: Academic) ch IV-1 ISBN 0-12-728250-5

I L, Juan W-T, Chiang C-H and Chu J H 1996 Microscopic particle motions in strongly coupled dusty plasmas *Science* **272** 1626–8

Kay E, Coburn J and Dolks A 1980 Plasma chemistry of fluorocarbons as related to plasma etching and plasma polymerization *Plasma Chemistry* vol III, ed S Veprek and M Venugopalan (New York: Springer) pp 1–42 ISBN 0-387-10166-7

Kobayashi H, Shen M and Bell A T 1974 Effects of reaction conditions on the plasma polymerization of ethylene *J. Macromol. Sci.* A **8** 373–91

Lieberman M A and Lichtenberg A J 1994 *Principles of Plasma Discharges and Materials Processing* (New York: Wiley) ISBN 0-471-00577-0

Morita S, Bell A T and Shen M 1979 The effect of frequency on the plasma polymerization of ethane *J. Polymer Sci.: Polymer Chem. Edition* **17** 2775–82

Morita S and Hattori S 1990 Applications of plasma polymers *Plasma Deposition, Treatment, and Etching of Polymers* ed R d'Agostino (New York: Academic) ch 6, pp 423–61 ISBN 0-12-200430-2

Morosoff N 1990 An introduction to plasma polymerization *Plasma Deposition, Treatment, and Etching of Polymers* ed R d'Agostino (New York: Academic) ch 1, pp 1–93 ISBN 0-12-200430-2

Mort J and Jansen F (ed) 1986 *Plasma Deposited Thin Films* (Boca Raton, FL: Chemical Rubber Company) ISBN 0-8493-5119-7

Poll H-U, Arzt M and Wickleder K-H 1976 Reaction kinetics in the polymerization of thin films on the electrodes of a glow-discharge gap *Eur. Polymer J.* **12** 505–12

Raizer Y P, Shneider M N and Yatsenko N A 1995 *Radio-Frequency Capacitive Discharges* (Boca Raton, FL: Chemical Rubber Company) ISBN 0-8493-8644-6

Ratner B D, Chilkoti A and Lopez G P 1990 Plasma deposition and treatment for biomaterial applications *Plasma Deposition, Treatment, and Etching of Polymers* ed R d'Agostino (New York: Academic) ch 7, pp 463–516 ISBN 0-12-200430-2

Rose P W, Kolluri O S and Cormia R D 1989 Continuous plasma polymerization onto carbon fiber *Proc. 34th Int. SAMPE Symp. (Reno, NV)*

Roth J R 1995 *Industrial Plasma Engineering Vol. 1: Principles* (Bristol: Institute of Physics Publishing) ISBN 0-7503-0318-2

Roth J R, Tsai P P and Liu C 1995a Steady-state, glow discharge plasma *US Patent* 5,387,842

Roth J R, Tsai P P, Liu C, Laroussi M and Spence P D 1995b One atmosphere, uniform glow discharge plasma *US Patent* 5,414,324

Shen M And Bell A T 1979 *Plasma Polymerization* (Washington, DC: American Chemical Society) ISBN 0-8412-0510-8

Smith D L 1995 *Thin-Film Deposition: Principles and Practice* (New York: McGraw-Hill) ISBN 0-07-058502-4

Tibbitt J M, Shen M and Bell A T 1976 Structural characterization of plasma-polymerized hydrocarbons *J. Macromol. Sci.* A **10** 1623–48

Veprek S and Venugopalan M (ed) 1980 *Plasma Chemistry* vol III (New York: Springer) ISBN 0-387-10166-7

Veprek-Heijman M G J and Boutard D 1991 The hydrogen content and properties of SiO_2 films deposited from tetraethoxysilane at 27 MHz in various gas mixtures *J. Electrochem. Soc.* **138** 2042–6

Vinzant J W, Shen M and Bell A T 1979 Polymerization of hydrocarbons in a pulsed plasma *Plasma Polymerization* ed M Shen and A T Bell (Washington, DC: American Chemical Society) ch 5, pp 79–85 ISBN 0-8412-0510-8

Wrobel A M and Wertheimer M R 1990 Plasma-polymerized organosilicones and organometallics *Plasma Deposition, Treatment, and Etching of Polymers* ed R d'Agostino (New York: Academic) ch 3, pp 163–268 ISBN 0-12-200430-2

Yasuda H 1978 Glow discharge polymerization *Thin Film Processes* ed J Vossen and W Kern (New York: Academic) ch IV-2 ISBN 0-12-728250-5

——1985 *Plasma Polymerization* (Orlando, FL: Academic) ISBN 0-12-768760-2

25

Plasma Etching

25.1 SURVEY OF PLASMA ETCHING

Plasma etching is a technology essential to microelectronic circuit fabrication. Its importance to this major industry stimulated the publication of several monographs in which the reader can find additional and more detailed information than can be included in this chapter. Useful references in this field include Hollahan and Bell (1974), Vossen and Kern (1978), Powell (1984), Sugano (1985), Morgan (1985), Manos and Flamm (1989), and finally Lieberman and Lichtenberg (1994). These monographs emphasize the heterogeneous reactions and chemical processes associated with etching; this chapter emphasizes its plasma and plasma reactor-related aspects.

25.1.1 History of Microelectronic Technology

The field of microelectronics originated with the first point contact transistor, invented by Bardeen, Brattain, and Shockley in 1947 at the Bell Telephone Laboratories, and reported in 1948 (Bardeen and Brattain 1948, Brattain and Bardeen 1948). A schematic diagram of their first solid state transistor circuit of 1947 is shown in figure 25.1. Over the next two decades, individual transistors were developed to replace vacuum tubes, and miniaturized for this and other electronic applications. In 1956, Bardeen, Brattain, and Shockley jointly received the Nobel Prize in Physics for their contribution.

The first well documented attempt to fabricate what is now known as an *integrated circuit* or *microelectronic chip* was made by Jack St Clair Kilby in 1958 (Reid 1984), while employed by the Texas Instruments Company. A sketch from Kilby's laboratory notebook of one of his early integrated circuits containing several solid state elements is shown in figure 25.2(*a*), and a photograph of one of these early devices is shown in figure 25.2(*b*). This important contribution

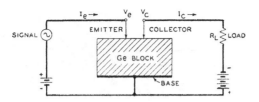

Figure 25.1. Schematic circuit containing the first point contact transistor, from Bardeen and Brattain (1948) and Brattain and Bardeen (1948).

was recognized by Kilby's sharing the Nobel Prize in Physics in 2000. Unlike modern integrated circuits, however, this early version was not monolithic and featured 'flying wires' that connected the solid state components to each other and to terminals at the boundary.

Subsequently, Robert N Noyce of the Fairchild Camera and Instrument Corporation independently conceived the monolithic fully integrated microelectronic chip containing several solid state circuit elements. On figure 25.3 is a photograph of an early (ca 1961) monolithic integrated circuit manufactured by the Fairchild Camera and Instrument Corporation. A circuit is fully monolithic if the components *and* their interconnecting wiring are either embedded in the surface or enclosed in the volume of the integrated circuit. Because of this distinction, Noyce (who died in 1990) is generally regarded as the primary inventor (in 1959) of the modern integrated circuit (Reid 1984).

Once the possibility of monolithic integrated circuits containing multiple solid state components was demonstrated, rapid progress was made in increasing the numbers and reducing the size of individual components in a single integrated circuit. This progress is illustrated in figure 25.4, modified from Tobey (1985). This figure shows the *design rule*, the dimension in microns of the smallest electronically active circuit element, plotted as a function of time from 1970 until recently. While the first monolithic circuits shown in figure 25.3 had a design rule comparable to 100 μm, very rapid progress reduced the design rule below 10 μm by 1971, and it is now below 0.3 μm at the time of writing.

At the same time that increasingly smaller circuit components were fabricated, this reduction in size made possible an enormous increase in the number of individual circuit elements in a single integrated circuit. This technology has reached the point where individual integrated circuits are now routinely made with many millions of individual circuit components. Inspection of figure 25.4 also reveals an interesting relation first noted by Tobey (1985). The progress of microelectronic technology has been such that the design rule of integrated circuits has decreased exponentially with time, with an e-folding time of about seven years. This exponential decrease in the design rule is indicated by the straight lines drawn in figure 25.4.

At some future time, progress to ever-smaller and faster integrated circuits

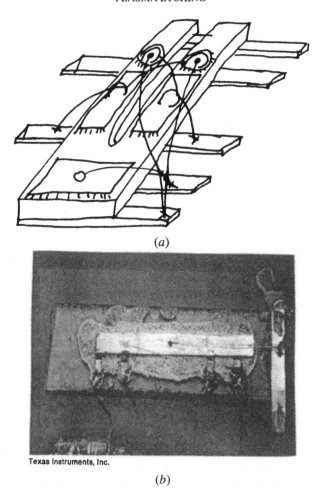

(a)

Texas Instruments, Inc.

(b)

Figure 25.2. Kilby's microelectronic chip of 1958: (a) sketch from Kilby's notebook; (b) photograph of an early microelectronic chip containing several circuit elements (Reid 1984).

will be halted by the laws of physics, including the speed of light, the uncertainty principle, mask diffraction effects, and the size of individual atoms. Other less fundamental technological limitations include removing the heat generated by large numbers of closely spaced circuit elements operating at high speed, illuminating photoresist at high powers and short wavelengths, and arranging reliable connections to external circuitry.

The early integrated circuits were fabricated on silicon wafers up to 10 cm in diameter, which formed their base and structural foundation. Prior to 1990, it was normal practice to simultaneously subject several such wafers to deposition

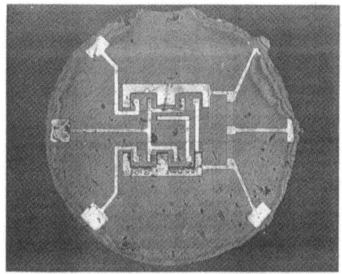

Figure 25.3. An early monolithic integrated circuit manufactured by Fairchild Camera and Instrument Corporation, ca 1961 (Reid 1984).

or etching processes in the same vacuum chamber. As the design rule shrank, the number of circuit elements per integrated circuit increased. Concurrently, it became possible to fit more integrated circuits onto a (larger diameter) wafer, but it also became harder to maintain uniformity of effect on multiple large wafers processed simultaneously. As a result, the microelectronic industry has moved to processing one wafer at a time, each with a diameter of 20 or 30 cm to accommodate the simultaneous processing of a large number of integrated circuits.

The remarkable decrease in design rule and increase in the number of components on an individual integrated circuit required a major change in the technology used to fabricate such circuits. In the earliest days of large-scale integration, isotropic wet chemical etching was sufficient to produce the patterns required. As commercial pressures forced the industry to ever smaller design rules however, the limitations of wet chemical etching became progressively less satisfactory.

By the early 1970s the etching of silicon by intermediate-pressure RF glow discharge plasma reactors was widespread in the microelectronic industry. This technology was developed to the point that when design rules reached approximately 2 μm in the early 1980s, 'wet' chemical etching was almost entirely superseded by 'dry' plasma etching processes. This transition was further motivated by the unwanted byproducts and adverse environmental impacts of wet

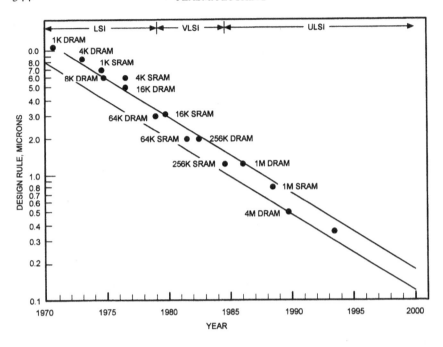

Figure 25.4. An illustration of Tobey's law, according to which the design rules of microelectronic circuits have decreased exponentially with time (modified from Tobey 1985).

chemical etching, compared to plasma-based etching methods (Perry 1993). If suitable photolithographic technology for mask fabrication can be developed, currently available RF glow discharge plasma reactor technology will probably be adequate for the next generation beyond the Pentium-class integrated circuit. Beyond that, more advanced plasma masking and etching technologies than those used in the late 1990s will be required.

The historically important 'wet' chemical etching process, still used in specialty applications, is illustrated in figure 25.5. In this technology, a *mask*, containing openings that define the material to be removed, covers the layer to be etched. The mask is sometimes called a *stencil*, a term better descriptive of its function. A chemical etching medium that attacks the etched layer, but preferably not the mask or the substrate, is allowed to react through openings in the mask. When complete, this process leaves behind the pattern to be transferred to the substrate. The etching medium attacks the etched layer isotropically, so that material is removed horizontally under the mask at the same rate that the etched layer is removed vertically. This results in poor pattern transfer. Such an etching process cannot be used if the width of the mask horizontally is less than twice the thickness of the layer to be etched, since the mask would be entirely undercut by

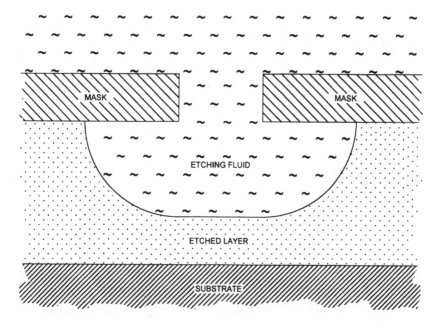

Figure 25.5. Schematic drawing of an isotropic wet chemical etching process.

the etching fluid before it reached the substrate.

The more recently developed 'dry' plasma etching is accomplished at pressures below 133 Pa (1 Torr) in RF glow discharge plasma reactors like those discussed in section 17.6, using the arrangement shown on figure 25.6. Active species from the plasma impinge on the mask and the exposed etched layer at equal rates. Because the mask is relatively unetchable, it remains in place to shield the layer below it from bombardment by active species originating in the plasma. An important requirement of a dry plasma etching process is that the etching occur vertically to the greatest extent possible: that is, that the etching be at least *anisotropic*, or ideally, *vertical*. To the extent that the ions responsible for promoting the etching reaction arrive vertically at the surface of the layer to be etched, a vertical trench with vertical sidewalls will result. In addition, the etching process should display *selectivity*, in which only the layer intended to be etched is actually etched, not the mask or the substrate.

Dry plasma etching is accomplished by starting with a relatively inert molecular gas such as carbon tetrafluoride (CF_4), and allowing it to interact with an RF glow discharge plasma to create active species capable of reacting chemically with the layer to be etched. As an example, CF_4 can be dissociated in a plasma to form carbon trifluoride and atomic fluorine, both very chemically reactive. These active species may react with the material to be etched, silicon for example. It is important to form a volatile reaction product that can leave

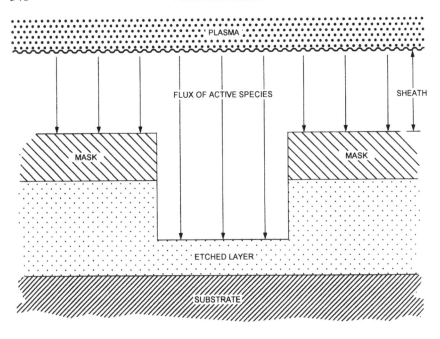

Figure 25.6. A directional, dry plasma etching process producing a trench in an etched layer.

the etching trench and be pumped away by the vacuum system. In the example of atomic fluorine reacting with silicon, the volatile reaction product is silicon tetrafluoride (SiF_4). The CF_4 working gas does not itself react chemically with the silicon being etched, and therefore does not attack the sidewalls of the trench. In this example, the etching process is the result of active species generated by the plasma, and not by the mere presence of the working gas. In many cases, the chemical reaction between the active species and the material to be etched is promoted by ions, electrons, or photons that originate from the plasma, or from external sources.

25.1.2   Characteristic Pattern Dimensions

In discussing plasma etching for microelectronics, it is useful to be aware of the units and size scales discussed in section 14.1.5, which are summarized on table 25.1. This table gives characteristic sizes of features relevant to plasma etching of integrated circuits. Dust and bacteria can be of the same order of size as the microelectronic circuit elements currently being etched, hence the importance of clean rooms. A single layer of a microelectronic circuit might range from 0.5–10 μm. For the best current microelectronic etching technology, the upper limit for the circuit feature size might be from 1–3 μm, and the smallest circuit feature

Table 25.1. Characteristic size scales relevant to microelectronics.

	Size	
Item	Microns (μm)	Ångstroms (Å)
Dust	0.1–10	1 000–100 000
Bacteria	0.5–5	5 000–50 000
Microelectronic circuit layer	0.5–10	5 000–100 000
Large circuit feature size	1–3	10 000–30 000
Small circuit feature size (design rule)	0.15–0.70	1 500–7 000
Wavelength of visible light	0.38–0.78	3 800–7 800
One monolayer ⟨111⟩ silicon	0.000 22	2.2
Diameter of argon atom	0.000 37	3.7
Diameter of oxygen atom	0.000 13	1.3

size (the design rule) might be from 0.20–0.4 μm. This design rule is comparable to the wavelength of visible light. Diffraction effects combined with this small design rule limit the use of visible lithography to produce the masks required for micro-electronic etching, and impose a requirement for shorter wavelength light sources, including plasma sparks and synchrotron sources for microlithography.

On the atomic scale, table 25.1 lists the thickness of one monolayer of the ⟨111⟩ plane of a silicon crystal. One of the larger atoms widely used in microelectronic plasma processing is argon, which has a diameter of 3.7 Å; and one of the smaller atoms is atomic oxygen, with a diameter of 1.3 Å. Several of the size scales listed in table 25.1 are further illustrated by the scaled drawing in figure 25.7, a characteristic microelectronic structure. The dimensions of the components are on the order of the wavelength of visible light. A monolayer of silicon in a silicon crystal would be smaller than the finest line shown in this figure.

25.1.3 The Plasma Etching Process

The plasma etching process is illustrated in figure 25.8, which shows a plasma etch in progress with a trench width of about one micron. We will proceed from the top to the bottom of this diagram, and discuss some key terms and processes encountered in plasma etching.

(1) The *plasma* is typically generated by an RF or microwave *glow discharge* source that provides the active species necessary for etching. The plasma also provides the energetic ions that may be required to promote a heterogeneous etching reaction between the surface of the etched layer and either the working gas or one or more active species. The etching rate

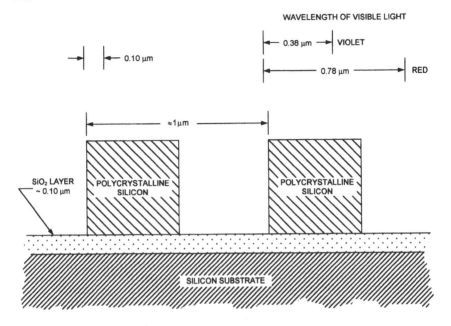

Figure 25.7. An illustration of some characteristic scale sizes relevant to microelectronic circuit fabrication.

is normally increased substantially by ion bombardment of the layer to be etched.

(2) The *sheath* consists of the region between the plasma and the mask. The sheath normally has an electric field pointing toward the workpiece, which accelerates ions to the wafer and helps to retain electrons in the plasma. The sheath is at least 10 to 100 times thicker than the dimensions of the microelectronic circuit elements being etched.

(3) The *mask* (sometimes referred to as the *stencil*) contains the pattern that is to be transferred to the layer beneath it. The mask must be highly resistant to etching and capable of being stripped off after etching the layer underneath it is complete, a process called *stripping* or *ashing*.

(4) The *etched layer* (or *film*) is the layer immediately under the mask that is to undergo etching. At the end of the etching process, the pattern formed by openings on the mask should be transferred without distortion to the etched layer. A typical layer thickness may be from one-half to a few microns. The term *layer* is often used to refer to the remainder of any etched layer after etching is complete.

(5) The *trench* is the cavity produced in the film or etched layer by the etching process. When complete, the trench extends from the mask to the substrate.

(6) The *substrate* is the material immediately under the etched layer or film, where etching proceeds at a greatly reduced rate or stops.

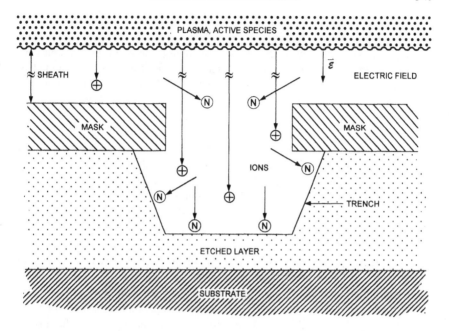

Figure 25.8. A directional plasma etch in progress, illustrating the major factors responsible for dry plasma etching.

(7) The *wafer* provides mechanical support and rigidity for all layers of a microelectronic circuit. It is the foundation on which all implantation, deposition, and etching operations are conducted. It is sometimes also referred to as a '*platelet*', '*specimen*', or '*workpiece*'.

(8) The *neutral working gas* may be a single input feed gas or a mixture of gases that reacts with the plasma to form the active species responsible for etching. The working gas may, but usually does not, react directly with the material to be etched.

(9) The *chemically reactive species* react with the material to be etched, or with other components of the microelectronic chip. Chemically reactive species may include the active species produced by the plasma, the neutral working gas fed into the reactor, and reaction products of etching.

(10) The *active species* are formed by the plasma and were discussed in section 14.2. They may include ions, electrons, free radicals, excited states, molecular fragments, atomic species, and photons. One or more of the active species, such as ions, may promote etching reactions at the bottom, but not at the sidewalls of the trench. Other active species may serve directly as an etching agent of the layer to be etched.

(11) *Energetic promoters of surface reactions*. The successful promotion of etching reactions at the bottom of a trench requires not only the arrival of

chemical precursors to the etching reaction, but also a significant amount of energy. Ions accelerated by the electric field in the sheath above the mask, as shown in figure 25.8, may have enough energy to serve as energetic promoters of surface reactions. Such energetic ions can be provided by beam sources originating outside the plasma. Photons that originate from laser illumination or electron beams can also provide the energy required to promote etching reactions.

The deposition of energetic ions, electrons, or photons at the bottom, but not at the sidewalls of the trench, is one of the principal factors that makes possible the directionality of plasma-assisted microelectronic etching. Energetic promoters of surface reactions also may be required to obtain useful etching rates for some relatively unreactive combinations of working gas and material to be etched. Beyond this, energetic promoters of surface reactions may also suppress the onset of unwanted polymerization on the surface being etched or on the surface of the mask.

(12) *Neutral reaction products* are the chemical reaction products of etching reactions. These products must be volatile so that they can be removed from the system, and not accumulate on, or contaminate, the surfaces being etched. The detailed heterogeneous chemical reaction processes responsible for etching begin with initial adsorption of the etching species on the surface. This may include disassociation, if the etching species is molecular. The next step in this surface chemical reaction is the formation of a product molecule from one or more atoms of the etched material and one or more atoms of the active species or the neutral working gas. Finally, the product molecule is desorbed from the surface, and if volatile, it will be pumped away by the vacuum system.

(13) *Etching tools* are the sophisticated vacuum systems in which plasma etching operations are conducted. These vacuum systems are made of expensive materials that resist chemical attack by working gases and etching agents. They must be outfitted with elaborate subsystems to control the gas flow rates, the gas pressure, and the plasma characteristics. The requirement that etching occur under vacuum enforces batch processing of wafers. Early fabrication practice processed many 7.5 or 10 cm diameter wafers simultaneously. As the design rule decreased, fabrication practice has shifted to processing one large (20 or 30 cm) wafer at a time containing many more integrated circuits than the smaller wafers.

25.1.4 Performance Issues in Plasma Etching

At the design rules currently used in the microelectronic industry, only plasma-based etching is capable of producing commercially viable integrated circuits. Nonetheless, it is important to look beyond the current situation and consider the advantages and disadvantages of plasma etching as it is currently practiced, because these performance issues will guide future research and development.

25.1.4.1   Positive Features of Plasma Etching

Especially by comparison with wet chemical etching, 'dry' plasma etching has a number of significant advantages. These include the ability to produce highly directional, even vertical, etching of trenches, and the potential for production line automation. In plasma etching, the minimum design rule is limited by masking technology, and not by the etching technology, as is true of wet chemical etching. Unlike the adverse environmental impacts of wet chemical etching discussed in section 14.1.4, plasma etching is a clean process that produces an absolute minimum of toxic or unwanted byproducts. Plasma etching requires much lower quantities of specialty chemicals/gases than are required for wet chemical etching, and the very small quantity of reaction products from the etching process results in minimal environmental impact.

25.1.4.2   Negative Features of Plasma Etching

A negative feature of plasma etching is the requirement that the etching process be conducted in a vacuum system. The etching 'tools' or vacuum systems needed to do plasma etching are expensive because of a requirement for automation, and a requirement for expensive materials that will not be attacked by the working gases or etching agents. A related difficulty has emerged as design rules in the microelectronic industry decreased below one micron. It has been necessary to go from the *intermediate-pressure* vacuum regime between 13.3 and 133 Pa (0.1– 1 Torr), which is relatively easy to produce and maintain, to the *low-pressure* vacuum regime below 1.3 Pa (10 mTorr), where the vacuum technology is more sophisticated and the pump-down times longer.

A further disadvantage of current plasma etching methods is the requirement for batch processing imposed by the use of vacuum systems. The automation of microelectronic circuit fabrication has proceeded in the direction of keeping the wafers in a vacuum system for the largest possible fraction of the steps required to produce an integrated circuit. However, for some operations the wafer must be taken in and out of the vacuum system, which slows the production process. Further, the requirement for batch processing may cause bottlenecks by either slow deposition or etching steps.

25.2   PATTERN TRANSFER BY PLASMA-RELATED ETCHING

Integrated circuits may contain millions of individual electronic circuit elements embedded in chips arranged in a planar array on a silicon wafer. In order to connect these resistors, capacitors, and semiconducting circuit elements in each chip, multiple layers of electrical insulators and conductors, patterned to avoid the 'flying wire' connections illustrated in figure 25.2, must be deposited on the wafer. At least six layers of conducting connections and up to several dozen deposition and etching operations may be required to produce the larger or more sophisticated integrated circuits.

A characteristic step in building up an integrated circuit involves the deposition of a thin layer of electrical conductor above an insulating layer. This is followed by etching away the unwanted material, to leave behind a pattern that forms the circuit connections appropriate to that layer. The two-dimensional pattern appropriate to the electrical circuitry of a given layer is carried by an unetchable *mask*, or *stencil*, which is formed by photolithographic techniques that are beyond the scope of this text.

25.2.1 Etching Terminology

In this section, we discuss the nomenclature and methods used to transfer a pattern from a mask to a layer to be etched by plasma-related etching processes.

25.2.1.1 Directionality

How well the pattern on a mask is transferred to the layer below it depends strongly on the *directionality* of the etching process, illustrated in figure 25.9. At the top of this figure is an unetchable mask that contains the pattern to be transferred to the layer to be etched. This layer is located on a substrate that is normally not to be etched. In figure 25.9(*b*) is shown the consequences of an *isotropic etch*, in which the etching agent acts isotropically in all directions. Such an agent etches horizontally under the mask at the same rate that it etches vertically downward into the layer to be etched. Obviously, this undercuts the mask and leads to poor pattern transfer, and it is not an acceptable approach when the feature size on the mask is comparable to the depth of the layer to be etched.

A *directional* or *anisotropic etch* is illustrated in figure 25.9(*c*), in which the etching rate is slower horizontally than vertically. This results in a trench that is etched vertically, but with some undercutting of the mask. The extent of the undercutting depends on the rate at which the etching process proceeds horizontally (parallel to the plane of the mask), relative to the vertical etch rate. Finally, the ideal pattern transfer occurs for a *vertical etch*, illustrated in figure 25.9(*d*). In this case, the etching process proceeds only in the vertical direction, and not at all horizontally. This results in a vertical trench under the openings in the mask, and the best possible transfer of the mask pattern to the etched layer. *Over-etching* consists of etching for longer than is necessary to expose a substrate or to remove a film, and may result in unwanted etching of a susceptible substrate (lack of *selectivity*).

25.2.1.2 Etching processes

The production of a layer of electrical interconnects among circuit elements on a wafer can be accomplished in one of two ways. The first is by *additive etching*,

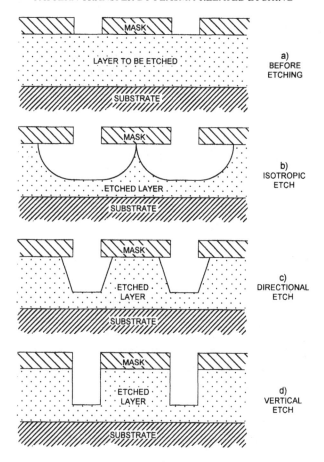

Figure 25.9. Various degrees of etching directionality: (*a*) an unetched layer; (*b*) result of an isotropic etch: (*c*) result of a directional etch; and (*d*) result of a vertical etch.

illustrated in the left-hand column of figure 25.10, and a second is by *subtractive etching*, illustrated in the right-hand column. In both cases, one starts at the top with a substrate on which a feature (a connecting conductor) is to be deposited, and finishes with the feature deposited on the substrate. The additive etching strategy is often preferred, since it requires one less processing step to achieve the result.

25.2.1.3   Endpoint of Etch

The *endpoint* of an etching process occurs when the trench reaches the intended depth in the etched layer. This usually occurs when the trench arrives at the substrate and completely transfers the pattern on the mask to the etched layer. In

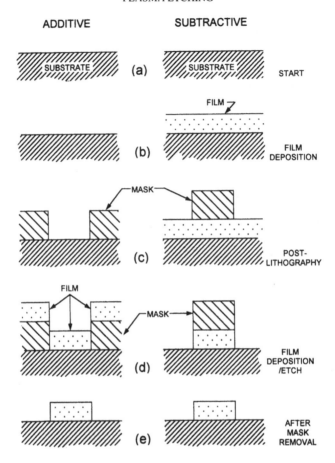

Figure 25.10. Additive and subtractive etching processes to produce a surface feature on a substrate.

actual manufacturing situations, however, the etched layer can be of non-uniform thickness over the surface of a wafer containing many individual microelectronic chips. In addition, non-uniformities in the plasma density or the flux of active species can result in different parts of a wafer being etched at different rates.

Both effects can lead to the situations illustrated in figure 25.11. At the top is illustrated an *incomplete etch*, in which the etching process has stopped short of the bottom of the etched layer. Figure 25.11(*b*) illustrates a *completed etch*, in which the pattern of the mask has been transferred to the etched layer. Figure 25.11(*c*) illustrates an *overetch*, in which the etching process has been continued long enough to etch into the substrate. In most situations, the substrate etches more slowly than the etched layer. Since the consequences for the final product of an incomplete etch are almost always worse than an overetch, the

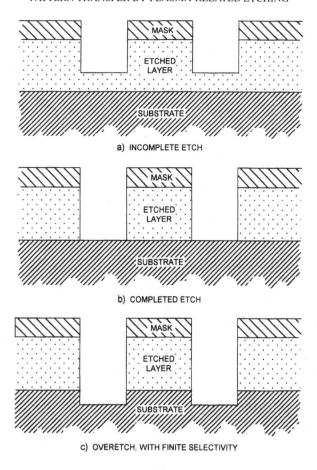

a) INCOMPLETE ETCH

b) COMPLETED ETCH

c) OVERETCH, WITH FINITE SELECTIVITY

Figure 25.11. Relative completion of the etching process: (*a*) an incomplete etch; (*b*) a completed etch; and (*c*) an over-etched trench with finite selectivity.

etching process is normally allowed to go on longer than is required to produce a nominal complete etch. This ensures that all of the material to be removed is, in fact, gone, even at the cost of having some over-etching of the kind illustrated in figure 25.11(*c*).

To prevent excessive over-etching, *endpoint detection* is used to determine when an etching operation is complete. Usually, when the etching process reaches the substrate, the flux of etching reaction products will suddenly decrease. This decrease can be detected by monitoring, with a mass spectrometer, the concentration of reaction products in the effluent gas pumped out of the vacuum system. At the same time that the primary etching reaction products decrease, a different set of secondary reaction products, obtained from etching the substrate, also will appear in the mass spectrometer data. Both sets of reaction products

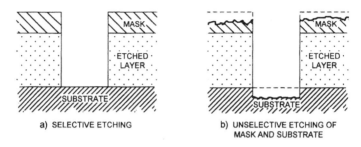

Figure 25.12. Relative selectivity of an etching process: (*a*) selective etching; (*b*) unselective etching of mask and substrate.

can indicate that the etching process has reached the bottom of a layer being etched. These and other procedures to monitor the completion of etching a layer are referred to as *endpoint detection*.

25.2.1.4 Etching Selectivity

The ideal etching agent would be *selective* and etch a layer vertically and rapidly, while leaving the mask and the substrate unaffected. This would produce the ideal *selective etch* in figure 25.12(*a*). A vertical etch is illustrated, but it might also be directional, with sloping sidewalls. A selective etching process transfers the pattern on the mask to the etched layer, while etching the mask and/or the substrate at greatly reduced rates.

The selectivity of an etch can be defined as

$$S \equiv \frac{\text{etching rate of layer}}{\text{etching rate of substrate}}. \qquad (25.1)$$

If the selectivity is finite, an *unselective etch* will occur, with the consequences shown in figure 25.12(*b*). In this case, the mask and/or the substrate will be etched in addition to the etched layer.

25.2.1.5 Detrimental Outcomes

Ideal etches and perfect pattern transfer may not occur. One of the possible detrimental outcomes is *mask erosion*, illustrated in figure 25.13. In this situation the mask is etched at a rate high enough compared to the rate at which the trench is advanced into the etched layer, that the quality of the pattern transfer is affected. The erosion of the mask at the edge of the trench can progressively remove its protective covering of the etched layer, and produce a sidewall angle somewhat less than vertical, as illustrated in figure 25.13.

In addition to opening up a larger than intended area of the mask for etching, mask erosion can lead to *faceting* of the mask, also illustrated in figure 25.13. The

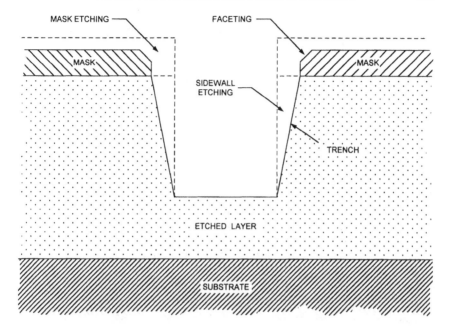

Figure 25.13. Consequences of unselective mask erosion.

phenomenon of faceting occurs because the edge of the mask sees a larger solid angle and a higher flux of etching active species than do the flat surfaces on the top or sidewalls of the mask. On the upper surface away from the edges of the mask, the flux of active species with a number density n_a is given by equation (2.28) from Volume 1 as

$$\Gamma = \tfrac{1}{4}n_a\bar{v} \qquad \text{particles/m}^2\text{-s} \qquad (25.2)$$

where $\bar{v}$ is the thermal velocity of the etching species.

Equation (25.2) is appropriate only for flat surfaces bombarded by incoming particles originating in the hemisphere above the point of interest. At the edge of a mask or etched layer, particles can not only approach the edge from the hemisphere above, but also from the half-hemisphere beside and below it. Thus, at the edges of masks and exposed etched layers, the particle flux is not given by equation (25.2), but by a flux that is up to half again as large, or

$$\Gamma_e = \tfrac{3}{8}n_a\bar{v} \qquad \text{particles/m}^2\text{-s.} \qquad (25.3)$$

This higher flux results in faster etching of edges than horizontal surfaces, and the phenomenon of faceting. If faceting continues long enough, a significant loss of definition of the pattern carried by the mask will result.

A common problem in plasma etching is *mask undercutting*, illustrated in figure 25.14. The most widely used etching agents are active species from

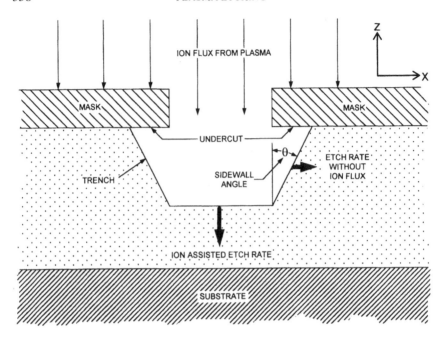

Figure 25.14. A directional etch with a finite undercutting rate producing sidewalls of half-angle θ.

the plasma that undergo chemical reactions with the layer to be etched. These reactions are promoted by vertical ion bombardment. Since such ions do not reach the sidewalls below the mask, ion-promoted etching reactions do not occur, and the sidewalls are subject only to the much lower purely chemical etching rate due to the active species or working gas. Any etching that occurs in the absence of an ion flux is sufficient to undercut the mask, as shown in figure 25.14, and thereby degrade the quality of the pattern transfer.

A second detrimental process occurs if the ion flux arrives at the mask with a small component of velocity parallel to the surface, as illustrated in figure 25.15. Such a horizontal component may be a remnant of the ion thermal velocity in the plasma, or result from ion scattering in the sheath between the plasma and the wafer. Incoming ion trajectories that depart from the vertical allow the ions to reflect from the sidewalls, and arrive at the bottom of the trench. This additional flux of reflected ions etches the bottom of the trench more rapidly at its edges than in the center, leading to the phenomenon of *trenching*, illustrated in figure 25.15. Trenching also occurs in *sputter etching*, when the trajectories of an incident beam of energetic ions are not parallel. Reflection of angled beam ions from the wall in the manner shown in figure 25.15 also will cause trenching.

Finally, figure 25.16 shows the consequences of *redeposition*, which may result either when a reaction product of etching is not volatile, or from a sputter

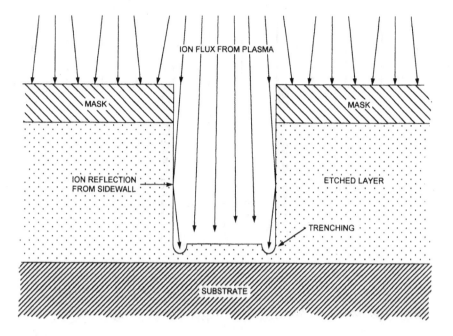

Figure 25.15. Trenching, resulting from a small ion velocity component parallel to the surface of the wafer.

etching process in which sputtered atoms redeposit on the sidewalls. The sputtered or non-volatile material builds up to a thickness that decreases with depth into the trench, since the top of the trench has a longer time to accumulate material. Such a layer of re-deposited material is normally detrimental, since it can form small particles that interfere with the deposition and etching of subsequent layers. In some cases, however, re-deposited material on the sidewalls can be a beneficial *passivating agent*, which shields the sidewalls from the etching agent. The etching agent then acts only on the bottom of the trench, producing a vertical etch.

25.2.1.6 *Performance Indices*

Many gases used in microelectronic etching are very expensive, greenhouse gases, toxic to humans, detrimental to the environment, flammable, or explosive. Some are several or most of these, so it is generally desirable to use up as much as possible of the working gases. The fraction of the input gas used for etching is the *utilization factor*, defined as

$$U \equiv \frac{\text{flow rate in exhaust}}{\text{input flow rate}}. \tag{25.4}$$

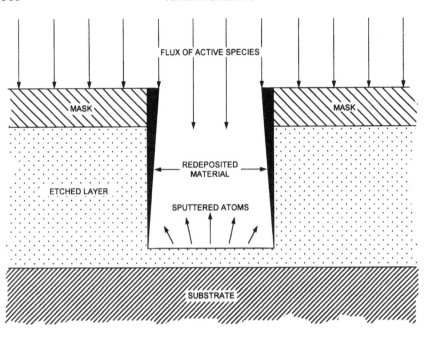

Figure 25.16. An example of etching with a buildup of re-deposited material, which may be either a non-volatile reaction product or sputtered material from the bottom of the trench.

Minimizing the utilization factor is a compromise between minimizing the total amount of gas required for the etching process, and ensuring uniformity of concentration and effect over the entire surface of the wafer undergoing etching.

A second performance index is the *loading effect*, in which the etching rate is reduced by having more than one wafer exposed in a plasma reactor at any given time. If there are m wafers undergoing etching simultaneously, then the loading effect equation is given by

$$\frac{R_0}{R_m} = 1 + m\phi_f \tag{25.5}$$

where R_0 is the etching rate in an empty reactor, R_m is the etching rate of the same reactor loaded with m wafers, and the parameter ϕ_f is the ratio of working gas consumed by etching a single wafer to that lost by recombination and reactions with the walls of the vacuum system. The parameter ϕ_f shown in equation (25.5) is the slope of the loading effect curve, obtained by plotting the left-hand side of equation (25.5) against the number, m, of wafers being etched. The parameter ϕ_f is proportional to the area of the wafers exposed to the working gas.

As the endpoint is approached and the substrate exposed, the area of film being etched is rapidly reduced. Equation (25.5) then predicts that the etching rate will be speeded up, finally reaching the etching rate of the unloaded reactor at the endpoint. Some etching agents are so reactive that the amount used in etching the wafers is small by comparison with that lost in chemically attacking the interior of the vacuum system. In such cases, the loading effect is negligible.

In etching integrated circuits, it is sometimes necessary to make deep trenches for capacitors or isolation barriers, with a ratio of depth/width of 10 or more. The low pressure at which such vertical etching is done may be in the free molecular flow regime. In this regime, the mean free paths of the reaction products and the working gas may be greater than the width of the trench. When this happens, a phenomenon called *microloading* may occur. *Microloading* is a reduction of the normal, open-surface etching rate due to the confined nature of the trench and free molecular flow effects.

25.2.2   Isotropic Etching Agents

Isotropic etching provides a mechanism for pattern transfer in printed circuit boards, and provided pattern transfer during the early development of integrated circuits prior to 1980. Several etching agents can produce the isotropic etch illustrated in figure 25.5. Isotropic *wet chemical etching* is used in the fabrication of large-scale printed circuit boards, and relies on the fact that the etching agent and the material to be etched undergo chemical reactions which lead to a rapid, isotropic etch.

One of the oldest isotropic etching processes is *stripping*, or *plasma ashing*, which consists of removing hydrocarbon-based layers, such as photoresist or insulating films, by oxidizing them with an intermediate-pressure oxygen plasma. Oxygen plasmas usually form atomic oxygen and ozone, and these active species chemically attack hydrocarbons to produce water vapor and CO_2, both of which are volatile. This form of isotropic etching can proceed very rapidly, if the process is conducted at elevated temperatures.

In current microelectronic etching, some neutral working gases, including Cl_2 and F_2, undergo chemical reactions with the material to be etched at room temperature, and without the reaction-promoting effect of energetic ions or other active species from the plasma. These reactions between the neutral working gas and the etched layer are isotropic in the absence of energetic promoters of the etching reaction (ions), and are responsible for much of the mask undercutting that occurs in currently practiced etching operations. Finally, plasmas can produce highly chemically reactive species such as ozone, atomic oxygen, atomic fluorine, and atomic chlorine. These active species can produce an isotropic etch if they arrive at the bottom and sidewalls of a trench at equal rates.

25.2.3   Characteristics of Isotropic Etching

If any of the isotropic etching agents discussed in the previous section are present, the result will be as illustrated in figure 25.17 after the etching process reaches the bottom of an etched layer of thickness d. In an isotropic etching process, the pattern on the mask will be undercut by a distance d, as shown. It is clear from

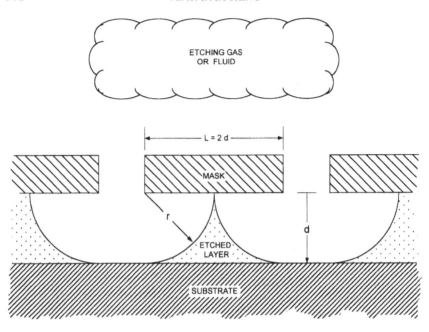

Figure 25.17. Relationship of mask dimensions to etched layer thickness in an isotropic etching process.

the geometry of figure 25.17 that the minimum feature size, L, which can produce a crude feature on the etched layer is given by

$$L_{min} = 2d. \tag{25.6a}$$

It is also clear from the geometry of figure 25.17 that if one wishes to leave behind a feature on the etched layer with a horizontal dimension equal to the depth of the etched layer, d, then the minimum dimension on the mask is given by

$$L = 3d. \tag{25.6b}$$

Since the etched layers are characteristically from 1–5 μm thick, with 2 μm being typical, it is clear that isotropic etching becomes progressively less useful as the feature size approaches 1 or 2 μm. The major technological shift from wet chemical etching to dry plasma etching in the late 1970s and early 1980s was motivated both by environmental considerations, and the increasingly smaller design rules indicated on figure 25.4 that were required to remain competitive.

There are several microelectronic applications, however, for which isotropic etching may still be useful. One of these is the removal of a contoured film, illustrated in figure 25.18. It sometimes happens in building up a complicated multilayer integrated circuit that it may be necessary to deposit a thin film over

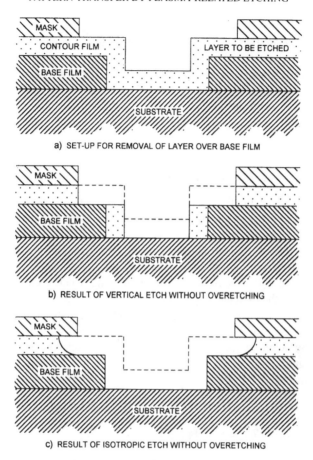

Figure 25.18. A situation in which isotropic etching may be preferable to a directional etching process: (*a*) initial setup for removal of a layer over a base film; (*b*) result of a vertical etch without over-etching; and (*c*) result of an isotropic etch without over-etching.

the entire wafer. This layer must then be selectively removed from some areas in the manner shown in figure 25.18. Figure 25.18(*a*) shows a setup for the removal of a layer that has been deposited over the substrate as shown. This layer is to be removed from the trench in the base film, within the area of the mask opening shown at the top.

As illustrated in figure 25.18(*b*), the result of a vertical etch without substantial over-etching is not satisfactory, because the vertical etch will remove only the horizontal portions of the deposited layer, leaving the vertical portions on the sidewalls of the trench in the base film still in place. An isotropic etch can remove the deposited film from the entire area as shown in figure 25.18(*c*), where the relatively small amount of undercutting of the mask is generally acceptable.

This leaves the rest of the integrated circuit with the required layer under the mask, but has cleared the deposited layer from the trench in the base film.

Other microelectronic uses of isotropic etching include removing dust, oxides, or contaminants from deposition or other etching operations from the surface. Finally, it may be necessary to treat the surface of a wafer or an integrated circuit isotropically to improve the adhesion of solder, potting agents, or even to ensure the printability of advertising logos, serial numbers, or inventory information.

25.2.4 Anisotropic Etching Processes

In the 1960s and 1970s, the limitations of isotropic etching motivated research and development programs intended to produce *directional* or *anisotropic etching* processes. These processes have a vertical etch rate, normal to the plane of the mask, higher than the horizontal etching rate into the sidewalls of the trench. These anisotropic etching processes have in common the preferential delivery of energy to the bottom of the trench (normally by ions) where etching takes place. The sidewalls of the trench do not receive a directed energy flux.

The delivery of an energy flux to the bottom of a trench is more important to producing an anisotropic etch than the medium or particle that carries the energy. Most anisotropic microelectronic etching currently is done by a beam or directed flux of energetic ions. However, it has been shown that a beam of electrons or photons delivered to the bottom of a trench in the presence of an appropriate etching gas will also produce a directional etch.

25.2.4.1 *Inert Ion-Beam Etching (IIBE)*

Historically the first directional etching was accomplished by sputtering with an inert ion-beam etching arrangement similar to that illustrated in figure 25.19. In *inert ion-beam etching*, a beam of ions is produced by a space-charge-limited source, usually a *Kaufman* (discussed in section 6.5) or a *Freeman ion source* (discussed in section 6.8). The ions from such a source impinge on a wafer to be etched in the manner illustrated in figure 25.19. In this relatively primitive form of directional etching, directionality is accomplished by sputter-induced erosion of the material at the bottom of the trench. This sputtered material may not react chemically with the working gas. If not, the material either re-deposits on the sidewalls of the trench, exits the trench and re-deposits on the surface of the mask, or re-deposits elsewhere in the interior of the vacuum system. Some of the difficulties with pure sputter etching were alluded to previously; figure 25.15 illustrates the phenomenon of trenching, and figure 25.16 illustrates the buildup of re-deposited material on the sidewalls of a trench.

Further information on the phenomenon of sputtering was presented in section 14.5, where additional information on sputtering yields for a selection of incident ions and target materials are listed. Some data relevant to sputter etching

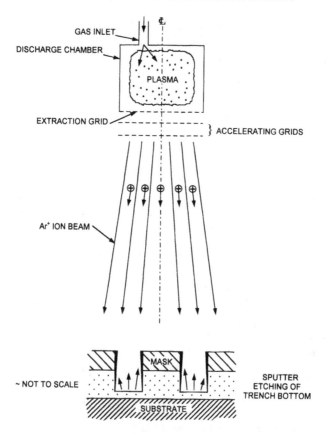

Figure 25.19. An inert ion-beam etching apparatus, in which the etching process is accomplished by sputtering due to monoenergetic ions.

taken from Coburn and Winters (1979b) are reproduced on figure 25.20. The measurements shown were made in a specialized laboratory apparatus illustrated in the upper left-hand side of the figure. A silicon wafer was subject to bombardment by 500 eV argon ions for 365 s, and then to a combination of argon ions and neutral fluorine gas after that time. The data prior to 365 s in figure 25.20 represent the sputter etching rate of 500 eV argon ions on silicon for the conditions measured; this rate was approximately 2.5 Å (0.25 nm)/min.

25.2.4.2 Reactive Sputter Etching

Reactive sputter etching is a process in which energetic ions react chemically with the material to be etched. This process is infrequently utilized, in part because it is more difficult to develop and maintain an ion source to produce chemically reactive ions, than it is a source that produces chemically inert ions.

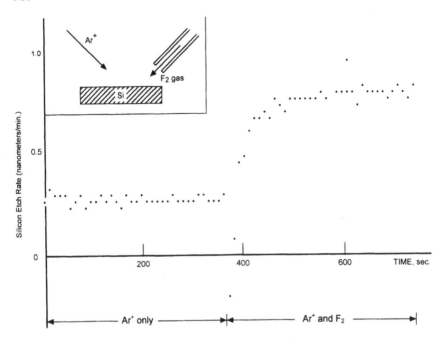

Figure 25.20. Data from Coburn and Winters (1979b) showing sputter etching of silicon by 500 eV argon ions prior to 365 s, and chemically assisted (by fluorine gas) ion-beam etching after that time.

25.2.4.3 Reactive Ion-Beam Etching (RIBE)

A third directional etching process is *reactive ion-beam etching* (RIBE), illustrated in figure 25.21 and discussed by Powell and Downey (1984). This process also uses an energetic ion beam, but pure physical sputtering is not the primary mechanism yielding a directional etch. In RIBE, the energetic beam ions promote chemical reactions between the working gas and the material to be etched.

An example of RIBE is illustrated by the data on the right-hand side of figure 25.20. The presence of fluorine working gas when a 500 eV argon ion beam hits the surface of the silicon results in etching the silicon at a rate at least three times higher than the pure sputter etching rate of the argon ions alone. Figure 25.20 represents a case of energetic ion-promoted chemical etching, since the argon ions themselves do not react chemically with either the silicon or the fluorine. They promote a chemical reaction between the fluorine and the silicon that produces SiF_4, a volatile reaction product.

The ion-promoted etching process illustrated in figure 25.21 shows an energetic argon ion beam impinging on a silicon layer in the presence of XeF_2 working gas. In this example, energetic argon ions promote a chemical reaction

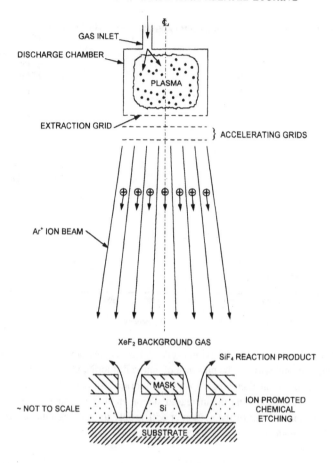

Figure 25.21. A schematic drawing showing RIBE, in which the energetic ions catalyze a reaction with a background etching gas to produce a directional etch.

that strips fluorine atoms from the xenon and causes them to react with the silicon, to produce silicon tetrafluoride (SIF_4), a volatile reaction product.

Some classic data on the interaction of an argon ion beam with silicon in the presence of xenon difluoride gas are taken from Coburn and Winters (1979a) and shown on figure 25.22. These data illustrate the result of a 450 eV argon ion beam impinging perpendicular to a silicon wafer, while xenon difluoride gas was made to flow from a small tube over the wafer. These data show that up to 200 s, with only xenon difluoride gas present, the purely isotropic, chemical etching of the silicon by xenon difluoride proceeds at the relatively low rate of approximately 5 Å/min. When the argon ion beam is turned on at 200 s, the argon-ion-promoted etching rate of silicon increases by more than a factor of 10, to levels of at least 55 Å/min. This greatly increased etching rate is due to the

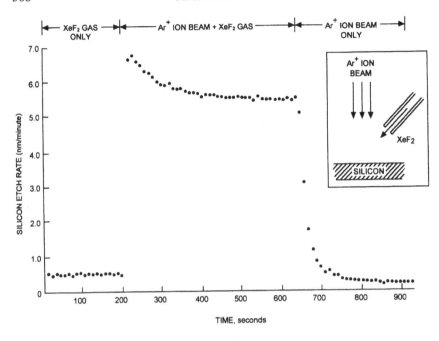

Figure 25.22. Data of Coburn and Winters (1979a) on the interaction of 450 eV argon ions with silicon, with and without the presence of xenon difluoride gas. The etching rate prior to 200 s is due to chemical etching by xenon difluoride gas. The 450 eV argon ion beam was turned at 200 s, producing the increased silicon etching rate shown. After the xenon difluoride was pumped out of the system at 650 s, only the etching rate caused by argon ion sputtering is evident.

effect of the energetic argon ions on the reaction rate of xenon difluoride with the silicon. Also interesting is that, after 660 s, the argon ion beam was left on, and the flow of xenon difluoride turned off. The etching rate of silicon then rapidly declined to a level of about 3 Å/min, characteristic of the sputtering rate of silicon by the argon ion beam alone.

25.2.4.4 Plasma-Assisted Etching

Plasma-assisted etching requires only the presence of a plasma to produce the ions and/or active species responsible for etching. It does not utilize an externally generated ion beam such as those illustrated in figures 25.19 and 25.21. This form of etching is also referred to by other names in the literature, some of which are confusing because they have related specialized meanings. *Plasma-assisted etching* is sometimes referred to simply as *plasma etching*, although this term also refers to a particular type of RF electrode arrangement in parallel-plate capacitive RF glow discharge plasma reactors (see section 17.6.3.1). Plasma-assisted etching is sometimes confused with reactive sputter etching. In this text, we will reserve 'reactive sputter etching' for the reactive-ion-beam-induced etching discussed in section 25.2.4.2. Plasma-assisted etching is also sometimes referred to as *'reactive ion etching'* which emphasizes the chemically active role of energetic ions accelerated across the sheath between the plasma and the wafer. Here, we reserve the term 'reactive ion etching' for its more specialized sense in describing a particular electrode arrangement in parallel-plate RF glow discharge plasma reactors (see section 17.6.3.1).

Plasma-assisted etching is accomplished in the manner illustrated previously in figure 19.1. The directionality of the etching process is achieved by energetic ions that originate in the glow discharge plasma, and are accelerated to high velocity across the sheath. In the example of figure 19.1, the ion-induced chemical etching is similar in principle to the RIBE discussed previously. The directionality of the energetic ions ensures a much higher rate of chemical reactions vertically than horizontally, because the horizontal etching rate is limited to chemical reactions that proceed without the benefit of promotion by energetic ions.

The ions responsible for directionality acquire energy by acceleration across the sheath between the plasma and the workpiece. The potential across this sheath may range from 10 to 120 V. When the ion energies exceed 50 eV, undesirable 'radiation damage' may occur to some features on the microelectronic chips. At energies between 10 and 50 eV, the ions have enough energy to promote chemical reactions between the working gas and the etched material. The directionality and etching rates of ions originating from plasmas are similar to the more carefully controlled laboratory data shown in figures 25.20 and 25.22. The latter data resulted from the impingement of low-energy ion beams on the silicon to be etched.

A second category of plasma-assisted etching results from chemical reactions induced by active species produced in the plasma other than ions. These active species will generally produce an isotropic etch, since they tend to be delivered at equal rates to both the sidewalls and bottom of the trench.

25.2.4.5   *Electron-Beam-Promoted Etching*

Finally, it is interesting to note that energetic ions are not the only possible promoters of directional etching reactions. Figure 25.23, taken from Coburn and Winters (1979b), shows the etching effect of an electron beam on SiO_2, a material normally difficult to etch, in the presence of xenon difluoride. Until 800 s elapsed, the silicon dioxide was exposed only to xenon difluoride, and no etching occurred. At 800 s, a 1500 eV electron beam with a total current of 45 μA and a current density of approximately 50 mA/cm^2 was directed on the silicon dioxide, in the presence of xenon difluoride. The electron beam promoted an etching reaction at the rate of 200 Å/min. An electron beam in the absence of xenon difluoride does not produce sustained etching of the silicon dioxide.

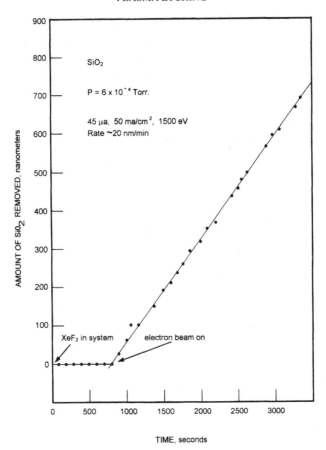

Figure 25.23. The etching of silicon dioxide in the presence of xenon difluoride, catalyzed by a 1500 eV electron beam turned on at 800 s, which served as an energetic promoter of the reaction. Taken from Coburn and Winters (1979b).

25.2.4.6 Photon-Beam-Promoted Etching

It has been found also that a high-intensity flux of laser photons is capable of etching materials of microelectronic interest in the presence of an appropriate working gas. There seems to be relatively little difference in the etching effect of energetic chemically inert ions, electrons, and even photons produced by a laser, although the microelectronic industry has opted to use plasma-generated ions as the preferred etching agent.

25.2.5 Characteristics of Anisotropic Etching

After discussing several mechanisms capable of producing anisotropic etching of microelectronic materials, we now look quantitatively at the production of a directional etch. With reference to figure 25.14, a directional etch with a sidewall angle θ results from a finite horizontal (in the x-direction) etching velocity, v_x m/s, perpendicular to the axis of the trench. This velocity is smaller than the vertical etching rate in the z-direction, v_z. Since etching of the sidewalls and bottom of the trench occur simultaneously and for the same duration, the sidewall angle θ shown on figure 25.14 is related to the horizontal and vertical etching velocities by

$$\tan \theta = \frac{v_x}{v_y} = \frac{\Delta x}{\Delta z}. \tag{25.7}$$

In the discussion of isotropic etching in figure 25.17, it was shown that the minimum mask feature size that will produce a significant feature in the etched layer is given by equation (25.6a) for isotropic etching,

$$L_{\min} = 2d. \tag{25.6a}$$

However, for the directional etch illustrated on figure 25.14, the minimum mask dimension is

$$L_{\min} = 2d \tan \theta. \tag{25.8}$$

Compared to isotropic etching, the minimum design rule can be reduced by the ratio of the horizontal to vertical etching velocities.

An example of one method by which vertical etching is accomplished is illustrated on figure 25.24, taken from Coburn and Winters (1979b). Plotted in this figure are the etching rates for silicon and silicon dioxide as functions of the DC bias voltage applied to a wafer. The energy of ions reaching the wafer is proportional to the bias voltage, and these ions promote etching in the presence of a suitable working gas. In the case of silicon in figure 25.24, the etching rate is finite when the ions have zero energy, i.e. when the ion flux is zero. Without an ion flux, the silicon sidewalls are etched at a finite but slower rate than the trench bottom, where an ion flux is present. A directional etch results, as shown below and on the left of figure 25.24. The sidewall angle of a silicon etch can be adjusted by changing the bias voltage on the wafer as shown in this figure. Silicon dioxide behaves differently, because no etching occurs when the ion energy is zero or no ion flux is present. Thus, the sidewalls of a trench in silicon dioxide will not be etched under ion bombardment, yielding the vertical etch illustrated on the right of figure 25.24.

The ratio of v_x to v_z and therefore the sidewall angle also can be adjusted by changing the percentage of hydrogen in the working gas. A vertical etch can be produced by depositing a *passivating layer*, an inert coating on the sidewall that prevents etching that occurs in its absence. A passivating layer may consist of material sputtered from the bottom of the trench, or material formed by a chemical reaction between the sidewalls, and plasma active species or the working gas.

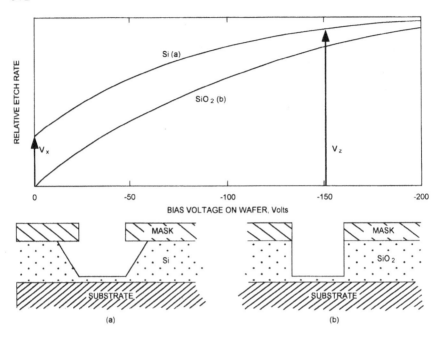

Figure 25.24. The effect of wafer bias on the etch rate of silicon and silicon dioxide. (*a*) Etching of silicon; (*b*) etching of SiO$_2$. See text for discussion.

25.2.6   Etching of Three-Dimensional Microstructures

Plasma etching is not confined to the microelectronic industry, although that is its most widespread application. Microelectronic etching techniques have been used to produce fine points for field emitters and/or corona discharges used in flat panel displays and lighting applications. Plasma etching has also been used to produce vacuum transistors, and even to fabricate moving parts on a scale of a few microns, including levers, wheels, and electrostatic motors.

Figure 25.25 is a scanning electron microscope (SEM) image of pyramids grown epitaxially at the end of small rods with the scale size indicated (Ugajin *et al* 1994). Such fine points are potentially useful as field emitters for flat panel displays. Some early work on the techniques required to fabricate three-dimensional microstructures is shown in figure 25.26 (Geis *et al* 1981). This figure illustrates the combined result of vertical etching by normally incident argon ions (shown in figure 25.26(*a*)) with an obliquely incident reactive gas flux flowing from the upper left. Although energetic argon ions did not directly impact some areas under the mask, etching occurred at an angle determined by the angle of the incident gas flux. The result of this etching process is shown in figure 25.26(*b*).

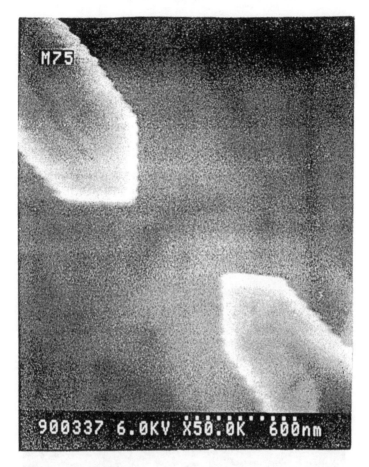

Figure 25.25. A scanning electron microscope (SEM) image of pyramids grown epitaxially at the ends of fine rods (Ugajin *et al* 1994).

25.3 CONTROL VARIABLES FOR PLASMA ETCHING

Control of the plasma etching process involves the input, plasma, and output variables illustrated in figure 25.27. Control is effected by adjusting the *independent input variables* listed on the left. The independent input variables determine the *plasma-related parameters* shown in the center of the diagram, and these in turn determine the *dependent* or *output variables* on the right-hand side of the diagram. The output variables define the nature and quality of the etching process being controlled. Causality in this diagram flows from left to right, and we proceed to consider the effects of these variables in that order.

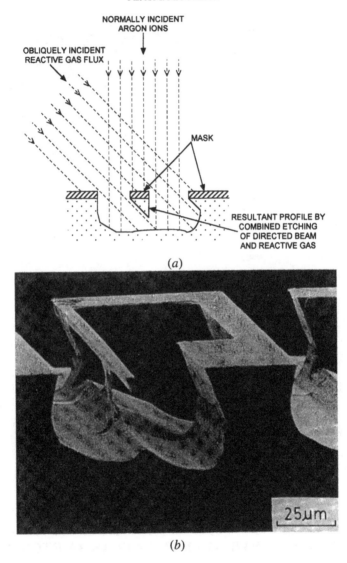

Figure 25.26. The production of three-dimensional structures by manipulation of the incident angle of neutral etching gas: (*a*) impingement direction of argon ions and etching gas flux on the wafer surface, in relation to the mask and the layer to be etched; (*b*) SEM micrograph of a structure etched in the manner shown in (*a*) (Geis *et al* 1981).

25.3.1 Independent (Input) Variables

The *independent input variables* are measured or determined by the process control system discussed in section 17.6.2.4. These input variables may be

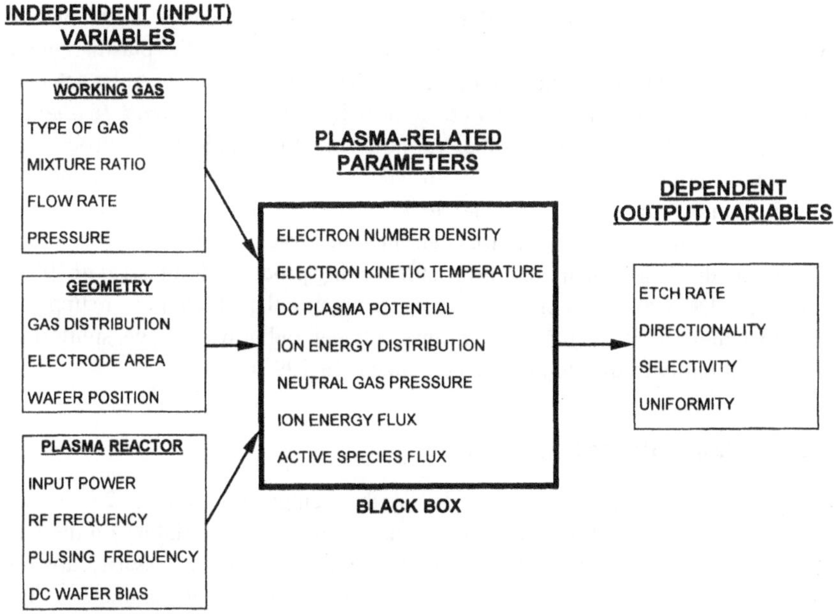

INDEPENDENT (INPUT)
VARIABLES

Figure 25.27. Independent input variables, plasma-related parameters, and dependent output variables of a plasma etching reactor.

grouped into three principal factors: the *working gases*, the *geometry*, and the *plasma source parameters*. The independent input variables relevant to the working gases are the type of gas or gases; the mixture ratio(s) or concentrations if more than one gas; the flow rate(s) of the gas(es) into the vacuum system; and the total pressure in the vacuum system.

The geometry is the second principal input variable. It is usually determined by a development program prior to a production run, rather than being a parameter available for adjustment during the etching process. Geometrical factors that can be adjusted for long production runs include the distribution of the working gas flow over the wafer surface, and the electrode gap spacing and electrode area ratio for RF parallel-plate glow discharge reactors.

The electrode area ratio determines the sheath voltage drop on the powered electrode, where wafers are usually located. This sheath voltage drop can cause radiation damage if it is too high, but it also promotes vertical etching by increasing the ratio of vertical to horizontal velocity of ions reaching the wafer. The wafer position within the plasma reactor or on the powered electrode may be adjusted to optimize the uniformity of active-species flux or etching effect. Finally, the plasma reactor configuration can be changed in order to optimize the impedance match between the RF power supply and the plasma reactor, or to ensure uniformity of the plasma and/or uniformity of etching effect.

There are additional independent input variables associated with the plasma source subsystem. These variables include the input power to the plasma, which can be adjusted during an etching operation. The RF frequency can be selected to lie at a standard frequency either above or below the *critical sheath frequency*, discussed in section 19.5.3. Another variable sometimes adjusted in production runs is the frequency and duration with which the plasma is turned on and off. This frequency can vary from hertz to kilohertz. Adjustment of this pulsing frequency and the duration of pulses can, in some cases, have a beneficial effect in speeding up or making more uniform the etching process (Bounasri *et al* 1993). Finally, the DC wafer bias may be adjusted with an independent power supply in order to achieve a satisfactory compromise between radiation damage, sputtering, and directionality of the etching ions (Reinke *et al* 1992).

25.3.2 Plasma-Related Parameters

The plasma-related parameters shown at the center of figure 25.27 are the consequence of particular settings of the independent input variables on the left-hand side of the diagram. There is usually, although not necessarily always, a unique single-valued relationship between these two sets of parameters. The independent input variables determine the electron number density, the electron kinetic temperature, the plasma potential, and the ion energy distribution function of ions in the plasma, and of ions reaching the wafer surface. The working gas pressure is also an input variable. The dwell time of a neutral gas atom in the vacuum system can be adjusted by changing the pressure and/or the flow rate of the working gases. The active-species flux on the wafer can be adjusted by changing the total pressure of the working gas, or the electron number density of the plasma.

Of these plasma-related parameters, only the working gas pressure and the energy distribution function of ions reaching the surface of the wafer are likely to be continuously monitored in production reactors. Two examples of the latter measurement may be found in the work of Hope *et al* (1993) and Olthoff *et al* (1994). The remainder of the plasma-related parameters are rarely measured outside an exploratory or basic research context, but for examples of such measurements, see Deshmukh and Cox (1988).

25.3.3 Dependent (Output) Variables

The independent input variables on the left-hand side of figure 25.27 determine the plasma-related parameters in the middle of the diagram, and these in turn determine the dependent output variables of a microelectronic etching process. These dependent output variables include:

(1) the *etch rate*, the velocity at which the etched layer will be removed in the vertical direction below the mask;

(2) the *directionality* of the etch, the relative rate of horizontal undercutting of the mask relative to the rate of etching in the vertical direction (for such measurements, see Zheng *et al* (1995));

(3) the *selectivity*, the ability of the process to etch only the layer to be etched, while leaving relatively undisturbed the material of the mask and the substrate below the layer to be etched; and

(4) the *uniformity* of etching, which results from a uniform flux of active species on the surface of the wafer.

25.3.4 System Control Strategies

When etching is conducted for routine production purposes, reproducibility of the process is essential. In order to achieve reproducibility, two general strategies may be adopted. One, which is almost universally used in the microelectronic industry, can be characterized as the *black box strategy* that, in terms of figure 25.27, regards the plasma as a black box into which wafers are put to be etched. One then adjusts the independent input variables to achieve a desired result in terms of the dependent output variables.

A second control strategy, rarely used in commercial practice, may be called the *plasma parameter strategy*. In this approach, the plasma-related parameters shown at the center of figure 25.27 are measured by plasma diagnostic instruments of the kind discussed in section 20.3. These parameters are kept constant by the control system to achieve a uniform and reproducible result, as indicated by the dependent output variables. In this strategy, the independent input variables are altered only in order to adjust the plasma-related parameters to known or specified set points that produce the desired result.

The plasma parameter strategy may be more difficult to implement than the black box strategy. However, it does offer the advantage of a more direct, causal relationship between the plasma-related parameters and the dependent output variables than is the case of a phenomenological correlation between the independent input variables and the dependent output variables. The plasma parameter strategy also is capable of leading to better understanding of physical processes in the plasma and in the etching process itself.

The faster results and lower short-term cost of the black box strategy have widely recommended it to the microelectronic industry. However, situations can arise in which the black box strategy can be costly in the long run. For example, it can happen that an integrated circuit manufacturer will receive an order for a batch of integrated circuits exactly like those fabricated a number of years previously. If the cluster tools and other hardware used for the original production run are no longer available, the settings used for the earlier production run may no longer be relevant on newer equipment with different characteristics. Without knowledge of the plasma-related parameters that produced the results of the previous production run, a costly developmental program may be needed to reproduce a previous production run on currently available equipment.

One concern about employing the plasma parameter strategy is whether a given set of plasma parameters, shown at the center of figure 25.27, is a necessary and sufficient condition to yield a reproducible, single-valued set of independent output variables. Recent studies have indicated that the particular type of plasma source employed to etch an integrated circuit is immaterial, so long as the plasmas produced by the various sources had the same set of plasma parameters (Hershkowitz *et al* 1996). It did not appear to matter whether the plasma was produced by a helicon reactor, an ECR reactor, or a helical flat-coil inductively coupled reactor, provided that the plasma parameters which affect etching were the same.

A further control strategy is the use of *in situ diagnostics* of the wafer to determine the thickness of deposited films or etched layers by direct measurement. Some of these diagnostics are discussed in section 20.8, and include quartz crystal microbalance measurements of the amount of material remaining in a sample of the etched layer exposed to the same active-species flux and etching environment as the wafer itself.

The most important event in an etching production run is the *endpoint* of the etching process, discussed in section 20.9. The endpoint occurs when the trench reaches the substrate over the entire wafer. Non-uniformities in the flux of active species, or variations in the thickness of the layer to be etched, may prevent the etching process from reaching the substrate simultaneously over the entire wafer. The endpoint can be detected in several ways. The most effective are based on monitoring, either spectroscopically or with a mass spectrometer, the density of the etching reaction products, or the density of reaction products that result only from etching the substrate.

A normal etching process will yield a more or less constant efflux of etching reaction products until the substrate is reached. At that point, the density of the etching products in the effluent gas will drop precipitously as the etched material at the bottom of the trench is exhausted. If the working gas attacks the substrate, then reaction products from the substrate will increase in density, and this increase and its leveling off can be used as an indication of the endpoint. When this happens, the system controller can shut off the working gas, and proceed to the next fabrication step.

25.4   THE CHEMISTRY OF PLASMA ETCHING

In this and section 25.5 that follows, neither the etching that results from sputtering by energetic ion beams, nor wet chemical etching will be considered. Only 'dry' plasma etching processes will be considered that result from reactive working gases or active species produced by glow discharge plasmas. These plasma-based etching processes now dominate the microelectronic industry, and are the only technology available for etching circuits with design rules below 0.5 μm.

25.4.1 Heterogeneous Processes in Plasma Etching

The plasma–chemical reactions that produce active species for microelectronic etching are initiated by inelastic collisions of energetic electrons, unlike the gas-phase reactions in a normal chemical reactor. Etching reactions utilize a variety of heterogeneous chemical reactions on a surface unlike the ordinary gas- or liquid-phase reactions chemists normally encounter. This section of the chapter is concerned with these etching processes.

As the result of several decades of applied research, a great deal is now known about the plasma–chemical and heterogeneous surface reactions responsible for the etching process. Here, we summarize only a small part of a very large literature on this subject. Those wishing for more information about the chemistry of plasma etching are referred to the classic papers of Coburn and Winters (1979), Coburn (1982), Winters *et al* (1983) and Coburn and Winters (1983). In addition, the full-length books by Powell (1984), Morgan (1985), Sugano (1985), Lieberman and Lichtenberg (1994), and Madou (1995) all include coverage of the chemistry of plasma etching.

25.4.1.1 *Characteristic Etching Reactions*

Microelectronic etching reactions occur at the heterogeneous state boundary between the gas filling the plasma reactor, and the solid surface of the layer being etched. Although the inputs and outputs of these heterogeneous reactions are well known, details of their intermediate steps are not. As examples, two characteristic etching reactions of silicon are shown in equations (25.9) and (25.10),

$$2XeF_2 + Si \xrightarrow{Ar^+,\,surface} 2Xe + SiF_4 \uparrow \qquad (25.9)$$

and

$$2Cl_2 + Si \xrightarrow{Ar^+,\,surface} SiCl_4 \uparrow . \qquad (25.10)$$

The first reaction between xenon difluoride and solid silicon occurs at the surface of the silicon and proceeds most rapidly in the presence of energetic argon ions. The argon ions do not participate in the chemistry as reactants or reaction products, but simply perform a 'catalytic' function as energetic promoters of the surface reaction. After the silicon surface is exposed to xenon difluoride in the presence of energetic argon ions, the xenon is released as an atomic gas, or, to a small extent, as xenon fluoride (XeF). Silicon tetrafluoride, a volatile gas, leaves the surface and is pumped away by the vacuum system.

A second characteristic etching reaction is shown in equation (25.10) between molecular chlorine and a solid silicon surface. In the presence of energetic argon ion bombardment, these reactants will combine to form silicon tetrachloride, a volatile gas, which removes the reacted silicon from the bottom of the trench. The details of how the diatomic or polyatomic etching reactants are disassembled on the silicon surface, and how their components reassemble themselves on the surface into the volatile reaction product, are poorly understood.

25.4.1.2   Role of Adsorbed Monolayers

As was discussed in section 21.2.2, the surface of solid materials, including the layer to be etched on a microelectronic wafer, may be covered with as many as several hundred adsorbed monolayers. These monolayers are normally molecules of the neutral working gas to which the surface has been most recently exposed for a long period. The lowermost monolayers, immediately adjacent to the surface, are tightly bound with a relatively high binding energy, and thus are difficult to dislodge by purely thermal or chemical means. These tightly bound near-surface monolayers may be hydrocarbon oils from machining operations, vacuum pump oils, or other similar contaminants.

In a normal ion-promoted etching process, the molecules of the etching gas and the energetic ions must penetrate the monolayers of adsorbed gases in order to interact with the silicon surface below. Exactly how the energetic ions and etching species interact with the adsorbed monolayers before and during the etching process is not well understood. However, it is likely that one of the functions of energetic ions is to scour and remove most, if not all, of the adsorbed monolayers from the surface in order to facilitate the etching reaction. As was pointed out in section 21.1.3.1, the removal of adsorbed monolayers is associated with a significant increase in surface energy, a decrease in the water contact angle, and other effects. These effects indicate the enhanced readiness of the surface to undergo surface etching reactions like those shown in equations (25.9) and (25.10).

A schematic representation of a partially etched trench before turning on the plasma is shown in figure 25.28(a). In this diagram, the silicon atoms at the bottom and the sidewalls of the trench are covered by multiple monolayers of adsorbed atoms. When the plasma is turned on in the absence of the working gas, as illustrated in figure 25.28(b), the adsorbed monolayers are scoured away by either chemical or physical means. This process leaves the silicon atoms at the bottom of the trench in a far more exposed state, ready to undergo etching reactions. Finally, when a working gas such as molecular chlorine (Cl_2) is introduced, the chlorine molecules arrive at the bottom of the trench, where they break up, react with the silicon, and leave the surface of the trench as a volatile reaction product, as indicated in figure 25.28(c).

25.4.1.3   Steps in an Etching Reaction

The steps in an etching reaction may be summarized as follows:

(1) *removal* or *penetration* of adsorbed monolayers by energetic ions and etching species;
(2) *adsorption* of etching species on the surface of the layer to be etched;

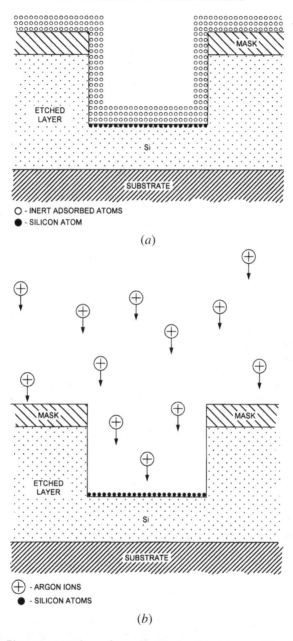

Figure 25.28. Phenomena at the wafer surface during three stages of the plasma etching process: (*a*) a partially etched trench before restarting plasma etching, covered with inert adsorbed monolayers; (*b*) impingement of energetic ions on the surface to be etched after removing adsorbed monolayers; and (*c*) initiation of etching by introduction of Cl_2 gas, producing silicon tetrachloride, $SiCl_4$.

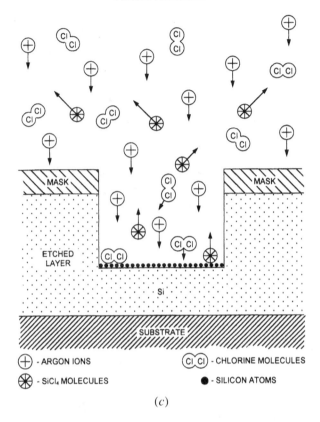

(c)

Figure 25.28. (Continued)

(3) *dissociation* of the molecules of the etching species;
(4) *formation of a reaction product* on the surface; and
(5) *desorption* of the product molecule.

The details of steps (3)–(5) are not well understood and are the subject of ongoing research.

25.4.1.4 Surface Reaction Mechanisms

Exothermic etching reactions that proceed in the manner discussed earlier normally require a 'catalytic' energy input from energetic ions or other such energetic promoters as electron or laser beams. Energetic ions may also be chemically active, and serve as a reactant in an etching reaction.

Etching can also be performed by ordinary heterogeneous chemical reactions that proceed on a surface without an input of energy from energetic ions. These reactions are usually undesirable, since they produce an isotropic etch. Nearly all directional etching reactions are the result of energy input from energetic ions, electrons, or photons, as discussed in section 25.2.4. It is relatively easy to produce inert or reactive ions in a glow discharge plasma reactor, and to accelerate those ions to the required energies (a few tens of eV) across the sheath above a wafer. A small proportion of etching is done by other means, including inert and reactive ion-beam etching (IIBE and RIBE), discussed in section 25.2.4.

Although energetic electron beams can promote etching reactions (see for example, figure 25.23), and energetic photon beams from lasers also have produced etching reactions, these have not been incorporated into microelectronic etching technology to any significant extent.

25.4.1.5 Polymerization

Polymerization of monomers on a surface to form polymeric layers is an undesirable heterogeneous chemical process when it occurs in etching applications. Polymerization is most likely to occur in plasma reactors that produce large concentrations of monomers such as CF_2. When monomers reach a surface and combine to form long molecular chains such as $(CF_2)_n$, the process can compete with, and even prevent, etching reactions on the surface.

Polymerization reactions are particularly likely to occur in plasma reactors containing hydrocarbon or fluorocarbon gases. In the case of fluorocarbon-based gases, whether the outcome is etching or polymerization is known to depend on the bias voltage applied to the surface being etched (i.e. on the energy of ions reaching the surface) and the fluorine-to-carbon ratio (F/C) of the gas-phase etching species. The fluorine-to-carbon ratio is calculated from the concentrations of the active species in the glow discharge that are likely to react chemically in the etching process, and does not include the concentrations of relatively inert working gases.

Figure 25.29, taken from Coburn and Winters (1979b), illustrates the formation of polymers in an etching reactor. Illustrated is a parameter space defined by the DC bias applied to the surface (approximately the energy of ions reaching the surface) on the ordinate, and the fluorine-to-carbon ratio (F/C) of the gas-phase etching species on the abscissa. The dotted line approximately defines the polymer point, below and to the left of which polymerization will occur on the surface, whereas above and to the right, etching will occur. It is significant that this curve bends over to the left near the top, suggesting that the higher energy ions which reach the surface under these conditions scour away polymers at a faster rate. This process thus promotes etching under conditions that, with lower energy ions at the bottom of the ordinate, would produce polymerization. As was discussed in chapter 24 however, the formation of polymeric films on the surfaces of microelectronic chips is often desired: here we are concerned only with polymerization as an unwanted by-product of the etching process.

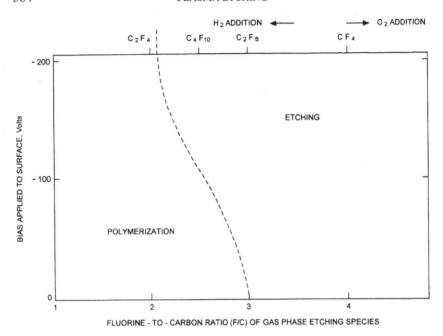

Figure 25.29. The effect of wafer bias and fluorine-to-carbon ratio of the gas-phase etching species on surface polymerization (Coburn and Winters 1979b).

25.4.1.6   Dust Formation

Solid dust particles can form in the plasma if the concentration of monomers is sufficiently dense, and the operating gas pressure sufficiently high. A striking example is shown in figure 25.30, from Garrity *et al* (1996). The polymeric material of these dust particles forms in a fractal-like manner, producing approximately spherical structures that look much like heads of cauliflower. The examples shown are a few microns in diameter. These dust particles can become electrically charged, and levitate above the surface of a wafer in the sheath electric field. When the RF power is turned off between operations in a cluster tool, these suspended dust particles may drop to the surface of the wafer, thereby contaminating it (Barnes *et al* 1992). In general, the operating conditions for microelectronic etching and deposition are adjusted to suppress this form of dust formation.

25.4.2   Characteristics of Etching Reactions

Although microelectronic etching reactions are heterogeneous, they share some characteristics with ordinary chemical reactions. They also have additional characteristics that result from a multistep process involving a gas and a solid. Here we examine some significant chemical characteristics of etching reactions.

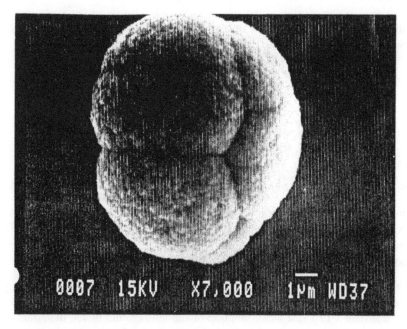

Figure 25.30. Scanning electron microscope image of a cauliflower-like dust particle collected in the exhaust during silicon etching (Garrity *et al* 1996).

25.4.2.1   Desirable Reaction Characteristics

Not every substance that reacts chemically with the layer to be etched will produce a result suitable for microelectronic circuit fabrication. Some characteristics of a good etching reaction include the following.

(1) The reaction should be *exothermic*, otherwise the reaction may not take place.

(2) The reaction should produce a *volatile reaction product*. If the reaction product is solid and remains on the bottom of the etching trench, the product will shield the layer to be etched, and interfere with the etching reaction.

(3) A desirable etching reaction should have *little or no reaction with the sidewalls of the trench*, in order to produce a vertical etch and avoid undercutting the mask.

(4) Finally, the etching reaction should be *selective*, such that it undergoes no significant reactions with either the mask or substrate. Selective etching allows the pattern to be transferred to the layer to be etched without eroding the mask, or damaging the substrate.

Table 25.2. Arrhenius etch rates for silicon and silicon dioxide at $T = 300$ K and a fluorine atom density $n_F = 3 \times 10^{21}/m^3$.

Material etched	C_1	E_a (eV/atom)	Etch rate (μm/min)
Si	2.86×10^{-22}	0.108	0.228
SiO$_2$	6.14×10^{-23}	0.163	0.0058

25.4.2.2 Temperature Dependence of Etching Rate

The temperature dependence of the chemical reaction rate between the etching agents and the layer to be etched is described by the *Arrhenius relation*. This relation states that the temperature-dependent *chemical rate constant* for a chemical reaction may be written

$$k(T) = C(T)\exp[-eE_a'/kT]. \tag{25.11}$$

The constant $C(T)$ is a weak function of the temperature of the material to be etched, and the parameter E_a' is the *activation energy* for the chemical reaction, which characteristically lies between 0.1 and 1 eV.

The etching rate of a solid material is given by the product of the reaction rate of equation (25.11), and the flux of etching species on the surface. This flux is the product of the thermal velocity of the etching species, and the number density of the etching species (taken to be fluorine in the example below), n_F. The product of the flux of working gas and the reaction rate of equation (25.11) therefore yields an etching rate given by the expression

$$\text{Etching rate}(ER) = C_1 n_F T^{1/2}\exp[-eE_a'/kT] \qquad \mu\text{m/min}. \tag{25.12}$$

Thus, when the logarithm of the etching rate is plotted as a function of T^{-1}, the resulting data fall along a straight line.

An example of the temperature dependence of plasma etching is shown in figure 25.31, taken from Manos and Flamm (1989). In this figure, data are shown for silicon and silicon dioxide etched by fluorine with a density $n_F = 3 \times 10^{21}$ fluorine atoms/m^3. The etching rate, in μm/min, is plotted as a function of inverse temperature for these two materials. A very strong dependence on temperature is evident. The etching rate constants appearing in equation (25.12) for silicon and silicon dioxide at a temperature of 300 K and a fluorine atom density $n_F = 3 \times 10^{21}/m^3$ are listed in table 25.2. For these conditions, the etching rate of silicon is about 230 nm/min, and for silicon dioxide 5.8 nm/min.

If a layer of silicon dioxide lies below a layer of polycrystalline silicon being etched to form a microelectronic circuit, the selectivity of etching silicon in preference to silicon dioxide is an issue. The selectivity, S, may be found by

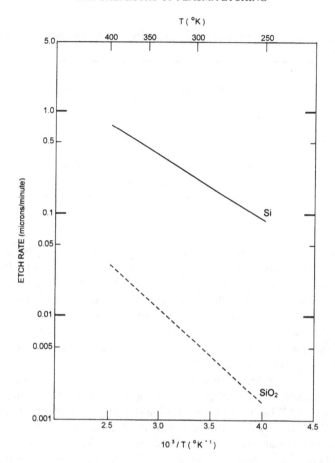

Figure 25.31. The etch rate, in μm/min, of silicon and silicon dioxide as functions of inverse temperature, as the result of etching in a CF_4 plasma. The $1/T$ dependence shown is a result of the Arrhenius relation of equation (25.11) (Manos and Flamm 1989, p 119).

taking the ratio of the etch rates of the two species expressed by equation (25.12). This ratio is, for the data of table 25.2,

$$S = \frac{ER(\text{Si})}{ER(\text{SiO}_2)} = 4.66e^{638.22/T}. \tag{25.13}$$

The number density of the etching species (fluorine) and the square of temperature have canceled out. This cancellation leaves the ratio of the etching constants C_1, and an exponential factor that is the difference of the activation energy of the two materials divided by their temperature.

The selectivity of etching of silicon and silicon dioxide by fluorine in equation (25.13) is listed for reference temperatures on table 25.3. It is evident

Table 25.3. Temperature dependence of selectivity for fluorine etching of silicon and silicon dioxide.

Temperature, T (K)	Comment	Relative etch rate (Si/SiO$_2$)
77	Liquid nitrogen	18 500
273	H$_2$O freezing	48.3
300	Lab temperature	39.1
373	H$_2$O boiling	25.8
500	Hot wafer	16.7
∞	Asymptotic limit	4.66

that as the material becomes very hot ($T \rightarrow \infty$), the selectivity approaches the constant appearing in front of the exponential, $S = 4.66$. At a laboratory temperature of 300 K, the selectivity is $S = 39$. If the wafer is cooled to lower temperatures, for example with liquid nitrogen at 77 K, the selectivity increases to 18 500. The selectivity of etching reactions with a marginally adequate or inadequate selectivity at room temperature can be increased by conducting the etching operation at a lower temperature, provided that the etching rates themselves remain high enough to be useful.

The temperature dependence of the etching rate implied by equation (25.12) produces a number of subtle effects. An example, from unpublished data taken at the University of Wisconsin's Center for Plasma Aided Manufacturing, is shown in figure 25.32 (Hershkowitz 1995). This figure shows a three-dimensional profile of the etching rate of a silicon wafer by carbon tetrafluoride working gas. The gross variations in etching rate are due to plasma effects, but there is a subtle cruciform indentation across two perpendicular diameters of the wafer that are the image of cooling lines imbedded in the wafer chuck. The presence of the cooling lines kept the surface of the wafer immediately above them slightly cooler, and this lower temperature slowed down the etching rate sufficiently to produce the pattern shown in figure 25.32.

25.4.2.3 Phenomenological Etching Correlation

Research at the University of Wisconsin's Center for Plasma Aided Manufacturing revealed an important correlation for the etching rate. This correlation addressed the relation between the etching rate, and the particle flux of the working gas and the energy flux arriving at the surface being etched (Hershkowitz et al 1996). Data are available for etching silicon dioxide in carbon tetrafluoride gas, or in mixtures of carbon tetrafluoride and argon. This relationship appears to be independent of the type of plasma source used to produce the etching species, and may apply to a wide range of microelectronic

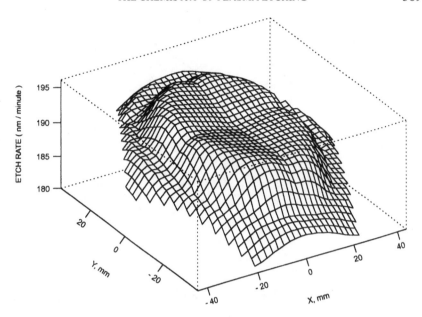

Figure 25.32. An isometric profile of the etch rate of a silicon wafer in a carbon tetrafluoride (CF_4) plasma (Hershkowitz 1995).

etching reactions.

The Wisconsin data are shown in figure 25.33. The contours of constant etching rate (in Å/min) are plotted parametrically against the number density of fluorine atoms on the ordinate, and the energy flux (in mW/cm^2) on the abscissa. The data on this plot can be correlated by the following semi-empirical expression for the etching rate (ER),

$$ER = \frac{C_2 E_i \Gamma_i \left[1 - c \left(\frac{\Gamma_p}{\Gamma_F} \right)^2 \right]}{1 + \frac{a E_i \Gamma_i}{\Gamma_F} + \frac{b \Gamma_p}{\Gamma_F}}. \tag{25.14}$$

Here the etching rate is written in terms of the following variables:

Γ_p precursor flux responsible for forming a polymer on the surface (precursors/cm^2-s);

Γ_F flux of fluorine atoms on surface (atoms/cm^2-s);

Γ_i ion flux on the surface to be etched (ions/cm^2-s);

E_i ion energy at the surface of the wafer (J); and

$\Gamma_i E_i$ ion energy flux on the surface (mW/cm^2).

Equation (25.14) applies to etching reactions for which polymeric films may form on the surface, a possibility in the carbon tetrafluoride etching reactors in

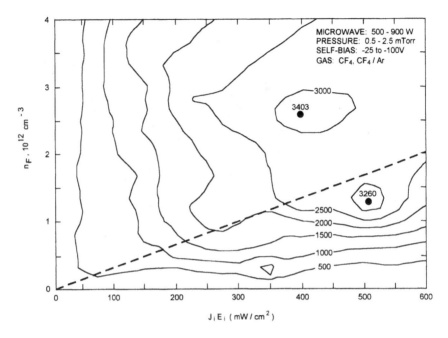

Figure 25.33. Data for the etching rate of silicon by carbon tetrafluoride, as a function of the ion power flux (in mW/cm^2) on the wafer surface (abscissa), and the fluorine number density above the wafer surface in atoms/cm^3 (ordinate) (Hershkowitz *et al* 1996).

which these data were taken. The term containing b in the denominator is usually negligible, and the remainder of equation (25.14) can be used to obtain a good approximation to figure 25.33.

If the flux of monomers on the surface is negligible, such that polymeric materials do not form, then $\Gamma_p \approx 0$. For this non-polymeric case, two situations can be distinguished. The first occurs when the second term in the denominator is much less than one, for which equation (25.14) becomes

$$ER \propto E_i\Gamma_i, \qquad \frac{aE_i\Gamma_i}{\Gamma_F} \ll 1. \qquad (25.15)$$

Thus, when the term containing the constant a in the denominator is small, the etching rate is proportional to the energy flux impinging on the wafer. This is the situation shown in figure 25.33 above the dotted line, where the contours of constant etching rate are approximately vertical, and the etching rate is proportional only to the energy flux on the wafer.

A second significant regime occurs when the polymer-forming particle flux is negligible and the second term in the denominator is much greater than one. In this case, the etching rate of equation (25.14) reduces to

$$ER \propto \Gamma_F, \qquad \frac{a E_i \Gamma_i}{\Gamma_F} \ll 1. \qquad (25.16)$$

In this second case, the etching rate is proportional to the flux of the etching species, atomic fluorine. This regime lies below the dotted line in figure 25.33, where the contours of constant etching rate are almost horizontal. This behavior implies that the etching rate is proportional to the flux, and hence to the number density, of the fluorine atoms.

The etching process in figure 25.33 was carried out with a ECR microwave generated plasma in the low-pressure regime between 0.07 and 0.33 Pa (0.5 and 2.5 mTorr). Similar experiments were carried out at the University of Wisconsin with other plasma sources, including inductively coupled and helicon etching tools. The ECR reactor was operated at 2.45 GHz, and the inductively coupled and helicon etching tools were operated at 13.56 MHz. The mean thermal velocities of the monomers and fluorine atoms that impinge on the surface are approximately the same. The ratio of the monomer to the fluorine particle fluxes on the surface that appears in equation (25.14) can therefore be written in terms of the relative number density of the two species,

$$\frac{\Gamma_p}{\Gamma_F} \approx \frac{n_p}{n_F}. \qquad (25.17)$$

Equation (25.14) was used to correlate a wide range of data on the etching of silicon dioxide by a carbon tetrafluoride plasma, using a variety of RF plasma sources (Hershkowitz et al 1996). The resulting correlation is shown in figure 25.34. The measured etching rate is plotted on the ordinate of figure 25.34, and the correlation parameter on the abscissa is equation (25.14), with the constants $a = 3.3 \times 10^{-3}$, $b = 0$, and $c = 0.13$,

$$ER = \frac{\Gamma_i E_i \left[1 - 0.13 \left(\frac{n_p}{n_F}\right)^2\right]}{1 + 3.3 \times 10^{-3} \frac{\Gamma_i E_i}{n_F}}. \qquad (25.18)$$

For this combination of substrate and working gas, the etching rate correlates with equation (25.18) over a factor of approximately 10 in etching rate and power flux to the surface. This correlation parameter was found to be useful for plasmas generated with four types of plasma source, including an electron cyclotron resonance source, a reactive ion etching (RIE) reactor, an inductively coupled plasma reactor, and a helicon plasma source. This good agreement indicates a wide range of etching conditions for which equation (25.14) holds. It also indicates that the etching rate does not depend on the plasma source used for the etching, only on the ionic power flux and the flux of etching species.

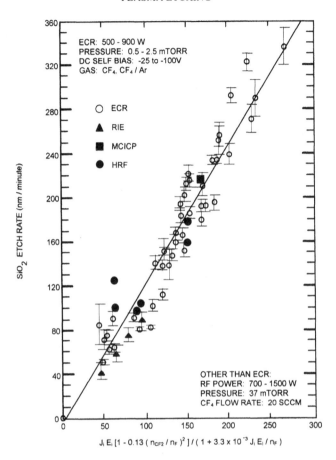

$$J_i E_i [1 - 0.13 (n_{CF2} / n_F)^2] / (1 + 3.3 \times 10^{-3} J_i E_i / n_F)$$

Figure 25.34. The data of figure 25.33 correlated by equation (25.18), for etching of silicon by carbon tetrafluoride plasmas using four kinds of plasma source (Hershkowitz *et al* 1996).

25.4.2.4 *Other Functional Dependences*

In the gas phase, the chemical reaction rate is a function of the working gas pressure, given by

$$\text{Reaction rate} \propto p^n \tag{25.19}$$

where the index n is the number of gaseous phase species participating in the reaction. An example is the polymerization reaction, given by

$$CF_2 + CF_2 + CF_2 = C_3F_6. \tag{25.20}$$

The production of C_3F_6, a polymer, requires a three-body process, for which the reaction rate proceeds with a cubic pressure dependence, $n = 3$. The reaction probability itself is not large at low pressures, but the cubic dependence can give rise to significant reaction rates at intermediate and atmospheric pressure. This is an important reason why the formation of dust and polymers is enhanced by operating at higher pressures. Since these reaction products are generally detrimental to the etching process, their suppression may provide another reason for operating at low or intermediate background neutral gas pressures.

Etching the interior walls of a plasma reactor is only one of the *wall effects* that can occur. In a characteristic etching plasma, most of the recombination reactions and some of the dissociation reactions also occur on the surrounding walls, rather than in the gas phase of the plasma or the sheath. The rate of such wall reactions is a strong function not only of the wall temperature, but also of the nature of the wall material. For this reason, the mix of active species that appears at a wafer surface may depend on processes that occur on the walls of the reactor. These processes can differ in different plasma etching reactors because of a difference in wall materials.

In particular, highly reactive etching species such as atomic fluorine, atomic chlorine and atomic oxygen may react chemically with the wall or not, depending on the chemical reactivity of the wall material. A chemically reactive wall material will provide a sink for the etching species that reduces their flux on the surface of the wafer. This lower etching flux results in a lower etching rate than a reactor in which the etching species is not lost by reactions on the wall. The reactions of the etching species on the wall are generally undesirable, since they irreversibly damage components of the vacuum system.

The agent responsible for etching can be removed from the plasma not only by interaction with surrounding walls of the vacuum system, but also by reaction with other wafers in the reactor. This is the *loading effect* discussed in section 25.2.1.6.

25.4.3 Species Concentrations in Etching Plasmas

The mean energy of electrons in RF glow discharge etching plasmas is at least several electronvolts, high enough to produce a variety of inelastic collisions, including the production of ion–electron pairs. These inelastic collisions also may produce neutral and ionized active species that participate directly in the etching reactions of interest, and the production of other neutral and ionized molecular fragments that do not participate directly in etching reactions. Some of these intermediate reaction products may, after further electron collisions, produce active species capable of etching reactions.

25.4.3.1 Neutral Active Species

The *ionization fraction* in RF glow discharges used for microelectronic etching appears in figure 25.35 as the parameter on the sloping lines. At the upper left of this diagram is a line which represents the fully ionized state of a background

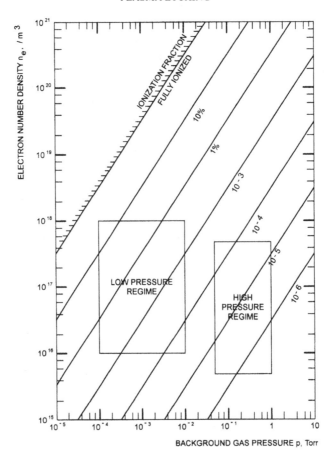

Figure 25.35. Plasma electron and active-species number densities as functions of the background gas pressure in Torr. A fully ionized background gas is shown as the last line to the upper left. The low- and intermediate- ('high'-) pressure regimes for microelectronic etching are indicated.

neutral gas, the number density of which is related to pressure by the perfect gas law. The number density is given in terms of the pressure in Torr for a gas with a temperature $T = 300$ K by

$$n = 3.22 \times 10^{22} p \text{ (Torr)} \qquad /\text{m}^3. \qquad (25.21)$$

If a glow discharge plasma contains only singly charged ions, the 'fully ionized' line at the upper left of figure 25.35 represents the maximum electron number density possible for singly ionized industrial glow discharge plasmas.

If the ordinate of figure 25.35 is identified with the electron number density, the intermediate-pressure regime, used for design rules above 0.5 μm in parallel-

Table 25.4. Some halocarbon gases for plasma etching and their principal fragments from mass spectrometer data (Vossen and Kern 1978).

Etching gas	Etchable materials	Principal fragments and relative intensities from mass spectra					
CF_4	Si, SiO_2, Si_3N_4, Ti, Mo, Ta, W	CF_3 100	CF_2 12	F 7	CF 5		
C_2F_6	SiO_2, Si_3N_4	CF_3 100	C_2F_5 41	CF 10	CF_2 10	C 2	F 1
$CClF_3$	Au, Ti	CF_3 100	$CClF_2$ 21	CF_2 10	Cl 7	CF 4	
$CBrF_3$	Ti, Pt	CF_3 100	$CBrF_3$ 15	$CBrF_2$ 14	Br 5		
C_3F_8	SiO_2, Si_3N_4	CF_3 100	CF 29	C_3F_7 25	CF_2 9	C_2F_5 9	C_2F_4 7
CCl_4	Al, Cr_2O_3, Cr	CCl_3 100	Cl 41	CCl 41	CCl_2 29		

plate RF glow discharge reactor technology, is shown on the lower right. Such plasma reactors characteristically have ionization fractions between 10^{-4} and 10^{-6}, a very slight degree of ionization. In such intermediate-pressure plasmas, the number density of neutral molecular fragments and other neutral active species may be many orders of magnitude higher than the electron or ion number densities. In the low-pressure regime below 1.3 Pa (10 mTorr), characteristic of design rules below 0.5 μm, the ionization fractions are significantly greater, ranging from approximately 10^{-4} to as high as 10% in some ECR reactors operating at pressures below 0.13 Pa (1 mTorr). In this low-pressure regime, the number density of etching species may be comparable to or greater than the electron number density.

The lower concentration of etching species in the low-pressure regime relative to the electron number density is in part due to their longer mean free paths. The etching species and their precursors collide more frequently with the walls than with electrons or other species inside the glow discharge plasma. When active species or etching precursors collide with the walls, heterogeneous reactions other than etching can occur, including recombination, polymerization, and chemical reactions with the wall material. Such direct interactions with the walls tend to lower the number density of active species and molecular fragments relative to the electron number density. In most RF glow discharge etching reactors, the working gas fed in by the gas supply subsystem comprises far more than 50% of the neutral gas present. The etching reaction product gas is usually the next most numerous gas, followed by the neutral and ionized active species.

A measurement of the molecular fragments that can be produced in a characteristic RF etching glow discharge with a fluorine-containing working gas is provided by table 25.4, taken from Vossen and Kern (1978). This table shows the concentration of principal fragments, inferred from the relative intensity of mass spectrometric peaks. Several combinations of working gas and materials to be etched were investigated. Each working gas was bombarded by 70 eV electrons. In each case, the carbon trifluoride (CF_3) mass spectrometric peak was taken as a normalizing concentration of 100 units, and the concentrations of other species are given relative to it. In many of these working gases, molecular fragments all the way down to atomic carbon and fluorine are evident above the noise level of the mass spectrometer.

25.4.3.2 Ionized Active Species

Some RF glow discharge plasmas, especially those containing *electrophilic* (electron attaching) species such as nitrogen (N_2) and sulfur hexafluoride (SF_6), produce a significant number of negative ions. Quasi-neutrality, which applies even in the low-pressure regime shown on figure 25.35, requires that the number density of positive and negative charges in the glow discharge plasma be approximately equal. For the ionic species S,

$$n(S^+) = n(e) + n(S^-). \qquad (25.22)$$

Only the most electrophilic species (such as sulfur hexafluoride, SF_6) will have a higher concentration of negative ions than of electrons. In RF glow discharge etching plasmas, the number density of positive ions will normally be much greater than the number density of negative ions,

$$n(S^+) > n(S^-). \qquad (25.23)$$

25.4.3.3 Magnetic Field Effects

The magnetization of electrons in a glow discharge plasma can have a significant effect on the relative concentrations of molecular fragments in it. This effect is illustrated by the work of Mayer and Barker (1982), discussed previously in section 19.7. In their low-pressure regime experiment, the electrons are *confined* at magnetic inductions above 0.4 mT, and the electrons are fully *magnetized* above 2 mT, where the *magnetization parameter* $\omega\tau$ is greater than unity. The work of Mayer and Barker (1982), showed that as electrons in the plasma make the transition from unmagnetized to magnetized confinement, the species balance of ions and neutrals in the plasma changes significantly. This change is from higher-molecular-weight and more complicated molecular fragments when the electrons are poorly confined at low magnetic fields, to simpler and lower-molecular-weight species as the electrons become better confined and more highly magnetized.

25.5 PLASMA ETCHING OF MICROELECTRONIC MATERIALS

Only a limited number of materials have the electrical, mechanical, thermal, and chemical properties suitable for fabricating microelectronic circuits. The requirements that the material undergo heterogeneous chemical reactions promoted by ion bombardment (to ensure directionality), and produce a volatile reaction product, greatly restricts the number of candidate materials available for etching.

Among the more commonly etched materials are aluminum, silicon and other semiconductors, silicon dioxide, hydrocarbons and photoresist, and refractory heavy metals used for transistor elements and interconnect vias. Many potentially attractive materials for microelectronic circuit fabrication, such as copper, silver, gold, platinum, and nickel, are easy to deposit in layers, but very difficult to etch in a practical and economic manner. We will briefly survey the etching of the materials more commonly used in integrated circuit fabrication.

25.5.1 Plasma Etching of Aluminum

Aluminum is widely used to form electrically conducting films in microelectronic circuits, but it is a difficult material to etch. Fluorine based gases cannot be used, because aluminum trifluoride (AlF_3) condenses as a solid at normal operating temperatures. A result of the condensability of aluminum trifluoride is illustrated in figure 25.36, taken from Coburn and Winters (1979a). In this experiment, aluminum was bombarded by a beam of 450 eV argon ions for the first 300 s. At that time, fluorine gas was introduced above the sample surface (an experimental setup similar to that shown in figure 25.21 was used.) A result of the formation of condensable aluminum trifluoride was to reduce the etching rate below the sputtering level initially observed. This reduction was due to the formation of a coating of aluminum trifluoride that protects the surface against argon ion sputtering.

Aluminum trichloride ($AlCl_3$) is volatile enough at room temperature to allow etching, but it is a reaction product with undesirable characteristics. It tends to condense on the walls of the reactor, and is *deliquescent* (i.e. it absorbs water during exposure to the normal atmosphere). The water vapor absorbed by the aluminum trichloride will consume working gas in the next etching cycle, and introduce a very undesirable dependence of the etching rate on the past history of the etching tool used. To prevent water absorption by condensed aluminum trichloride, airlock or dry box access is needed for aluminum etching, and often flushing of the vacuum system with dry air or an inert gas is required in addition. Unless such precautions are taken, rapid corrosion of the etched aluminum occurs following exposure of the wafer to air. To prevent such corrosion, all traces of residual chlorine and chlorine compounds must be removed, and conditions established which allow the aluminum to restore its normal protective surface

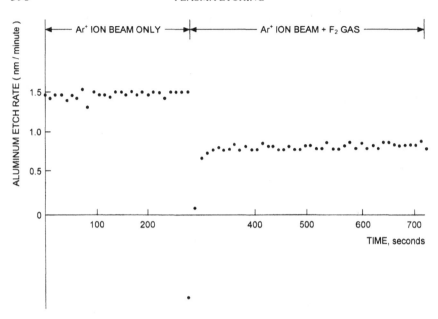

Figure 25.36. The etching rate of aluminum by 450 eV argon ions only, and 450 eV argon ions in the presence of fluorine gas. Note that the etching rate decreases in the presence of fluorine gas due to the formation of an non-volatile reaction product (Coburn and Winters 1979a).

layer of aluminum oxide (Al_2O_3).

The etching of aluminum may require some form of post-treatment not required by other etching processes, because of a requirement to clean away chlorine and chlorine compounds. Post-treatments of aluminum may include either exposure to a carbon tetrafluoride (CF_4) glow discharge, exposure to a hydrogen or hydrogen-forming glow discharge, immersion in distilled water, a brief wet etch, or exposure to an oxygen glow discharge. The latter procedure must be done with caution, because the presence of oxygen and chlorine in a glow discharge plasma can produce toxic compounds such as phosgene.

The characteristics of some chlorine-based etching reactions for aluminum are listed in table 25.5. The first three reactions on this table are most widely used. Inert gases such as argon or helium may be used as carrier gases for aluminum etching. The etching species is atomic chlorine in most cases, although molecular chlorine can play a direct role in the etching reaction.

The aluminum trichloride molecule has a heat of formation of 6.07 eV/molecule, whereas aluminum oxide (Al_2O_3) has a heat of formation of 17.4 eV/molecule. As a result, whenever oxygen is present during an etching operation, aluminum tends to form the more stable solid oxide layer, rather than the volatile aluminum trichloride reaction product. The aluminum oxide coating

Table 25.5. Characteristics of etching reactions for aluminum.

Material to be etched	Primary etching gas	Additive etching gas	Probable etching species	Volatile etching reaction product
Al	BCl_3	Ar, He	Cl atoms	$AlCl_3$, Al_2Cl_6
Al	CCl_4	Ar, He	Cl atoms	$AlCl_3$, Al_2Cl_6
Al	Cl_2	Ar, He	Cl_2 molecules	$AlCl_3$, Al_2Cl_6
Al	$SiCl_4$	—	Cl atoms	$AlCl_3$, Al_2Cl_6
Al	$CHCl_3$	—	Cl atoms	$AlCl_3$, Al_2Cl_6

normally present on aluminum must be removed before the etching process can proceed. The variable characteristics of this oxide film can result in a relatively long and often irregularly variable period of 'conditioning' prior to the onset of steady state etching. For this reason, the etching of aluminum (and aluminum oxide) may require better and more extensive process monitoring than other etching reactions. Such monitoring may be done by emission spectroscopy using one of the emission lines of aluminum, or aluminum chloride ($AlCl_3$).

The etching of aluminum oxide layers may also be required in the fabrication of integrated circuits, as well as in the cases for which an aluminum oxide surface coating has to be removed before an aluminum etch takes place. In general, because of its high heat of formation, aluminum oxide etching processes tend to be very slow in chlorine-based gases.

25.5.2   Plasma Etching of Silicon

Silicon is the most frequently etched material in microelectronic circuit fabrication, and there is a very large literature on the subject. Silicon is normally etched with either chlorine or fluorine based gases, although other halogens (such as bromine) may also be used. Some of the more widely used etching reactions of silicon are listed in table 25.6, in no particular order. Figure 25.37, taken from Coburn and Winters (1979a), shows the etching of silicon with molecular chlorine (Cl_2), in the presence of an energetic beam of 450 eV argon ions. This etching reaction for silicon is an analog of that for aluminum shown in figure 25.36, with fluorine gas. Unlike the aluminum etch, however, immediately after molecular chlorine was admitted to the apparatus, a transient buildup of reaction products occurred on the surface (this is the significance of the negative data points in figure 25.37). In addition, more than three minutes were required for the etching process to stabilize to a steady state, a relatively long duration in microelectronic etching operations.

The etching of silicon by fluorine gas in the presence of a 500 eV argon

Table 25.6. Characteristics of etching reactions for silicon.

Material to be etched	Primary etching gas	Additive etching gas	Probable etching species	Volatile etching reaction product
Si	CF_4	O_2	F atoms	SiF_4
Si	SF_6	O_2	F atoms	SiF_4
Si	XeF_2	None	F atoms	SiF_4
Si	F_2	None	F atoms	SiF_4
Si	Cl_2	None	Cl atoms	$SiCl_4$
Si	CF_3Cl	—	F, Cl atoms	SiF_4, $SiCl_4$
Si	CF_3Br	—	F, Br atoms	SiF_4, $SiBr_4$
Si	CCl_4	O_2	Cl atoms	$SiCl_4$
Si	$SiCl_4$	O_2	Cl atoms	$SiCl_4$
Si	C_2F_6	O_2	F atoms	SiF_4

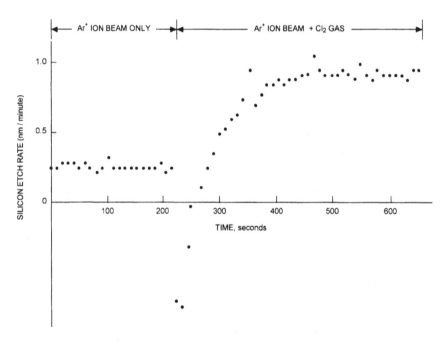

Figure 25.37. The etch rate of silicon in the presence of a 450 eV argon ion beam, with and without chlorine gas as etching agent (Coburn and Winters 1979a).

ion beam is shown in figure 25.20, also taken from Coburn and Winters. In this case, the time required for the etching process to stabilize after the introduction

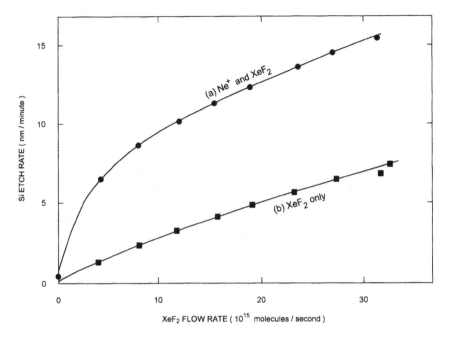

Figure 25.38. The etching rate of silicon as a function of the flow rate of XeF_2 with and without the catalytic effect of energetic neon ions (Gerlach-Meyer *et al* 1981).

of fluorine gas is significantly shorter than that for the chlorine etch shown on figure 25.37.

Figure 25.38, taken from Gerlach-Meyer *et al* (1981), shows the etching rate of silicon by xenon difluoride (XeF_2) on the lower curve, and a combination of xenon difluoride and energetic neon ions in the upper curve, both as functions of the xenon difluoride flow rates. These two curves illustrate how a directional but not a fully vertical etch is produced. The chemical etching of silicon by xenon difluoride, represented by the lower curve, affects the sidewalls as well as the bottom of the trench. When the xenon ion beam is present, the ions do not promote the reactions on the sidewalls of the trench, but they do increase the etching rate at bottom of the trench to that shown by the upper curve. Using the xenon difluoride silicon etching process represented on figure 25.38 will result in a trench, the sidewalls of which have been etched at the rate indicated by the lower curve, and the bottom of which has been etched into the silicon at the rate given by the upper curve.

This relative etching rate is illustrated also in the classic data of Coburn and Winters (1979a) in figure 25.22. These data illustrate the separate effects of the presence of xenon difluoride gas, the argon ion beam, and the synergistic presence of both the argon ion beam and the xenon difluoride gas. During the initial phase (up to approximately 200 s) on figure 25.22, silicon was chemically etched isotropically by the xenon difluoride gas only, at a rate of approximately 5 Å (0.5 nm)/min. After an energetic 450 eV argon ion beam was turned on in the presence of the xenon difluoride gas, the etching rate stabilized to approximately 55 Å/min, approximately 11 times the purely chemical initial etching rate. At approximately 650 s, the xenon difluoride was removed from the vacuum system, leaving only the sputter-etching rate of the argon ion beam, which removed silicon at the low rate of approximately 3 Å/min. In an etching reactor, the sidewalls would be etched at the initial rate appropriate to the presence of xenon difluoride only, and the bottom of the trench etched at the 11 times faster rate appropriate to the combination of the energetic ion beam and xenon difluoride.

The addition of both oxygen and hydrogen can affect the etching rate of silicon in a fluorine or chlorine based plasma reactor. The presence of either can, by a complicated chain of chemical reactions, add or remove competitors for the active species responsible for etching the silicon. An example of this in a carbon tetrafluoride (CF_4) plasma etching silicon with the addition of oxygen is shown in figure 25.39, taken from Mogab *et al* (1978). These data show an almost tenfold increase in the relative etching rate of silicon as the percentage of oxygen increases from zero to about 15%, beyond which it again falls off as the fraction of carbon tetrafluoride and its molecular fragments declines with an increasing percentage of oxygen.

25.5.3 Plasma Etching of Silicon Dioxide (SiO_2)

Silicon dioxide is ordinary glass in its amorphous form, and is used as an insulator and dielectric layer material in microelectronic circuits. Silicon dioxide is difficult to etch, and only a few fluorine-based compounds are able to etch silicon dioxide at acceptable rates. The etching is accomplished through chemical reactions of fluorine atoms or sometimes HF molecular fragments.

With the exception of the electron-beam-induced etching of silicon dioxide using xenon difluoride as the working gas, discussed previously in connection with figure 25.23, nearly all etching of silicon dioxide is done with carbon tetrafluoride or the related compounds listed in table 25.7. The carbon tetrafluoride is often mixed with argon; with hydrogen, if it is desired to increase the etching rate by the presence of HF molecular fragments; or with oxygen, if it is desired to improve the selectivity of the silicon dioxide etch with respect to a hydrocarbon such as photoresist.

In a mixture of carbon tetrafluoride and argon, the etching rate of silicon dioxide decreases monotonically as the percentage of argon in the carbon tetrafluoride increases from zero. In a plasma containing only carbon tetrafluoride, the etching rate of silicon dioxide is a monotone increasing function of the RF input power. If the etching rate of silicon dioxide in pure carbon tetrafluoride gas is plotted as a function of pressure, there is often a maximum. This maximum etching rate as a function of pressure for silicon dioxide etched in an ECR plasma characteristically occurs at pressures between 0.13 and 0.26 Pa (1 and 2 mTorr).

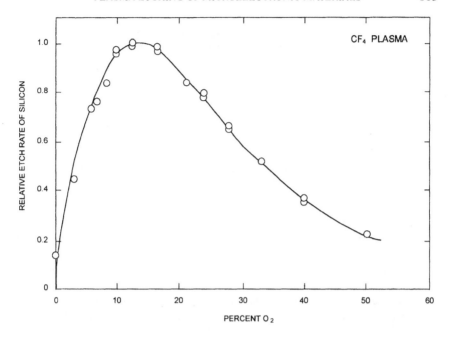

Figure 25.39. The relative etch rate of silicon in a carbon tetrafluoride (CF_4) plasma, as a function of the percentage of oxygen mixed with the carbon tetrafluoride (from Mogab *et al* 1978).

Table 25.7. Characteristics of etching reactions for silicon dioxide.

Material to be etched	Primary etching gas	Additive etching gas	Probable etching species	Volatile etching reaction product
SiO_2	CF_4	Ar, H_2	F, CF_3 atoms	SiF_4
SiO_2	CF_3	Ar	F, CF_3 atoms	SiF_4
SiO_2	C_3F_8	—	F, CF_3 atoms	SiF_4
SiO_2	XeF_2	Electron beam	F, CF_3 atoms	SiF_4
SiO_2	$C_2H_2F_4$	O_2	F, CF_3	SiF_4

25.5.4   Plasma Etching of Photoresist and Hydrocarbons

During the fabrication of microelectronic circuits, it is necessary to remove hydrocarbon-based masking, photoresist, or polymeric layers. This process is also known as *plasma ashing*, and is normally done in an oxygen plasma. Many of the reactions involved resemble the ordinary combustion of hydrocarbons to

Table 25.8. Characteristics of etching reactions for photoresist and hydrocarbons.

Material to be etched	Primary etching gas	Additive etching gas	Probable etching species	Volatile etching reaction product
Carbon, hydrocarbons	O_2	None	O atoms	CO, CO_2, H_2
Carbon, hydrocarbons	O_2	CF_4	O, F atoms	CO, CO_2, HF
Photoresist	CF_4	O_2	O, F atoms	CO, CO_2, HF
Photoresist	CF_4	H_2	F, H atom	HF, CH_4

yield carbon dioxide and water vapor, but in some cases, in order to achieve the selectivity desired, carbon tetrafluoride (CF_4) is used instead of oxygen. Some etching or ashing reactions for photoresist and hydrocarbons are listed in table 25.8.

The etching rates of several proprietary photoresists are shown as functions of the oxygen fraction and RF input power in figures 25.40(a) and (b), which have been taken from Melliar-Smith and Mogab (1978). Figure 25.40(a) shows the etching rate of photoresist in a carbon tetrafluoride–oxygen plasma, operated at an RF input power of 200 W, in the intermediate-pressure regime at a pressure of 47 Pa (350 mTorr). The rate of removal of photoresist is a direct function of the percentage of oxygen in the working gas, and is very low when oxygen is not present. This suggests that chemical reactions between the photoresist and carbon tetrafluoride are relatively inconsequential, and that the dominant reaction in etching the photoresist depends upon the amount of oxygen available.

Figure 25.40(b), from the same source, shows several types of proprietary photoresist, each represented by a different symbol, etched by a carbon tetrafluoride–oxygen plasma that contains 8% oxygen at a pressure of 40 Pa (300 mTorr). The etching rate is shown as a function of RF power and is almost a linear function of input power. This relationship is expected, since the electron number density and the rate of formation of active species are approximately proportional to the RF power delivered to the plasma.

25.5.5 Plasma Etching of Refractory Metals and Semiconductors

In addition to the materials discussed earlier, it is also necessary to etch layers of silicon doped with refractory or rare metals; semiconductors; or vias and interconnects made with rare or refractory metals from the third to fifth rows of the periodic table of the elements. A few of these reactions are listed in table 25.9. These reactions characteristically use a fluorine-containing gas that produces a volatile hexafluoride reaction product that allows the reaction products to be removed from the etching site. In some cases, chlorine is the preferred

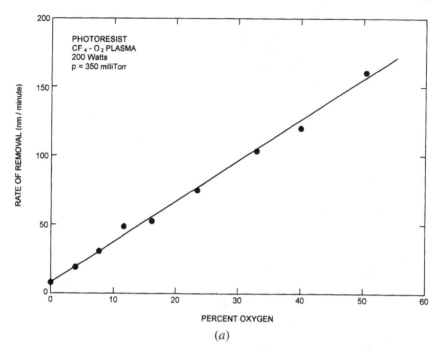

Figure 25.40. The etch rate of photoresist in a carbon tetrafluoride–oxygen plasma, taken from Melliar-Smith and Mogab (1978): (*a*) the etching rate of photoresist as a function of percentage of oxygen gas at a power level of 200 W and a background pressure $p = 47$ Pa (350 mTorr); (*b*) the etching rate of photoresist at 8% oxygen at $p = 40$ Pa (300 mTorr) as a function of RF power for several proprietary photoresists.

working gas, and will produce a volatile reaction product. Further information concerning these reactions is available in the references cited at the beginning of this section.

25.6 TECHNICAL ISSUES IN PLASMA ETCHING

The plasma etching processes discussed in this chapter have been developed since about 1970. As a result of the relatively recent development of this technology, a number of issues in plasma etching remain to be satisfactorily resolved. Here we survey some of these issues.

25.6.1 Desirable Etching Reactions That Do Not Work

The number of gases available to etch the materials with which microelectronic circuits are fabricated is relatively limited. Some etching gases create safety and regulatory difficulties when they are used, they may require purging the working

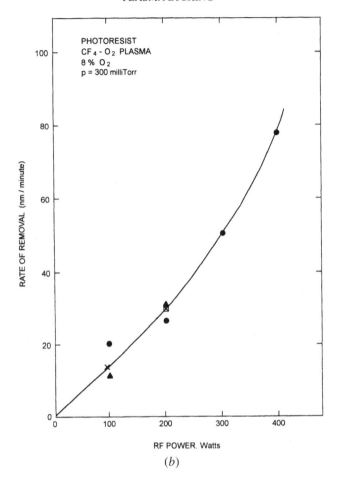

gas from previous steps in the fabrication process, or they may require moving a wafer from one plasma reactor to another in a cluster tool. It would, in many cases, be desirable to increase the flexibility of the etching process by having available a larger variety of etching reactions from which to choose. Some desirable chemical reactions that do not work are discussed below.

The etching of aluminum by fluorine would be very convenient if it were possible, and would proceed according to the reaction

$$Al + 3F \rightarrow AlF_3. \tag{25.24}$$

This reaction is exothermic, but produces a solid reaction product, which does not leave the etching site.

The etching of insulating layers of aluminum oxide is frequently required

Table 25.9. Characteristics of etching reactions for refractory/rare metals.

Material to be etched	Primary etching gas	Additive etching gas	Probable etching species	Volatile etching reaction product
W	CF_4, SF_6	—	F atoms	WF_6
Ta	NF_3, F_2	—	F atoms	TaF_6
Nb	NF_3, F_2	—	F atoms	NbF_6
Mo	NF_3, F_2	—	F atoms	MoF_6
GaAs	BCl_3, Cl_2	—	Cl atoms	$GaCl_3$
InP	CCl_4	—	Cl atoms	$AsCl_5$
$MoSi_2$	Cl_2	O_2	Cl atoms	$SiCl_4$

in etching operations. It would be convenient if this could be done in a chlorine discharge, which may be the gas of choice for operations that precede or follow the etching of aluminum oxide. This aluminum oxide etching reaction would proceed according to the chemical equation

$$2Al_2O_3 + 6Cl_2 \rightarrow 4AlCl_3 + 3O_2. \tag{25.25}$$

This reaction does not happen, however, because it is endothermic.

A similar case is the etching of silicon dioxide in a chlorine discharge, which would be convenient if it were possible. This reaction would proceed according to the equation

$$SiO_2 + Cl_2 \rightarrow SiCl_2 + O_2. \tag{25.26}$$

This reaction also is endothermic, and does not proceed at a rate rapid enough to be of practical use.

Finally, layers of metallic nickel would be more widely used if they were easier to etch. The etching of nickel would proceed in chlorine-based gases according to the equation

$$Ni + Cl_2 \rightarrow NiCl_2. \tag{25.27}$$

However, this is another instance in which the reaction product, nickel dichloride, is not volatile, and it therefore will not produce an acceptable etch.

There are many other similar examples of potential etching reactions that are not available, either because they are endothermic, because they produce a nonvolatile reaction product, or they simply do not produce the anticipated chemical reactions for reasons that are poorly understood. The etching reactions discussed in section 25.6 of this chapter are satisfactory and do work. They have been culled from a very large number of candidate reactions, the large majority of which do not work.

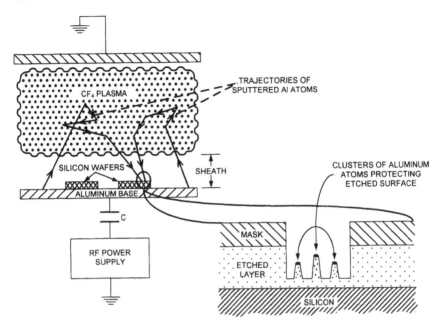

Figure 25.41. Illustration of sputtered aluminum atoms or clusters that can deposit on the surface of an etched layer to produce spikes and surface roughness.

25.6.2   Interference from Sputtered Atomic Clusters

One problem that can ruin an etched wafer is the deposition of sputtered atomic clusters on the surface of the layer during the etching process. This process is illustrated in figure 25.41. In this example, silicon wafers are being etched on an aluminum base plate in the presence of a carbon tetrafluoride plasma. The same ions that strike the wafer and are responsible for promoting the etching process also impact the aluminum base plate. The aluminum can be sputtered either individually as atoms, or as atomic clusters of aluminum that become electrically charged, and levitate in the electric field of the sheath and plasma.

After being sputtered off the aluminum base, the aluminum particles suffer a number of random collisions in the plasma, some of which can result in their being deposited on the surface of the material to be etched, as shown in the enlargement to the lower right of figure 25.41. When a cluster lands on the surface to be etched, or an individual atom impinges on the surface and then accumulates other atoms, the aluminum cluster can provide a protective shield ('umbrella') for the etched material underneath it. The etching process can proceed around it, leaving a tall, columnar structure under the sputtered aluminum cluster. These vertical spires of un-etched material can ruin the etched layer, and must be avoided.

25.6.3 Interference from Microscopic Dust

Particles of dust can be introduced by poor housekeeping, or can form by condensation from monomers of polymeric species in the plasma discharge. The dust can become electrically charged in the plasma, and levitate above the surface of the wafers. Upon shutting off the RF power, levitated dust particles can fall onto the surface (Barnes *et al* 1992). If they acquire a positive charge, the dust particles can be accelerated onto the surface of the wafer by the sheath electric field during RF operation. In either case, dust particles with dimensions on the order of the microelectronic circuit elements being etched may interfere with the etched or deposited layers to the point at which the circuit elements being processed are no longer functional. In the microelectronic industry, strenuous attempts are made to minimize the formation of dust in the plasma by chemical polymerization processes, and also to prevent the contamination of the plasma reactor by outside dust, using cleanroom techniques.

25.6.4 Unwanted Surface Polymerization

As was discussed in section 25.4.1.5, some etching reactions produce monomers, which can polymerize on surrounding surfaces. An example of this is shown in figure 25.29. As this figure indicates, the process can be transformed from polymerization to etching by increasing the negative bias on the wafer, or by changing the carbon-to-fluorine ratio in the etching reactor.

25.6.5 Large Number of Process Variables

As was pointed out in section 25.3.1, a large number of independent process variables affect the results of etching, some of which cannot be easily varied. Often there is an incomplete understanding of why varying some parameters produces the observed results. It appears that the most productive way to minimize the number of relevant process variables for process monitoring is to measure the plasma parameters, rather than black box process variables.

The value of this approach is emphasized by results from the University of Wisconsin's Center for Plasma Aided Manufacturing, in which the data presented in figure 25.33 indicate that plasma etching rates are insensitive to the type of plasma reactor. They are, however, a function of fundamental, relatively easily measured variables such as the ion power flux on the surface being etched, and the concentration and flux of etching species. Additional studies need to be made of the etching process, to further identify parameters relevant to plasma etching, and to understand the effects of each significant parameter on the etching process.

25.6.6 Inadequate Diagnostics

Additional and more adequate diagnostic methods are needed of the plasma parameters in the discharge chamber, of the surface being etched, and for *in situ* process monitoring. Some combinations of the material to be etched and substrate require better means of endpoint detection. In addition, virtually all etching processes would benefit by a more precise method of assessing the uniformity of the etching process during the etching process itself.

25.6.7 Understanding Physical Processes

Better understanding of physical processes that occur during the heterogeneous chemical reactions on the surface of the material being etched would almost certainly explain why some desirable etching reactions do not take place. More important, such understanding may allow the identification of new and additional etching reactions that will increase the flexibility and compatibility of sequential etching operations in a fabrication sequence.

25.6.8 Plasma-Induced Damage to Surface

Many production processes are limited by plasma-induced damage to semiconductor films and devices. One of the disadvantages of plasma etching relative to wet chemical etching is that plasma-induced damage is an issue, although wet chemical etching has its own, different set of problems.

Plasma-induced damage often affects oxide coatings. Three types of oxide damage can result from plasma processing. These include *radiation damage*, *particle bombardment*, and *electrostatic effects*. *Radiation damage* can occur during plasma processing as the result of bombardment by ions with energies above a few tens of eV, or by soft x-rays or UV radiation. *Particle bombardment* of an oxide by these active species can break chemical bonds on the oxide layer, thus producing electron and hole pairs. This form of radiation damage can be minimized by reducing the potential through which the ions fall in the sheath between the plasma and the wafer. *Electrostatic damage* to an oxide layer also can result from an electric field across the oxide layer.

25.6.9 Health and Safety Concerns

Although the microelectronic industry uses relatively small amounts of input materials, many of the working gases used are hazardous, and can be either explosive, flammable, toxic, or poisonous, or sometimes all of these. An excellent discussion of health and safety concerns may be found in Herb (1989), from which tables 25.10 and 25.11 are taken.

In table 25.10 the flammability of some common working gases is shown. The gas is listed on the left, its chemical symbol is in the central column, and its flammability ratio in air is listed in the right hand column. Many of these gases are flammable over a relatively wide range of concentrations. Silane and phosphine are particularly hazardous, as they are *pyrophoric*, that is they burst spontaneously into flame when exposed to air.

Table 25.10. Flammability of common etching gases (Herb 1989).

Gas	Chemical symbol	Flammability ratio in air
Silane	SiH_4	Pyrophoric
Phosphine	PH_3	Pyrophoric
Hydrocarbon	C_3H_8	2.2–9.5%
Hydrogen	H_2	4–75%
Carbon monoxide	CO	13–74%
Ammonia	NH_3	15–28%

Table 25.11. Hazardous concentrations of common etching gases (Herb 1989).

Gas	Chemical symbol	Designation	PEL	IDLH
			(Concentration in ppm)	
Arsenine	AsH_3	Poison	0.05	6
Phosphine	PH_3	Poison	0.3	200
Chlorine	Cl_2	Poison	1.0	25
Hydrochloric acid	HCl		5.0	100
Nitrogen trifluoride	NF_3		10.0	2000
Carbon tetrafluoride	CCl_4		10.0	300

In table 25.11, also taken from Herb (1989), the hazardous concentrations of some common etching gases are shown. The first three listed are classified as poisons, because of their ability to produce a fatal effect at very low concentrations. The *permissible exposure level* (PEL), also known as the threshold limit value (TLV) is the concentration, in parts per million (ppm), which can be tolerated on a continuous basis by an 'average' worker in an 8 hr day, over a normal 40 hr working week, without any adverse effect.

The last column in table 25.11 is the concentration, in parts per million, that is *immediately dangerous to life and health* (IDLH). This is the concentration that the 'average' worker can tolerate during a 30 min exposure, and be able to escape from a contaminated area without any help, either human assistance or with equipment, and suffer no biological effect. The reader must be warned that the gases listed in tables 25.10 and 25.11 are not the only hazardous gases widely used in the microelectronic industry. Any new or unfamiliar gas that appears in the workplace should be carefully researched in one of several databases of hazardous materials.

REFERENCES

Bardeen J and Brattain W H 1948 The transistor, a semi-conductor triode *Phys. Rev.* **74** 230–1

Barnes M S, Keller J H, Forster J C, O'Nell J A and Coultas D K 1992 Transport of dust particles in glow-discharge plasmas *Phys. Rev. Lett.* **68** 313–16

Bounasri F, Moisan M, Sauve G and Pelletier J 1993 Influence of the frequency of a periodic biasing voltage upon the etching of polymers *J. Vac. Sci. Technol.* B **11** 1859–67

Brattain W H and Bardeen J 1948 Nature of the forward current in germanium point contacts *Phys. Rev.* **74** 231–2

Coburn J W 1982 Plasma-assisted etching *Plasma Chem. Plasma Process.* **2** 1–41

Coburn J W and Winters H F 1979a Ion- and electron-assisted gas–surface chemistry—an important effect in plasma etching *J. Appl. Phys.* **50** 3189–96

——1979b Plasma etching—a discussion of mechanisms *J. Vac. Sci. Technol.* **16** 391–403

——1983 Plasma-assisted etching in microfabrication *Annu. Rev. Mater. Sci.* **13** 91–116

Deshmukh V G I and Cox T I 1988 Physical characterization of dry etching plasmas used in semiconductor fabrication *Plasma Phys. Control. Fusion* **30** 21–33

Garrity M P, Peterson T W and O'Hanlon J F 1996 Particle formation rates in sulfur hexafluoride plasma etching of silicon *J. Vac. Sci. Technol.* A **14** 550–5

Geis M W, Lincoln G A, Efremow N and Piacentini W J 1981 A novel anisotropic dry etching technique *J. Vac. Sci. Technol.* **19** 1390–3

Gerlach-Meyer U, Coburn J W and Key E 1981 Ion-enhanced gas–surface chemistry: the influence of the mass of the incident ion *Surf. Sci.* **103** 177–88

Herb G K 1989 Safety, health, and engineering considerations for plasma processing *Plasma Etching* ed D M Manos and D L Flamn (San Diego, CA: Academic) pp 425–70 ISBN 0-12-469370-9

Hershkowitz N, Ding J, Breun R A, Chen R T S, Meyer J and Quick A K 1996 Does high density–low pressure etching depend on the type of plasma source? *Phys. Plasmas* **3** 2197–202

Hollahan J R and Bell A T (ed) 1974 *Techniques and Applications of Plasma Chemistry* (New York: Wiley) ISBN 0-471-40628-7

Hope D A D, Monnington G J, Gill S S, Borsing N, Smith J A and Rees J A 1993 Ion energy distributions in $SiCl_4$ and Ar/O_2 dry etching discharges *Vacuum* **44** 245–8

Lieberman M A and Lichtenberg A J 1994 *Principles of Plasma Discharges and Materials Processing* (New York: Wiley) ISBN 0-471-00577-0

Madou M 1995 *Fundamentals of Microfabrication* (Boca Raton, FL: Chemical Rubber Company) ISBN 0-8493-9451-1

Manos D M and Flamm D L (ed) 1989 *Plasma Etching* (New York: Academic) ISBN 0-12-469370-9

Mayer T M and Barker R A 1982 Reactive ion beam etching with CF_4—characterization of a Kaufman ion source and details of SiO_2 etching *J. Electrochem. Soc.* **129** 585–91

Melliar-Smith C M and Mogab C J 1978 Plasma-assisted etching techniques for pattern delineation *Thin Film Processes* ed J L Vossen and W Kern (New York: Academic) ISBN 0-12-728250-5

Mogab C J, Adams A C and Flamm D L 1978 Plasma etching of Si and SiO_2—the effect of oxygen additions to CF_4 plasmas *J. Appl. Phys.* **49** 3796–803

Morgan R A 1985 *Plasma Etching in Semiconductor Fabrication* (New York: Elsevier)

ISBN 0-444-42419-9

Olthoff J K, Van Brunt R J, Radovanov S B, Rees J A and Surowiec R 1994 Kinetic-energy distributions of ions sampled from argon plasma in a parallel-plate, radio frequency reference cell *J. Appl. Phys.* **75** 115–25

Perry T S 1993 Coming clean *IEEE Spectrum* **30** 20–6

Powell R A (ed) 1984 *Dry Etching for Microelectronics* (New York: Elsevier) ISBN 0-444-86905-0

Powell R A and Downey D F 1984 Reactive ion-beam etching *Dry Etching for Microelectronics* ed R A Powell (Amsterdam: North-Holland) ch 4 ISBN 0-444-86905-0

Reid T R 1984 *The Chip* (New York: Simon and Schuster) ISBN 0-671-45393-9

Reinke P, Schelz S, Jacob W and Moller W 1992 Influence of a direct current bias on the energy of ions from an electron cyclotron resonance plasma *J. Vac. Sci. Technol.* A **10** 434–8

Sugano T (ed) 1985 *Applications of Plasma Processes to VLSI Technology* (New York: Wiley) ISBN 0-471-86960-0

Tobey A C 1985 Semiconductor microlithography through the eighties *Microelectron. Manufact. Test.* **19**

Ugajin R, Ishibashi A and Mori Y 1994 Advanced fabrication techniques of three-dimensional microstructures for future electronic devices *J. Vac. Sci. Technol.* B **12** 3160–5

Vossen J L and Kern W (ed) 1978 *Thin Film Processes* (New York: Academic) ISBN 0-12-728250-5

Winters H F, Coburn J W and Chuang T J 1983 Surface processes in plasma-assisted etching environment *J. Vac. Sci. Technol.* B **1** 469–80

Zheng J, Brinkmann R P and McVittie J P 1995 The effect of the presheath on the ion angular distribution at the wafer surface *J. Vac. Sci. Technol.* A **13** 859–64

Appendix A

Nomenclature

LATIN SYMBOLS

A	Area (m^2)
	Arbitrary constant
	Nuclear mass number
	Richardson emission constant (A/m^2-K^2)
a	Plasma or electrode radius (m)
	Similarity scaling factor
	Effective atomic raidus (m), equation (14.43)
a_0	Bohr radius, 5.2918×10^{-11} m
B, $\boldsymbol{B}$	Magnetic induction: scalar, vector quantities (T)
B	Arbitrary constant
b	Distance, radius (m)
	Distance of closest approach (m)
C	Capacitance (F)
	Arbitrary constant
	First Fowler–Nordheim constant (A/V^2)
C_v	Specific heat capacity (J/kg-K)
c	Speed of light, 2.998×10^8 m/s
D	Arbitrary constant
	Diffusion coefficient (m^2/s)
	Second Fowler–Nordheim constant (V/m)
	Displacement vector
	Beam diameter (m)
	Duty cycle, equation (24.5)
	Dose (ions/m^2)
DR	Deposition rate (particles/m^2-s)

614

D	Diffusion matrix
d	Distance (m)
$\mathcal{E}, \mathcal{E}; E, \boldsymbol{E}$	Electric field intensity: scalar, vector quantities (V/m)
E	Incomplete elliptic integral
E	Energy (J)
E'	Energy (eV)
E_0	Dimensionless sputtering energy, equation (14.18)
ER	Etching rate (m/s)
e	Electronic charge, 1.602×10^{-19} C
$F, \boldsymbol{F}$	Force: scalar, vector quantities (N)
F	Body force (N/m^3)
	Form factor
	Incomplete elliptic integral
	Ionization fraction
	Fraction of neutral gas in active species
$f, \boldsymbol{f}$	Force per particle: scalar, vector quantities (N)
f	Particle distribution function
g	Gravitational acceleration (m/s^2)
	Statistical weight of atomic state
H	Magnetic field strength (A/m)
	Latent enthalpy of vaporization (Joules/mole)
h	Planck's constant, 6.626×10^{-34} J s
$\hbar$	$h/2\pi$
I	Electric current (A)
	Luminous intensity, power flux (W/m^2)
i	Half angle of lens aperture
$J, \boldsymbol{J}$	Electric current density: scalar, vector quantities (A/m^2)
J	Bessel function
$\mathcal{J}$	Relativistic dimensionless current density
K	Complete elliptic integral
	Wavenumber (m^{-1})
	Dimensionless wavenumber
	Phenomenological constant, equation (14.7)
k	Boltzmann's constant, 1.380×10^{-23} J/K
	Exponent
	Wave number (m^{-1})
	Chemical rate constant
	Integer index
L	Distance, separation (m)
	Density scale length (m)
	Inductance (H)
	Extraordinary wave parameter, equation (13.22)
l	Length (m)

l (cont.)	Electron mean free path (m)
	Resolution distance (m)
M	Extraordinary wave parameter, equation (13.23)
	Magnetic moment (J/T)
	Ionic, neutral mass (kg)
	Volumetric momentum (kg m/s-m^3)
M_e	Radiant exitance (W/m^2)
m	Particle mass (kg)
	Electron mass (kg)
	Exponent
	Surface roughness factor
m_r	Reduced mass, equation (14.50)
N	Number of events, etc
	Dimensionless electron number density
	Number of electrodes per RF cycle
	Atomic density (atoms/m^3)
N_A	Avogadro's number, 6.023×10^{26}/kg-mole
n	Number density (particles/m^3)
	Index of refraction
P	Total power (W)
	Power flux (W/m^2)
P, p	Power density (W/m^3)
p	Pressure (Pa or N/m^2)
	Power per particle (W/charge)
$\mathcal{P}$	Perveance (A/V$^{3/2}$)
Q	Charge (C)
	Phenomenological constant
q	Charge, a signed quantity (C)
	Parametric exponent
r	Radial coordinate (m)
	Relative off time, equation (24.6)
R	Gas constant
	Radius of curvature (m)
	Reaction rate (reactions/m^3-s)
	Radius of gyration (m)
	Resistance (Ω)
	Reflection coefficient, equation (14.4)
S	Source rate (particles/m^3-s)
	Sheath thickness
	Power flux (W/m^2)
	Pumping speed (liters/s, m^3/s)
	Formation rate (particles/s)
	Spreading coefficient, equation (20.40)

S (cont.)	Selectivity of etch, equation (25.1)
s	Distance along trajectory (m)
T	Kinetic temperature (K)
	Time duration (s)
	Dimensionless time
	Trapping coefficient, equation (14.7)
T'	Kinetic temperature (eV)
T_b	Absorption coefficient, equation (14.5)
T_d	Adsorption coefficient, equation (14.6)
T	Transport matrix
t	Time (s)
U	Work (J)
	Potential, stored energy (J)
	Utilization factor, equation (25.4)
V	Volume (m^3, liters)
	Potential (V)
V^*	Effective ionization potential (eV)
$v, \boldsymbol{v}$	Speed: scalar, vector quantities (m/s)
W	Energy or work done (J)
	Dimensionless frequency
w	Particle kinetic energy (J)
	Width of etched trench (m)
	Dimensionless range, equation (22.7)
X	Dimensionless value of pd
	Mole fraction of implanted species
x	Cartesian coordinate (m)
Y	Dimensionless potential
Y_n	Phenomenological energy function, equation (14.9)
y	Cartesian coordinate (m)
z	Cartesian coordinate (m)
Z	Ionic charge number

GREEK SYMBOLS

α	Attenuation coefficient
	Radius ratio, equation (8.102)
	Townsend's first ionization coefficient (ion–electron pairs/m)
β	Relativistic velocity ratio, v/c
	Dimensionless cylindrical diode parameter
	Propagation constant
	Dimensionless ion drift velocity
	Townsend's second ionization coefficient

Γ, $\boldsymbol{\Gamma}$	Particle flux: scalar, vector quantities (particles/m^2-s)
γ	Ratio of specific heats
	Relativistic parameter
	Dimensionless neutral drift velocity
	Secondary electron emission coefficient (electrons/ion)
	Momentum transfer parameter, equation (14.16)
Δ	Small increment of quantity prefixed
δ	Direction cosine
	Small departure
	Interference thickness period, equation (20.47) (m)
	Atmospheric density factor, equation (8.105)
	Grid geometry ratio, equation (8.49)
	Ionization fraction
	Skin depth (m)
ε_0	Permittivity of free space, 8.854×10^{-12} F/m
ε	Emissivity of a plasma
	Small parameter
	Dimensionless energy
	Sputtering yield (atoms/ion)
η	Resistivity (Ω-m)
	Amplitude modulation factor
	Energy cost of ion–electron pair (eV)
	Energy coupling parameter, equation (11.34)
	Efficiency
	Coefficient of viscosity (N s/m^3)
θ	Angle with respect to magnetic field (rad)
	Azimuthal angle (rad)
	Vertex half-angle of escape cone
	Contact angle of droplet (rad)
κ	Thermal conductivity (W/m-K)
Λ	Argument of Coulomb logarithm
	Characteristic diffusion length (m)
λ	Debye shielding length (m)
	Wavelength (m)
	Mean free path (m)
μ, $\boldsymbol{\mu}$	Mobility: scalar, tensor quantities (C s/kg)
	Magnetic moment (J/T)
	Index of refraction (real)
μ_0	Permeability of free space, $4\pi \times 10^{-7}$ H/m
ν	Frequency of an event/particle (Hz)
ν_c	Collision frequency/particle (Hz)
ν_i	Ionization frequency/particle (Hz)
ξ	Dimensionless distance, equation (3.165b)
	Penetration parameter, equation (14.48)

π	3.14159
ρ	Charge density (C/m^3)
	Dimensionless radial parameter, equation (3.196a)
	Mass density (kg/m^3)
	Resistivity (Ω-m)
	Areal mass density (kg/m^2)
σ	Cross section (m^2)
	Electrical conductivity (S/m $\equiv$ mho/m)
	Stefan–Boltzmann constant, 5.671×10^{-8} W/m^2-K^4
	Surface charge density (C/m^2)
$\langle \sigma v \rangle$	Reaction rate coefficient (m^3/s)
τ	Characteristic time (s)
	Particle containment time (s)
	Time between collisions (s)
	Dwell time in vacuum system (s)
ϕ	Potential (V)
	Angle of rotation (rad)
	Magnetic flux (Wb)
	Maximum allowable power flux (W/m^2)
	Dimensionless potential
	Phase angle (rad)
	Particle flux (particles/m^2-s)
χ	'Child' coefficient, equation (3.144)
	Attenuation index
	Quantum scale parameter, equation (14.52)
Ω	Solid angle (rad)
ω	Angular frequency (rad/s)
	Relativistic parameter, equation (3.175)
ω_p	Plasma frequency (rad/s)

SUBSCRIPTS

0, o	Neutral gas
	At rest value
	Outside surface value
	Free space value
	Axial value
	Maximum value
	Initial value
	Operating value
A	Ambipolar value
A, a	Anode
B	Bohm value

B, b	Breakdown
b	Beam value
	Outer radius
C	Child Law
	Compton wavelength
c	Critical
	Cathode region
	Gyrofrequency
	Collision
	Centrifugal
	Child-law critical current
	Charged particle
cl	Classical
d	Deflection
	Debye length
	Electrical discharge
	Particle drift
E	Electric field
	Efficiency
	Energy
	Electrostatic
	Erosion
e	Electrons
eff	Effective value
ex	External
	Extraordinary wave
F	Floating potential
f	Filament
	Final value
	Floating value
g	Gravitational
	Radius of gyration
	Relating to gas
H	Heating
	Hall direction
i	Ions
	Ionization
	Injection
in	Inside
L	Left circularly polarized
	Load
lim	Limiting value
m	Magnetic
	Most probable value

mob	Mobility
max	Maximum value
min	Minimum value
n	Neutral gas
ne	Electron–neutral ionization
op	Operating
oh	Ohmic heating
ord	Ordinary wave
out	Outside
P	Pedersen direction
p	Plasma
	Proton
	Particles
R	Right circularly polarized
r	Radial
	Relativistic
rms	Root mean square value
S	Standing
	Source
s	Sheath
	Surface value
	Species
	Scattering
	Gasdynamic stagnation
sc	Space charge
T	Trapping
t, tot	Total
uh	Upper hybrid
v	Volume, constant volume
w	Pertaining to kinetic energy
	Wall, wall value
x	Cartesian direction
y	Cartesian direction
z	Cartesian direction
$\parallel$	Parallel to magnetic field
$\perp$	Perpendicular to magnetic field
*	Associated with energy transfer frequency
1	Perturbed value

SUPERSCRIPTS

N	Angular dependence exponent, equation (17.2)
n	Pressure exponent, equation (19.15)

$'$ (prime)	Indicates energy $\mathcal{E}'$, E', and kinetic temperature T' in electronvolts
	Derivative
	Moving coordinate system
	Value per unit axial length
*	Associated with eV/ion
o	Time derivative

OTHER NOTATION

| $\langle\,\rangle$ | Implies averaging over a Maxwellian distribution for the quantity between angle brackets |
| $-$ | Mean value |

Appendix B

Physical Constants

c	Velocity of light	2.998×10^8 m/s
e	Electronic charge	1.602×10^{-19} C
m_e	mass of an electron	9.109×10^{-31} kg
e/m_e	Electron charge to mass ratio	1.759×10^{11} C/kg
	Electron rest mass	0.511 MeV
m_p	Mass of a proton	1.673×10^{-27} kg
e/m_p	Proton charge to mass ratio	9.564×10^7 C/kg
	Proton rest mass	938.21 MeV
m_p/m_e	Proton to electron mass ratio	1836
h	Planck's constant	6.626×10^{-34} J s
k	Boltzmann's constant ($k = R/N_A$)	1.381×10^{-23} J/K
N_A	Avogadro's number	6.022×10^{26} /kg-mole
σ	Stefan–Boltzmann constant	5.671×10^{-8} J/m^2-s-K^4
g	Standard gravity	9.807 m/s^2
ε_0	Permittivity of free space	8.854×10^{-12} F/m
μ_0	Permeability of free space	$4\pi \times 10^{-7}$ H/m
R	Molar gas constant	8.315 J/K-mol

Appendix C

Units and Conversion Factors

This appendix lists the quantities and units that are relevant to plasma physics and engineering. All units are SI except those marked with an asterisk (*). The following quantities have their own named units.

Quantity	Name	Symbol	Equivalent units
Mass	kilogram	kg	
Length	meter	m	
Time	second	s	
Temperature	kelvin	K	
	*electronvolt	eV	11 604 K
Electrical current	ampere	A	C/s
Amount of substance	mole	mol	*kg mole
Plane angle	radian	rad	
Frequency	hertz	Hz	/s
Force	newton	N	$m\,kg/s^2$
Energy	joule	J	N m
	*electronvolt	eV	
Power	watt	W	$J/s \equiv V\,A$
Pressure	pascal	Pa	N/m^2
	*torr	Torr	*mm Hg
	*standard atmosphere	atm	
Electric charge	coulomb	C	A s
Electric potential, voltage	volt	V	$J/C \equiv W/A$
Electric resistance	ohm	Ω	V/A
Electric conductance	siemens	S	$A/V \equiv /\Omega \equiv$ *mho
Capacitance	farad	F	C/V
Magnetic flux	weber	Wb	V s
Magnetic induction	tesla	T	Wb/m^2
Inductance	henry	H	Wb/A

The following quantities are expressed in terms of the units listed above.

Quantity	Name	Symbol
Area, cross section	square meter	m^2
Volume	cubic meter	m^3
Mass density	kilogram per cubic meter	kg/m^3
Velocity	meter per second	m/s
Acceleration	meter per second squared	m/s^2
Wave number	per meter	/m
Electric current density	ampere per square meter	A/m^2
Electric field strength	volt per meter	V/m
Power flux	watt per square meter	W/m^2
Power density	watt per cubic meter	W/m^3
Resistivity	ohm meter	Ω-m
Electrical conductivity	siemens per meter	S/m
Magnetic field strength	ampere per meter	A/m
Magnetic moment	joule per tesla	J/T
Permittivity	farad per meter	F/m
Permeability	henry per meter	H/m
Mobility	coulomb second per kilogram	C s/kg
Specific heat capacity	joule per kilogram kelvin	J/kg K
Thermal conductivity	watt per meter kelvin	W/m K
Angular frequency	radian per second	rad/s
Plasma frequency	radian per second	rad/s
Number density	particles per cubic meter	$particles/m^3$

CONVERSION FACTORS

Energy	$1\text{ eV} = 1.602 \times 10^{-19}$ J
pages 56, 57	$\mathcal{E}(\text{J}) = e\mathcal{E}'(\text{eV})$
Temperature	$1\text{ eV} = e/k = 11\,604$ K
page 40	$kT(\text{K}) = eT'(\text{eV})$
Pressure	1 atm = 760 Torr = 101 325 Pa
pages 34, 35	1 Torr = 101 325/760 Pa

Conversion formulae for neutral particle density (in $particles/m^3$) and pressure (in Pa, Torr and atm) are given in Appendix D.

Appendix D

Useful Formulae

SI units are used except as indicated. The kinetic temperatures T', T_e' and T_i' are in electronvolts throughout, whereas T is in kelvin; m_e, m_i and m_p are the electron, ion and proton masses, respectively; n_e and n_i are the electron and ion number densities, respectively; B is the magnetic induction, e is the electronic charge; Z is the charge of an ion; and $A \approx m_i/m_p$ is the atomic mass number and is approximately the ratio of ion to proton mass.

FREQUENCIES

Electron cyclotron (or gyro) frequency:

$$\nu_{ce} \equiv \frac{eB}{2\pi m_e} = 28.00B \qquad \text{(GHz)} \tag{D.1}$$

Ion cyclotron (or gyro-) frequency:

$$\nu_{ci} \equiv \frac{ZeB}{2\pi m_i} = 15.22\frac{ZB}{A} \qquad \text{(MHz)} \tag{D.2}$$

Electron plasma frequency:

$$\nu_{pe} \equiv \frac{1}{2\pi}\left(\frac{n_e e^2}{\varepsilon_0 m_e}\right)^{1/2} = 8.980(n_e)^{1/2} \qquad \text{(Hz)} \tag{D.3}$$

Ion plasma frequency:

$$\nu_{pi} \equiv \frac{1}{2\pi}\left(\frac{n_i Z^2 e^2}{\varepsilon_0 m_i}\right)^{1/2} = 0.2095Z\left(\frac{n_i}{A}\right)^{1/2} \qquad \text{(Hz)} \tag{D.4}$$

CHARACTERISTIC LENGTHS

Electron Debye length:

$$\lambda_{de} \equiv \left(\frac{\varepsilon_0 T'}{n_e e}\right)^{1/2} = 7434 \left(\frac{T'}{n_e}\right)^{1/2} \quad \text{(m)} \qquad \text{(D.5)}$$

Electron gyroradius (based on mean thermal velocity):

$$r_{ge} \equiv \frac{m_e v_e}{eB} = \frac{1}{B}\left(\frac{8m_e T'_e}{\pi e}\right)^{1/2} = 3.805 \times 10^{-6} \frac{(T'_e)^{1/2}}{B} \quad \text{(m)} \qquad \text{(D.6)}$$

Ion gyroradius (based on mean thermal velocity):

$$r_{gi} \equiv \frac{m_i v_i}{ZeB} = \frac{1}{ZB}\left(\frac{8m_p A T'_i}{\pi e}\right)^{1/2} = 1.632 \times 10^{-4} \frac{(AT'_i)^{1/2}}{ZB} \quad \text{(m)} \qquad \text{(D.7)}$$

VELOCITIES

Electron mean thermal velocity:

$$v_e \equiv \left(\frac{8eT'_e}{\pi m_e}\right)^{1/2} = 6.693 \times 10^5 (T'_e)^{1/2} \quad \text{(m/s)} \qquad \text{(D.8)}$$

Ion mean thermal velocity:

$$v_i \equiv \left(\frac{8eT'_i}{\pi m_i}\right)^{1/2} = 1.561 \times 10^4 \left(\frac{T'_i}{A}\right)^{1/2} \quad \text{(m/s)} \qquad \text{(D.9)}$$

NUMBER DENSITY/PRESSURE CONVERSIONS

At 300 K, the number density of a neutral gas, as derived from equation (2.3), is given by differing conversion formulae, depending on the units in which the pressure is given:

$$n \equiv P/kT \qquad \text{(2.3)}$$
$$= 2.415 \times 10^{20} p(\text{Pa}) \qquad (\text{atoms/m}^3) \qquad \text{(D.10)}$$
$$= 2.447 \times 10^{25} p'(\text{atm}) \qquad (\text{atoms/m}^3) \qquad \text{(D.11)}$$
$$= 3.220 \times 10^{22} p'(\text{Torr}) \qquad (\text{atoms/m}^3) \qquad \text{(D.12)}$$
$$= 3.220 \times 10^{16} p'(\text{Torr}) \qquad (\text{atoms/cm}^3). \qquad \text{(D.13)}$$

Index

Page numbers in *italic* typeface refer to volume 1.